BASIC COMPLEX ANALYSIS

SECOND EDITION

BASIC COMPLEX ANALYSIS

JERROLD E. MARSDEN

University of California, Berkeley

MICHAEL J. HOFFMAN

California State University, Los Angeles

W. H. Freeman and Company   New York

The computer-generated cover illustration representing the dynamics of complex analytic mappings is courtesy of R. Devany of Boston University with the assistance of C. Mayberry, C. Small, and S. Smith. See pp. 446–448 for an introduction to this topic.

LIBRARY OF CONGRESS CATALOGING-IN-PUBLICATION DATA

Marsden, Jerrold E.
  Basic complex analysis.

  Includes index.
  1. Functions of complex variables.   I. Hoffman,
Michael J.   II. Title.
QA331.M378   1987   515.9   86-18413
ISBN 0-7167-1814-6

Printed in the United States of America

   5 6 7 8 9   VB   6 5 4 3 2

FOR ISOBEL PHILLIPS
AND LENORE HARRIS

CONTENTS

CHAPTER 3

SERIES REPRESENTATION OF ANALYTIC FUNCTIONS   205

CHAPTER 4

CALCULUS OF RESIDUES   266

CHAPTER 5

CONFORMAL MAPPINGS   345

PREFACE

This text is intended for undergraduates in mathematics, the physical sciences, and engineering who are taking complex analysis for the first time. Two years of calculus, up through calculus of several variables and Green's theorem, are adequate preparation for the course. The text contains some references to linear algebra and basic facts about ϵ-δ analysis, but these are not central to the course and may be considered optional.

The book is generous in the number of examples, exercises, and applications provided. We have made a special effort to motivate students by making the book readable for self-study and have provided plenty of material to help students gain an intuitive understanding of the subject. For example, Section 2.2 presents an intuitive version of Cauchy's theorem using Green's theorem so that the main content of the theorem quickly becomes apparent. Section 2.3 then presents a more careful proof based on the bisection method and deformation of curves. This arrangement enables application-oriented students to skip the more technical parts without sacrificing an understanding of the main theoretical points. The applications include electric potentials, heat conduction, and hydrodynamics — studied with the aid of harmonic functions and conformal mappings, and Laplace transforms, asymptotic expansions, the Gamma function, and Bessel functions.

Our experience teaching complex variables has been that most of Chapters 1 to 6 can be taught in a one-semester course for mathematics majors. In applied mathematics courses, if the technical parts of Chapter 2 and parts of Chapter 6 are omitted, then parts of Chapters 7 and 8 can be covered in one semester. In our opinion it is healthy for mathematics majors to see as many of the applications as possible, for they are an integral part of the cultural and historical heritage of mathematics.

The symbols used in this text are, for the most part, standard. $\mathbf{R}$ denotes the set of real numbers, while $\mathbf{C}$ denotes the set of complex numbers. "iff" stand for "if and only if" (except in definitions, where we write only "if"). The end of a proof is marked ■, while the end of the proof of a lemma in the middle of a proof of a theorem is marked ▼. The notation $]a, b[$ represents the open interval consisting of all real numbers x satisfying $a < x < b$. This is to avoid confusion with the ordered pair notation (a, b). The symbols $f: A \subset \mathbf{C} \to \mathbf{C}$ mean that the mapping f maps the domain A into $\mathbf{C}$, and we write $z \mapsto f(z)$ for the effect of f on the point $z \in A$. Occasionally, $\Rightarrow$ is used to mean "implies." The set theoretic difference of the sets A and B is denoted by $A \backslash B$, while their union and intersection are denoted by $A \cup B$ and $A \cap B$. The definitions, theorems, propositions, lemmas, and examples are numbered consecutively for easy cross-reference; for example, Definition 6.2.3 refers to the third item in Section 6.2.

This second edition represents a substantial revision of the first edition in detail, but not in spirit. Many passages have been rewritten for clarity, and many corrections have been made. For example, the deformation or homotopy version of Cauchy's theorem has been completely rewritten, as has the presentation of the method of stationary phase. Supplementary material on topics such as integrals along continuous curves, normal families and the proof of the Riemann mapping theorem, dynamics of complex mappings, and functions of bounded variation has also been provided. Throughout the text more examples have been added. Answers are now given for just the odd-numbered exercises, which should help the instructor in making up homework assignments.

Despite the large number of texts written in recent years, some of the older classics remain the best. A few that are worth looking at are A. Hurwitz and R. Courant, *Vorlesungen über allgemeine Functionentheorie und elliptische Functionen* (Berlin: Julius Springer, 1925); E. T. Whittaker and G. N. Watson, *A Course of Modern Analysis* (London: Cambridge University Press, 1927); E. T. Titchmarch, *The Theory of Functions,* 2d ed. (New York: Oxford University Press, 1939, reprinted 1985); and K. Knopp, *Theory of Functions* (New York: Dover, 1947). The reader who wishes further information on various of the more advanced topics can also profitably consult E. Hille, *Analytic Function Theory* (2 volumes), (Boston: Ginn, 1959); L. V. Ahlfors, *Complex Analysis* (New York: McGraw-Hill, 1966); W. Rudin, *Real and Complex Analysis* (New York: McGraw-Hill, 1969); S. Lang, *Complex Analysis,* 2d ed. (New York: Springer Verlag, 1985); and P. M. Morse and H. Feshbach, *Methods of Theoretical Physics* (New York: McGraw-Hill, 1953). More information on complex dynamics (see p. 446) may be found in R. Devany, *An Introduction to Chaotic Dynamical Systems* (Reading, Mass.: Addison Wesley, 1985) and H. O. Peitgen and P. H. Richter, *The Beauty of Fractals: Images of Complex Dynamical Systems* (Berlin, Heidelberg: Springer-Verlag, 1986). Some additional references are given throughout the text.

The modern treatment of complex analysis did not evolve rapidly or smoothly. The numerous creators of this area of mathematics travel over many rough roads and encounter many blind alleys before the superior routes are found. An appreciation of the history of mathematics and its intimate connection to the physical sciences is important to every student's education. We recommend looking at M. Klein's *Mathematical Thought from Ancient to Modern Times* (London: Oxford University Press, 1972).

We are grateful for the many readers who supplied corrections and comments for this edition. There are too many to all be thanked individually, but we would like to especially mention (more or less chronologically) M. Buchner (who helped significantly with the first edition), C. Risk, P. Roeder, W. Barker, G. Hill, J. Seitz, J. Brudowski, H. O. Cordes, M. Choi, W. T. Stallings, E. Green, R. Iltis, N. Starr, D. Fowler, L. L. Campbell, D. Goldschmidt, T. Kato, J. Mesirov, P. Kenshaft, K. L. Teo, G. Bergmann, J. Harrison and C. Daniels.

Jerry Marsden   Michael Hoffman
Berkeley, California   *Los Angeles, California*

BASIC COMPLEX ANALYSIS

CHAPTER 1

ANALYTIC FUNCTIONS

In this chapter the basic ideas about complex numbers and analytic functions are introduced. The organization of the text is analogous to that of an elementary calculus textbook, which begins with the real line $\mathbf{R}$ and a function $f(x)$ of a real variable x and then studies differentiation of f. Similarly, in complex analysis we begin with complex numbers z and study differentiable functions $f(z)$. (These are called analytic functions.) The analogy, however, is deceptive, because complex analysis is a much richer theory; a lot more can be said about an analytic function than about a differentiable function of a real variable. The properties of analytic functions will be fully developed in subsequent chapters.

In addition to becoming familiar with the theory, the student should gain some facility with the standard (or "elementary") functions—such as polynomials, e^z, log z, sin z— used in calculus. These functions are studied in Sec. 1.3 and appear frequently throughout the text.

1.1 INTRODUCTION TO COMPLEX NUMBERS

Historical Sketch

The following discussion will assume some familiarity with the main properties of real numbers. The real number system resulted from the search for a system (an abstract set together with certain rules) that included the rationals but that also provided solutions to such polynomial equations as $x^2 - 2 = 0$.

Historically, a similar consideration gave rise to an extension of the real numbers. As early as the sixteenth century, Geronimo Cardano considered quadratic (and cubic) equations such as $x^2 + 2x + 2 = 0$, which is satisfied by no real number x. The quadratic formula $(-b \pm \sqrt{b^2 - 4ac})/2a$ yields "formal" expressions for the two solutions of the equation $ax^2 + bx + c = 0$. But this formula may involve square roots of negative numbers; for example, $-1 \pm \sqrt{-1}$ for the equation $x^2 + 2x + 2 = 0$. Cardano noticed that if these "complex numbers" were treated as ordinary numbers with the added rule that $\sqrt{-1} \cdot \sqrt{-1} = -1$, they did indeed solve the equations.

The important expression $\sqrt{-1}$ is now given the widely accepted designation $i = \sqrt{-1}$. (An alternative convention is followed by many electrical engineers, who prefer the symbol $j = \sqrt{-1}$ since they wish to reserve the symbol i for electric current.) However, in the past it was felt that no meaning could actually be assigned to such expressions, which were therefore termed "imaginary." Gradually, especially as a result of the work of Leonhard Euler in the eighteenth century, these imaginary quantities came to play an important role. For example, Euler's formula $e^{i\theta} = \cos\theta + i\sin\theta$ revealed the existence of a profound relationship between complex numbers and the trigonometric functions. The rule $e^{i(\theta_1 + \theta_2)} = e^{i\theta_1}e^{i\theta_2}$ was found to summarize the rules for expanding sine and cosine of a sum of two angles in a neat way, and this result alone indicated that some meaning should be attached to these "imaginary" numbers.

However, not until the work of Casper Wessel (ca. 1797), Jean Robert Argand (1806), Karl Friedrich Gauss (1831), Sir William R. Hamilton (1837), and others was the meaning of complex numbers clarified, and was it realized that there is nothing "imaginary" about them at all (although this term is still used).

The complex analysis that is the subject of this book was developed in the nineteenth century, mainly by Augustin Cauchy (1789–1857). Later his theory was made more rigorous and extended by such mathematicians as Peter Dirichlet (1805–1859), Karl Weierstrass (1815–1897), and Georg Friedrich Bernhard Riemann (1826–1866).

The search for a method to describe heat conduction influenced the development of the theory, which has found many uses outside mathematics. Subsequent chapters will discuss some of these applications to problems in physics and engineering, such as hydrodynamics and electrostatics. The theory also has mathematical applications to problems that at first do not seem to involve complex numbers. For example, the proof that

$$\int_0^\infty \frac{\sin^2 x}{x^2}\, dx = \frac{\pi}{2}$$

or that

$$\int_0^\infty \frac{x^{\alpha-1}}{1+x}\, dx = \frac{\pi}{\sin(\alpha\pi)} \qquad \text{for } 0 < \alpha < 1$$

or that

$$\int_0^{2\pi} \frac{d\theta}{a + \sin\theta} = \frac{2\pi}{\sqrt{a^2 - 1}}$$

may be difficult or impossible using elementary calculus, but these identities can be readily proved using the techniques of complex variables.

Complex analysis has become an indispensable and standard tool of the working mathematician, physicist, and engineer. Neglect of it can prove to be a severe handicap in most areas of research and application involving mathematical ideas and techniques.

Definition of the Complex Numbers

The first task in this section will be to define complex numbers and to show that they possess properties that are adequate for the usual algebraic manipulations to hold. The basic idea of complex numbers is credited to Jean Robert Argand, who suggested using points in the plane to represent complex numbers. The student will recall that the xy plane, denoted by $\mathbf{R}^2$, consists of all ordered pairs (x, y) of real numbers. Let us begin with the formal definition, followed by a discussion of why complex multiplication is defined the way it is.

1.1.1 DEFINITION *The system of complex numbers, denoted $\mathbf{C}$, is the set $\mathbf{R}^2$ together with the usual rules of vector addition and scalar multiplication by a real number a, namely,*

$$(x_1, y_1) + (x_2, y_2) = (x_1 + x_2, y_1 + y_2)$$
$$a(x, y) = (ax, ay)$$

and with the operation of complex multiplication, defined by

$$(x_1, y_1)(x_2, y_2) = (x_1 x_2 - y_1 y_2, x_1 y_2 + y_1 x_2)$$

Rather than using (x, y) to represent a complex number, we will find it more convenient to return to more standard notation as follows. Let us identify *real* numbers x with points on the x axis; thus x and $(x, 0)$ stand for the same point $(x, 0)$ in $\mathbf{R}^2$. The y axis will be called the *imaginary axis,* and the unit point $(0, 1)$ will be denoted i. Thus, by definition, $i = (0, 1)$. Then

$$(x, y) = x + yi$$

because the right side of the equation stands for $(x, 0) + y(0, 1) = (x, 0) + (0, y) = (x, y)$. Using $y = (y, 0)$ and Definition 1.1.1 of complex multiplication, we get $iy = (0, 1)(y, 0) = (0 \cdot y - 1 \cdot 0, y \cdot 1 + 0 \cdot 0) = (0, y) = y(0, 1) = yi$, and so we can also write $(x, y) = x + iy$. A single symbol such as $z = a + ib$ is generally used to indicate a complex number. The notation $z \in \mathbf{C}$ means that z belongs to the set of complex numbers.

Note that $i^2 = i \cdot i = (0, 1) \cdot (0, 1) = (0 \cdot 0 - 1 \cdot 1, (1 \cdot 0 + 0 \cdot 1)) = (-1, 0) = -1$, so we do have the property we want:

$$i^2 = -1$$

If we remember this equation, then the rule for multiplication of complex numbers is also easy to remember and motivate:

$$(a + ib)(c + id) = ac + iad + ibc + i^2bd$$
$$= (ac - bd) + i(ad + bc)$$

Thus, for example, $2 + 3i$ is the complex number $(2, 3)$, and $(2 + 3i)(1 - 4i) = 2 - 12i^2 + 3i - 8i = 14 - 5i$ is another way of saying that $(2, 3)(1, -4) = (2 \cdot 1 - 3(-4), 3 \cdot 1 + 2(-4)) = (14, -5)$. The reason for using the expression $a + bi$ is twofold. First, it is conventional. Second, the rule $i^2 = -1$ is easier to use than the rule $(a, b)(c, d) = (ac - bd, bc + ad)$, although both rules produce the same result.

Because multiplication of real numbers is associative, commutative, and distributive, multiplication of complex numbers is also; that is, for all complex numbers, z, w, s we have $(zw)s = z(ws)$, $zw = wz$, and $z(w + s) = zw + zs$. Let us verify the first of these properties; the others can be similarly verified.

Let $z = a + ib$, $w = c + id$, and $s = e + if$. Then $zw = (ac - bd) + i(bc + ad)$, and so $(zw)s = e(ac - bd) - f(bc + ad) + i[e(bc + ad) + f(ac - bd)]$. Similarly,

$$z(ws) = (a + bi)[(ce - df) + i(cf + de)]$$
$$= a(ce - df) - b(cf + de) + i[a(cf + de) + b(ce - df)]$$

Comparing these expressions and accepting the usual properties of real numbers, we conclude that $(zw)s = z(ws)$. Thus we can write, without ambiguity, an expression like $z^n = z \cdots z$ (n times).

Note that $a + ib = c + id$ means $a = c$ and $b = d$ (since this is what equality means in $\mathbf{R}^2$) and that 0 stands for $0 + i0 = (0, 0)$. Thus $a + ib = 0$ means that *both* $a = 0$ and $b = 0$.

In what sense are these complex numbers an extension of the reals? We have already said that if a is real we also write a to stand for $a + 0i = (a, 0)$. In other words, the reals $\mathbf{R}$ are identified with the x axis in $\mathbf{C} = \mathbf{R}^2$; we are thus regarding the real numbers as those complex numbers $a + bi$ for which $b = 0$. If, in the expression $a + bi$, the term $a = 0$, we call $bi = 0 + bi$ a *pure imaginary number*. In the expression $a + bi$ we say that a is the *real part* and b is the *imaginary part*. This is sometimes written Re $z = a$, Im $z = b$, where $z = a + bi$. Note that Re z and Im z are always real numbers (see Fig. 1.1.1).

Actually, $\mathbf{C}$ obeys all the algebraic rules that ordinary real numbers do. For example, it will be shown in the following discussion that multiplicative inverses exist for nonzero elements. This means that if $z \neq 0$, then there is a (complex) number z' such that $zz' = 1$, and we write $z' = z^{-1}$. We can write this expression unambiguously (in other words, z' is uniquely determined), because if $zz'' = 1$ as well, then $z' = z' \cdot 1 = z'(zz'') = (z'z)z'' = 1 \cdot z'' = z''$, and so $z'' = z'$. To show that z' exists, suppose that $z = a + bi \neq 0$. Then at least one of $a \neq 0$, $b \neq 0$ holds, and so $a^2 + b^2 \neq 0$. To find z', we set $z' = a' + b'i$. The condition $zz' = 1$ imposes conditions that will enable us to compute a' and b'. Computing the product gives $zz' = (aa' - bb') + (ab' + a'b)i$. The linear equations $aa' - bb' = 1$ and $ab' + a'b = 0$ can be solved for a' and b' giving $a' = a/(a^2 + b^2)$ and $b' = -b/(a^2 + b^2)$, since $a^2 + b^2 \neq 0$. Thus for $z = a + ib \neq 0$, we may set

$$z^{-1} = \frac{a}{a^2 + b^2} - \frac{ib}{a^2 + b^2}$$

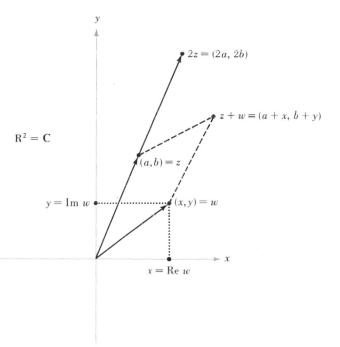

FIGURE 1.1.1 The geometry of complex numbers.

If z and w are complex numbers with $w \neq 0$, then the symbol z/w means zw^{-1}; we call z/w the *quotient* of z by w. Thus $z^{-1} = 1/z$. To compute z^{-1}, the following series of equations is common and is a useful way to remember the preceding formula for z^{-1}:

$$\frac{1}{a + ib} = \frac{a - ib}{(a + ib)(a - ib)} = \frac{a - ib}{a^2 + b^2} = \frac{a}{a^2 + b^2} - \frac{b}{a^2 + b^2} i$$

In short, all the usual algebraic rules for manipulating real numbers, fractions, polynomials, and so on, hold in complex analysis.

Formally, the system of complex numbers is an example of a *field*. The crucial rules for a field, stated here for reference, are:

Addition rules:
  (i) $z + w = w + z$
  (ii) $z + (w + s) = (z + w) + s$
  (iii) $z + 0 = z$
  (iv) $z + (-z) = 0$
Multiplication rules:
  (i) $zw = wz$
  (ii) $(zw)s = z(ws)$

(iii) $1z = z$

(iv) $z(z^{-1}) = 1$ for $z \neq 0$

Distributive law: $z(w + s) = zw + zs$

In summary, we have:

1.1.2 THEOREM   *The complex numbers $\mathbf{C}$ form a field.*

The student is cautioned that we generally do not define the symbol $\leq$, as in $z \leq w$, for complex z and w. If one requires the usual ordering properties for reals to hold, then *such an ordering is impossible* for complex numbers. This statement can be proved as follows. Suppose that such an ordering exists. Then either $i \geq 0$ or $i \leq 0$. Suppose that $i \geq 0$. Then $i \cdot i \geq 0$ and so $-1 \geq 0$, which is absurd. Alternatively, suppose that $i \leq 0$. Then $-i \geq 0$, so $(-i)(-i) \geq 0$, or $-1 \geq 0$, again absurd. If $z = a + ib$ and $w = c + id$, we could say that $z \leq w$ iff $a \leq c$ and $b \leq d$. This is an ordering of sorts, but it does not satisfy all the rules that might be required, such as those obeyed by real numbers. Thus in this text the notation $z \leq w$ will be avoided unless z and w happen to be real.

Roots of Quadratic Equations

As mentioned previously, one of the reasons for using complex numbers is to enable us to take square roots of negative real numbers. That this can, in fact, be done for all complex numbers is verified in the next proposition.

1.1.3 PROPOSITION   *Let $z \in \mathbf{C}$. Then there exists a $w \in \mathbf{C}$ such that $w^2 = z$. (Notice that $-w$ also satisfies this equation.)*

We shall give a purely algebraic proof here; another proof, based on polar coordinates, is given in Sec. 1.2.

PROOF   Let $z = a + bi$. We want to find $w = x + iy$ such that $a + bi = (x + iy)^2 = (x^2 - y^2) + (2xy)i$, and so we must simultaneously solve $x^2 - y^2 = a$ and $2xy = b$. The existence of such solutions is geometrically clear from examination of the graphs of the two equations. These are shown in Fig. 1.1.2 for the case in which both a and b are positive. From the graphs it is clear that there should be two solutions which are negatives of each other. In the following paragraph, these will be obtained algebraically.
  We know that $(x^2 + y^2)^2 = (x^2 - y^2)^2 + 4x^2y^2 = a^2 + b^2$. Hence $x^2 + y^2 = \sqrt{a^2 + b^2}$, and so $x^2 = (a + \sqrt{a^2 + b^2})/2$ and $y^2 = (-a + \sqrt{a^2 + b^2})/2$. If we let

$$\alpha = \sqrt{\frac{a + \sqrt{a^2 + b^2}}{2}} \quad \text{and} \quad \beta = \sqrt{\frac{-a + \sqrt{a^2 + b^2}}{2}}$$

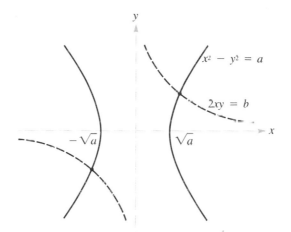

FIGURE 1.1.2 Graphs of the curves $x^2 - y^2 = a$ and $2xy = b$.

where $\sqrt{}$ denotes the positive square root of positive real numbers, then, in the event that b is positive, we have either $x = \alpha$, $y - \beta$ or $x = -\alpha$, $y = -\beta$; in the event that b is negative, we have either $x = \alpha$, $y = -\beta$ or $x = -\alpha$, $y = \beta$. We conclude that the equation $w^2 = z$ has solutions $\pm(\alpha + \mu\beta i)$, where $\mu = 1$ if $b \geq 0$ and $\mu = -1$ if $b < 0$. ∎

From the expressions for α and β we can conclude three things: (1) the square roots of a complex number are real if and only if the complex number is real and positive; (2) the square roots of a complex number are purely imaginary if and only if the complex number is real and negative; and (3) the two square roots of a number coincide if and only if the complex number is zero. (The student should check these conclusions.)

We can easily check that the quadratic equation $az^2 + bz + c = 0$ for complex numbers a, b, c has solutions $z = (-b \pm \sqrt{b^2 - 4ac})/2a$, where now the symbol $\sqrt{u}$ denotes the square root of the complex number u, as constructed in Proposition 1.1.3.

Uniqueness of the Complex Numbers

The following discussion will not be fully rigorous; rather it will, at times, be somewhat informal. An absolutely precise explanation would be tantamount to a short course in abstract algebra. However, the student should nevertheless be able to grasp the important points of the presentation.

We have constructed the field **C**, which contains the reals and in which every quadratic equation has a solution. It is only natural to ask whether there are any other fields containing the reals in which every quadratic equation has a solution.

The answer is that any other such field called, say, F, must already contain the complex numbers; therefore, the complex numbers are the *smallest* field containing $\mathbf{R}$ in which all quadratic equations are solvable. In that sense it is unique. The reason for this is quite simple. Let j be any solution in F to the equation $z^2 + 1 = 0$. Consider, in F, all numbers of the form $a + jb$ for real numbers a and b. This set is, algebraically, the same as $\mathbf{C}$, because of the simple fact that since $j^2 = -1$, j plays the role of and may be identified with i. We must also check that $a + jb = c + jd$ implies that $a = c$ and $b = d$ to be sure that equality in this set coincides with that in $\mathbf{C}$. Indeed, $(a - c) + j(b - d) = 0$, so we must prove that $e + jf = 0$ implies that $e = 0$ and $f = 0$ (where $e = a - c$ and $f = b - d$). If $f = 0$, then clearly $e = 0$ as well. But if $f \neq 0$, then $j = -e/f$, which is real, and no *real* number satisfies $j^2 = -1$, because the square of any real number is nonnegative, and so f must be zero. This proves our claim. We can rephrase our result by saying that $\mathbf{C}$ *is the smallest field extension of* $\mathbf{R}$ *in which quadratic equations are solvable.*

Another question arises at this point. We made $\mathbf{R}^2$ into a field. For what other n can $\mathbf{R}^n$ be made into a field? Let us demand at the outset that the algebraic operations agree with those on $\mathbf{R}$, assuming that $\mathbf{R}$ is the x axis. The answer is, Only in the case $n = 2$. A fieldlike structure, called the *quaternions,* can be obtained for $n = 4$, except that the rule $zw = wz$ fails. Such a structure is called a *noncommutative field.* The proof of these facts can be found in an advanced abstract algebra text.

Worked Examples

1.1.4   *Prove that* $1/i = -i$ *and that* $1/(i + 1) = (1 - i)/2.$

Solution. First,

$$\frac{1}{i} = \frac{1}{i} \cdot \frac{-i}{-i} = -i$$

because $i \cdot -i = -(i^2) = -(-1) = 1$. Also,

$$\frac{1}{i+1} = \frac{1}{i+1} \frac{1-i}{1-i} = \frac{1-i}{2}$$

since $(1 + i)(1 - i) = 1 + 1 = 2$.

1.1.5   *Find the real and imaginary parts of* $(z + 2)/(z - 1)$ *where* $z = x + iy.$

Solution.

$$\frac{z+2}{z-1} = \frac{(x+2)+iy}{(x-1)+iy} = \frac{(x+2)+iy}{(x-1)+iy} \cdot \frac{(x-1)-iy}{(x-1)-iy}$$

$$= \frac{(x+2)(x-1)+y^2+i[y(x-1)-y(x+2)]}{(x-1)^2+y^2}$$

Hence,

$$\text{Re}\,\frac{z+2}{z-1} = \frac{x^2 + x - 2 + y^2}{(x-1)^2 + y^2}$$

and

$$\text{Im}\,\frac{z+2}{z-1} = \frac{-3y}{(x-1)^2 + y^2}$$

1.1.6 *Solve $z^4 + i = 0$ for z.*

Solution. Let $z^2 = w$. Then $w^2 + i = 0$. Substituting in the formula $\sqrt{a+ib} = \pm(\alpha + \mu\beta i)$ we developed for taking square roots, and letting $a = 0$ and $b = -1$, we get

$$w = \pm\left(\frac{1}{\sqrt{2}} - \frac{1}{\sqrt{2}}\,i\right)$$

Consider the equation $z^2 = (1-i)/\sqrt{2}$. Again substituting in the formula, now letting $a = 1/\sqrt{2}$ and $b = -1/\sqrt{2}$, we obtain the two solutions

$$z = \pm\left(\frac{\sqrt{2+\sqrt{2}}}{2} - \frac{\sqrt{2-\sqrt{2}}}{2}\,i\right)$$

From the other value for w we obtain two further solutions:

$$z = \pm\left(\frac{\sqrt{2-\sqrt{2}}}{2} + \frac{\sqrt{2+\sqrt{2}}}{2}\,i\right)$$

Remark. In the next section, de Moivre's formula will be developed, which will enable us to find, quite simply, the nth root of any complex number.

1.1.7 *Prove that, for complex numbers z and w,*

$$\text{Re}\,(z+w) = \text{Re}\,z + \text{Re}\,w$$

and

$$\text{Im}\,(z+w) = \text{Im}\,z + \text{Im}\,w$$

Solution. Let $z = x + iy$ and $w = a + ib$. Then $z + w = (x+a) + i(y+b)$, and so $\text{Re}\,(z+w) = x + a = \text{Re}\,z + \text{Re}\,w$. Similarly, $\text{Im}\,(z+w) = y + b = \text{Im}\,z + \text{Im}\,w$.

Exercises

1. Express the following complex numbers in the form $a + ib$:

(a) $(2 + 3i) + (4 + i)$

(b) $\dfrac{2 + 3i}{4 + i}$

(c) $\dfrac{1}{i} + \dfrac{3}{1 + i}$

2. Express the following complex numbers in the form $a + bi$:

(a) $(2 + 3i)(4 + i)$

(b) $(8 + 6i)^2$

(c) $\left(1 + \dfrac{3}{1 + i}\right)^2$

3. Find the solutions to $z^2 = 3 - 4i$.

4. Find the solutions to:

(a) $(z + 1)^2 = 3 + 4i$

(b) $z^4 - i = 0$

5. Find the real and imaginary parts of the following, where $z = x + iy$:

(a) $\dfrac{1}{z^2}$

(b) $\dfrac{1}{3z + 2}$

6. Find the real and imaginary parts of the following, where $z = x + iy$:

(a) $\dfrac{z + 1}{2z - 5}$

(b) z^3

7. Is it true that $\mathrm{Re}\,(zw) = (\mathrm{Re}\,z)(\mathrm{Re}\,w)$?

8. If a is real and z is complex, prove that $\mathrm{Re}\,(az) = a\,\mathrm{Re}\,z$ and that $\mathrm{Im}\,(az) = a\,\mathrm{Im}\,z$. Generally, show that $\mathrm{Re}\colon \mathbf{C} \to \mathbf{R}$ is a real linear map; that is, that $\mathrm{Re}\,(az + bw) = a\,\mathrm{Re}\,z + b\,\mathrm{Re}\,w$ for a, b real, z, w complex.

9. Show that $\mathrm{Re}\,(iz) = -\,\mathrm{Im}\,(z)$ and that $\mathrm{Im}\,(iz) = \mathrm{Re}\,(z)$ for any complex number z.

10. (a) Fix a complex number $z = x + iy$ and consider the linear mapping $\phi_z\colon \mathbf{R}^2 \to \mathbf{R}^2$ (that is, of $\mathbf{C} \to \mathbf{C}$) defined by $\phi_z(w) = z \cdot w$ (that is, multiplication by z). Prove that the matrix of ϕ_z in the standard basis $(1, 0)$, $(0, 1)$ of $\mathbf{R}^2$ is given by

$$\begin{pmatrix} x & -y \\ y & x \end{pmatrix}$$

(b) Show that $\phi_{z_1 z_2} = \phi_{z_1} \circ \phi_{z_2}$.

11. Assuming that they work for real numbers, show that the nine rules given for a field also work for complex numbers.

12. Using only the axioms for a field, give a formal proof (including all details) for the following:

(a) $\dfrac{1}{z_1 z_2} = \dfrac{1}{z_1} \cdot \dfrac{1}{z_2}$

(b) $\dfrac{1}{z_1} + \dfrac{1}{z_2} = \dfrac{z_1 + z_2}{z_1 z_2}$

13. Let $(x - iy)/(x + iy) = a + ib$. Prove that $a^2 + b^2 = 1$.

14. Prove the binomial theorem for complex numbers; that is, letting z, w be complex numbers and n be a positive integer,

$$(z + w)^n = z^n + \binom{n}{1} z^{n-1} w + \binom{n}{2} z^{n-2} w^2 + \cdots + \binom{n}{n} w^n$$

where

$$\binom{n}{r} = \frac{n!}{r!(n - r)!}$$

Use induction on n.

15. Show that z is real if and only if Re $z = z$.

16. Prove that, for any integer k,

$$i^{4k} = 1,\ i^{4k+1} = i,\ i^{4k+2} = -1,\ i^{4k+3} = -i$$

Show how this result gives a formula for i^n for all n by writing $n = 4k + j, 0 \le j \le 3$.

17. Simplify the following:

(a) $(1 + i)^4$

(b) $(-i)^{-1}$

18. Simplify the following:

(a) $(1 - i)^{-1}$

(b) $\dfrac{1 + i}{1 - i}$

19. Simplify the following:

(a) $\sqrt{1 + \sqrt{i}}$

(b) $\sqrt{1 + i}$

(c) $\sqrt{\sqrt{-i}}$

20. Show that the following rules uniquely determine complex multiplication on $\mathbf{C} = \mathbf{R}^2$.

(a) $(z_1 + z_2)w = z_1 w + z_2 w$.

(b) $z_1 z_2 = z_2 z_1$.

(c) $i \cdot i = -1$.

(d) $z_1(z_2 z_3) = (z_1 z_2)z_3$.

(e) If z_1 and z_2 are real, $z_1 \cdot z_2$ is the usual product of real numbers.

1.2 PROPERTIES OF COMPLEX NUMBERS

In mathematics it is very important to be able to picture the concepts being studied and to develop what is called geometric intuition. This ability is particularly valuable in dealing with complex numbers. In studying them we shall also encounter the important concepts of the absolute value, argument, polar representation, and complex conjugate of a complex number. These concepts have simple geometric interpretations that are important to understand.

Properties of Complex Numbers

In the preceding section a complex number was defined to be a point in the plane $\mathbf{R}^2$. Besides addition and multiplication by a real scalar, the operation of complex multiplication was defined. A complex number may be thought of geometrically as a (two-dimensional) vector and pictured as an arrow from the origin to the point in $\mathbf{R}^2$ given by the complex number (see Fig. 1.2.1).

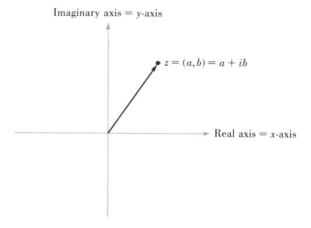

FIGURE 1.2.1   Vector representation of complex numbers.

Because the points $(x, 0) \in \mathbf{R}^2$ correspond to real numbers, the horizontal or x axis is called the *real axis*. Similarly, the vertical axis (the y axis) is called the *imaginary axis,* because points on it have the form $iy = (0, y)$ for y real. Addition of complex numbers can thus be pictured as addition of vectors (see Figs. 1.1.1 and 1.2.2).

It would be helpful if we could picture complex multiplication in the same manner. To this end, we shall first write complex numbers in what is called polar coordinate form. To do this, recall that the *length* of the vector $(a, b) = a + ib$ is defined as $r = \sqrt{a^2 + b^2}$. Suppose the vector makes an angle θ with the positive direction of the real axis,

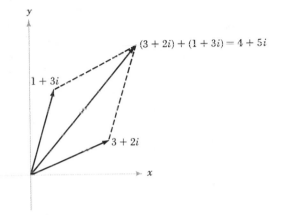

FIGURE 1.2.2 Addition of complex numbers.

where $0 \le \theta < 2\pi$ (see Fig. 1.2.3). Thus $\tan \theta = b/a$. Since $a = r \cos \theta$ and $b = r \sin \theta$, we thus have $a + bi = r \cos \theta + (r \sin \theta)i = r(\cos \theta + i \sin \theta)$. This way of writing the complex number is called the *polar coordinate representation*. The length of the vector $z = (a, b) = a + ib$ is denoted $|z|$ and is called the *norm*, or *modulus*, or *absolute value* of z. The angle θ is called the *argument* of the complex number and is denoted $\theta = \arg z$.

If we restrict our θ, as we actually did, to the interval $[0, 2\pi[$—that is, $0 \le \theta < 2\pi$— then each nonzero complex number has an unambiguously defined argument. (We accept this as known from trigonometry.) However, it is clear that we can add integral multiples of 2π to θ and still obtain the same complex number. In fact, we shall find it

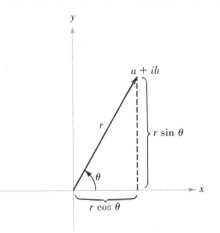

FIGURE 1.2.3 Polar coordinate representation of complex numbers.

convenient to be flexible in our requirements for the values that θ is to assume. For example, we could equally well allow the range of θ to be $-\pi < \theta \le \pi$. Such an interval will always be specified in subsequent chapters. The student should remember that once this interval is specified for each $z \ne 0$, a unique θ is determined that lies within that specified interval. It is clear that any $\theta \in \mathbf{R}$ can be brought into our specified interval ($[0, 2\pi[$or$] -\pi, \pi]$, for example) by addition of some (positive or negative) integral multiple of 2π. For these reasons it is sometimes best to think of arg z as the set of possible values of the angle. If θ is one possible value, then so is $\theta + 2\pi n$ for any integer n, and we can sometimes think of arg z as $\{\theta + 2\pi n: n$ is an integer$\}$. Specification of a particular range for the angle is known as choosing a *branch of the argument*.

Use of the polar representation of complex numbers simplifies the task of describing geometrically the product of two complex numbers. Let $z_1 = r_1(\cos \theta_1 + i \sin \theta_1)$ and $z_2 = r_2(\cos \theta_2 + i \sin \theta_2)$. Then $z_1 z_2 = r_1 r_2 [(\cos \theta_1 \cdot \cos \theta_2 - \sin \theta_1 \cdot \sin \theta_2) + i[\cos \theta_1 \cdot \sin \theta_2 + \cos \theta_2 \cdot \sin \theta_1)] = r_1 r_2 [\cos (\theta_1 + \theta_2) + i \sin (\theta_1 + \theta_2)]$, by the addition formula for the sine and cosine functions used in trigonometry. Thus we have proven

1.2.1 PROPOSITION   *For any complex numbers z_1 and z_2,*

$$|z_1 z_2| = |z_1| \cdot |z_2|$$

and

$$\arg (z_1 z_2) = \arg z_1 + \arg z_2 \quad (\bmod 2\pi)$$

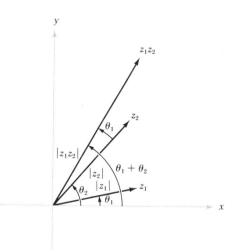

FIGURE 1.2.4   Multiplication of complex numbers.

Translated into words, the product of two complex numbers is the complex number that has a length equal to the product of the lengths of the two complex numbers and an argument equal to the sum of the arguments of those numbers. This is the basic geometric representation of complex multiplication (see Fig. 1.2.4).

The second equality in Proposition 1.2.1 means that the sets of possible values for the left and right sides are the same, that is, that the two sides can be made to agree by the addition of the appropriate multiple of 2π to one side. If a particular branch is desired and arg z_1 + arg z_2 lies outside the interval that we specify, we should adjust it by a multiple of 2π to bring it within that interval. For example, if our interval is $[0, 2\pi[$ and $z_1 = -1$ and $z_2 = -i$, then arg $z_1 = \pi$ and arg $z_2 = 3\pi/2$ (see Fig. 1.2.5), but $z_1 z_2 = i$, so arg $z_1 z_2 = \pi/2$, and arg z_1 + arg $z_2 = \pi + 3\pi/2 = 2\pi + \pi/2$. We can obtain the correct answer by subtracting 2π to bring it within the interval $[0, 2\pi[$.

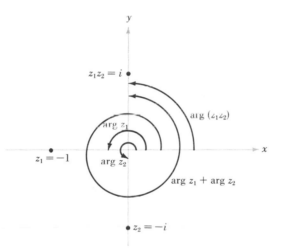

FIGURE 1.2.5   Multiplication of the complex numbers -1 and $-i$.

There is also a geometric construction using similar triangles that enables us to construct the product of two complex numbers. The two shaded triangles in Figure 1.2.6 are similar. The construction of the point $z_1 z_2$ is effected by drawing the two angles shown, each with magnitude θ_1 and the angle σ. That the constructed point is the correct one follows from Proposition 1.2.1 and properties of similar triangles.

Multiplication of complex numbers can be analyzed in another useful way. Let $z \in \mathbf{C}$ and define $\psi_z : \mathbf{C} \to \mathbf{C}$ by $\psi_z(w) = wz$; that is, ψ_z is the map "multiplication by z." By Proposition 1.2.1, it is clear that *the effect of this map is to rotate a complex number through an angle equal to* arg z *in the counterclockwise direction and to stretch its length by the factor* $|z|$. For example, ψ_i (multiplication by i) simply rotates complex numbers by $\pi/2$ in the counterclockwise direction (see Figure 1.2.7).

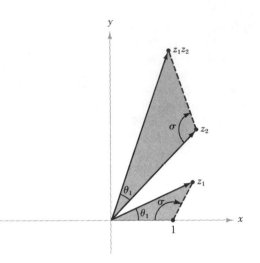

FIGURE 1.2.6   Geometric construction of the product of two complex numbers.

The map ψ_z is a linear transformation on the plane, in the sense that $\psi_z(\lambda w_1 + \mu w_2) = \lambda\psi_z(w_1) + \mu\psi_z(w_2)$, where λ, μ are real numbers and w_1, w_2 are complex numbers. Any linear transformation of the plane to itself can be represented by a matrix. If $z = a + ib = (a, b)$, then the matrix of ψ_z is

$$\begin{pmatrix} a & -b \\ b & a \end{pmatrix}$$

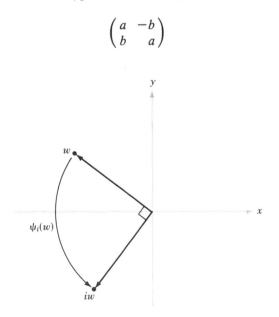

FIGURE 1.2.7   Multiplication by i.

since

$$\begin{pmatrix} a & -b \\ b & a \end{pmatrix}\begin{pmatrix} x \\ y \end{pmatrix} = \begin{pmatrix} ax - by \\ ay + bx \end{pmatrix}$$

(see Exercise 10, Sec. 1.1).

de Moivre's Formula

The formula we derived for multiplication, using the polar coordinate representation, provides more than geometric intuition. We can use it to obtain a formula that enables us to find the nth roots of any complex number.

1.2.2 DE MOIVRE'S FORMULA If $z = r(\cos \theta + i \sin \theta)$ and n is a positive integer, then

$$z^n = r^n(\cos n\theta + i \sin n\theta)$$

PROOF By Proposition 1.2.1, $z^2 = r^2[\cos(\theta + \theta) + i \sin (\theta + \theta)] = r^2(\cos 2\theta + i \sin 2\theta)$. Multiplying again by z gives $z^3 = z \cdot z^2 = r \cdot r^2[\cos (2\theta + \theta) + i \sin (2\theta + \theta)] = r^3(\cos 3\theta + i \sin 3\theta)$. This procedure may be continued by induction to obtain the desired result for any integer n. ■

Let w be a complex number; that is, let $w \in \mathbb{C}$. Using de Moivre's formula we will solve the equation $z^n = w$ for z when w is given. Suppose that $w = r(\cos \theta + i \sin \theta)$ and $z = \rho(\cos \psi + i \sin \psi)$. Then, by de Moivre's formula, $z^n = \rho^n(\cos n\psi + i \sin n\psi)$. It follows that $\rho^n = r = |w|$ by uniqueness of the polar representation and $n\psi = \theta + k(2\pi)$, where k is some integer. Thus

$$z = \sqrt[n]{r}\left[\cos\left(\frac{\theta}{n} + \frac{k}{n} 2\pi\right) + i \sin\left(\frac{\theta}{n} + \frac{k}{n} 2\pi\right)\right]$$

Each value of $k = 0, 1, \ldots, n - 1$ gives a different value of z. Any other value of k merely repeats one of the values of z corresponding to $k = 0, 1, 2, \ldots, n - 1$. Thus there are exactly n nth roots of any complex number.

For example, the preceding formula shows the three solutions to the equation $z^3 = 1 = 1(\cos 0 + i \sin 0)$ to be

$$z = \cos\frac{k2\pi}{3} + i \sin\frac{k2\pi}{3} \qquad k = 0, 1, 2$$

That is,

$$z = 1, -\frac{1}{2} + \frac{i\sqrt{3}}{2}, -\frac{1}{2} - \frac{i\sqrt{3}}{2}$$

This procedure for finding roots is summarized as follows.

1.2.3 COROLLARY   *Let w be a nonzero complex number with polar representation* $w = r(\cos\theta + i\sin\theta)$. *Then the nth roots of w are given by the n complex numbers*

$$z_k = \sqrt[n]{r}\left[\cos\left(\frac{\theta}{n} + \frac{2\pi k}{n}\right) + i\sin\left(\frac{\theta}{n} + \frac{2\pi k}{n}\right)\right] \quad k = 0, 1, \ldots, n-1$$

As a special case of this formula we note that the n roots of 1 (that is, the nth roots of unity) are 1 and $n-1$ other points equally spaced around the unit circle, as illustrated in Fig. 1.2.8 for the case $n = 8$.

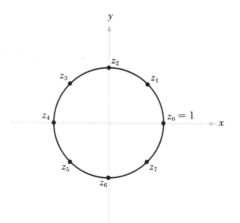

FIGURE 1.2.8   The eighth roots of unity.

Complex Conjugation

Subsequent chapters will include many references to the simple idea of conjugation, which is defined as follows: If $z = a + ib$, then $\bar{z}$, the *complex conjugate* of z, is defined by $\bar{z} = a - ib$. Complex conjugation can be pictured geometrically as reflection in the real axis (see Fig. 1.2.9).

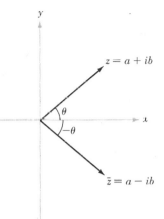

FIGURE 1.2.9 Complex conjugation.

Proposition 1.2.4 summarizes the main properties of complex conjugation.

1.2.4 PROPOSITION

(i) $\overline{z + z'} = \bar{z} + \bar{z}'$.

(ii) $\overline{zz'} = \bar{z}\,\bar{z}'$.

(iii) $\overline{z/z'} = \bar{z}/\bar{z}'$ for $z' \neq 0$.

(iv) $z\bar{z} = |z|^2$ and hence if $z \neq 0$, we have $z^{-1} = \bar{z}/|z|^2$.

(v) $z = \bar{z}$ if and only if z is real.

(vi) Re $z = (z + \bar{z})/2$ and Im $z = (z - \bar{z})/2i$.

(vii) $\bar{\bar{z}} = z$.

PROOF

(i) Let $z = a + ib$ and let $z' = a' + ib'$. Then $z + z' = a + a' + i(b + b')$, and so $\overline{z + z'} = (a + a') - i(b + b') = a - ib + a' - ib' = \bar{z} + \bar{z}'$.

(ii) Let $z = a + ib$ and let $z' = a' + ib'$. Then

$$\overline{zz'} = \overline{(aa' - bb') + i(ab' + a'b)} = (aa' - bb') - i(ab' + a'b)$$

On the other hand, $\bar{z}\,\bar{z}' = (a - ib)(a' - ib') = (aa' - bb') - i(ab' + a'b)$.

(iii) By (ii) we have $\overline{z'}\,\overline{z/z'} = \overline{z'z/z'} = \bar{z}$. Hence $\overline{z/z'} = \bar{z}/\bar{z}'$.

(iv) $z\bar{z} = (a + ib)(a - ib) = a^2 + b^2 = |z|^2$.

(v) If $a + ib = a - ib$, then $ib = -ib$, and so $b = 0$.

(vi) This assertion is clear by the definition of $\bar{z}$.

(vii) This assertion is also clear by the definition of complex conjugation. ∎

The absolute value of a complex number $|z| = |a + ib| = \sqrt{a^2 + b^2}$, which is merely the usual Euclidean length of the vector representing the complex number, has already been defined. From Proposition 1.2.4(iv), we note that $|z|$ is also given by $|z|^2 = z\bar{z}$. The absolute value of a complex number is encountered throughout complex analysis; the following properties of the absolute value are quite basic.

1.2.5 PROPOSITION

(i) $|zz'| = |z| \cdot |z'|$.

(ii) *If $z' \neq 0$, then* $|z/z'| = |z|/|z'|$.

(iii) $-|z| \leq \mathrm{Re}\ z \leq |z|$ *and* $-|z| \leq \mathrm{Im}\ z \leq |z|$; *that is,* $|\mathrm{Re}\ z| \leq |z|$ *and* $|\mathrm{Im}\ z| \leq |z|$.

(iv) $|\bar{z}| = |z|$.

(v) $|z + z'| \leq |z| + |z'|$.

(vi) $|z - z'| \geq ||z| - |z'||$.

(vii) $|z_1 w_1 + \cdots + z_n w_n| \leq \sqrt{|z_1|^2 + \cdots + |z_n|^2} \sqrt{|w_1|^2 + \cdots + |w_n|^2}$.

Statement (iv) is clear geometrically from Fig. 1.2.9, (v) is the usual *triangle inequality* for vectors in $\mathbf{R}^2$ (see Fig. 1.2.10) and (vii) is referred to as *Cauchy's inequality*. By repeated application of (v) we get the general statement $|z_1 + \cdots + z_n| \leq |z_1| + \cdots + |z_n|$.

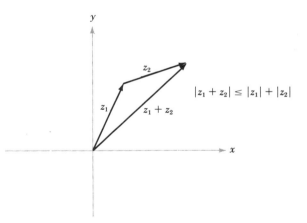

FIGURE 1.2.10 Triangle inequality.

PROOF

(i) This equality was shown in Proposition 1.2.1.

(ii) By (i), $|z'||z/z'| = |z' \cdot (z/z')| = |z|$, so $|z/z'| = |z|/|z'|$.

(iii) If $z = a + ib$, then $-\sqrt{a^2 + b^2} \leq a \leq \sqrt{a^2 + b^2}$ since $b^2 \geq 0$. The other inequality asserted in (iii) is similarly proved.

(iv) If $z = a + ib$, then $\bar{z} = a - ib$, and we clearly have $|z| = \sqrt{a^2 + b^2} = \sqrt{a^2 + (-b)^2} = |\bar{z}|$.

(v) By Proposition 1.2.4 (iv),

$$|z + z'|^2 = (z + z')\overline{(z + z')}$$
$$= (z + z')(\bar{z} + \bar{z}')$$
$$= z\bar{z} + z'\bar{z}' + z'\bar{z} + z\bar{z}'$$

But $z\bar{z}'$ is the conjugate of $z'\bar{z}$ (Why?), so by Proposition 1.2.4(vi) and (iii) in this proof, $|z|^2 + |z'|^2 + 2\,\mathrm{Re}\,z'\bar{z} \leq |z|^2 + |z'|^2 + 2|z'\bar{z}| = |z|^2 + |z'|^2 + 2|z||z'|$. But this equals $(|z| + |z'|)^2$, so we get our result.

(vi) By applying (v) to z' and $z - z'$ we get $|z| = |z' + (z - z')| \leq |z'| + |z - z'|$, so $|z - z'| \geq |z| - |z'|$. By interchanging the roles of z and z', we similarly get $|z - z'| \geq |z'| - |z| = -(|z| - |z'|)$, which is what we originally claimed.

(vii) This inequality is less evident, and the proof of it requires a slight mathematical trick (see Exercise 22 for a different proof). Let us suppose that not all the $w_k = 0$ (or else the result is clear). Let

$$v = \sum_{k=1}^{n} |z_k|^2 \qquad t = \sum_{k=1}^{n} |w_k|^2 \qquad s = \sum_{k=1}^{n} z_k\overline{w}_k \qquad \text{and} \qquad c = s/t$$

Now consider

$$\sum_{k=1}^{n} |z_k - c\overline{w}_k|^2$$

which is ≥ 0 and equals

$$v + |c|^2 t - c\sum_{k=1}^{n} \overline{z}_k\overline{w}_k - \bar{c}\sum_{k=1}^{n} z_k w_k = v + |c|^2 t - 2\,\mathrm{Re}\,\bar{c}s$$

$$= v + \frac{|s|^2}{t} - 2\,\mathrm{Re}\,\frac{\overline{s}s}{t}$$

Since t is real and $s\overline{s} = |s|^2$ is real, $v + (|s|^2/t) - 2(|s|^2/t) = v - |s|^2/t \geq 0$. Hence $|s|^2 \leq vt$, which is the desired result. ∎

Worked Examples

1.2.6 *Solve $z^8 = 1$ for z.*

Solution. Since $1 = \cos k2\pi + i\sin k2\pi$ when k equals any integer, Corollary 1.2.3 gives

$$z = \cos\frac{k2\pi}{8} + i\sin\frac{k2\pi}{8} \qquad k = 0, 1, 2, \ldots, 7$$

$$= 1, \frac{1}{\sqrt{2}} + \frac{i}{\sqrt{2}}, i, \frac{-1}{\sqrt{2}} + \frac{i}{\sqrt{2}}, -1, \frac{-1}{\sqrt{2}} - \frac{i}{\sqrt{2}}, -i, \frac{1}{\sqrt{2}} - \frac{i}{\sqrt{2}}$$

These may be pictured as points evenly spaced on the circle in the complex plane (see Fig. 1.2.11).

1.2.7   *Show that* $\overline{\left[\dfrac{(3+7i)^2}{(8+6i)}\right]} = \dfrac{(3-7i)^2}{(8-6i)}.$

Solution. The point here is that it is not necessary first to work out $(3+7i)^2/(8+6i)$ if we simply use the properties developed in the text, namely, $\overline{z^2} = (\bar{z})^2$ and $\overline{z/z'} = \bar{z}/\bar{z}'$. Thus we obtain

$$\overline{\left[\frac{(3+7i)^2}{(8+6i)}\right]} = \frac{\overline{(3+7i)^2}}{\overline{(8+6i)}} = \frac{\overline{(3+7i)}^2}{8-6i} = \frac{(3-7i)^2}{8-6i}$$

1.2.8   *If* $|z| = 1$, *prove that*

$$\left|\frac{az+b}{\bar{b}z+\bar{a}}\right| = 1$$

for any complex numbers a and b.

Solution. Since $|z| = 1$, we have $z = \bar{z}^{-1}$. Thus

$$\frac{az+b}{\bar{b}z+\bar{a}} = \frac{az+b}{\bar{b}+\bar{a}\bar{z}} \cdot \frac{1}{z}$$

$z \cdot \bar{z} = |z|^2$

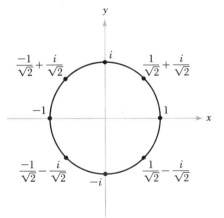

FIGURE 1.2.11   The eight 8th roots of unity.

Because of the properties of the absolute value and because $|z| = 1$,

$$\left|\frac{az + b}{\bar{b}z + \bar{a}}\right| = \left|\frac{az + b}{\bar{a}\bar{z} + \bar{b}}\right| \cdot \frac{1}{|z|} = 1$$

since $|az + b| = |\overline{az + b}| = |\bar{a}\bar{z} + \bar{b}|$.

1.2.9 *Show that the maximum absolute value of $z^2 + 1$ on the unit disk $|z| \leq 1$ is 2.*

Solution. By the triangle inequality, $|z^2 + 1| \leq |z^2| + 1 = |z|^2 + 1 \leq 1^2 + 1 = 2$, since $|z| < 1$ thus $|z^2 + 1|$ does not exceed 2 on the disk. Since the value 2 is achieved at $z = 1$, the maximum is 2.

1.2.10 *Express $\cos 3\theta$ in terms of $\cos \theta$ and $\sin \theta$ using de Moivre's formula.*

Solution. De Moivre's formula for $r = 1$ and $n = 3$ gives the identity

$$(\cos \theta + i \sin \theta)^3 = \cos 3\theta + i \sin 3\theta$$

The left side of this equation, when expanded (see Exercise 14 of Sec. 1.1), becomes

$$\cos^3 \theta + i3 \cos^2 \theta \sin \theta - 3 \cos \theta \sin^2 \theta - i \sin^3 \theta$$

By equating real and imaginary parts, we get

$$\cos 3\theta = \cos^3 \theta - 3 \cos \theta \sin^2 \theta$$

and the additional formula

$$\sin 3\theta = -\sin^3 \theta + 3 \cos^2 \theta \sin \theta$$

1.2.11 *Write, in complex notation, the equation of a straight line; of a circle; of an ellipse.*

Solution.

The straight line is most conveniently expressed in parametric form: $z = a + bt$, $a, b \in \mathbf{C}, t \in \mathbf{R}$, which represents a line in the direction of b and passing through the point a.

The circle can be expressed as $|z - a| = r$ (radius r, center a).

The ellipse can be expressed as $|z - d| + |z + d| = 2a$; the foci are located at $\pm d$ and the semimajor axis equals a.

These equations, in which $|\cdot|$ is interpreted as length, are the usual geometric definitions of these loci.

1.2.12   *Suppose $u = a + ib$ and $v = c + id$ are fixed complex numbers and that μ is a positive real number. Show that the equation*

$$\left| \frac{z - u}{z - v} \right|^2 = \mu$$

describes a circle or a straight line in the complex plane.

Solution. If we set $z = x + iy$, the equation becomes

$$\mu = \left| \frac{(x - a) + i(y - b)}{(x - c) + i(y - d)} \right|^2 = \frac{(x - a)^2 + (y - b)^2}{(x - c)^2 + (y - d)^2}$$

Cross-multiplying, expanding, and gathering terms involving x and y on the left-hand side gives

$$(1 - \mu)x^2 + (1 - \mu)y^2 - 2(a - \mu c)x - 2(b - \mu d)y = \mu(c^2 + d^2) - (a^2 + b^2)$$

If $\mu = 1$, this is the equation of a line. In fact, it is the perpendicular bisector of the line segment between u and v. If $\mu \neq 1$, completing the square gives

$$\left(x - \frac{a - \mu c}{1 - \mu} \right)^2 + \left(y - \frac{b - \mu d}{1 - \mu} \right)^2 = \frac{\mu(c^2 + d^2) - (a^2 + b^2)}{1 - \mu} + \left(\frac{a - \mu c}{1 - \mu} \right)^2 + \left(\frac{b - \mu d}{1 - \mu} \right)^2$$

$$= \frac{\mu}{(1 - \mu)^2} [(a - c)^2 + (b - d)^2]$$

which is the equation of a circle centered at $((a - \mu c)/(1 - \mu), (b - \mu d)/(1 - \mu))$ and with radius $(\sqrt{\mu}/|1 - \mu|)\sqrt{(a - c)^2 + (b - d)^2}$.

Exercises

1. Solve the following equations:

(a) $z^5 - 2 = 0$ (b) $z^4 + i = 0$

2. Solve the following equations:

(a) $z^6 + 8 = 0$ (b) $z^3 - 4 = 0$

3. What is the complex conjugate of $(3 + 8i)^4/(1 + i)^{10}$?

4. What is the complex conjugate of $(8 - 2i)^{10}/(4 + 6i)^5$?

5. Express $\cos 5x$ and $\sin 5x$ in terms of $\cos x$ and $\sin x$.

6. Express $\cos 6x$ and $\sin 6x$ in terms of $\cos x$ and $\sin x$.

7. Find the absolute value of $[i(2 + 3i)(5 - 2i)]/(-2 - i)$

8. Find the absolute value of $(2 - 3i)^2/(8 + 6i)^2$.

9. Let w be an nth root of unity, $w \neq 1$. Show that $1 + w + w^2 + \cdots + w^{n-1} = 0$.

10. Show that the roots of a polynomial with real coefficients occur in conjugate pairs.

11. If $a, b \in \mathbb{C}$, prove the *parallelogram identity*: $|a - b|^2 + |a + b|^2 = 2(|a|^2 + |b|^2)$.

12. Interpret the identity in Exercise 11 geometrically.

13. When does equality hold in the triangle inequality $|z_1 + z_2 + \cdots + z_n| \leq |z_1| + |z_2| + \cdots + |z_n|$? Interpret your result geometrically.

14. Assuming either $|z| = 1$ or $|w| = 1$ and $\bar{z}w \neq 1$, prove that

$$\left| \frac{z - w}{1 - \bar{z}w} \right| = 1$$

15. Does $z^2 = |z|^2$? If so, prove this equality. If not, for what z is it true?

16. Letting $z = x + iy$, prove that $|x| + |y| \leq \sqrt{2}|z|$.

17. Let $z = a + ib$ and $z' = a' + ib'$. Prove that $|zz'| = |z||z'|$ by evaluating each side.

18. Prove the following:
(a) $\arg \bar{z} = -\arg z \pmod{2\pi}$ (b) $\arg (z/w) - \arg z - \arg w \pmod{2\pi}$
(c) $|z| = 0$ if and only if $z = 0$

19. What is the equation of the circle with radius 3 and center $8 + 5i$ in complex notation?

20. Using the formula $z^{-1} = \bar{z}/|z|^2$, show how to construct z^{-1} geometrically.

21. Describe the set of all z such that $\text{Im}\,(z + 5) = 0$.

22. Prove Lagrange's identity:

$$\left| \sum_{k=1}^{n} z_k w_k \right|^2 = \left(\sum_{k=1}^{n} |z_k|^2 \right) \left(\sum_{k=1}^{n} |w_k|^2 \right) - \sum_{k<j} |z_k \bar{w}_j - z_j \bar{w}_k|^2$$

Deduce the Cauchy inequality from your proof.

23. Find the maximum of $|z^n + a|$ for those z with $|z| \leq 1$.

24. Compute the least upper bound (that is, supremum) of the following set of real numbers: $\{\mathrm{Re}\,(iz^3 + 1)$ such that $|z| < 2\}$.

25. Prove Lagrange's trigonometric identity:

$$1 + \cos\theta + \cos 2\theta + \cdots + \cos n\theta = \frac{1}{2} + \frac{\sin\left(n + \frac{1}{2}\right)\theta}{2\sin\frac{\theta}{2}}$$

(Assume that $\sin(\theta/2) \neq 0$.)

26. Suppose that the complex numbers z_1, z_2, z_3 satisfy the equation

$$\frac{z_2 - z_1}{z_3 - z_1} = \frac{z_1 - z_3}{z_2 - z_3}$$

Prove that $|z_2 - z_1| = |z_3 - z_1| = |z_2 - z_3|$. (*Hint.* Argue geometrically, interpreting the meaning of each statement.)

27. Give a necessary and sufficient condition for
(a) z_1, z_2, z_3 to lie on a straight line.
(b) z_1, z_2, z_3, z_4 to lie on a straight line or a circle.

28. Prove the identity

$$\sin\frac{\pi}{n}\sin\frac{2\pi}{n}\cdots\sin\frac{(n-1)\pi}{n} = \frac{n}{2^{n-1}}$$

(*Hint.* The given product can be written as $1/2^{n-1}$ times the product of the nonzero roots of the polynomial $(1 - z)^n - 1$.)

29. Let w be an nth root of unity, $w \neq 1$. Evaluate $1 + 2w + 3w^2 + \cdots + nw^{n-1}$.

30. The correspondence of the complex number $z = a + bi$ with the matrix $\begin{pmatrix} a & -b \\ b & a \end{pmatrix} = \psi_z$ noted in the text preceding item 1.2.2 gives another way of displaying the complex numbers. Show that
(a) $\psi_{zw} = \psi_z\psi_w$.
(b) $\psi_{z+w} = \psi_z + \psi_w$.
(c) $\psi_1 = \begin{pmatrix} 1 & 0 \\ 0 & 1 \end{pmatrix}$.
(d) $\lambda\psi_z = \psi_{\lambda z}$ if λ is real.
(e) $\psi_{\bar{z}} = (\psi_z)^t$ (the transposed matrix).
(f) $\psi_{1/z} = (\psi_z)^{-1}$.
(g) z is real if and only if $\psi_z = (\psi_z)^t$.
(h) $|z| = 1$ if and only if ψ_z is an orthogonal matrix.

1.3 SOME ELEMENTARY FUNCTIONS

The trigonometric functions sine and cosine, as well as the exponential function and the logarithmic function, are studied in elementary calculus. Recall that the trigonometric functions may be defined in terms of the ratios of sides of a right-angled triangle. The definition of "angle" may be extended to include any real value, and thus $\cos \theta$ and $\sin \theta$ become real-valued functions of the real variable θ. It is a basic fact that $\cos \theta$ and $\sin \theta$ are differentiable, with derivatives given by $d(\cos \theta)/d\theta = -\sin \theta$ and $d(\sin \theta)/d\theta = \cos \theta$. Alternatively, $\cos \theta$ and $\sin \theta$ can be defined by their power series:

$$\sin x = x - \frac{x^3}{3!} + \frac{x^5}{5!} - \cdots$$

$$\cos x = 1 - \frac{x^2}{2!} + \frac{x^4}{4!} - \cdots$$

Convergence of these series, of course, must also be proved; such a proof can be found in Chap. 3 and in any calculus text.* Alternatively, $\sin x$ can be defined as the unique solution $f(x)$ to the differential equation $f''(x) + f(x) = 0$ satisfying $f(0) = 0, f'(0) = 1$; and $\cos x$ can be defined as the unique solution to $f''(x) + f(x) = 0, f(0) = 1, f'(0) = 0$ (again, see a calculus text for proofs).

The Exponential Function

The exponential function, denoted e^x, may be defined as the unique solution to the differential equation $f'(x) = f(x)$, subject to the initial condition that $f(0) = 1$. (It can be shown that a unique solution exists.) The exponential function can also be defined by its power series:

$$e^x = 1 + x + \frac{x^2}{2!} + \frac{x^3}{3!} + \cdots$$

We accept from calculus the fact that e^x is a positive, strictly increasing function of x. Therefore, for $y > 0$, $\log y$ can be defined as the inverse function of e^x; that is, $e^{\log y} = y$. Another approach that is often used in calculus books is to begin by defining

$$\log y = \int_1^y \frac{1}{t}\, dt$$

---

* Such as J. Marsden and A. Weinstein, *Calculus,* 2d ed. (New York: Springer-Verlag, 1985), Chap. 12.

for $y > 0$ and then to define e^x as the inverse function of log y. (*Note:* Many calculus books write ln y for the logarithm to the base e. As in most advanced mathematics, throughout this book we will write log y for ln y.)

In this section these functions will be extended to the complex plane. In other words, the functions sin z, cos z, e^z, and log z will be defined for complex z, and their restrictions to the real line will be the usual sin x, cos x, e^x, and log x. The extension to complex numbers should be natural in that many of the familiar properties of sin, cos, exp, and log are retained. These functions, as well as the power function z^n, will be studied geometrically later in this section.

We first extend the exponential function. We know from calculus that for real x, e^x can be represented by its Maclaurin series:

$$e^x = 1 + \frac{x}{1!} + \frac{x^2}{2!} + \frac{x^3}{3!} + \cdots$$

Thus it would be most natural to define e^{iy} by

$$1 + \frac{(iy)}{1!} + \frac{(iy)^2}{2!} + \cdots$$

for $y \in \mathbf{R}$. Of course, this definition is not quite legitimate, as convergence of series in $\mathbf{C}$ has not yet been discussed. Chapter 3 will show that this series does indeed represent a well-defined complex number for each y, but for the moment the series is used informally as the basis for the definition that follows, which will be precise. A slight rearrangement of the series (using Exercise 16, Sec. 1.1) shows that

$$e^{iy} = \left(1 - \frac{y^2}{2!} + \frac{y^4}{4!} - \cdots\right) + i\left(y - \frac{y^3}{3!} + \frac{y^5}{5!} - \cdots\right)$$

But we recognize this as being simply cos $y + i$ sin y. Thus we *define*

$$e^{iy} = \cos y + i \sin y.$$

So far, we have defined e^z for z along both the real and imaginary axes. How do we define $e^z = e^{x+iy}$? We desire our extension of the exponential to retain the familiar properties, and among these is the law of exponents: $e^{a+b} = e^a \cdot e^b$. This requirement forces us to define $e^{x+iy} = e^x \cdot e^{iy}$. This can be stated in a formal definition:

1.3.1 DEFINITION   *If $z = x + iy$, then e^z is defined by $e^x(\cos y + i \sin y)$.*

Note that if z is real (that is, if $y = 0$), this definition agrees with the usual exponential function e^x. The student is cautioned that we are not, at this stage, justified in thinking of

e^z as "e raised to the 'power' of z," since the concept of complex exponents has not yet been defined. The notation e^z is merely shorthand for the function defined by $e^x(\cos y + i \sin y)$.

There is another, again purely formal, reason for defining $e^{iy} = \cos y + i \sin y$. If we write $e^{iy} = f(y) + ig(y)$, we note that since we want $e^0 = 1$, we should have $f(0) = 1$, and $g(0) = 0$. Differentiating with respect to y gives us $ie^{iy} = f'(y) + ig'(y)$, so when $y = 0$ we get $f'(0) = 0, g'(0) = 1$. Differentiating again gives us $-e^{iy} = f''(y) + ig''(y)$. Comparing this equation with $e^{iy} = f(y) + ig(y)$, we get $f''(y) + f(y) = 0, f(0) = 1, f'(0) = 0$, so that $f(y) = \cos y$; and $g''(y) + g(y) = 0, g(0) = 0, g'(0) = 1$, so that $g(y) = \sin y$, by the definition of $\cos y$ and $\sin y$ in terms of differential equations. Thus we would obtain $e^{iy} = \cos y + i \sin y$ as in Definition 1.3.1.

Some of the important properties of e^z are summarized in the following proposition. To state it, we need to recall the definition of a periodic function; namely, a function $f: \mathbf{C} \to \mathbf{C}$ is called *periodic* if there exists a $w \in \mathbf{C}$ (called a *period*) such that $f(z + w) = f(z)$ for all $z \in \mathbf{C}$.

1.3.2 PROPOSITION

(i) $e^{z+w} = e^z e^w$ for all $z, w \in \mathbf{C}$.
(ii) e^z is never zero.
(iii) If x is real, then $e^x > 1$ when $x > 0$ and $e^x < 1$ when $x < 0$.
(iv) $|e^{x+iy}| = e^x$.
(v) $e^{\pi i/2} = i, e^{\pi i} = -1, e^{3\pi i/2} = -i, e^{2\pi i} = 1$.
(vi) e^z is periodic; any period for e^z, has the form $2\pi n i$, n an integer.
(vii) $e^z = 1$ iff $z = 2n\pi i$ for some integer n (positive, negative, or zero).

PROOF

(i) Let $z = x + iy$, and let $w = s + it$. By our definition of e^z,

$$
\begin{aligned}
e^{z+w} &= e^{(x+s)+i(y+t)} \\
&= e^{x+s}[\cos(y+t) + i \sin(y+t)] \\
&= [e^x(\cos y + i \sin y)][e^s(\cos t + i \sin t)]
\end{aligned}
$$

using the addition formulas for sine and cosine and the property, well known from elementary calculus, that $e^{x+s} = e^x \cdot e^s$ for real numbers x and s. Thus $e^{z+w} = e^z \cdot e^w$ for all complex numbers z and w.

(ii) For any z, we have $e^z \cdot e^{-z} = e^0 = 1$ since we know that the usual exponential satisfies $e^0 = 1$. Thus e^z can never be zero, because if it were, then $e^z \cdot e^{-z}$ would be zero, which is not true.

(iii) This follows from elementary calculus. For example, obviously

$$
e^x = 1 + \frac{x}{1!} + \frac{x^2}{2!} + \frac{x^3}{3!} + \cdots > 1 \qquad \text{when } x > 0
$$

(Another proof utilizing the definition of e^x in terms of differential equations is as follows. Recall that e^x is the unique solution to $f'(x) = f(x)$ with $e^0 = 1$ (x real). Since e^x is continuous and is never zero, it must be strictly positive. Hence $(e^x)' = e^x$ is always positive and consequently e^x is strictly increasing. Thus for $x > 0$, $e^x > 1$; and for $x < 0$, $e^x < 1$.)

(iv) Using $|zz'| = |z||z'|$ (see Proposition 1.2.5), we get

$$\begin{aligned}
|e^{x+iy}| = |e^x e^{iy}| &= |e^x||e^{iy}| \\
&= e^x |\cos y + i \sin y| && \text{since } e^x > 0 \\
&= e^x && \text{since } \cos^2 y + \sin^2 y = 1
\end{aligned}$$

(v) By definition, $e^{\pi i/2} = \cos(\pi/2) + i \sin(\pi/2) = i$. The proofs of the other formulas are similar.

(vi) Suppose that $e^{z+w} = e^z$ for all $z \in \mathbf{C}$. Setting $z = 0$, we get $e^w = 1$. If $w = s + ti$, then, using (iv), $e^w = 1$ implies that $e^s = 1$ and so $s = 0$. Hence any period is of the form ti, $t \in \mathbf{R}$. Suppose that $e^{ti} = 1$, that is, that $\cos t + i \sin t = 1$. Then $\cos t = 1$, $\sin t = 0$; and so $t = 2\pi n$ for some integer n.

(vii) $e^0 = 1$, as we have seen, and $e^{2\pi ni} = 1$ because e^z is periodic, by (vi). Conversely, $e^z = 1$ implies that $e^{z+z'} = e^{z'}$ for all z'; so by (vi), $z = 2\pi ni$ for some integer n. ■

How can we picture e^{iy}? As y goes from 0 to 2π, e^{iy} moves along the unit circle in a counterclockwise direction, reaching i at $y = \pi/2$, -1 at π, -1 at $3\pi/2$, and 1 again at 2π. Thus e^{iy} is the point on the unit circle with argument y (see Fig. 1.3.1).

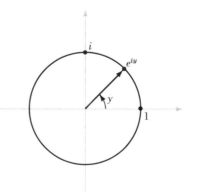

FIGURE 1.3.1   Points on the unit circle.

Note that in exponential form, the polar representation of a complex number becomes

$$z = |z|e^{i(\arg z)}$$

which is sometimes abbreviated to $z = re^{i\theta}$.

The Trigonometric Functions

Next we wish to extend the definitions of cosine and sine to the complex plane. The extension of the exponential to the complex plane suggests a way to extend the definitions of sine and cosine. We have $e^{iy} = \cos y + i \sin y$, and $e^{-iy} = \cos y - i \sin y$, which implies that

$$\sin y = \frac{e^{iy} - e^{-iy}}{2i} \quad \text{and} \quad \cos y = \frac{e^{iy} + e^{-iy}}{2}$$

But since e^{iz} is now defined for any $z \in \mathbf{C}$, we are led to formulate the following definition:

1.3.3 DEFINITION

$$\sin z = \frac{e^{iz} - e^{-iz}}{2i} \quad \text{and} \quad \cos z = \frac{e^{iz} + e^{-iz}}{2}$$

for any complex number z.

Again, if z is real, these definitions agree with the usual definitions of sine and cosine learned in elementary calculus.

The next proposition lists some of the properties of the sine and cosine functions that have now been defined over the whole of $\mathbf{C}$ and not merely on $\mathbf{R}$.

1.3.4. PROPOSITION

(i) $\sin^2 z + \cos^2 z = 1$.
(ii) $\sin(z + w) = \sin z \cdot \cos w + \cos z \cdot \sin w$ *and*
$\cos(z + w) = \cos z \cdot \cos w - \sin z \cdot \sin w$.

Again the student is cautioned that these formulas, although plausible, must be proved, since at this stage we know their validity only when w and z are real.

PROOF

(i) $\sin^2 z + \cos^2 z = \left(\dfrac{e^{iz} - e^{-iz}}{2i}\right)^2 + \left(\dfrac{e^{iz} + e^{-iz}}{2}\right)^2$

$$= \frac{e^{2iz} - 2 + e^{-2iz}}{-4} + \frac{e^{2iz} + 2 + e^{-2iz}}{4} = 1$$

(ii) $\sin z \cdot \cos w + \cos z \cdot \sin w$

$$= \frac{e^{iz} - e^{-iz}}{2i} \cdot \frac{e^{iw} + e^{-iw}}{2} + \frac{e^{iz} + e^{-iz}}{2} \cdot \frac{e^{iw} - e^{-iw}}{2i}$$

which, using $e^{iz}e^{iw} = e^{i(z+w)}$, simplifies to

$$\frac{e^{i(z+w)} - e^{-i(z+w)}}{4i} + \frac{e^{i(z+w)} - e^{-i(z+w)}}{4i} = \frac{e^{i(z+w)} - e^{-i(z+w)}}{2i} = \sin(z+w)$$

The student can similarly check the addition formula for $\cos(z+w)$. ∎

In addition to $\cos z$ and $\sin z$, we can define $\tan z = (\sin z)/(\cos z)$ when $\cos z \neq 0$, and similarly obtain the other trigonometric functions.

The Logarithm Function

We now shall define the logarithm in such a way that our definition will agree with the usual definition of $\log x$ when x is located on the positive part of the real axis. In the real case we can define the logarithm as the inverse of the exponential (that is, $\log x = y$ is the solution of $e^y = x$). When we allow z to range over **C**, we must be more careful, because, as has been shown, the exponential is periodic and thus does not have an inverse. Furthermore, the exponential is never zero and so we cannot expect to be able to define the logarithm at zero. Thus we must be more careful in our choice of the domain in **C** on which we can define the logarithm. The next proposition indicates how this may be done.

1.3.5 PROPOSITION   *Let A_{y_0} denote the set of complex numbers $x + iy$ such that $y_0 \leq y < y_0 + 2\pi$; symbolically,*

$$A_{y_0} = \{x + iy \mid x \in \mathbf{R} \text{ and } y_0 \leq y < y_0 + 2\pi\}$$

Then e^z maps A_{y_0} in a one-to-one manner onto the set $\mathbf{C}\backslash\{0\}$.

Recall that a map is *one-to-one* when the map takes every two distinct points to two distinct points; in other words, two distinct points never get mapped to the same point. The statement that a map is *onto* a set B means that every point of B is the image of some point under the mapping. The notation $\mathbf{C}\backslash\{0\}$ means the whole plane $\mathbf{C}$ minus the point 0; that is, the plane with the origin removed.

PROOF If $e^{z_1} = e^{z_2}$, then $e^{z_1 - z_2} = 1$, and so $z_1 - z_2 = 2\pi in$ for some integer n, by Proposition 1.3.2. But because z_1 and z_2 both lie in A_{y_0}, where the difference between the imaginary parts of any points is less than 2π, we must have $z_1 = z_2$. This argument shows that e^z is one-to-one. Let $w \in \mathbf{C}$ with $w \neq 0$. We claim the equation $e^z = w$ has a solution z in A_{y_0}. The equation $e^{x+iy} = w$ is equivalent to the two equations $e^x = |w|$ and $e^{iy} = w/|w|$. (Why?) The solution of the first equation is $x = \log|w|$, where "log" is the ordinary logarithm defined on the positive part of the real axis. The second equation has infinitely many solutions y, each differing by integral multiples of 2π, but exactly one of these is in the interval $[y_0, y_0 + 2\pi[$. This y is merely arg w, where the specified range for the arg function is $[y_0, y_0 + 2\pi[$. Thus e^z is onto $\mathbf{C}\backslash\{0\}$. ∎

The sets defined in this Proposition are shown in Fig. 1.3.2. Here e^z maps the horizontal strip between $y_0 i$ and $(y_0 + 2\pi)i$ one to one onto $\mathbf{C}\backslash\{0\}$. (The notation $z \mapsto f(z)$ is used to indicate that z is sent to $f(z)$ under the mapping f.)

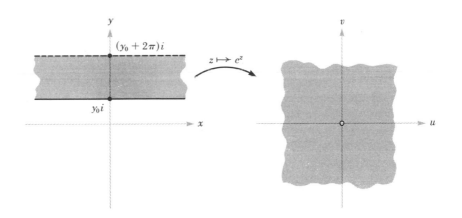

FIGURE 1.3.2 e^z as a one-to-one function onto $\mathbf{C}\backslash\{0\}$.

In the proof of Proposition 1.3.5 an explicit expression was also derived for the inverse of e^z restricted to the strip $y_0 \leq \operatorname{Im} z < y_0 + 2\pi$, and this expression is stated formally in the following definition.

1.3.6 DEFINITION   *The function log: $\mathbf{C}\backslash\{0\} \to \mathbf{C}$, with range $y_0 \le \text{Im log } z < y_0 + 2\pi$, is defined by*

$$\log z = \log |z| + i \text{ arg } z,$$

where arg z takes values in the interval $[y_0, y_0 + 2\pi[$ and $\log |z|$ is the usual logarithm of the positive real number $|z|$.

This function is sometimes referred to as the "*branch* of the logarithm function lying in $\{x + iy \,|\, y_0 \le y < y_0 + 2\pi\}$." But we must remember that the function log z is well defined only when we specify an interval of length 2π in which arg z takes its values, that is, when a specific branch is chosen. For example, suppose that the specified interval is $[0, 2\pi[$. Then $\log (1 + i) = \log \sqrt{2} + i\pi/4$. However, if the specified interval is $[\pi, 3\pi[$, then $\log (1 + i) = \log \sqrt{2} + i9\pi/4$. Any particular branch of the logarithm defined in this way undergoes a sudden jump as z moves across the ray arg $z = y_0$. To avoid this one can restrict the domain to $y_0 < y < y_0 + 2\pi$. This idea will be important in Sec. 1.6.

1.3.7 PROPOSITION   log z *is the inverse of e^z in the following sense: For any branch of* log z, *we have $e^{\log z} = z$, and if we choose the branch lying in $y_0 \le y < y_0 + 2\pi$, then* $\log (e^z) = z$ *for $z = x + iy$ and $y_0 \le y < y_0 + 2\pi$.*

PROOF   Since $\log z = \log |z| + i \text{ arg } z$,

$$e^{\log z} = e^{\log|z|} e^{i \text{ arg } z} = |z| e^{i \text{ arg } z} = z$$

Conversely, suppose that $z = x + iy$ and $y_0 \le y < y_0 + 2\pi$ By definition, $\log e^z = \log |e^z| + i \text{ arg } e^z$. But $|e^z| = e^x$ and arg $e^z = y$ by our choice of branch. Thus $\log e^z = \log e^x + iy = x + iy = z$. ∎

The logarithm defined on $\mathbf{C}\backslash\{0\}$ behaves the same way with respect to products as the logarithm restricted to the positive part of the real axis.

1.3.8 PROPOSITION   *If $z_1, z_2 \in \mathbf{C}\backslash\{0\}$, then $\log (z_1 z_2) = \log z_1 + \log z_2$ (up to the addition of integral multiples of $2\pi i$).*

PROOF $\log z_1 z_2 = \log |z_1 z_2| + i \text{ arg } (z_1 z_2)$, where an interval $[y_0, y_0 + 2\pi[$ has been chosen for the values of the arg function. We know that $\log |z_1 z_2| = \log |z_1||z_2| = $

$\log |z_1| + \log |z_2|$ and $\arg (z_1 z_2) = \arg z_1 + \arg z_2$ (up to integral multiples of 2π). Thus $\log z_1 z_2 = (\log |z_1| + i \arg z_1) + (\log |z_2| + i \arg z_2) = \log z_1 + \log z_2$ (up to integral multiples of $2\pi i$). ∎

To illustrate: Let us find $\log [(-1 - i)(1 - i)]$ where the range for the arg function is chosen as, for instance, $[0, 2\pi[$. Thus $\log [(-1 - i)(1 - i)] - \log (-2) = \log 2 + \pi i$. On the other hand, $\log (-1 - i) = \log \sqrt{2} + i5\pi/4$ and $\log (1 - i) = \log \sqrt{2} + i7\pi/4$. Thus $\log (-1 - i) + \log (1 - i) = \log 2 + i3\pi = (\log 2 + \pi i) + 2\pi i$; so in this case, when $z_1 = -1 - i$ and $z_2 = 1 - i$, $\log z_1 z_2$ differs from $\log z_1 + \log z_2$ by $2\pi i$.

The basic property in Proposition 1.3.8 can help one remember the definition of $\log z$ by writing $\log z = \log (re^{i\theta}) = \log r + \log e^{i\theta} = \log |z| + i \arg z$.

Complex Powers

We are now in a position to define the term a^b where $a, b \in \mathbf{C}$ and $a \neq 0$ (read "a raised to the power b"). Of course, however we define a^b, the definition should reduce to the usual one in which a and b are real numbers. The trick is to notice that a can also be written $e^{\log a}$ by Proposition 1.3.7. If b is an integer, we have $a^b = (e^{\log a})^b = e^{b \log a}$. This last equality holds since if n is an integer and z is any complex number, $(e^z)^n = e^z \cdots e^z = e^{nz}$ by Proposition 1.3.2(i). Thus we are led to formulate the following definition.

1.3.9 DEFINITION a^b (where $a, b \in \mathbf{C}$ and $a \neq 0$) is defined to be $e^{b \log a}$; it is understood that some interval $[y_0, y_0 + 2\pi[$ (that is, some branch of log) has been chosen within which the arg function takes its values.

It is of the utmost importance to understand precisely what this definition involves. Note especially that in general $\log z$ is "multiple-valued"; that is, $\log z$ can be assigned many different values because different intervals $[y_0, y_0 + 2\pi[$ can be chosen. This is not surprising, for if $b = 1/q$, where q is an integer, then our previous work with de Moivre's formula would lead us to expect that a^b is one of the qth roots of a and thus should have q distinct values. The following theorem elucidates this point.

1.3.10 PROPOSITION   Let $a, b \in \mathbf{C}$, $a \neq 0$. Then a^b is single-valued (that is, the value of a^b does not depend on the choice of branch for log) if and only if b is an integer. If b is a real, rational number, and if $b = p/q$ is in its lowest terms (in other words, if p and q have no common factor), then a^b has exactly q distinct values; namely, the q roots of a^p. If b is real and irrational or if b has a nonzero imaginary part, then a^b has infinitely many values. Where a^b has distinct values, these values differ by factors of the form $e^{2\pi nbi}$.

PROOF   Choose some interval, for example, $[0, 2\pi[$, for the values of the arg function. Let log z be the corresponding branch of the logarithm. If we were to choose any other branch of the log function, we would obtain $\log a + 2\pi ni$ rather than $\log a$, for some integer n. Thus $a^b = e^{b\log a + 2\pi nbi} = e^{b\log a} \cdot e^{2\pi nbi}$, where the value of n depends on the branch of the logarithm (that is, on the interval chosen for the values of the arg function). But by Proposition 1.3.2, $e^{2\pi nbi}$ remains the same for different values of n if and only if b is an integer. Similarly, it can be shown that $e^{2\pi nip/q}$ has q distinct values if p and q have no common factor. If b is irrational, and if $e^{2\pi nbi} = e^{2\pi mbi}$, it follows that $e^{(2\pi bi)(n-m)} = 1$ and hence $b(n-m)$ is an integer; since b is irrational, this implies that $n - m = 0$. Thus if b is irrational, $e^{2\pi nbi}$ has infinitely many distinct values. If b is of the form $x + iy, y \neq 0$, then $e^{2\pi nbi} = e^{-2\pi ny} \cdot e^{2\pi nix}$, which also has infinitely many distinct values. ∎

To repeat: When we write $e^{b\log a}$, it is understood that some branch of log has been chosen, and accordingly $e^{b\log a}$ has a single well-defined value. But as we change the branch of log, we get values for $e^{b\log a}$ that differ by factors of $e^{2\pi inb}$. This is what we mean when we say that $a^b = e^{b\log a}$ is "multiple-valued."

An example should make this clear. Let $a = 1 + i$ and let b be some real irrational number. Then the infinitely many different possible values of a^b are given by $e^{b[\log(1+i)+2\pi ni]} = e^{b(\log\sqrt{2}+i\pi/4+2\pi ni)} = (e^{b\log\sqrt{2}+ib\pi/4}) e^{b2\pi ni}$ as n takes on all integral values (corresponding to different choices of the branch). For instance, if we used the branch corresponding to $[-\pi, \pi[$ or $[0, 2\pi[$, we would set $n = 0$.

Some general properties of a^b are found in the exercises at the end of this section, but we are now interested in the special case when b is of the form $1/n$, because this gives the nth root.

The nth Root Function

We know that $\sqrt[n]{z}$ has exactly n values for $z \neq 0$. To make it a specific function we must again single out a branch of log as described in the preceding paragraphs.

1.3.11 DEFINITION   *The nth root function is defined by*

$$\sqrt[n]{z} = z^{1/n} = e^{(\log z)/n}$$

for a specific choice of branch of log z; *with this choice,* $\sqrt[n]{z} = e^{(\log z)/n}$ *is called a **branch** of the nth root function.*

The following theorem should not surprise us.

1.3.12 PROPOSITION $\sqrt[n]{z}$ *so defined is an nth root of z. It is obtained as follows. If* $z = re^{i\theta}$, *then*

$$\sqrt[n]{z} = \sqrt[n]{r}e^{i\theta/n}$$

where θ is chosen so that it lies within a particular interval corresponding to the branch choice. As we add multiples of 2π to θ, we run through the n nth roots of z. On the right, $\sqrt[n]{r}$ is the usual positive real nth root of the positive real number r.

PROOF By our definition,

$$\sqrt[n]{z} = e^{(\log z)/n}$$

But $\log z = \log r + i\theta$; so

$$e^{(\log z)/n} = e^{(\log r)/n} \cdot e^{i\theta/n} = \sqrt[n]{r}e^{i\theta/n}$$

The assertion is then clear. ∎

The reader should now take the time to become convinced that this way of describing the n nth roots of z is the same as that described in Corollary 1.2.3.

Geometry of the Elementary Functions

To further understand the functions z^n, $\sqrt[n]{z}$, e^z, and $\log z$, we shall consider the geometric interpretation of each in the remainder of this section. Let us begin with the power function z^n and let $n = 2$. We know that z^2 has length $|z|^2$ and argument $2 \arg z$. Thus the map $z \mapsto z^2$ squares lengths and doubles arguments (see Fig. 1.3.3).

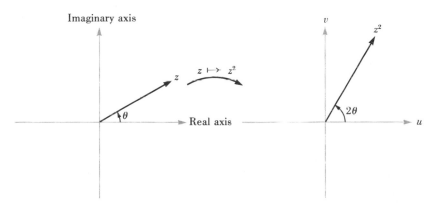

FIGURE 1.3.3 Squaring function.

It should be clear from this doubling of angles that the power function z^2 maps the first quadrant to the whole upper half plane (see Fig. 1.3.4). Similarly, the upper half plane is mapped to the whole plane.

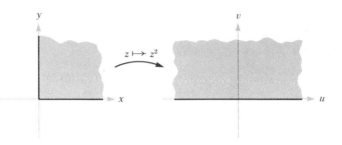

FIGURE 1.3.4   Effect of the squaring function on the first quadrant.

Now consider the square root function $\sqrt{z} = \sqrt{r}e^{i\theta/2}$. Suppose that we choose a branch by using the interval $0 \le \theta < 2\pi$. Then $0 \le \theta/2 < \pi$, so $\sqrt{z}$ will always lie in the upper half plane, and the angles thus are cut in half. The situation is similar to that involving the exponential function in that $z \mapsto \sqrt{z}$ is the inverse for $z \mapsto z^2$ when the latter is restricted to a region on which it is one-to-one. In like manner, if we choose the branch $-\pi \le \theta < \pi$, we have $-\pi/2 \le \theta/2 < \pi/2$, so $\sqrt{z}$ takes its values in the right half plane instead of the upper half plane. (Generally, any "half plane" could be used—see Fig. 1.3.5.) If we choose a specific branch of $\sqrt{z}$, we also choose which of the two possible square roots we shall obtain.

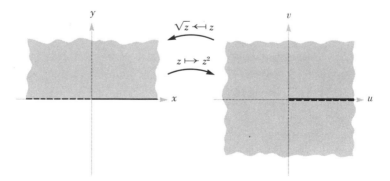

FIGURE 1.3.5   The squaring function and its inverse.

Various geometric statements can be made concerning the map $z \mapsto z^2$ that also give information about the inverse, $z \mapsto \sqrt{z}$. For example, a circle of radius r, described by the

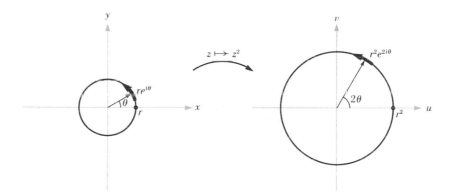

FIGURE 1.3.6 Effect of the squaring function on a circle of radius r.

set of points $re^{i\theta}$, $0 \le \theta < 2\pi$, is mapped to $r^2 e^{i2\theta}$, a circle of radius r^2; as $re^{i\theta}$ moves once around the first circle, the image point moves twice around (see Fig. 1.3.6). The inverse map does the opposite: as z moves along the circle $re^{i\theta}$ of radius r, $\sqrt{z}$ moves half as fast along the circle $\sqrt{r}e^{i\theta/2}$ of radius $\sqrt{r}$.

The correct domains on which $z \mapsto e^z$ and $z \mapsto \log z$ are inverses have already been discussed (see Fig. 1.3.2). Let us note that a line $y = $ constant, described by the points $x + iy$ as x varies, are mapped by the function $z \mapsto e^z$ to points $e^x e^{iy}$, which is a ray with argument y. As x ranges from $-\infty$ to $+\infty$, the image point on the ray goes from 0 out to infinity (see Fig. 1.3.7). Similarly, the vertical line $x = $ constant is mapped to a circle of radius e^x. If we restrict y to an interval of length 2π, the image circle is described once, but if y is unrestricted, the image circle is described infinitely many times as y ranges from $-\infty$ to $+\infty$. The logarithm, being the inverse of e^z, maps points in the opposite direction to e^z, as shown in Fig. 1.3.7. Because of the special nature of the striplike

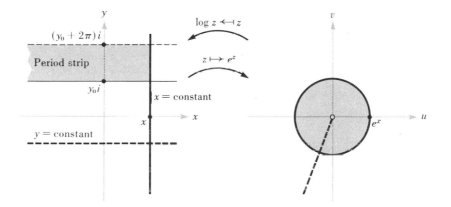

FIGURE 1.3.7 Geometry of e^z and $\log z$.

regions in Figs. 1.3.2 and 1.3.7 (on them e^z is one-to-one) and because of the periodicity of e^z, these regions deserve a name. They are usually called *period strips* of e^z.

Worked Examples

1.3.13   *Find the real and imaginary parts of* exp e^z. (exp w *is another way of writing* e^w.)

Solution. Let $z = x + iy$; then $e^z = e^x \cos y + ie^x \sin y$. Thus exp $e^z = e^{e^x \cos y}[\cos (e^x \sin y) + i \sin (e^x \sin y)]$. Hence Re $(\exp e^z) = (e^{e^x \cos y}) \cos (e^x \sin y)$ and Im $(\exp (e^z)) = (e^{e^x \cos y}) \times \sin (e^x \sin y)$.

1.3.14   *Show that* $\sin (iy) = i \sinh y$, *where by definition,*

$$\sinh y = \frac{e^y - e^{-y}}{2}$$

Solution.

$$\sin (iy) = \frac{e^{i(iy)} - e^{-i(iy)}}{2i} = i\frac{e^y - e^{-y}}{2} = i \sinh y.$$

1.3.15   *Find all the values of* i^i.

Solution.

$$i^i = e^{i \log i} = e^{i[\log 1 + (i\pi/2) + (2\pi n)i]} = (e^{-2\pi n})e^{-\pi/2} = e^{-2\pi(n + 1/4)}$$

All the values of i^i are given by the last expression as n takes integral values, $n = 0, \pm 1, \pm 2, \ldots$.

1.3.16   *Solve* $\cos z = \frac{1}{2}$.

Solution. We know that $z_n = \pm(\pi/3 + 2\pi n)$, n an integer, solves $\cos z = \frac{1}{2}$; we shall show that z_n, $n = 0, \pm 1, \ldots$, are the *only* solutions; that is, there are no solutions off the real axis. We are given

$$\cos z = \frac{e^{iz} + e^{-iz}}{2} = \frac{1}{2}$$

Hence $e^{2iz} - e^{iz} + 1 = 0$, so $e^{iz} = (1 \pm \sqrt{-3})/2 = \frac{1}{2} \pm \sqrt{3}i/2$. Therefore $iz = \log (\frac{1}{2} \pm \sqrt{3}i/2) = \pm \log (\frac{1}{2} + \sqrt{3}i/2)$, since $\frac{1}{2} + \sqrt{3}i/2$ and $\frac{1}{2} - \sqrt{3}i/2$ are reciprocals of one another. We thus obtain

$$z = \pm i \log \left(\frac{1}{2} + \frac{\sqrt{3}}{2} i\right) = \pm i \left(\log 1 + \frac{\pi}{3} i + 2\pi n i\right) = \pm \left(\frac{\pi}{3} + 2\pi n\right)$$

1.3.17   *Consider the mapping* $z \mapsto \sin z$. *Show that lines parallel to the real axis are mapped to ellipses and that lines parallel to the imaginary axis are mapped to hyperbolas.*

Solution. Using Proposition 1.3.4 (also see Example 1.3.14), we get

$$\sin z = \sin (x + iy) = \sin x \cos (iy) + \sin (iy) \cos x$$
$$= \sin x \cosh y + i \sinh y \cos x$$

where

$$\cosh y = \frac{e^y + e^{-y}}{2} \quad \text{and} \quad \sinh y = \frac{e^y - e^{-y}}{2}$$

Suppose that $y = y_0$ is constant; then if we write

$$\sin z = u + iv$$

we have

$$\frac{u^2}{\cosh^2 y_0} + \frac{v^2}{\sinh^2 y_0} = 1$$

since $\sin^2 x + \cos^2 x = 1$. This is an ellipse.

Similarly, if $x = x_0$ is constant, from $\cosh^2 y - \sinh^2 y = 1$ we obtain

$$\frac{u^2}{\sin^2 x_0} - \frac{v^2}{\cos^2 x_0} = 1$$

which is a hyperbola.

1.3.18 *Let $f(z) = z^2$ and suppose $f(z_0) = a + bi$. Describe the curves defined implicitly in the xy plane by the equations $\operatorname{Re} f(z) = a$ and $\operatorname{Im} f(z) = b$. Show that these curves are perpendicular to each other at z_0.*

Solution. If $z = x + iy$, then $f(z) = x^2 - y^2 + 2xyi$. The curves desired are the hyperbolas $x^2 - y^2 = a$ and $xy = b/2$, which are sketched in Fig. 1.3.8 for the case in which $a > 0$ and $b > 0$. These curves are the level curves of the functions $u(x, y) = x^2 - y^2$ and $v(x, y) = 2xy$. Normal vectors to them are, from calculus, given by their gradients

$$\nabla u = (2x, -2y)$$

and

$$\nabla v = (2y, 2x)$$

At a point $z_0 = x_0 + iy_0$ on both curves, their dot product is $\nabla u \cdot \nabla v = 4x_0 y_0 - 4y_0 x_0 = 0$. The normal vectors are thus perpendicular to each other at z_0, and so the curves are also.

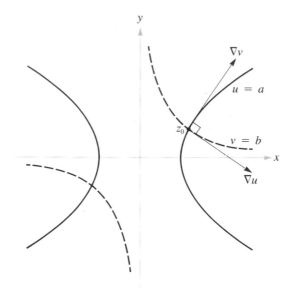

FIGURE 1.3.8 The level sets of Re $f(z)$ and Im $f(z)$ for $f(z) = z^2$.

Exercises

1. Express in the form $a + bi$:
(a) e^{2+i} (b) $\sin (1 + i)$

2. Express in the form $a + bi$:
(a) e^{3-i} (b) $\cos (2 + 3i)$

3. Solve:
(a) $\cos z = \frac{3}{4} + i/4$ (b) $\cos z = 4$

4. Solve:
(a) $\sin z = \frac{3}{4} + i/4$ (b) $\sin z = 4$

5. Find all the values of:
(a) $\log 1$ (b) $\log i$

6. Find all the values of:
(a) $\log (-i)$ (b) $\log (1 + i)$

7. Find all the values of:
(a) $(-i)^i$ (b) $(1 + i)^{1+i}$

8. Find all the values of:

(a) $(-1)^i$ (b) 2^i

9. For what values of z is $\overline{(e^{iz})} = e^{i\bar{z}}$?

10. Let $\sqrt{\ }$ denote the particular square root defined by $\sqrt{r}(\cos\theta + i\sin\theta) = r^{1/2}[\cos(\theta/2) + i\sin(\theta/2)]$, $0 \le \theta < 2\pi$; the other square root is $r^{1/2}\{\cos[(\theta + 2\pi)/2] + i\sin[(\theta + 2\pi)/2]\}$. For what values of z does the equation $\sqrt{z^2} = z$ hold?

11. Along which rays through the origin (a ray is determined by arg z = constant) does $\lim\limits_{z \to \infty} |e^z|$ exist?

12. Prove that

$$z = \tan\left[\frac{1}{i}\log\left(\frac{1 + iz}{1 - iz}\right)^{1/2}\right]$$

13. Simplify e^{z^2}, e^{iz}, and $e^{1/z}$, where $z = x + iy$. For $e^{1/z}$ we specify that $z \neq 0$.

14. Examine the behavior of e^{x+iy} as $x \to \pm\infty$ and the behavior of e^{x+iy} as $y \to \pm\infty$.

15. Prove that $\sin(-z) = -\sin z$; that $\cos(-z) = \cos z$; and that $\sin(\pi/2 - z) = \cos z$.

16. Define sinh and cosh on all of **C** by $\sinh z = (e^z - e^{-z})/2$ and $\cosh z = (e^z + e^{-z})/2$. Prove that:

(a) $\cosh^2 z - \sinh^2 z = 1$
(b) $\sinh(z_1 + z_2) = \sinh z_1 \cosh z_2 + \cosh z_1 \sinh z_2$
(c) $\cosh(z_1 + z_2) = \cosh z_1 \cosh z_2 + \sinh z_1 \sinh z_2$
(d) $\sinh(x + iy) = \sinh x \cos y + i \cosh \sin y$
(e) $\cosh(x + iy) = \cosh x \cos y + i \sinh x \sin y$

17. Use the equation $\sin z = \sin x \cosh y + i \sinh y \cos x$ where $z = x + iy$ to prove that $|\sinh y| \le |\sin z| \le |\cosh y|$.

18. If b is real, prove that $|a^b| = |a|^b$.

19. Is it true that $|a^b| = |a|^{|b|}$ for all a, $b \in$ **C**?

20. (a) For complex numbers a, b, c, prove that $a^b a^c = a^{b+c}$, using a fixed branch of log.
(b) Show that $(ab)^c = a^c b^c$ if we choose branches so that $\log(ab) = \log a + \log b$ (with no extra $2\pi n i$).

21. Using polar coordinates, show that $z \mapsto z + 1/z$ maps the circle $|z| = 1$ to the interval $[-2, 2]$ on the x axis.

22. (a) The map $z \mapsto z^3$ maps the first quadrant onto what?

(b) Discuss the geometry of $z \mapsto \sqrt[3]{z}$ as was done in the text for $\sqrt{z}$.

23. The map $z \mapsto 1/z$ takes the exterior of the unit circle to the interior (excluding zero) and vice versa. What are lines arg $z = $ constant mapped to?

24. What is the image of vertical and horizontal lines under $z \mapsto \cos z$?

25. Under what conditions does $\log a^b = b \log a$ for complex numbers a, b? (Use the branch of log with $-\pi \le \theta < \pi$)

26. (a) Show that under the map $z \mapsto z^2$, lines parallel to the real axis are mapped to parabolas.

(b) Show that under (a branch of) $z \mapsto \sqrt{z}$, lines parallel to the real axis are mapped to hyperbolas.

27. Show that the n nth roots of unity are $1, w, w^2, w^3, \ldots, w^{n-1}$, where $w = e^{2\pi i/n}$.

28. Show that the trigonometric identities can be deduced if $e^{i(x_1+x_2)} = e^{ix_1} \cdot e^{ix_2}$ is assumed.

29. Show that $\sin z = 0$ iff $z = k\pi$, $k = 0, \pm 1, \pm 2, \ldots$.

30. Show that the sine and cosine are periodic with minimum period 2π; that is, that

(a) $\sin (z + 2\pi) = \sin z$ for all z.

(b) $\cos (z + 2\pi) = \cos z$ for all z.

(c) $\sin (z + \omega) = \sin z$ for all z implies $\omega = 2\pi n$ for some integer n.

(d) $\cos (z + \omega) = \cos z$ for all z implies $\omega = 2\pi n$ for some integer n.

31. Find the maximum of $|\cos z|$ on the square $0 \le \text{Re } z \le 2\pi, 0 \le \text{Im } z \le 2\pi$.

32. Show that $\log z = 0$ iff $z = 1$, using the branch with $-\pi < \arg z \le \pi$.

33. Compute the following numerically to two significant figures: $e^{3.2+6.1i}$, $\log (1.2 - 3.0i)$, $\sin (8.1i - 3.2)$.

34. Show that $\sin z$ maps the strip $-\pi/2 < \text{Re } z < \pi/2$ onto $\mathbb{C} \backslash \{z \,|\, \text{Im } z = 0 \text{ and } |\text{Re } z| \ge 1\}$.

35. Discuss the inverse functions $\sin^{-1} z$, $\cos^{-1} z$. For example, is $\sin z$ one-to-one on $0 \le \text{Re } z < 2\pi$?

1.4 CONTINUOUS FUNCTIONS

In this section and the next, the fundamental notions of continuity and differentiability for complex-valued functions of a complex variable will be analyzed. The results are quite similar to those learned in the calculus of functions of real variables. These sections will be concerned mostly with the underlying theory. This theory will be applied to the elementary functions in Sec. 1.6.

Since $\mathbf{C}$ is $\mathbf{R}^2$ with the extra structure of complex multiplication, many geometric concepts can be translated from $\mathbf{R}^2$ into complex notation. This has already been done for the absolute value, $|z|$, which is the same as the norm, or length, of z regarded as a vector in $\mathbf{R}^2$. Furthermore, the student will soon be able to use knowledge of calculus for functions of two variables in the study of functions of a complex variable.

Open Sets

It is necessary first to have available the notion of an open set. A set $A \subset \mathbf{C} = \mathbf{R}^2$ is *open* when, for each point z_0 in A, there is a real number $\epsilon > 0$ such that $z \in A$ whenever $|z - z_0| < \epsilon$. (See Fig. 1.4.1.) Note that the value of ϵ depends on z_0; as z_0 gets close to the "edge" of A, ϵ gets smaller. Intuitively, a set is open if it does not contain any of its "boundary" or "edge" points.

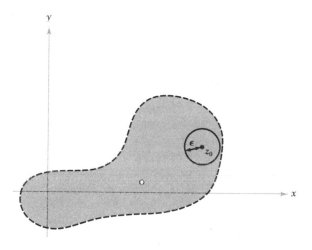

FIGURE 1.4.1 An open set.

For a number $r > 0$, the *r neighborhood* or *r disk* around a point z_0 in $\mathbf{C}$ is defined to be the set $D(z_0; r) = \{z \in \mathbf{C}$ such that $|z - z_0| < r\}$. For practice, the student should prove that for each $w_0 \in \mathbf{C}$ and $r > 0$, the disk $A = \{z \in \mathbf{C}$ such that $|z - w_0| < r\}$ is itself open. A *deleted r neighborhood* is an r neighborhood whose center point has been removed. Thus a deleted neighborhood has the form $D(z_0; r)\backslash\{z_0\}$, which stands for the set $D(z_0; r)$ minus the singleton set $\{z_0\}$. (See Fig. 1.4.2.). A *neighborhood* of a point z_0 is, by definition, an open set containing z_0. Thus we can rephrase the definition of "open" as follows: a set A is open iff for each z_0 in A, there is an r neighborhood of z_0 wholly contained in A.

The basic properties of open sets are collected in Proposition 1.4.1.

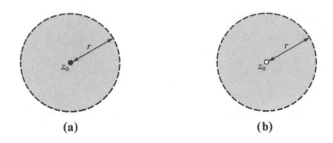

FIGURE 1.4.2   Neighborhood (a) and deleted neighborhood (b).

1.4.1 PROPOSITION

(i) **C** *is open.*

(ii) *The empty set* $\varnothing$ *is open.*

(iii) *The union of any collection of open subsets of* **C** *is open.*

(iv) *The intersection of any finite collection of open subsets of* **C** *is open.*

PROOF   The first two assertions hold almost by definition; the first because any ϵ will work for any point z_0, and the second because there are no points for which we are required to find such an ϵ. The reader is asked to supply proofs of the last two in Exercises 19 and 20 at the end of this section. ∎

Mappings, Limits, and Continuity

Let A be a subset of **C**. We recall that a mapping $f: A \rightarrow$ **C** is an assignment of a specific point $f(z)$ in **C** to each point z in A. The set A is called the *domain* of f, and we say f is *defined on A*. When the domain and the range (the set of values f assumes) are both subsets of **C**, as here, we speak of f as a *complex function of a complex variable*. Alternatively, we can think of f as a map $f: A \subset \mathbf{R}^2 \rightarrow \mathbf{R}^2$; then f is called a vector-valued function of two real variables. For $f: A \subset \mathbf{C} \rightarrow \mathbf{C}$, we can let $z = x + iy = (x, y)$ and define $u(x, y) = \text{Re } f(z)$ and $v(x, y) = \text{Im } f(z)$. Then u and v are merely the components of f thought of as a vector function. Hence we may write, in a unique way, $f(x + iy) = u(x, y) + iv(x, y)$, where u and v are real-valued functions defined on A.

Let us next transcribe the notion of limit to the notation of complex numbers.

1.4.2 DEFINITION   *Let f be defined on a set containing some deleted r neighborhood of* z_0. *The expression*

$$\lim_{z \to z_0} f(z) = a$$

means that for every $\epsilon > 0$, there is a $\delta > 0$ such that $z \in D(z_0; r)$, $z \neq z_0$, and $|z - z_0| < \delta$ imply that $|f(z) - a| < \epsilon$.

The expression in this definition has the same intuitive meaning as it has in calculus; namely, $f(z)$ is close to a whenever z is close to z_0. It is not necessary to define f on a *whole* deleted neighborhood to have a valid theory of limits, but deleted neighborhoods are used here for the sake of simplicity and also because such usage will be appropriate later in the text.

Just as with real numbers and real-valued functions, a function can have no more than one limit at a point, and limits behave well with respect to algebraic operations. This is the content of the next two propositions.

1.4.3 PROPOSITION *Limits are unique if they exist.*

PROOF Suppose that $\lim_{z \to z_0} f(z) = a$ and $\lim_{z \to z_0} f(z) = b$ with $a \neq b$. Let $2\epsilon = |a - b|$, so that $\epsilon > 0$. There is a $\delta > 0$ such that $0 < |z - z_0| < \delta$ implies that $|f(z) - a| < \epsilon$ and $|f(z) - b| < \epsilon$. Choose such a point $z \neq z_0$ (because f is defined in a deleted neighborhood of z_0). Then, by the triangle inequality, $|a - b| \leq |a - f(z)| + |f(z) - b| < 2\epsilon$, a contradiction. Thus $a = b$. ∎

1.4.4 PROPOSITION *If $\lim_{z \to z_0} f(z) = a$ and $\lim_{z \to z_0} g(z) = b$, then*

(i) $\lim_{z \to z_0} [f(z) + g(z)] = a + b$.

(ii) $\lim_{z \to z_0} [f(z)g(z)] = ab$.

(iii) $\lim_{z \to z_0} [f(z)/g(z)] = a/b$ if $b \neq 0$.

PROOF Only assertion (ii) will be proved here. The proof of assertion (i) is easy, and proof of assertion (iii) is slightly more challenging, but the reader can get the necessary clues from the corresponding real-variable case. To prove assertion (ii), we write

$$|f(z)g(z) - ab| \leq |f(z)g(z) - f(z)b| + |f(z)b - ab| \quad \text{(triangle inequality)}$$
$$= |f(z)||g(z) - b| + |f(z) - a||b| \quad \text{(factoring)}$$

We want to estimate each term. To do so, we choose $\delta_1 > 0$ so that $0 < |z - z_0| < \delta_1$ implies that $|f(z) - a| < 1$, and thus $|f(z)| < |a| + 1$, since $|f(z) - a| \geq |f(z)| - |a|$, by Proposition 1.2.5(vi). Given $\epsilon > 0$, we can choose $\delta_2 > 0$ such that $0 < |z - z_0| < \delta_2$ implies that

$$|f(z) - a| < \begin{cases} \dfrac{\epsilon}{2|b|} & \text{if } b \neq 0 \\ 1 & \text{if } b = 0 \end{cases}$$

and $\delta_3 > 0$ such that if $0 < |z - z_0| < \delta_3$, then

$$|g(z) - b| < \frac{\epsilon}{2(|a| + 1)}$$

Let δ be the smallest of $\delta_1, \delta_2, \delta_3$. Then, if $0 < |z - z_0| < \delta$,

$$|f(z)g(z) - ab| \leq |f(z)||g(z) - b| + |f(z) - a||b|$$

$$< \frac{\epsilon}{2(|a| + 1)}|f(z)| + \frac{\epsilon}{2|b|}|b| \qquad \text{(if } b \neq 0; \text{ if } b = 0, \text{ replace the second term by 0)}$$

$$\leq \frac{\epsilon}{2} + \frac{\epsilon}{2} = \epsilon \qquad \blacksquare$$

1.4.5 DEFINITION   *Let $A \subset \mathbf{C}$ be an open set and let $f : A \to \mathbf{C}$ be a function. We say f is **continuous** at $z_0 \in A$ if and only if*

$$\lim_{z \to z_0} f(z) = f(z_0)$$

*and that f is **continuous on** A if f is continuous at each point z_0 in A.*

This definition has the same intuitive meaning as it has in elementary calculus: If z is close to z_0, then $f(z)$ is close to $f(z_0)$. From Proposition 1.4.4 we can immediately deduce that if f and g are continuous on A, then so are the sum $f + g$ and the product fg, and so is f/g if $g(z_0) \neq 0$ for all points z_0 in A. It is also true that a composition of continuous functions is continuous.

1.4.6 PROPOSITION
(i) *If $\lim\limits_{z \to z_0} f(z) = a$ and h is a function defined on a neighborhood of a and is continuous at a, then $\lim\limits_{z \to z_0} h(f(z)) = h(a)$.*

(ii) *If f is a continuous function on an open set A in $\mathbf{C}$ and h is continuous on $f(A)$, then the composite function $(h \circ f)(z) = h(f(z))$ is continuous on A.*

PROOF   Given $\epsilon > 0$, there is a $\delta_1 > 0$ such that $|h(w) - h(a)| < \epsilon$ whenever $|w - a| < \delta_1$ and a $\delta > 0$ such that $|f(z) - a| < \delta_1$ whenever $0 < |z - z_0| < \delta$. Therefore we get

$|h(f(z)) - h(a)| < \epsilon$ whenever $0 < |z - z_0| < \delta$, which establishes (i). A proof of (ii) follows from (i) and is requested in Exercise 22 at the end of this section. ∎

Sequences

The concept of convergent sequences of complex numbers is analogous to that for sequences of real numbers studied in calculus. A sequence z_n, $n = 1, 2, 3, \ldots$, of points of **C** *converges to* z_0 if and only if for every $\epsilon > 0$, there is an integer N such that $n \geq N$ implies $|z_n - z_0| < \epsilon$. The limit of a sequence is expressed as

$$\lim_{n \to \infty} z_n = z_0 \quad \text{or} \quad z_n \to z_0$$

Limits of sequences have the same properties, obtained by the same proofs, as limits of functions. For example, the limit is unique if it exists; and if $z_n \to z_0$ and $w_n \to w_0$, then

(i) $z_n + w_n \to z_0 + w_0$

(ii) $z_n w_n \to z_0 w_0$

and

(iii) $z_n/w_n \to z_0/w_0$ (if w_0 and w_n are not 0)

Also, $z_n \to z_0$ iff Re $z_n \to$ Re z_0 and Im $z_n \to$ Im z_0. A proof of this for functions is requested in Exercise 2 at the end of this section.

A sequence z_n is called a *Cauchy sequence* if for every $\epsilon > 0$ there is an integer N such that $|z_n - z_m| < \epsilon$ whenever both $n \geq N$ and $m \geq N$. A basic property of real numbers which we will accept without proof is that every Cauchy sequence in **R** converges. More precisely, if $\{x_n\}_{n=1}^{\infty}$ is a Cauchy sequence of real numbers, then there is a real number x_0 such that $\lim_{n \to \infty} x_n = x_0$. This is equivalent to the *completeness* of the real number system.* From the fact that $z_n \to z_0$ iff Re $z_n \to$ Re z_0 and Im $z_n \to$ Im z_0, we can conclude that

*every Cauchy sequence in **C** converges.*

This is a technical point, but is useful in convergence proofs, as we shall see in Chap. 3.

It should be noted that a link exists between sequences and continuity; namely, $f: A \subset \mathbf{C} \to \mathbf{C}$ is continuous iff for every convergent sequence $z_n \to z_0$ of points in A (that is, $z_n \in A$ and $z_0 \in A$), we have $f(z_n) \to f(z_0)$. The student is requested to prove this in Exercise 18 at the end of this section.

---

* See, for example, J. Marsden, *Elementary Classical Analysis* (New York: W. H. Freeman and Company, 1974).

Closed Sets

A subset F of $\mathbf{C}$ is said to be *closed* if its complement, $\mathbf{C} \backslash F = \{z \in \mathbf{C} \mid z \notin F\}$, is open. By taking complements and using Proposition 1.4.1, one discovers the following properties of closed sets.

1.4.7 PROPOSITION

  (i)   *The empty set is closed.*
 (ii) $\mathbf{C}$ *is closed.*
 (iii) *The intersection of any collection of closed subsets of* $\mathbf{C}$ *is closed.*
 (iv)   *The union of any finite collection of closed subsets of* $\mathbf{C}$ *is closed.*

Closed and open sets are important for their relationships to continuous functions and to sequences and for other constructions we will see later.

1.4.8 PROPOSITION   *A set* $F \subset \mathbf{C}$ *is closed iff whenever* $z_1, z_2, z_3, \ldots$ *is a sequence of points in* F *such that* $w = \lim_{n \to \infty} z_n$ *exists, then* $w \in F$.

PROOF   Suppose F is closed and z_n is a sequence of points in F. If $D(w; r)$ is any disk around w, then by the definition of convergence, z_n is in $D(w; r)$ for large enough n. Thus, $D(w; r)$ cannot be contained in the complement of F. Since that complement is open, w must not be in the complement of F. It must be in F.

If F is not closed, then the complement is not open. In other words, there is a point w in $\mathbf{C} \backslash F$ such that no neighborhood of w is contained in $\mathbf{C} \backslash F$. In particular, we may pick points z_n in $F \cap D(w; 1/n)$; this yields a convergent sequence of points of F whose limit is not in F. ∎

1.4.9 PROPOSITION   *If* $f: \mathbf{C} \to \mathbf{C}$, *the following are equivalent:*

  (i) f *is continuous.*
 (ii)   *The inverse image of every closed set is closed.*
 (iii) *The inverse image of every open set is open.*

PROOF   To show that (i) implies (ii), suppose f is continuous and that F is closed. Let $z_1, z_2, z_3, \ldots$ be a sequence of points in $f^{-1}(F)$ and suppose that $z_n \to w$. Since f is continuous, $f(z_n) \to f(w)$. But the points $f(z_n)$ are in the closed set F, and so $f(w)$ is also in F. That is, w is in $f^{-1}(F)$. Proposition 1.4.8 shows that $f^{-1}(F)$ is closed.

To show that (ii) implies (iii), let U be open. Then $F = \mathbf{C} \setminus U$ is closed. If (ii) holds, then $f^{-1}(F)$ is closed. Therefore $\mathbf{C} \setminus f^{-1}(F) = f^{-1}(\mathbf{C} \setminus F) = f^{-1}(U)$ is open.

To show that (iii) implies (i), fix z_0 and let $\epsilon > 0$. Then z_0 is a member of the open set $f^{-1}(D(f(z_0); \epsilon))$. Hence there is a $\delta > 0$ with $D(z_0; \delta) \subset f^{-1}(D(f(z_0); \epsilon))$. This says precisely that $|f(z) - f(z_0)| < \epsilon$ whenever $|z - z_0| < \delta$. We thus get exactly the inequality needed to establish continuity. ∎

To handle continuity on a subset of $\mathbf{C}$, it is convenient to introduce the notion of relatively open and closed sets. If $A \subset \mathbf{C}$, a subset B of A is called *open relative to A* if $B = A \cap U$ for some open set U. It is said to be *closed relative to A* if $B = A \cap F$ for some closed set F. This leads to the following proposition, whose proof is left to the reader.

1.4.10 PROPOSITION   *If $f : A \to \mathbf{C}$, the following are equivalent:*

(i) *f is continuous.*
(ii) *The inverse image of every closed set is closed relative to A.*
(iii) *The inverse image of every open set is open relative to A.*

Connected Sets

This subsection and the next study two important classes of sets which occupy to some extent the place in the theory of complex variables held by intervals and by closed bounded intervals in the theory of functions of a real variable. These are the connected sets and the compact sets.

A connected set should be one which "consists of one piece." This may be approached from a positive point of view: "Any point can be connected to any other"; or from a negative: "The set cannot be split into two parts." This leads to two possible definitions.

1.4.11 DEFINITION   *A set C in $\mathbf{C}$ is **path-connected** if for every pair of points a, b in C there is a continuous map $\gamma: [0, 1] \to C$ with $\gamma(0) = a$ and $\gamma(1) = b$. We call γ a **path** joining a and b.*

One can often easily tell if a set is path-connected, as is shown in Fig. 1.4.3. The negative point of view suggests a slightly different definition.

1.4.12 DEFINITION   *A set $C \subset \mathbf{C}$ is **not connected** if there are open sets U and V such that*

(i) *$C \subset U \cup V$*
(ii) *$C \cap U \neq \varnothing$ and $C \cap V \neq \varnothing$*

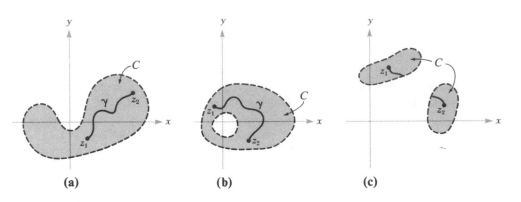

FIGURE 1.4.3   Connectedness. (a) and (b) are connected regions; (c) is not connected.

and

(iii) $(C \cap U) \cap (C \cap V) = \varnothing$
(See Fig. 1.4.4.)

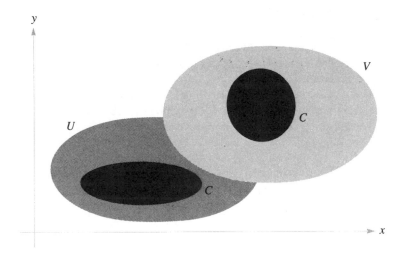

FIGURE 1.4.4   The set C is not connected.

The notions of relatively open and closed sets allow this to be rephrased in terms of subsets of C. Since the intersection of C with U is the same as its intersection with the complement of V, the set $C \cap U$ is both open and closed relative to C, as is $C \cap V$. This proves the next result.

1.4.13 PROPOSITION *A set C is connected if and only if the only subsets of C which are both open and closed relative to C are the empty set and C itself.*

The next two propositions give the relationship between the two definitions. The two notions are not in general equivalent, but they are for open sets. The proof of this last assertion (given below in Proposition 1.4.15) illustrates a fairly typical way of using the notion of connectivity. One shows that a certain property holds everywhere in C by showing that the set of places where it holds is not empty and is both relatively open and relatively closed.

1.4.14 PROPOSITION *A path-connected set is connected.*

PROOF Suppose C is a path-connected set and that D is a nonempty subset of C which is both open and closed relative to C. If $C \neq D$, there is a point z_1 in D and a point z_2 in $C \backslash D$. Let $\gamma: [a, b] \to C$ be a continuous path joining z_1 to z_2. Let $B = \gamma^{-1}(D)$. Then B is a subset of the interval $[a, b]$ which is both open and closed relative to $[a, b]$, since γ is continuous. (See Proposition 1.4.10). Since a is in B, B is not empty, and $[a, b] \backslash B$ is not empty since it contains b.

This argument shows that it is sufficient to prove the theorem for the case of an interval $[a, b]$. We thus need to establish the following lemma:

Intervals of the real line are connected.

A proof requires the use of the least upper bound property (or some other characterization of the fact that the system of real numbers is complete). Let $x = \sup B$ (that is, the least upper bound of B). We find that x is in B since B is closed. But since B is open there is a neighborhood around x lying in B (remember that $x \neq b$, since b is in $[a, b] \backslash B$). This means that for some $\epsilon > 0$, the point $x + \epsilon$ is in B. Thus x cannot be the least upper bound. This contradiction shows that such a set B cannot exist. ∎

A connected set need not be path-connected,* but if it is open it must be. In fact, more is true.

---

* A standard example is given by letting C be the union of the graph of $y = \sin 1/x$, where $x > 0$, and the line segment $-1 \leq y \leq 1$, $x = 0$. This set is connected but not path-connected.

1.4.15 PROPOSITION   *If C is an open connected set and a and b are in C, then there is a differentiable path γ: [0, 1] $\rightarrow$ C with $\gamma(0) = a$ and $\gamma(1) = b$.*

PROOF   Let a be in C. If z_0 is in C, then since C is open, there is an $\epsilon > 0$ such that the disk $D(z_0; \epsilon)$ is contained in C. By combining a path from a to z_0 with one from z_0 to z which stays in this disk, we see that z_0 can be connected to a by a differentiable path if and only if the same is true for every point z in $D(z_0; \epsilon)$. This shows that both the sets

$$A = \{z \in \mathbf{C} \,|\, z \text{ can be connected to } a \text{ by a differentiable path}\}$$

and

$$B = \{z \in \mathbf{C} \,|\, z \text{ cannot be so connected to } a\}$$

are open. Since C is connected, one of A or B must be empty. Obviously it is B. See Fig. 1.4.5. ∎

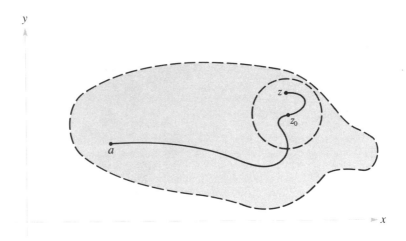

FIGURE 1.4.5   An open connected set is path-connected.

Because of the importance of open connected sets, they are often designated by a special term. Although the usage is not completely standard in the literature, the words *region* and *domain* are often used. In this text these terms will be used synonymously to mean an open connected subset of **C**. The reader should be careful to check the meanings when these words are encountered in other texts.

The notion of connected sets will be of use to us several times. One observation is that a continuous function cannot break apart a connected set.

1.4.16 PROPOSITION *If f is a continuous function defined on a connected set C, then the image set f(C) is also connected.*

PROOF If U and V are open sets which disconnect $f(C)$, then $f^{-1}(U)$ and $f^{-1}(V)$ are open sets disconnecting C. ∎

Be careful. This proposition works in the opposite direction from the one about open and closed sets. For continuous functions, the inverse images of open sets are open and the inverse images of closed sets are closed. But it is the direct images which are guaranteed to be connected and *not* the inverse images of connected sets. (Can you think of an example?) The same sort of thing will happen with the class of sets studied in the next subsection, the compact sets.

Compact Sets

The next special class of sets we wish to introduce is that of the compact sets. These will turn out to be those subsets K of $\mathbf{C}$ which are bounded in the sense that there is a number M such that $|z| \le M$ for every z in K and which are closed. One of the nice properties of such sets is that every sequence of points in the set must have a subsequence which converges to some point in the set. For example, the sequence $1, \frac{1}{2}, \frac{2}{3}, \frac{1}{3}, \frac{3}{5}, \frac{1}{4}, \frac{7}{4}, \frac{1}{5}, \frac{2}{5}, \ldots$ of points in $]0, 1[$ has the subsequence $1, \frac{1}{2}, \frac{1}{3}, \ldots$, which converges to the point 0, which is not in the open interval $]0, 1[$ but is in the closed interval $[0, 1]$. Note that in the claimed property, the sequence itself is not asserted to converge. All that is claimed is that some subsequence does; the example shows that this is necessary.

As often happens in mathematics, the study consists of three parts:

(i) An easily recognized characterization: closed and bounded
(ii) A property we want: the existence of convergent subsequences
(iii) A technical definition useful in proofs and problems

In the case at hand, the technical definition involves the relationship between compactness and open sets. A collection of open sets U_α for α in some index set $\mathscr{A}$ is called a *cover* (or an *open cover*) of a set K if K is contained in their union: $K \subset \bigcup_{\alpha \in \mathscr{A}} U_\alpha$. For example, the collection of all open disks of radius 2 is an open cover of $\mathbf{C}$:

$$U_z = D(z; 2) \qquad \mathbf{C} \subset \bigcup_{z \in \mathbf{C}} D(z; 2)$$

It may be, as here, that the covering process has been wasteful, using more sets than needed. In that case we may use only some of the sets and talk of a *subcover*, for example, $\mathbf{C} \subset \bigcup_{n,m \in \mathbf{Z}} D(n + mi; 2)$, where $\mathbf{Z}$ denotes the set of integers.

1.4.17 DEFINITION   *A set K is **compact** if every open cover of K has a finite subcover.*

That is, if U_α is any collection of open sets whose union contains K, then there is a finite subcollection $U_{\alpha_1}, U_{\alpha_2}, \ldots, U_{\alpha_k}$ such that $K \subset U_{\alpha_1} \cup U_{\alpha_2} \cup \cdots \cup U_{\alpha_k}$.

1.4.18 PROPOSITION   *The following conditions are equivalent for a subset K of* $\mathbf{C}$ *(or of* $\mathbf{R}$):

 (i) *K is closed and bounded.*
 (ii) *Every sequence of points in K has a subsequence which converges to some point in K.*
(iii) *K is compact.*

This proposition requires a deeper study of the completeness properties of the real numbers than it is necessary for us to go into here, so the proof is omitted. It may be found in most advanced calculus or analysis texts (such as J. Marsden, *Elementary Classical Analysis* (New York: W. H. Freeman and Co., 1974)). It is easy to see why (i) is necessary for (ii) and (iii). If K is not bounded we can select z_1 in K and then successively choose z_2 with $|z_2| > |z_1| + 1$ and, in general, z_n with $|z_n| > |z_{n-1}| + 1$. This gives a sequence with no convergent subsequence. The open disks $D(0; n)$, $n = 1, 2, 3, \ldots$, would be an open cover with no finite subcover.

If K is a set in $\mathbf{C}$ which is not closed, then there is a point w in $\mathbf{C} \backslash K$ and a sequence z_1, $z_2, \ldots$ of points in K which converges to w. Since the sequence converges, w is the only possible limit of a subsequence, so no subsequence can converge to a point of K. The sets $\{z$ such that $|z - w| > 1/n\}$ for $n = 1, 2, 3, \ldots$ form an open cover of K with no finite subcover.

The utility of the technical definition 1.4.17 is illustrated in the following results.

1.4.19 PROPOSITION   *If K is a compact set and f is a continuous function defined on K, then the image set $f(K)$ is also compact.*

PROOF   If U_α is an open cover of $f(K)$, then the sets $f^{-1}(U_\alpha)$ form an open cover of K. Selection of a finite subcover gives

$$K \subset f^{-1}(U_{\alpha_1}) \cup \cdots \cup f^{-1}(U_{\alpha_k})$$

so that $f(K) \subset U_{\alpha_1} \cup \cdots \cup U_{\alpha_k}$.   ∎

1.4.20 EXTREME VALUE THEOREM *If K is a compact set and $f : K \to \mathbf{R}$ is continuous, then f attains finite maximum and minimum values.*

PROOF The image $f(K)$ is compact, hence closed and bounded. Since it is bounded, the numbers $M = \sup \{f(z) \mid z \in K\}$ and $m = \inf \{f(z) \mid z \in K\}$ are finite. Since $f(K)$ is closed, m and M are included in $f(K)$. ∎

Another illustration of the use of compactness is given by the following lemma, which asserts that the distance from a compact set to a closed set is positive. That is, there must be a definite gap between the two sets.

1.4.21 DISTANCE LEMMA *Suppose K is compact, C is closed, and $K \cap C = \varnothing$. Then the distance $d(K, C)$ from K to C is greater than 0. That is, there is a number $\rho > 0$ such that $|z - w| > \rho$ whenever z is in K and w is in C.*

PROOF The complement of C, $U = \mathbf{C} \backslash C$, is an open set, and $K \subset U$, so that each point z in K is the center of some disk $D(z; \rho(z)) \subset U$. The family of smaller disks $D(z; \rho(z)/2)$ also cover K, and by compactness there is a finite collection of them, which we denote by $D_k = D(z_k; \rho(z_k)/2)$, $k = 1, 2, 3, \ldots, N$ which cover K. (See Fig. 1.4.6.) Let $\rho_k = \rho(z_k)/2$ and $\rho = \min(\rho_1, \rho_2, \ldots, \rho_N)$. If z is in K and w is in C, then z is in D_k for some k, and so $|z - z_k| < \rho_k$. But $|w - z_k| > \rho(z_k) = 2\rho_k$. Thus $|z - w| > \rho_k \geq \rho$. ∎

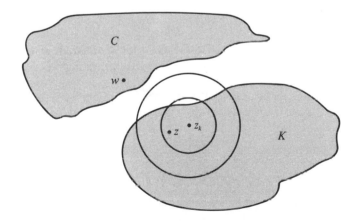

FIGURE 1.4.6 The distance between a closed set C and a compact set K is greater than zero.

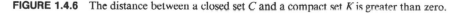

Uniform Continuity

Remember that a function is said to be continuous on a set K if it is continuous at each point of K. This is an example of what is called a *local property*. It is defined in terms of the behavior of the function at or near each point and can be determined for each point by looking only near the point and not at the whole set at once. This is in contrast to *global properties* of a function, which depend on its behavior on the whole set.

An example of a global property is boundedness. Saying that a function f is bounded by some number M on a set K is an assertion which depends on the whole set at once. If the function is continuous it is certainly bounded near each point, but that would not automatically say that it is bounded on the whole set. For example, the function $f(x) = 1/x$ is continuous on the open interval $]0, 1[$ but is certainly not bounded there. We have seen that if a function f is continuous on a compact set K, then it is bounded on K, and in fact that the bounds are attained. Thus compactness of K allowed us to carry the local boundedness near each point given by continuity over to the whole set. Compactness often can be used to make such a shift from a local property to a global one. The following is a global version of the notion of continuity.

1.4.22 DEFINITION *A function $f : A \to C$ (or R) is **uniformly continuous** on A if for every choice of $\epsilon > 0$ there is a $\delta > 0$ such that $|f(s) - f(t)| < \epsilon$ whenever s and t are in A and $|s - t| < \delta$.*

Notice that the difference between this and the definition of ordinary continuity is that now the choice of δ can be made so that the same δ will work everywhere in the set A. Obviously, uniformly continuous functions are continuous. On a compact set the opposite is true as well.

1.4.23 PROPOSITION *A continuous function on a compact set is uniformly continuous.*

PROOF Suppose f is a continuous function on a compact set K, and let $\epsilon > 0$. For each point t in K there is a number $\delta(t)$ such that $|f(s) - f(t)| < \epsilon/2$ whenever $|s - t| < \delta$. The open sets $D(t; \delta(t)/2)$ cover K, and so by compactness there are a finite number of points $t_1, t_2, \ldots, t_N$ such that the sets $D_k = D(t_k; \delta(t_k)/2)$ cover K. Let $\delta_k = \delta(t_k)/2$ and set δ equal to the minimum of $\delta_1, \delta_2, \ldots, \delta_N$. If $|s - t| < \delta$, then t is in D_k for some k, and so $|t - t_k| < \delta_k$. Thus $|f(t) - f(t_k)| < \epsilon/2$. But also,

$$
\begin{aligned}
|s - t_k| &= |s - t + t - t_k| \\
&\leq |s - t| + |t - t_k| \\
&\leq \delta + \delta_k \\
&\leq \delta(t_k)
\end{aligned}
$$

and so $|f(s) - f(t_k)| \le \epsilon/2$. Thus

$$
\begin{aligned}
|f(s) - f(t)| &= |f(s) - f(t_k) + f(t_k) - f(t)| \\
&\le |f(s) - f(t_k)| + |f(t_k) - f(t)| \\
&< \epsilon/2 + \epsilon/2 = \epsilon
\end{aligned}
$$

We have produced a single δ which works everywhere in K, and so f is uniformly continuous. ∎

Path Covering Lemma

The notion of uniform continuity is a very powerful one which will be useful to us several times. We use it first in conjunction with the distance lemma and some of the properties of compact sets to establish a useful geometric lemma about curves in open subsets of the complex plane. This lemma will be useful later in the text, particularly for studying integrals along such curves. It says that the curve can be covered by a finite number of disks centered along the curve in such a way that each disk is contained in the open set and each contains the centers of both the preceding and the succeeding disks along the curve. (See Fig. 1.4.7.)

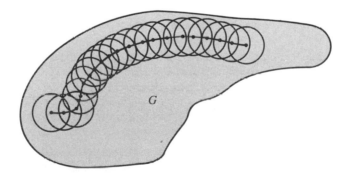

FIGURE 1.4.7 A continuous path in an open set can be covered by a finite number of well-overlapping disks.

1.4.24 PATH COVERING LEMMA *Suppose $\gamma: [a, b] \to G$ is a continuous path from the interval $[a, b]$ into an open subset G of $\mathbf{C}$. Then there are a number $\rho > 0$ and a subdivision of the interval $a = t_0 < t_1 < t_2 \cdots < t_n = b$ such that*

(i) $D(\gamma(t_k); \rho) \subset G$ *for all k*
(ii) $\gamma(t) \in D(\gamma(t_0); \rho)$ *for $t_0 \le t \le t_1$*
(iii) $\gamma(t) \in D(\gamma(t_k); \rho)$ *for $t_{k-1} \le t \le t_{k+1}$*

and

(iv) $\gamma(t) \in D(\gamma(t_n); \rho)$ for $t_{n-1} \leq t \leq t_n$

PROOF Since γ is continuous and the closed interval $[a, b]$ is compact, the image curve $K = \gamma([a, b])$ is compact. By the distance lemma, 1.4.21, there is a number ρ such that each point on the curve is a distance at least ρ from the complement of G. Therefore, $D(\gamma(t); \rho) \subset G$ for every t in $[a, b]$. Also, since γ is continuous on the compact set $[a, b]$, it is uniformly continuous, and there is a number $\delta > 0$ such that $|\gamma(t) - \gamma(s)| < \rho$ whenever $|s - t| < \delta$. Thus if the subdivision is chosen fine enough so that $t_k - t_{k-1} < \delta$ for all $k = 1, 2, 3, \ldots, N$, then the conclusions of the theorem hold. ∎

The Riemann Sphere and the "Point at Infinity"

For some purposes it is convenient to introduce a point "∞" in addition to the points $z \in \mathbf{C}$. One must be careful in doing so, since it can lead to confusion and abuse of the symbol ∞. But with care it can be useful, and we certainly want to be able to talk intelligently about infinite limits and limits at infinity.

In contrast to the real line, to which $+\infty$ and $-\infty$ can be added, we have only one ∞ for $\mathbf{C}$. The reason is that $\mathbf{C}$ has no natural ordering as $\mathbf{R}$ does. Formally we add a symbol "∞" to $\mathbf{C}$ to obtain the *extended complex plane,* $\overline{\mathbf{C}}$, and define operations with ∞ by the "rules"

$$z + \infty = \infty$$
$$z \cdot \infty = \infty \qquad \text{provided } z \neq 0$$
$$\infty + \infty = \infty$$
$$\infty \cdot \infty = \infty$$
$$\frac{z}{\infty} = 0$$

for $z \in \mathbf{C}$. Notice that some things are not defined: ∞/∞, $0 \cdot \infty$, $\infty - \infty$, and so forth, are *indeterminate forms* for essentially the same reasons that they are in the calculus of real numbers. We also define appropriate limit concepts:

$\lim\limits_{z \to \infty} f(z) = z_0$ *means: For any $\epsilon > 0$, there is an $R > 0$ such that $|f(z) - z_0| < \epsilon$ whenever $|z| \geq R$.*

$\lim\limits_{z \to z_0} f(z) = \infty$ *means: For any $R > 0$, there is a $\delta > 0$ such that $|f(z)| > R$ whenever $|z - z_0| < \delta$*

and for sequences:

$\lim\limits_{z \to \infty} z_n = \infty$ *means: For any $R > 0$, there is an $N > 0$ such that $|z_n| > R$ whenever $n \geq N$.*

Thus a point $z \in \mathbb{C}$ is "close to ∞" when it lies outside a large circle. This type of closeness can be pictured geometrically by means of the *Riemann sphere* shown in Fig. 1.4.8. By the method of *stereographic projection* illustrated in this figure, a point z' on the sphere is associated with each point z in $\mathbb{C}$. Exactly one point on the sphere S has been omitted—the "north" pole. We assign ∞ in $\overline{\mathbb{C}}$ to the north pole of S. We see geometrically that z is close to ∞ if and only if the corresponding points are close on the Riemann sphere in the usual sense of closeness in $\mathbb{R}^3$. Proof of this is requested in Exercise 24.

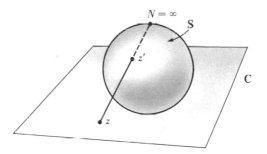

FIGURE 1.4.8 The Riemann sphere.

The Riemann sphere S represents a convenient geometric picture of the extended plane $\overline{\mathbb{C}} = \mathbb{C} \cup \{\infty\}$. The sphere does point up one fact about the extended plane which is sometimes useful in further theory. Since S is a closed bounded subset of $\mathbb{R}^3$, it is compact. Therefore every sequence in it has a convergent subsequence. Since stereographic projection makes convergence on S coincide with convergence of the sequence of corresponding points in $\overline{\mathbb{C}}$, the same is true there. That is, $\overline{\mathbb{C}}$ is compact. Every sequence of points in $\overline{\mathbb{C}}$ must have a subsequence convergent in $\overline{\mathbb{C}}$. *Caution:* Since the convergence is in the extended plane, the limit might be ∞, in which case we would normally say that the limit does not exist. Basically we have thrown in the point at infinity as another available limit so that sequences which did not formerly have a limit now have one. The sphere can be used both to help visualize and to make precise some notions about the behavior of functions "at infinity" that we will meet in future chapters.

Worked Examples

1.4.25   *Where is the function*

$$f(z) = \frac{z^3 + 2z + 1}{z^3 + 1}$$

continuous?

Solution. Since sums, products, and quotients of continuous functions are continuous except where the denominator is 0, this function is continuous on the whole plane except at the cube roots of -1. That is, on $\mathbf{C}\backslash\{e^{\pi i/3}, e^{5\pi i/3}, -1\}$.

1.4.26 *Show that $\{z \mid \operatorname{Re} z > 0\}$ is open.*

Solution. A proof can be based on the following properties of complex numbers (see Exercise 1): *If $w \in \mathbf{C}$, then*

(i) $|\operatorname{Re} w| \leq |w|$
(ii) $|\operatorname{Im} w| \leq |w|$

and

(iii) $|w| \leq |\operatorname{Re} w| + |\operatorname{Im} w|$

Let $U = \{z \mid \operatorname{Re} z > 0\}$ and let z_0 be in U. Suppose that z is in the disk $D(z_0; \operatorname{Re} z_0)$. Then $|\operatorname{Re} z - \operatorname{Re} z_0| = |\operatorname{Re} (z - z_0)| \leq |z - z_0| < \operatorname{Re} z_0$, and so $\operatorname{Re} z > 0$ and z is in U. Thus $D(z_0; \operatorname{Re} z_0)$ is a neighborhood of z_0 which is contained in U. Since this can be done for any point z_0 which is in U, the set U is open. See Fig. 1.4.9.

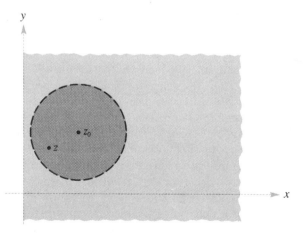

FIGURE 1.4.9   The open right half plane.

1.4.27 *Prove the following statement: Let $A \subset \mathbf{C}$ be open and $z_0 \in A$, and suppose that $D_r = \{z \text{ such that } |z - z_0| \leq r\} \subset A$. Then there is a number $\rho > r$ such that $D (z_0; \rho) \subset A$.*

Solution. We know from the extreme value theorem (1.4.20) that a continuous real-valued function on a closed bounded set in $\mathbf{C}$ attains its maximum and minimum at some point of the set. For z in D_r, let $f(z) = \inf \{|z - w| \text{ such that } w \in \mathbf{C}\backslash A\}$. (Here "inf" means the greatest lower bound.) In other words, $f(z)$ is the distance from z to the complement of A. Since A is open, $f(z) > 0$ for each z in D_r. We can also verify that f is continuous. Thus f assumes its minimum at some point z_1 in D_r. Let $\rho = f(z_1) + r$, and check that this ρ has the desired properties. See Fig. 1.4.10.

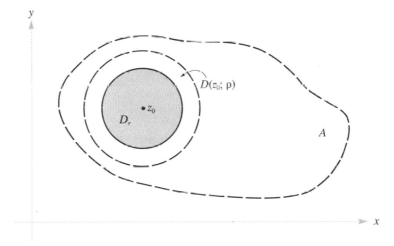

FIGURE 1.4.10 A closed disk in an open set may be enlarged.

1.4.28 *Find*

$$\lim_{z \to \infty} \frac{3z^4 + 2z^2 - z + 1}{z^4 + 1}$$

Solution.

$$\lim_{z \to \infty} \frac{3z^4 + 2z^2 - z + 1}{z^4 + 1} = \lim_{z \to \infty} \frac{3 + 2z^{-2} - z^{-3} + z^{-4}}{1 + z^{-4}} = 3$$

using $\lim_{z \to \infty} z^{-1} = 0$, and the basic properties of limits.

Exercises

1. Show that if $w \in \mathbf{C}$, then

(a) $|\text{Re } w| \leq |w|$ (b) $|\text{Im } w| \leq |w|$

(c) $|w| \leq |\text{Re } w| + |\text{Im } w|$

2. (a) Show that $|\text{Re } z_1 - \text{Re } z_2| \leq |z_1 - z_2| \leq |\text{Re } z_1 - \text{Re } z_2| + |\text{Im } z_1 - \text{Im } z_2|$ for any two complex numbers z_1 and z_2.

(b) If $f(z) = u(x, y) + iv(x, y)$, show that

$$\lim_{z \to z_0} f(z) = \lim_{\substack{x \to x_0 \\ y \to y_0}} u(x, y) + i \lim_{\substack{x \to x_0 \\ y \to y_0}} v(x, y)$$

exists if both limits on the right of the equation exist. Conversely, if the limit on the left exists, show that both limits on the right exist as well and equality holds. Show that $f(z)$ is continuous iff u and v are.

3. Prove: If f is continuous and $f(z_0) \neq 0$, there is a neighborhood of z_0 on which f is $\neq 0$.

4. If $z_0 \in \mathbf{C}$, show that the set $\{z_0\}$ is closed.

5. Prove: The complement of a finite number of points is an open set.

6. Use the fact that a function is continuous if and only if the inverse image of every open set is open to show that a composition of two continuous functions is continuous.

7. Show that $f(z) = \bar{z}$ is continuous.

8. Show that $f(z) = |z|$ is continuous.

9. What is the largest set on which the function $f(z) = 1/(1 - e^z)$ is continuous?

10. Prove or give an example if false: If $\lim\limits_{z \to z_0} f(z) = a$, h is defined at the points $f(z)$, and $\lim\limits_{w \to a} h(w) = c$, then $\lim\limits_{z \to z_0} h(f(z)) = c$. [*Hint:* We could have $h(a) \neq c$].

11. For what z does the sequence $z_n = nz^n$ converge?

12. Define $f : \mathbf{C} \to \mathbf{C}$ by setting $f(0) = 0$ and by setting $f(r[\cos \theta + i \sin \theta]) = \sin \theta$ if $r > 0$. Show that f is discontinuous at 0 but is continuous everywhere else.

13. For each of the following sets, state (i) whether or not it is open and (ii) whether or not it is closed.
(a) $\{z \text{ such that } |z| < 1\}$ (b) $\{z \mid 0 < |z| \leq 1\}$
(c) $\{z \mid 1 \leq \operatorname{Re} z \leq 2\}$

14. For each of the following sets, state (i) whether or not it is open and (ii) whether or not it is closed.
(a) $\{z \mid \operatorname{Im} z > 2\}$ (b) $\{z \mid 1 \leq |z| \leq 2\}$
(c) $\{z \mid -1 < \operatorname{Re} z \leq 2\}$

15. For each of the following sets, state (i) whether or not it is connected and (ii) whether or not it is compact.
(a) $\{z \mid 1 \leq |z| \leq 2\}$ (b) $\{z \text{ such that } |z| \leq 3 \text{ and } |\operatorname{Re} z| \geq 1\}$
(c) $\{z \text{ such that } |\operatorname{Re} z| \leq 1\}$ (d) $\{z \text{ such that } |\operatorname{Re} z| \geq 1\}$

16. For each of the following sets, state (i) whether or not it is connected and (ii) whether or not it is compact.

(a) $\{z \mid 1 < \mathrm{Re}\ z \leq 2\}$ (b) $\{z \mid 2 \leq |z| \leq 3\}$
(c) $\{z \text{ such that } |z| \leq 5 \text{ and } |\mathrm{Im}\ z| \geq 1\}$

17. If $A \subset \mathbf{C}$ and $f: \mathbf{C} \to \mathbf{C}$, show that $\mathbf{C} \backslash f^{-1}(A) = f^{-1}(\mathbf{C} \backslash A)$.

18. Show that $f: A \subset \mathbf{C} \to \mathbf{C}$ is continuous if and only if $z_n \to z_0$ in A implies that $f(z_n) \to f(z_0)$.

19. Show that the union of any collection of open subsets of $\mathbf{C}$ is open.

20. Show that the intersection of any finite collection of open subsets of $\mathbf{C}$ is open.

21. Give an example to show that the statement in Exercise 20 is false if the word "finite" is omitted.

22. Prove part (ii) of Proposition 1.4.6 by using part (i).

23. Show that if $|z| > 1$, then $\lim\limits_{n \to \infty} (z^n/n) = \infty$.

24. Introduce the *chordal metric* ρ on $\overline{\mathbf{C}}$ by setting $\rho(z_1, z_2) = d(z_1', z_2')$ where z_1' and z_2' are the corresponding points on the Riemann sphere and d is the usual distance between points in $\mathbf{R}^3$.
(a) Show that $z_n \to z$ in $\mathbf{C}$ if and only if $\rho(z_n, z) \to 0$.
(b) Show that $z_n \to \infty$ if and only if $\rho(z_n, \infty) \to 0$.
(c) If $f(z) = (az + b)/(cz + d)$ and $ad - bc \neq 0$, show that f is continuous at ∞.

1.5 ANALYTIC FUNCTIONS

Although continuity is an important concept, its importance in complex analysis is overshadowed by that of the complex derivative. There are several approaches to the theory of complex differentiation. We shall begin by defining the derivative as the limit of difference quotients exactly as is done for real variables in calculus. Many properties of the derivative and the computation rules in particular follow from the properties of limits just as they do in the calculus of functions of real variables. However, there are some surprising and beautiful results special to the complex theory.

Several different words are used to describe functions which are differentiable in the complex sense; for example: "regular," "holomorphic," and "analytic." We will use the term "analytic," the same word used in calculus to describe functions for which the Taylor series converges to the value of the function. One of the elegant results of complex analysis justifies this choice of language. Indeed, we will see in Chap. 3 that, in sharp distinction from the case of a single real variable, the assumption that a function is differentiable in the sense of complex variables guarantees the validity of the Taylor expansion of that function.

Differentiability

1.5.1 DEFINITION   *Let $f : A \to \mathbf{C}$ where $A \subset \mathbf{C}$ is an open set. Then f is said to be* ***differentiable (in the complex sense)*** *at $z_0 \in A$ if*

$$\underset{z \to z_0}{\text{limit}} \frac{f(z) - f(z_0)}{z - z_0}$$

*exists. This limit is denoted by $f'(z_0)$, or sometimes by $df/dz\,(z_0)$. Thus $f'(z_0)$ is a complex number. f is said to be **analytic** on A if f is complex-differentiable at each $z_0 \in A$. The word "**holomorphic**," which is sometimes used, is synonymous with the word "analytic." The phrase "**analytic at z_0**" means analytic on a neighborhood of z_0.*

Note that

$$\frac{f(z) - f(z_0)}{z - z_0}$$

is undefined at $z = z_0$, and this is the reason why deleted neighborhoods were used in the definition of limit.

The student is cautioned that although the definition of the derivative $f'(z_0)$ is similar to that of the usual derivative of a function of a real variable and although many of their properties are similar, the complex case is much richer. Note also that in the definition of $f'(z_0)$, we are dividing by the complex number $z - z_0$ and the special nature of division by complex numbers is a key consideration. The limit as $z \to z_0$ is taken for an arbitrary z approaching z_0 but not along any particular direction.

The existence of f' implies a great deal about f. It will be proven in Sec. 2.4 that if f' exists, then all the derivatives of f exist (that is, f'', the (complex) derivative of f', exists, and so on). This is in marked contrast to the case of a function $g(x)$ of the real variable x, in which $g'(x)$ can exist without the existence of $g''(x)$.

The analysis of what are called the Cauchy-Riemann equations in Theorem 1.5.8 will show how the complex derivative of f is related to the usual partial derivatives of f as a function of the real variables (x, y) and will supply a useful criterion for determining the existence of $f'(z_0)$. As in elementary calculus, continuity of f does not imply differentiability; for example, $f(z) = |z|$ is continuous but is not differentiable (see Exercise 10 at the end of this section). However, as in one-variable calculus, a differentiable function must be continuous:

1.5.2 PROPOSITION   *If $f'(z_0)$ exists, then f is continuous at z_0.*

PROOF By the sum rule for limits, we need only show that

$$\lim_{z \to z_0} [f(z) - f(z_0)] = 0$$

But

$$\lim_{z \to z_0} [f(z) - f(z_0)] = \lim_{z \to z_0} \left[\frac{f(z) - f(z_0)}{z - z_0} (z - z_0) \right]$$

which, by the product rule for limits, equals $f'(z_0) \cdot 0 = 0$. ∎

The usual rules of calculus—the product rule, the quotient rule, the chain rule, and the inverse function rule—can be used when differentiating analytic functions. We now explore these rules in detail.

1.5.3 PROPOSITION *Suppose that f and g are analytic on A, where $A \subset \mathbf{C}$ is an open set. Then*

(i) *$af + bg$ is analytic on A and $(af + bg)'(z) = af'(z) + bg'(z)$ for any complex numbers a and b.*

(ii) *fg is analytic on A and $(fg)'(z) = f'(z)g(z) + f(z)g'(z)$.*

(iii) *If $g(z) \neq 0$ for all $z \in A$, then f/g is analytic on A and*

$$\left(\frac{f}{g} \right)'(z) = \frac{f'(z)g(z) - g'(z)f(z)}{[g(z)]^2}$$

(iv) *Any polynomial $a_0 + a_1 z + \cdots + a_n z^n$ is analytic on all of $\mathbf{C}$ with derivative $a_1 + 2a_2 z + \cdots + na_n z^{n-1}$.*

(v) *Any rational function*

$$\frac{a_0 + a_1 z + \cdots + a_n z^n}{b_0 + b_1 z + \cdots + b_m z^m}$$

is analytic on the open set consisting of all z except those (at most, m) points where the denominator is zero. (See Review Exercise 24 for Chapter 1.)

PROOF The proofs of (i), (ii), and (iii) are all similar to the proofs of the corresponding results found in calculus. The procedure can be illustrated with a proof of (ii). Applying the limit theorems and the fact that $\lim_{z \to z_0} f(z) = f(z_0)$ (Proposition 1.5.2), we get

$$
\begin{aligned}
\operatorname*{limit}_{z \to z_0} \frac{f(z)g(z) - f(z_0)g(z_0)}{z - z_0} &= \operatorname*{limit}_{z \to z_0} \left[\frac{f(z)g(z) - f(z)g(z_0)}{z - z_0} + \frac{f(z)g(z_0) - f(z_0)g(z_0)}{z - z_0} \right] \\
&= \operatorname*{limit}_{z \to z_0} \left[f(z) \frac{g(z) - g(z_0)}{z - z_0} \right] + \operatorname*{limit}_{z \to z_0} \left[\frac{f(z) - f(z_0)}{z - z_0} g(z_0) \right] \\
&= f(z_0)g'(z_0) + f'(z_0)g(z_0)
\end{aligned}
$$

To prove (iv) we must first show that $f' = 0$ if f is constant. This is immediate from the definition of derivative because $f(z) - f(z_0) = 0$. It is equally easy to prove that $dz/dz = 1$. Then, using (ii), we can prove that

$$
\frac{d}{dz} z^2 = 1 \cdot z + z \cdot 1 = 2z
$$

and

$$
\frac{d}{dz} z^3 = \frac{d}{dz} (z \cdot z^2) = 1 \cdot z^2 + z \cdot 2z = 3z^2
$$

In general, we see by induction that $dz^n/dz = nz^{n-1}$. Then (iv) follows from this and (i), and (v) follows from (iv) and (iii). ∎

For example,

$$
\frac{d}{dz} (z^2 + 8z - 2) = 2z + 8
$$

and

$$
\frac{d}{dz} \left(\frac{1}{z+1} \right) = -\frac{1}{(z+1)^2}
$$

The student will also recall that one of the most important rules for differentiation is the chain rule, or "function of a function" rule. To illustrate,

$$
\frac{d}{dz} (z^3 + 1)^{10} = 10(z^3 + 1)^9 \cdot 3z^2
$$

$$
= 30 \, z^2 (z^3 + 1)^9
$$

This procedure for differentiating should be familiar; it is justified by the next result.

1.5.4 CHAIN RULE *Let $f. A \to C$ and $g. B \to C$ be analytic (A, B are open sets) and let $f(A) \subset B$. Then $g \circ f : A \to C$ defined by $g \circ f(z) = g(f(z))$ is analytic and*

$$\frac{d}{dz} g \circ f(z) = g'(f(z)) \cdot f'(z)$$

The basic idea of the proof of this theorem is that if $w = f(z)$ and $w_0 = f(z_0)$, then

$$\frac{g(f(z)) - g(f(z_0))}{z - z_0} = \frac{g(w) - g(w_0)}{w - w_0} \cdot \frac{f(z) - f(z_0)}{z - z_0}$$

and if we let $z \to z_0$, we also have $w \to w_0$, and the right side of the preceding equation thus becomes $g'(w_0)f'(z_0)$. The trouble is that even if $z \neq z_0$, we could have $w = w_0$. Because of this possibility, we give a more careful proof. (Although the chain rule here can be deduced from the chain rule for the usual derivative for functions of several variables—see the proof of 1.5.8—a separate proof is instructive.)

PROOF Let $w_0 = f(z_0)$, and define, for $w \subset B$,

$$h(w) = \begin{cases} \dfrac{g(w) - g(w_0)}{w - w_0} - g'(w_0) & w \neq w_0 \\ 0 & w = w_0 \end{cases}$$

Since $g'(w_0)$ exists, h is continuous. Since the composite of continuous functions is continuous,

$$\lim_{z \to z_0} h(f(z)) = h(w_0) = 0$$

From the definition of h and letting $w = f(z)$, we get $g \circ f(z) - g(w_0) = [h(f(z)) + g'(w_0)][f(z) - w_0]$. Note that this still holds if $f(z) = w_0$. For $z \neq z_0$, we get

$$\frac{g \circ f(z) - g \circ f(z_0)}{z - z_0} = [h(f(z)) + g'(w_0)] \frac{f(z) - f(z_0)}{z - z_0}$$

As $z \to z_0$, the right side of the equation converges to $[0 + g'(w_0)] \cdot [f'(z_0)]$, so the theorem is proved. ∎

An argument similar to this shows that if $\gamma:]a, b[\to C$ is differentiable, we can differentiate the curve $\sigma(t) = f(\gamma(t))$ and obtain $\sigma'(t) = f'(\gamma(t)) \cdot \gamma'(t)$. Here $\gamma'(t)$ is the derivative of γ as a function $]a, b[\rightarrow \mathbf{R}^2$; that is, if $\gamma(t) = (x(t), y(t))$, then $\gamma'(t) = (x'(t), y'(t)) = x'(t) + iy'(t)$.

This last version of the chain rule allows us to develop a complex version of the theorem from calculus which states that a function whose derivative is identically 0 must be constant. The result illustrates the importance of *regions,* or open connected sets, in which we may, by Proposition 1.4.15, connect any two points by a differentiable path.

1.5.5 PROPOSITION   *Let $A \subset \mathbf{C}$ be open and connected and let $f : A \rightarrow \mathbf{C}$ be analytic. If $f'(z) = 0$ on A, then f is constant on A.*

PROOF   Let $z_1, z_2 \in A$. We want to show that $f(z_1) = f(z_2)$. Let $\gamma(t)$ by a path joining z_1 to z_2. By the chain rule, $df(\gamma(t))/dt = f'(\gamma(t)) \cdot \gamma'(t) = 0$, since $f' = 0$. Thus if $f = u + iv$, we have $du(\gamma(t))/dt = 0$ and $dv(\gamma(t))/dt = 0$. From calculus, we know this implies that $u(\gamma(t))$ and $v(\gamma(t))$ are constant functions of t. Comparing the values at $t = a$ and $t = b$ gives us $f(z_1) = f(z_2)$. ∎

Clearly, connectedness is needed because if A consisted of two disjoint pieces, we could let $f = 1$ on one piece and $f = 0$ on the other. Then $f'(z)$ would equal 0 but f would not be constant on A.

Conformal Maps

The existence of the complex derivative f' places severe, but very useful, restrictions on f. The first of these restrictions will be briefly discussed here. Another restriction will be mentioned when the Cauchy-Riemann equations are analyzed in Theorem 1.5.8.

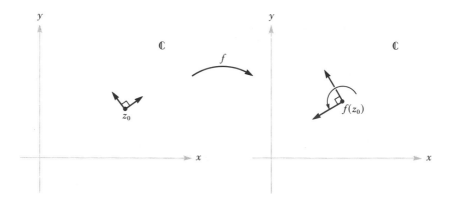

FIGURE 1.5.1   Conformal map at z_0.

It will be shown that "infinitesimally" near a point z_0 at which $f'(z_0) \neq 0$, f is a rotation by $\arg f'(z_0)$ and a magnification by $|f'(z_0)|$. The term "infinitesimally" is defined more precisely below, but intuitively it means that locally f is approximately a rotation together with a magnification (see Fig. 1.5.1). If $f'(z_0) = 0$, the structure of f is more complicated. (This point will be studied further in Chap. 6.)

1.5.6 DEFINITION *A map $f : A \to \mathbf{C}$ is called **conformal** at z_0 if there exist a $\theta \in [0, 2\pi[$ and an $r > 0$ such that for any curve $\gamma(t)$ which is differentiable at $t = 0$, for which $\gamma(t) \in A$ and $\gamma(0) = z_0$, and which satisfies $\gamma'(0) \neq 0$, the curve $\sigma(t) = f(\gamma(t))$ is differentiable at $t = 0$ and, setting $u = \sigma'(0)$ and $v = \gamma'(0)$, we have $|u| = r|v|$ and $\arg u = \arg v + \theta$ (mod 2π).*
*In this text a map will be called **conformal** when it is conformal at every point.*

Thus a conformal map merely rotates and stretches tangent vectors to curves. This is the precise meaning of "infinitesimal" as previously used. It should be noted that a conformal map *preserves angles* between intersecting curves. (By definition, the angle between two curves is the angle between their tangent vectors—see Fig. 1.5.2.)

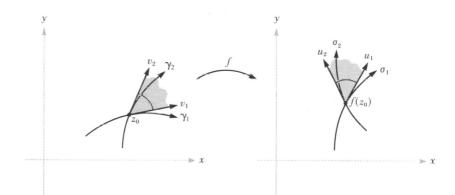

FIGURE 1.5.2 Preservation of angles by a conformal map.

1.5.7 CONFORMAL MAPPING THEOREM *If $f : A \to \mathbf{C}$ is analytic and if $f'(z_0) \neq 0$, then f is conformal at z_0 with $\theta = \arg f'(z_0)$ and $r = |f'(z_0)|$ fulfilling Definition 1.5.6.*

The proof of this theorem is remarkably simple:

PROOF   Using the preceding notation and the chain rule, we get $u = \sigma'(0) = f'(z_0) \cdot \gamma'(0) = f'(z_0) \cdot v$. Thus $\arg u = \arg f'(z_0) + \arg v \pmod{2\pi}$ and $|u| = |f'(z_0)| \cdot |v|$, as required. ∎

The point of this proof is that the tangent vector v to any curve is multiplied by a fixed complex number, namely, $f'(z_0)$, no matter in which direction v is pointing. This is because, in the definition of $f'(z_0)$, $\lim\limits_{z \to z_0}$ is "taken through all possible directions" as $z \to z_0$.

Cauchy-Riemann Equations

Let us recall that if $f: A \subset \mathbf{C} = \mathbf{R}^2 \to \mathbf{R}^2$ and if $f(x, y) = (u(x, y), v(x, y)) = u(x, y) + iv(x, y)$, then the *Jacobian matrix* of f is defined as the matrix of partial derivatives given by

$$
Df(x, y) = \begin{pmatrix} \dfrac{\partial u}{\partial x} & \dfrac{\partial u}{\partial y} \\[2mm] \dfrac{\partial v}{\partial x} & \dfrac{\partial v}{\partial y} \end{pmatrix}
$$

at each point (x, y). We shall relate these partial derivatives to the complex derivative. From the point of view of real variables, f is called *differentiable* with derivative the matrix $Df(x_0, y_0)$ at (x_0, y_0) iff for any $\epsilon > 0$, there is a $\delta > 0$ such that $|(x_0, y_0) - (x, y)| < \delta$ implies

$$
|f(x, y) - f(x_0, y_0) - Df(x_0, y_0)[(x, y) - (x_0, y_0)]| \leq \epsilon |(x, y) - (x_0, y_0)|
$$

where $Df(x_0, y_0) \cdot [(x, y) - (x_0, y_0)]$ means the matrix $Df(x_0, y_0)$ applied to the (column) vector

$$
\begin{pmatrix} x - x_0 \\ y - y_0 \end{pmatrix}
$$

and $|w|$ stands for the length of a vector w.

If f is differentiable, then the usual partials $\partial u/\partial x$, $\partial u/\partial y$, $\partial v/\partial x$, and $\partial v/\partial y$ exist and $Df(x_0, y_0)$ is given by the Jacobian matrix. The expression $Df(x_0, y_0)[w]$ represents the derivative of f in the direction w. If the partials exist and are continuous, then f is differentiable. Generally, then, differentiability is a bit stronger than existence of the individual partials. (Proofs of the preceding statements are not included here but can be found in any advanced calculus text, such as J. Marsden, *Elementary Classical Analysis* (New York: W. H. Freeman and Co., 1974), Chap. 6.) The main result connecting the partial derivatives and analyticity is stated in the next theorem.

1.5.8 CAUCHY-RIEMANN THEOREM *Let $f: A \subset \mathbf{C} \to \mathbf{C}$ be a given function, with A an open set. Then $f'(z_0)$ exists if and only if f is differentiable in the sense of real variables and, at $(x_0, y_0) = z_0$, u, v satisfy*

$$\frac{\partial u}{\partial x} = \frac{\partial v}{\partial y} \quad \text{and} \quad \frac{\partial u}{\partial y} = -\frac{\partial v}{\partial x}$$

*(called the **Cauchy-Riemann equations**).*

Thus, if $\partial u/\partial x$, $\partial u/\partial y$, $\partial v/\partial x$, and $\partial v/\partial y$ exist, are continuous on A, and satisfy the Cauchy-Riemann equations, then f is analytic on A.

If $f'(z_0)$ does exist, then

$$f'(z_0) = \frac{\partial u}{\partial x} + i\frac{\partial v}{\partial x} = \frac{\partial f}{\partial x}$$

$$= \frac{\partial v}{\partial y} - i\frac{\partial u}{\partial y} = \frac{1}{i}\frac{\partial f}{\partial y}$$

PROOF Let us first show that if $f'(z_0)$ exists, then u and v satisfy the Cauchy-Riemann equations. In the limit

$$f'(z_0) = \lim_{z \to z_0} \frac{f(z) - f(z_0)}{z - z_0}$$

let us take the special case that $z = x + iy_0$. Then

$$\frac{f(z) - f(z_0)}{z - z_0} = \frac{u(x, y_0) + iv(x, y_0) - u(x_0, y_0) - iv(x_0, y_0)}{x - x_0}$$

$$= \frac{u(x, y_0) - u(x_0, y_0)}{x - x_0} + i\frac{v(x, y_0) - v(x_0, y_0)}{x - x_0}$$

As $x \to x_0$, the left side of the equation converges to the limit $f'(z_0)$. Thus both the real and imaginary parts of the right side must converge to a limit (see Exercise 2 of Sec. 1.4). From the definition of partial derivatives, this limit is $(\partial u/\partial x)(x_0, y_0) + i(\partial v/\partial x)(x_0, y_0)$. Thus $f'(z_0) = \partial u/\partial x + i\,\partial v/\partial x$ evaluated at (x_0, y_0).

Next let $z = x_0 + iy$. Then we similarly have

$$\frac{f(z) - f(z_0)}{z - z_0} = \frac{u(x_0, y) + iv(x_0, y) - u(x_0, y_0) - iv(x_0, y_0)}{i(y - y_0)}$$

$$= \frac{u(x_0, y) - u(x_0, y_0)}{i(y - y_0)} + \frac{v(x_0 y) - v(x_0, y_0)}{y - y_0}$$

As $y \to y_0$, we get

$$\frac{1}{i}\frac{\partial u}{\partial y} + \frac{\partial v}{\partial y} = \frac{\partial v}{\partial y} - i\frac{\partial u}{\partial y}$$

Thus, since $f'(z_0)$ exists and has the same value regardless of how z approaches z_0, we get

$$f'(z_0) = \frac{\partial u}{\partial x} + i\frac{\partial v}{\partial x} = \frac{\partial v}{\partial y} - i\frac{\partial u}{\partial y}$$

By comparing real and imaginary parts of these equations, we derive the Cauchy-Riemann equations as well as the two formulas for $f'(z_0)$.

Another argument for this direction of the proof and one for the opposite implication may be based on the matrix representation for complex multiplication developed in Exercise 10 of Sec. 1.1:

1.5.9 LEMMA   *A matrix*

$$\begin{pmatrix} a & b \\ c & d \end{pmatrix}$$

represents, under matrix multiplication, multiplication by a complex number iff $a = d$ and $b = -c$. The complex number in question is $a + ic = d - ib$.

PROOF   First, let us consider multiplication by the complex number $a + ic$. It sends $x + iy$ to $(a + ic)(x + iy) = ax - cy + i(ay + cx)$, which is the same as

$$\begin{pmatrix} a & -c \\ c & a \end{pmatrix}\begin{pmatrix} x \\ y \end{pmatrix} = \begin{pmatrix} ax - cy \\ cx + ay \end{pmatrix}$$

Conversely, let us suppose that

$$\begin{pmatrix} a & b \\ c & d \end{pmatrix}\begin{pmatrix} x \\ y \end{pmatrix} = z \cdot (x + iy)$$

for a complex number $z = \alpha + i\beta$. Then we get

$$ax + by = \alpha x - \beta y$$

and

$$cx + dy = \alpha y + \beta x$$

for all x, y. This implies (setting $x = 1$, $y = 0$, then $x = 0$, $y = 1$) that $a = \alpha$, $b = -\beta$, $c = \beta$, and $d = \alpha$, and so the proof is complete. ▼

We can now complete the proof of Theorem 1.5.8

From the definition of f', the statement that $f'(z_0)$ exists is equivalent to the following statement: For any $\epsilon > 0$, there is a $\delta > 0$ such that $0 < |z - z_0| < \delta$ implies

$$|f(z) - f(z_0) - f'(z_0)(z - z_0)| < \epsilon|z - z_0|$$

First let us suppose that $f'(z_0)$ exists. By definition, $Df(x_0, y_0)$ is the unique matrix with the property that for any $\epsilon > 0$ there is a $\delta > 0$ such that, setting $z = (x, y)$ and $z_0 = (x_0, y_0)$, $0 < |z - z_0| < \delta$ implies

$$|f(z) - f(z_0) - Df(z_0)(z - z_0)| < \epsilon|z - z_0|$$

If we compare this equation with the preceding one and recall that multiplication by a complex number is a linear map, we conclude that f is differentiable in the sense of real variables and that the matrix $Df(z_0)$ represents multiplication by the complex number $f'(z_0)$. Thus, applying the lemma to the matrix

$$Df(z_0) = \begin{pmatrix} \dfrac{\partial u}{\partial x} & \dfrac{\partial u}{\partial y} \\ \dfrac{\partial v}{\partial x} & \dfrac{\partial v}{\partial y} \end{pmatrix}$$

with $a = \partial u/\partial x$, $b = \partial u/\partial y$, $c = \partial v/\partial x$, $d = \partial v/\partial y$, we have $a = d$, $b = -c$, which are the Cauchy-Riemann equations.

Conversely, if the Cauchy-Riemann equations hold, $Df(z_0)$ represents multiplication by a complex number (by the lemma) and then as above, the definition of differentiability in the sense of real variables reduces to that for the complex derivative.

The formula for $f'(z_0)$ follows from the last statement of the lemma. ∎

We can also express the Cauchy-Riemann equations in terms of polar coordinates, but care must be exercised because the change of coordinates defined by $r = \sqrt{x^2 + y^2}$ and $\theta = \arg(x + iy)$ is a differentiable change only if θ is restricted to the *open* interval $]0, 2\pi[$ or any other open interval of length 2π and if the origin ($r = 0$) is omitted. Without such a restriction θ is discontinuous, because it jumps by 2π on crossing the x axis. Using $\partial x/\partial r = \cos\theta$, $\partial y/\partial r = \sin\theta$, we see that the Cauchy-Riemann equations are equivalent to saying that

$$\frac{\partial u}{\partial r} = \frac{1}{r}\frac{\partial v}{\partial \theta} \qquad \frac{\partial v}{\partial r} = \frac{-1}{r}\frac{\partial u}{\partial \theta}$$

on a region contained in a region such as those shown in Fig. 1.5.3. Here we are employing standard abuse of notation by writing $u(r, \theta) = u(r \cos \theta, r \sin \theta)$. (For a more precise statement, see Exercise 12 at the end of this section.)

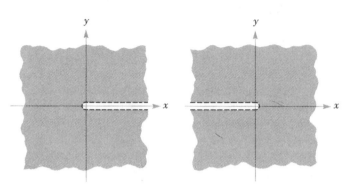

FIGURE 1.5.3   Two regions of validity of polar coordinates.

Inverse Functions

A basic theorem of real analysis is the inverse function theorem: *A continuously differentiable function is one-to-one and onto an open set and has a differentiable inverse in some neighborhood of a point where the Jacobian determinant of the derivative matrix is not* 0. We will give a proof here of the complex counterpart of this result, which assumes that the derivative f' is continuous and depends on the corresponding theorem for functions of real variables. After we have proved Cauchy's theorem in Chap. 2, we will see that the continuity of f' is automatic, and in Chap. 6 we will prove the theorem in another way which does not depend on the real-variable theorem. The proof given here, however, illustrates the relationship between real and complex variables and the relevance of the Cauchy-Riemann equations.

1.5.10 INVERSE FUNCTION THEOREM   *Let* $f: A \to \mathbf{C}$ *be analytic (with f' continuous) and assume that $f'(z_0) \neq 0$. Then there exist a neighborhood U of z_0 and a neighborhood V of $f(z_0)$ such that $f: U \to V$ is a bijection (that is, is one-to-one and onto) and its inverse function f^{-1} is analytic with derivative given by*

$$\frac{d}{dw}f^{-1}(w) = \frac{1}{f'(z)} \qquad where \qquad w = f(z)$$

The student is cautioned that application of the inverse function theorem allows one only to conclude the existence of a *local* inverse for f. For example, let us consider $f(z) = z^2$ defined on $A - \mathbf{C}\backslash\{0\}$. Then $f'(z) = 2z \neq 0$ at each point of A. The inverse function theorem says that f has a unique local analytic inverse, which is, in fact, merely some branch of the square root function. But f is not one-to-one on all of A, since, for example, $f(1) = f(-1)$. Thus f will be one-to-one only within sufficiently small neighborhoods surrounding each point.

To prove this theorem, let us recall the statement for real variables in two dimensions:

REAL-VARIABLE INVERSE FUNCTION THEOREM *If $f : A \subset \mathbf{R}^2 \to \mathbf{R}^2$ is continuously differentiable and $Df(x_0, y_0)$ has a nonzero determinant, then there are neighborhoods U of (x_0, y_0) and V of $f(x_0, y_0)$ such that $f : U \to V$ is a bijection, $f^{-1} : V \to U$ is differentiable, and*

$$Df^{-1}(f(x, y)) = [Df(x, y)]^{-1}$$

(this is the inverse of the matrix of partials).

The proof of this theorem may be found in advanced calculus texts. See, for instance, J. Marsden, *Elementary Classical Analysis* (New York: W. H. Freeman and Co., 1974), Chap. 7. Accepting this statement and assuming that f' in Theorem 1.5.10 is continuous, we can complete the proof.

PROOF OF THEOREM 1.5.10 For analytic functions such as $f(z)$, we have seen that the matrix of partial derivatives is

$$Df = \begin{pmatrix} \dfrac{\partial u}{\partial x} & \dfrac{\partial u}{\partial y} \\ \dfrac{\partial v}{\partial x} & \dfrac{\partial v}{\partial y} \end{pmatrix} = \begin{pmatrix} \dfrac{\partial u}{\partial x} & -\dfrac{\partial v}{\partial x} \\ \dfrac{\partial v}{\partial x} & \dfrac{\partial u}{\partial x} \end{pmatrix}$$

which has determinant

$$\left(\frac{\partial u}{\partial x}\right)^2 + \left(\frac{\partial v}{\partial x}\right)^2 = |f'(z)|^2$$

since $f'(z) = \partial u/\partial x + i\,\partial v/\partial x$. All these functions are to be evaluated at the point $(x_0, y_0) = z_0$. Now $f'(z_0) \neq 0$, so $\text{Det } Df(x_0, y_0) = |f'(z_0)|^2 \neq 0$. Thus the real-variable inverse function theorem applies. By the Cauchy-Riemann theorem we need only verify that the entries of $[Df(x, y)]^{-1}$ satisfy the Cauchy-Riemann equations and give $(f^{-1})'$ as stated.

As we have just seen,

$$
Df = \begin{pmatrix} \dfrac{\partial u}{\partial x} & \dfrac{\partial u}{\partial y} \\[2mm] \dfrac{\partial v}{\partial x} & \dfrac{\partial v}{\partial y} \end{pmatrix}
$$

and the inverse of this matrix is

$$
\frac{1}{\text{Det } Df} \begin{pmatrix} \dfrac{\partial v}{\partial y} & \dfrac{-\partial u}{\partial y} \\[2mm] -\dfrac{\partial v}{\partial x} & \dfrac{\partial u}{\partial x} \end{pmatrix}
$$

Thus if we write $f^{-1}(x, y) = t(x, y) + is(x, y)$, then, comparing

$$
D(f^{-1}) = \begin{pmatrix} \dfrac{\partial t}{\partial x} & \dfrac{\partial t}{\partial y} \\[2mm] \dfrac{\partial s}{\partial x} & \dfrac{\partial s}{\partial y} \end{pmatrix}
$$

with the inverse matrix for Df, we get

$$
\frac{\partial t}{\partial x} = \frac{1}{\text{Det } Df} \frac{\partial v}{\partial y} = \frac{1}{\text{Det } Df} \frac{\partial u}{\partial x}
$$

and

$$
\frac{\partial s}{\partial x} = \frac{1}{\text{Det } Df} \frac{-\partial v}{\partial x} = \frac{1}{\text{Det } Df} \frac{\partial u}{\partial y} \qquad \text{evaluated at } f(x_0, y_0)
$$

Similarly,

$$
\frac{\partial t}{\partial y} = \frac{1}{\text{Det } Df} \frac{\partial v}{\partial x} \qquad \text{and} \qquad \frac{\partial s}{\partial y} = \frac{1}{\text{Det } Df} \frac{\partial v}{\partial y}
$$

Thus the Cauchy-Riemann equations hold for t and s since they hold for u and v. Therefore, f^{-1} is complex-differentiable. From the Cauchy-Riemann theorem we see that at the point $f(z_0)$,

$$
(f^{-1})' = \frac{\partial t}{\partial x} + i \frac{\partial s}{\partial x} = \frac{1}{\text{Det } Df} \left(\frac{\partial u}{\partial x} - i \frac{\partial v}{\partial x} \right) = \frac{\overline{f'(z_0)}}{|f'(z_0)|^2} = \frac{1}{f'(z_0)} \qquad \blacksquare
$$

An alternative way to show that $(f^{-1})'(f(z)) = 1/f'(z)$ is outlined in Exercise 7.

Harmonic Functions and Harmonic Conjugates

The real and imaginary parts of an analytic function must satisfy the Cauchy-Riemann equations. Manipulation of these equations leads directly to another very important property, which we now isolate. A twice continuously differentiable function $u: A \to \mathbf{R}$ defined on an open set A is called *harmonic* if

$$\nabla^2 u = \frac{\partial^2 u}{\partial x^2} + \frac{\partial^2 u}{\partial y^2} = 0 \tag{1}$$

The expression $\nabla^2 u$ is called the *Laplacian of u* and is one of the most basic operations in mathematics and physics. Harmonic functions play a fundamental role in the physical examples discussed later in Chaps. 5 and 8. For the moment let us study them from the mathematical point of view. For Eq. (1) to make sense, the function u must be twice differentiable. In Chap. 3 an analytic function will be shown to be *infinitely* differentiable. Thus its real and imaginary parts are infinitely differentiable. Let us accept (or assume) these properties here. In particular, the second partial derivatives are continuous, and so a standard result of calculus says that the mixed partials are equal. The Cauchy-Riemann equations may then be used to show that the functions are harmonic.

1.5.11 PROPOSITION *If f is analytic on an open set A and f = u + iv (that is, if $u = \operatorname{Re} f$ and $v = \operatorname{Im} f$), then u and v are harmonic on A.*

PROOF We use the Cauchy-Riemann equations, $\partial u/\partial x = \partial v/\partial y$ and $\partial u/\partial y = -\partial v/\partial x$. Differentiating the first equation with respect to x and the second equation with respect to y, we get

$$\frac{\partial^2 u}{\partial x^2} = \frac{\partial^2 v}{\partial x\, \partial y} \quad \text{and} \quad \frac{\partial^2 u}{\partial y^2} = -\frac{\partial^2 v}{\partial y\, \partial x} \tag{2}$$

We know from calculus that the second partials are symmetric:

$$\frac{\partial^2 v}{\partial x\, \partial y} = \frac{\partial^2 v}{\partial y\, \partial x}$$

Adding the equations in equation set (2) gives us

$$\frac{\partial^2 u}{\partial x^2} + \frac{\partial^2 u}{\partial y^2} = \frac{\partial^2 v}{\partial x\, \partial y} - \frac{\partial^2 v}{\partial y\, \partial x} = 0$$

The equation for v is proved in the same way. ∎

If u and v are real-valued functions defined on an open subset A of $\mathbf{C}$ such that the complex-valued function $f = u + iv$ is analytic on A, we say that u and v are *harmonic conjugates* on A. For example,

$$u(x, y) = x^2 - y^2 \quad \text{and} \quad v(x, y) = 2xy$$

are harmonic conjugates of each other since they are the real and imaginary parts of $f(z) = z^2$. In Worked Example 1.3.18 we saw that the level curves of u and v intersect at right angles. This is no accident; the proof of its general validity for harmonic functions uses the same idea as that in 1.3.18 and also the Cauchy-Riemann equations.

1.5.12 PROPOSITION   *Let u and v be harmonic conjugates on a region A. Suppose that*

$$u(x, y) = constant = c_1$$

and

$$v(x, y) = constant = c_2$$

define smooth curves. Then these curves intersect orthogonally (see Fig. 1.5.4).

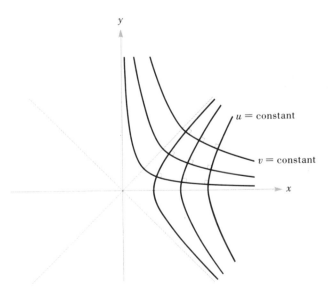

FIGURE 1.5.4   Harmonic conjugates in the case $u = x^2 - y^2$, $v = 2xy$, $f(z) = z^2$.

We shall accept from calculus the fact that $u(x, y) = c_1$ defines a smooth curve if the gradient grad $u(x, y) = (\partial u/\partial x, \partial u/\partial y) = (\partial u/\partial x) + i(\partial u/\partial y)$ is nonzero for x and y satisfying $u(x, y) = c_1$. (The student should be aware of this fact even though it is a technical point that does not play a major role in concrete examples.) It is also true that the vector grad u is perpendicular to that curve (see Fig. 1.5.5). This can be explained as follows. If $(x(t), y(t))$ is the curve, then $u(x(t), y(t)) = c_1$, so that

$$\frac{d}{dt}[u(x(t), y(t))] = 0$$

and so by the chain rule

$$\frac{\partial u}{\partial x} \cdot x'(t) + \frac{\partial u}{\partial y} \cdot y'(t) = 0$$

That is,

$$\left(\frac{\partial u}{\partial x}, \frac{\partial u}{\partial y}\right) \cdot (x'(t), y'(t)) = 0$$

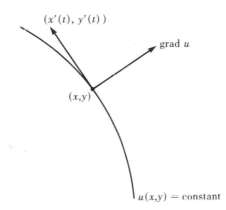

FIGURE 1.5.5 Gradients are orthogonal to level sets.

PROOF OF PROPOSITION 1.5.12 By the above remarks, it suffices to show that grad u and grad v are perpendicular. Their inner product is

$$\text{grad } u \cdot \text{grad } v = \frac{\partial u}{\partial x} \cdot \frac{\partial v}{\partial x} + \frac{\partial u}{\partial y} \cdot \frac{\partial v}{\partial y}$$

which is zero by the Cauchy-Riemann equations. ∎

This orthogonality property of harmonic conjugates has an important physical interpretation which will be used in Chap. 5. Another way to see why this property should hold is to consider conformal maps and the inverse function theorem. This is illustrated in Fig. 1.5.6. If $f = u + iv$ is analytic and $f'(z_0) \neq 0$, then f^{-1} is analytic on a neighborhood V of $w_0 = f(z_0)$ and $(f^{-1})'(w_0) \neq 0$ by the inverse function theorem. If $w_0 = c_1 + ic_2$, then the curves $u(x, y) = c_1$ and $v(x, y) = c_2$ are the images of the vertical and horizontal lines through w_0 under the mapping f^{-1}. They should cross at right angles since f^{-1} is conformal by Theorem 1.5.7.

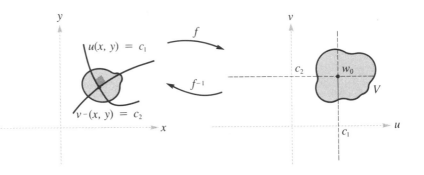

FIGURE 1.5.6 Since f and f^{-1} are analytic, they are conformal and so preserve orthogonality.

Proposition 1.5.11 says that the real part of an analytic function is harmonic. A natural question is the opposite one: Is every harmonic function the real part of an analytic function? More precisely: Given a harmonic function u on a set A, need there be a harmonic conjugate v such that $f = u + iv$ is analytic on A? The full answer is a little tricky and depends on the set A. However, the answer is simpler if we confine ourselves to small neighborhoods. Indeed, the property of being harmonic is what is called a *local property*. The function u is harmonic on a set if Eq. (1) holds at each point of that set, and the validity of Eq. (1) depends only on the behavior of u near that point.

1.5.13 PROPOSITION   *If u is a twice continuously differentiable harmonic function on an open set A and $z_0 \in A$, then there is some neighborhood of z_0 on which u is the real part of an analytic function.*

In other words, there exist an $r > 0$ and a function v defined on the open disk $D(z_0; r)$ such that u and v are harmonic conjugates on $D(z_0; r)$. In fact, $D(z_0; r)$ may be taken to be the largest disk centered at z_0 and contained in A. A direct proof of this is outlined in Exercise 32. A different proof of a slightly stronger result will be given in Chap. 2. Since the Cauchy-Riemann equations must hold, v is uniquely determined up to the addition of a constant. These equations may be used as a method for finding v when u is given. (See Worked Example 1.5.20.)

Worked Examples

1.5.14 *Where is*

$$f(z) = \frac{z^3 + 2z + 1}{z^3 + 1}$$

analytic? Compute the derivative.

Solution. By Proposition 1.5.3(iii) f is analytic on the set $A = \{z \in \mathbb{C} \mid z^3 + 1 \neq 0\}$; that is, f is analytic on the whole plane except the cube roots of $-1 = e^{\pi i}$: namely, the points $e^{\pi i/3}$, $e^{\pi i}$, and $e^{5\pi i/3}$. By the formula for differentiating a quotient, the derivative is

$$f'(z) = \frac{(z^3 + 1)(3z^2 + 2) - (z^3 + 2z + 1)(3z^2)}{(z^3 + 1)^2}$$

$$= \frac{(2 - 4z^3)}{(z^3 + 1)^2}$$

1.5.15 *Consider $f(z) = z^3 + 1$. Study the infinitesimal behavior of f at $z_0 = i$.*

Solution. We use the conformal mapping theorem (1.5.7). In this case $f'(z_0) = 3i^2 = -3$. Thus f rotates locally by $\pi = \arg(-3)$ and multiplies lengths by $3 = |f'(z_0)|$. More precisely, if c is any curve through $z_0 = i$, the image curve will, at $f(z_0)$, have its tangent vector rotated by π and stretched by a factor 3.

1.5.16 *Show that $f(z) = \bar{z}$ is not analytic.*

Solution. Let $f(z) = u(x, y) + iv(x, y) = x - iy$ where $z = (x, y) = x + iy$. Thus $u(x, y) = x$, $v(x, y) = -y$. But $\partial u / \partial x = 1$ and $\partial v / \partial y = -1$ and hence $\partial u / \partial x \neq \partial v / \partial y$, and so the Cauchy-Riemann equations do not hold. Therefore, $f(z) = \bar{z}$ cannot be analytic, by the Cauchy Riemann theorem (1.5.8).

1.5.17 *We know by Proposition 1.5.3 that $f(z) = z^3 + 1$ is analytic. Verify the Cauchy-Riemann equations for this function.*

Solution. If $f(z) = u(x, y) + iv(x, y)$ when $z = (x, y) = x + iy$, then in this case $u(x, y) = x^3 - 3xy^2 + 1$ and $v(x, y) = 3x^2y - y^3$. Therefore, $\partial u / \partial x = 3x^2 - 3y^2$, $\partial u / \partial y = -6xy$, $\partial v / \partial x = 6xy$, and $\partial v / \partial y = 3x^2 - 3y^2$, from which we see that $\partial u / \partial x = \partial v / \partial y$ and $\partial u / \partial y = -\partial v / \partial x$.

1.5.18 *Let A be an open subset of $\mathbb{C}$ and $A^* = \{z \mid \bar{z} \in A\}$. Suppose f is analytic on A, and define a function g on A^* by $g(z) = \overline{f(\bar{z})}$. Show that g is analytic on A^*.*

Solution. If $f(z) = u(x, y) + iv(x, y)$, then $g(z) = \overline{f(\bar{z})} = u(x, -y) - iv(x, -y)$. We check the Cauchy-Riemann equations for g as follows:

$$\frac{\partial}{\partial x}(\operatorname{Re} g) = \frac{\partial}{\partial x} u(x, -y) = \frac{\partial u}{\partial x}\bigg|_{(x,-y)} = \frac{\partial v}{\partial y}\bigg|_{(x,-y)}$$

$$= \frac{\partial}{\partial y}[-v(x, -y)] = \frac{\partial}{\partial y}(\operatorname{Im} g)$$

and

$$\frac{\partial}{\partial y}(\operatorname{Re} g) = \frac{\partial}{\partial y} u(x, -y) = -\frac{\partial u}{\partial y}\bigg|_{(x,-y)} = \frac{\partial v}{\partial x}\bigg|_{(x,-y)}$$

$$= -\frac{\partial}{\partial x}[-v(x, -y)] = -\frac{\partial}{\partial x}(\operatorname{Im} g)$$

Since the Cauchy-Riemann equations hold and g is differentiable in the sense of real variables (why?), it is analytic on A^* by the Cauchy-Riemann theorem. (One could also solve this exercise by direct appeal to the definition of complex differentiability.)

1.5.19   *Suppose f is an analytic function on a region (an open connected set) A and that $|f(z)|$ is constant on A. Show that $f(z)$ is constant on A.*

Solution. We use the Cauchy-Riemann equations to show that $f'(z) = 0$ everywhere in A. Let $f = u + iv$. Then $|f|^2 = u^2 + v^2 = c$ is constant. If $c = 0$, then $|f(z)| = 0$ and so $f(z) = 0$ for all z in A. If $c \neq 0$, we take derivatives of $u^2 + v^2 = c$ with respect to x and y to obtain

$$2u\frac{\partial u}{\partial x} + 2v\frac{\partial v}{\partial x} = 0 \qquad \text{and} \qquad 2u\frac{\partial u}{\partial y} + 2v\frac{\partial v}{\partial y} = 0$$

By the Cauchy-Riemann equations these become

$$u\frac{\partial u}{\partial x} - v\frac{\partial u}{\partial y} = 0 \qquad \text{and} \qquad v\frac{\partial u}{\partial x} + u\frac{\partial u}{\partial y} = 0$$

As a system of equations for the two unknowns $\partial u/\partial x$ and $\partial u/\partial y$, the matrix of coefficients has determinant $u^2 + v^2 = c$, which is not 0. Thus the only solution is $\partial u/\partial x = \partial u/\partial y = 0$ at all points of A. Therefore $f'(z) = \partial u/\partial x + i(\partial v/\partial x) = 0$ everywhere in A. Since A is connected, f is constant (by Proposition 1.5.5).

1.5.20.   *Find the harmonic conjugates of the following harmonic functions on $\mathbf{C}$:*
(a) $u(x, y) = x^2 - y^2$
(b) $u(x, y) = \sin x \cosh y$

Remark. The student might recognize $x^2 - y^2$ and $\sin x \cosh y$ as the real parts of z^2 and $\sin z$, $z = x + iy$. From this observation it follows that the conjugates, up to addition of constants, are $2xy$ and $\sinh y \cos x$. (We shall see in the next section that $\sin z$ is analytic.)

It is instructive, however, to solve the problem directly using the Cauchy-Riemann equations, because the student might not always recognize an appropriate analytic $f(z)$ by inspection.

Solution.

(a) If v is a harmonic conjugate of u, then $\partial v/\partial x = 2y$ and $\partial v/\partial y = 2x$. Therefore, $v = 2yx + g_1(y)$ and $v = 2xy + g_2(x)$. Hence $g_1(y) = g_2(x) =$ constant, and so $v(x, y) = 2yx +$ constant.

(b) $\partial v/\partial x = -\sin x \sinh y$ and $\partial v/\partial y = \cos x \cosh y$. The first equation implies that $v = \cos x \sinh y + g_1(y)$ and the second equation implies that $v = \cos x \sinh y + g_2(x)$. Hence $g_1(y) = g_2(x) =$ constant. Therefore $v(x, y) = \cos x \sinh y +$ constant.

1.5.21 *Show that $u(x, y) = \log \sqrt{x^2 + y^2}$ is harmonic on $\mathbf{C}\backslash\{0\}$. Discuss possible harmonic conjugates.*

Solution. Direct computation shows that

$$\frac{\partial^2 u}{\partial x^2} + \frac{\partial^2 u}{\partial y^2} = \frac{y^2 - x^2}{(x^2 + y^2)^2} + \frac{x^2 - y^2}{(x^2 + y^2)^2} = 0$$

Laplace's equation, Eq. (1), is satisfied, and so the function is harmonic.

If $v(x, y)$ is a harmonic conjugate, then

$$\frac{\partial v}{\partial x} = -\frac{\partial u}{\partial y} = -\frac{y}{x^2 + y^2} \quad \text{and} \quad \frac{\partial v}{\partial y} = \frac{\partial u}{\partial x} = \frac{x}{x^2 + y^2}$$

As a result of integrating the first equation, $v = -\tan^{-1}(x/y) + g_1(y)$, and from the second, $v = \tan^{-1}(y/x) + g_2(x)$. But $\tan^{-1}(y/x) + \tan^{-1}(x/y) = \pi/2$ (why?), and so $v = \tan^{-1}(y/x) +$ constant.

Notice that if $z = x + iy$, then $|z| = \sqrt{x^2 + y^2}$, and $u(x, y)$ is the real part of $\log z = \log|z| + i \arg z$. If $z_0 \neq 0$, we can choose the interval for the argument in such a way that the corresponding branch of the logarithm is defined and continuous on a neighborhood of z_0. Our function $u(x, y) = \log|z|$ is the real part of this. The function $v(x, y) = \tan^{-1}(x/y) = \arg z$ is a harmonic conjugate for u on a neighborhood of z_0, and the Cauchy-Riemann equations hold, and so $\log z$ is analytic. (See Fig. 1.5.7.) However, even though u is harmonic on all of $\mathbf{C}\backslash\{0\}$, there is no way to choose a harmonic conjugate which will work on this whole set at once, since a choice of $\tan^{-1}(y/x)$ must be made. Such a conjugate would have to be a choice of branch of the argument, and there is no way to do this without a jump discontinuity somewhere. (We shall study $\log z$ more systematically in subsequent sections.)

Exercises

1. Determine the sets on which the following functions are analytic, and compute their derivatives:

(a) $(z + 1)^3$

(b) $z + \dfrac{1}{z}$

(c) $\left(\dfrac{1}{z - 1}\right)^{10}$

(d) $\dfrac{1}{(z^3 - 1)(z^2 + 2)}$

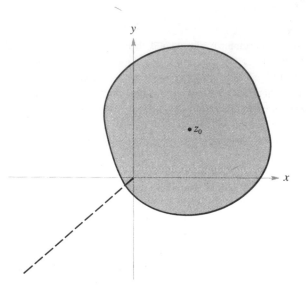

FIGURE 1.5.7 A branch of logarithm is analytic in a neighborhood of z_0.

2. Determine the sets on which the following functions are analytic, and compute their derivatives:

(a) $3z^2 + 7z + 5$ (b) $(2z + 3)^4$

(c) $\dfrac{3z - 1}{3 - z}$

3. On what sets are the following functions analytic? Compute the derivative for each.

(a) z^n, n being an integer (positive or negative)   (b) $\dfrac{1}{(z + 1/z)^2}$

(c) $z/(z^n - 2)$, n being a positive integer.

4. For $\gamma: \,]a, b[\to \mathbf{C}$ differentiable and $f: A \to \mathbf{C}$ analytic with $\gamma(]a, b[) \subset A$, prove that $\sigma = f \circ \gamma$ is differentiable with $\sigma'(t) = f'(\gamma(t)) \cdot \gamma'(t)$ by imitating the proof of the chain rule (1.5.4).

5. Study the infinitesimal behavior of the following functions at the indicated points:

(a) $f(z) = z + 3$, $z_0 = 3 + 4i$ (b) $f(z) = z^6 + 3z$, $z_0 = 0$

(c) $f(z) = \dfrac{z^2 + z + 1}{z - 1}$, $z_0 = 0$

6. Study the infinitesimal behavior of the following functions at the indicated points:

(a) $f(z) = 2z + 5$, $z_0 = 5 + 6i$ (b) $f(z) = z^4 + 4z$, $z_0 = i$

(c) $f(z) = 1/(z - 1)$, $z_0 = i$

7. Prove that $df^{-1}/dw = 1/f'(z)$ where $w = f(z)$ by differentiating $f^{-1}(f(z)) = z$, using the chain rule. Assume that f^{-1} is defined and is analytic.

8. Use the inverse function theorem to show that if $f: A \to \mathbf{C}$ is analytic and $f'(z) \neq 0$ for all $z \in A$, then f maps open sets in A to open sets.

9. Verify the Cauchy-Riemann equations for the function $f(z) = z^2 + 3z + 2$.

10. Prove that $f(z) = |z|$ is not analytic.

11. Show, by changing variables, that the Cauchy-Riemann equations in terms of polar coordinates become

$$\frac{\partial u}{\partial r} = \frac{1}{r} \frac{\partial v}{\partial \theta} \qquad \frac{\partial v}{\partial r} = -\frac{1}{r} \frac{\partial u}{\partial \theta}$$

12. Carefully perform the computations in Exercise 11 by the following procedure. Let f be defined on the open set $A \subset \mathbf{C}$ (that is, $f: A \subset \mathbf{C} \to \mathbf{C}$) and suppose that $f(z) = u(z) + iv(z)$. Let $T: \,] 0, 2\pi[\times \mathbf{R}^+ \to \mathbf{R}^2$, where $\mathbf{R}^+ = \{x \in \mathbf{R} \mid x > 0\}$, be given by $T(\theta, r) = (r \cos \theta, r \sin \theta)$. Thus T is one-to-one and onto the set $\mathbf{R}^2 \backslash \{(x, 0) \mid x \geq 0\}$. Define $\tilde{u}(\theta, r) = u(r \cos \theta, r \sin \theta)$ and $\tilde{v}(\theta, r) = v(r \cos \theta, r \sin \theta)$. Show that
(a) T is continuously differentiable and has a continuously differentiable inverse.
(b) f is analytic on $A \backslash \{x + iy \mid y = 0, x \geq 0\}$ if and only if $(\tilde{u}, \tilde{v}): T^{-1}(A) \to \mathbf{R}^2$ is differentiable and

$$\frac{\partial \tilde{u}}{\partial r} = \frac{1}{r} \frac{\partial \tilde{v}}{\partial \theta} \qquad \frac{\partial \tilde{v}}{\partial r} = -\frac{1}{r} \frac{\partial \tilde{u}}{\partial \theta}$$

on $T^{-1}(A)$

13. Define the symbol $\partial f / \partial \bar{z}$ by

$$\frac{\partial f}{\partial \bar{z}} = \frac{1}{2} \left(\frac{\partial f}{\partial x} - \frac{\partial f}{i \partial y} \right)$$

Show that the Cauchy-Riemann equations are equivalent to $\partial f / \partial \bar{z} = 0$. *Note:* It is sometimes said, because of this result, that analytic functions are not functions of $\bar{z}$ but of z alone. This statement can be made more precise as follows. Given $f(x, y)$, write $x = \frac{1}{2}(z + \bar{z})$ and $y = (1/2i)(z - \bar{z})$. Then f becomes a function of z and $\bar{z}$ and the chain rule gives

$$\frac{\partial f}{\partial \bar{z}} = \frac{\partial f}{\partial x} \frac{\partial x}{\partial \bar{z}} + \frac{\partial f}{\partial y} \frac{\partial y}{\partial \bar{z}} = \frac{1}{2} \left(\frac{\partial f}{\partial x} - \frac{1}{i} \frac{\partial f}{\partial y} \right)$$

14. Define the symbol $\partial f / \partial z$ by

$$\frac{\partial f}{\partial z} = \frac{1}{2} \left(\frac{\partial f}{\partial x} + \frac{1}{i} \frac{\partial f}{\partial y} \right)$$

(a) Show that if f is analytic, then $f' = \partial f / \partial z$.

(b) If $f(z) = z$, show that $\partial f / \partial z = 1$ and $\partial f / \partial \bar{z} = 0$.

(c) If $f(z) = \bar{z}$, show that $\partial f / \partial z = 0$ and $\partial f / \partial \bar{z} = 1$.

(d) Show that the symbols $\partial / \partial z$ and $\partial / \partial \bar{z}$ obey the sum, product, and scalar multiple rules for derivatives.

(e) Show that the expression $\sum_{n=0}^{N} \sum_{m=0}^{M} a_{nm} z^n \bar{z}^m$ is an analytic function of z if and only if $a_{nm} = 0$ whenever $m \neq 0$.

15. Suppose that f is an analytic function on the disk $D = \{z \text{ such that } |z| < 1\}$ and that $\operatorname{Re} f(z) = 3$ for all z in D. Show that f is constant on D.

16. (a) Let $f(z) = u(x, y) + iv(x, y)$ be an analytic function in a connected open set A. If $au(x, y) + bv(x, y) = c$ in A, where a, b, c are real constants not all 0, prove that $f(z)$ is constant in A.

(b) Is the result obtained in (a) still valid if a, b, c are complex constants?

17. Suppose f is analytic on $A = \{z \mid \operatorname{Re} z > 1\}$ and that $\partial u / \partial x + \partial v / \partial y = 0$ on A. Show that there are a real constant c and a complex constant d such that $f(z) = -icz + d$ on A.

18. Let $f(z) = z^5 / |z|^4$ if $z \neq 0$ and $f(0) = 0$.

(a) Show that $f(z)/z$ does not have a limit as $z \to 0$.

(b) If $u = \operatorname{Re} f$ and $v = \operatorname{Im} f$, show that $u(x, 0) = x$, $v(0, y) = y$, $u(0, y) = v(x, 0) = 0$.

(c) Conclude that the partials of u, v exist and that the Cauchy-Riemann equations hold, but that $f'(0)$ does not exist. Does this conclusion contradict the Cauchy-Riemann theorem?

(d) Repeat exercise (c), letting $f = 1$ on the x and y axes and 0 elsewhere.

(e) Repeat exercise (c), letting $f(z) = \sqrt{|xy|}$.

19. Let $f(z) = (z + 1)/(z - 1)$.

(a) Where is f analytic?

(b) Is f conformal at $z = 0$?

(c) What are the images of the x and y axes under f?

(d) At what angle do these images intersect?

20. Let f be an analytic function on an open connected set A and suppose that $f^{(n+1)}(z)$ (the $n + 1$st derivative) exists and is zero on A. Show that f is a polynomial of degree $\leq n$.

21. On what set is

$$u(x, y) = \operatorname{Re} \frac{z}{z^3 - 1}$$

harmonic?

22. Verify directly that the real and imaginary parts of $f(z) = z^4$ are harmonic.

23. On what sets are each of the following functions harmonic?

(a) $u(x, y) = \text{Im} (z^2 + 3z + 1)$

(b) $u(x, y) = \dfrac{x-1}{x^2 + y^2 - 2x + 1}$

24. On what sets are each of the following functions harmonic?

(a) $u(x, y) = \text{Im} (z + 1/z)$

(b) $u(x, y) = \dfrac{y}{(x-1)^2 + y^2}$

25. Let $f: A \to \mathbf{C}$ be analytic and let $w: B \to \mathbf{R}$ be harmonic with $f(A) \subset B$. Show that $w \circ f: A \to \mathbf{R}$ is harmonic.

26. If u is harmonic, show that, in terms of polar coordinates,

$$r^2 \frac{\partial^2 u}{\partial r^2} + r \frac{\partial u}{\partial r} + \frac{\partial^2 u}{\partial \theta^2} = 0$$

(*Hint.* Use the Cauchy-Riemann equations in polar form (Exercise 11).)

27. (a) Show that $u(x, y) = e^x \cos y$ is harmonic on $\mathbf{C}$.

(b) Find a harmonic conjugate $v(x, y)$ for u on $\mathbf{C}$ such that $v(0, 0) = 0$.

(c) Show that $f(z) = e^z$ is analytic on $\mathbf{C}$.

28. Show that $u(x, y) = x^3 - 3xy^2$ is harmonic on $\mathbf{C}$ and find a harmonic conjugate v such that $v(0, 0) = 2$.

29. If $u(z)$ is harmonic and $v(z)$ is harmonic, then

(a) Is $u(v(z))$ harmonic?

(b) Is $u(z) \cdot v(z)$ harmonic?

(c) Is $u(z) + v(z)$ harmonic?

30. Consider the function $f(z) = 1/z$. Draw the contours $u = \text{Re } f = \text{constant}$ and $v = \text{Im } f = \text{constant}$. How do they intersect? Is it *always* true that grad u is parallel to the curve $v = \text{constant}$?

31. Let u have continuous second partials on an open set A and let $\partial^2 u/\partial x^2 + \partial^2 u/\partial y^2 = 0$. Let $f = \partial u/\partial x - i \, \partial u/\partial y$. Show that f is analytic.

32. Suppose u is a twice continuously differentiable real-valued function on a disk $D(z_0; r)$ centered at $z_0 = x_0 + iy_0$. For $(x_1, y_1) \in D(z_0; r)$, show that the equation

$$v(x_1, y_1) = c + \int_{y_0}^{y_1} \frac{\partial u}{\partial x}(x_1, y) \, dy - \int_{x_0}^{x_1} \frac{\partial u}{\partial y}(x, y_0) \, dx$$

defines a harmonic conjugate for u on $D(z_0; r)$ with $v(x_0, y_0) = c$.

1.6   DIFFERENTIATION OF THE ELEMENTARY FUNCTIONS

This section will discuss the differentiability properties of the elementary functions discussed in Sec. 1.3.

Exponential Function and Logarithm

1.6.1 PROPOSITION   *The map $f : \mathbf{C} \to \mathbf{C}$, $z \mapsto e^z$, is analytic on $\mathbf{C}$ and*

$$\frac{de^z}{dz} = e^z$$

PROOF   By definition, $f(z) = e^x (\cos y + i \sin y)$, and so the real and imaginary parts are $u(x, y) = e^x \cos y$ and $v(x, y) = e^x \sin y$. These are C^∞ (infinitely often differentiable) functions, and so f is differentiable in the sense of real variables. To show that f is analytic, we must verify the Cauchy-Riemann equations. But

$$\frac{\partial u}{\partial x} = e^x \cos y \qquad \frac{\partial u}{\partial y} = -e^x \sin y$$

$$\frac{\partial v}{\partial x} = e^x \sin y \qquad \frac{\partial v}{\partial y} = e^x \cos y$$

Thus $\partial u / \partial x = \partial v / \partial y$ and $\partial u / \partial y = -\partial v / \partial x$, and so by the Cauchy-Riemann theorem, f is analytic. Also, since

$$\frac{df}{dz} = \frac{\partial u}{\partial x} + i \frac{\partial v}{\partial x} = e^x (\cos y + i \sin y) = e^z$$

the proposition is proved. ∎

A function that is defined and analytic on the whole plane $\mathbf{C}$ is called *entire*. Thus, $f(z) = e^z$ is an entire function.

Using the differentiation rules (Proposition 1.5.3) as they are used in elementary calculus, we can differentiate e^z in combination with various other functions. For instance, $e^{z^2} + 1$ is entire because $z \mapsto z^2$ and $w \mapsto e^w$ are analytic, and so by the chain rule, $z \mapsto e^{z^2}$ is analytic. By the chain rule and the sum rule, $d(e^{z^2} + 1)/dz = 2ze^{z^2}$.

We recall that $\log z : \mathbf{C}\backslash\{0\} \to \mathbf{C}$ is an inverse for e^z when e^z is restricted to a period strip $\{x + iy \mid y_0 \le y < y_0 + 2\pi\}$. However, for differentiability of $\log z$ we must restrict $\log z$ to a set that is smaller than $\mathbf{C}\backslash\{0\}$. The reason is simple: $\log z = \log|z| + i \arg z$, for, say, $0 \le \arg z < 2\pi$. But the arg function is discontinuous; it jumps by 2π as we cross the real axis. If we remove the real axis, then we are excluding the usual positive reals on which we want log z defined. Therefore, it is convenient to use the branch $-\pi < \arg z < \pi$. Then an appropriate set on which log z is analytic is given as follows.

1.6.2 PROPOSITION *Let A be the open set that is $\mathbf{C}$ minus the negative real axis including zero (that is, $\mathbf{C}\backslash\{x + iy \mid x \le 0 \text{ and } y = 0\}$). Define a branch of log on A by*

$$\log z = \log |z| + i \arg z \qquad -\pi < \arg z < \pi$$

*which is called the **principal branch of the logarithm**. Then log z is analytic on A with*

$$\frac{d}{dz} \log z = \frac{1}{z}$$

Analogous statements hold for other branches.

FIRST PROOF (using the inverse function theorem). From Sec. 1.3 we know that log z is the unique inverse of the function $f(z) = e^z$ restricted to the set $\{z \mid z = x + iy, -\pi < y < \pi\}$. Since $de^z/dz = e^z \neq 0$, the inverse function theorem implies that locally, e^z has an analytic inverse. Since the inverse is unique, it must be log z. The derivative of $f^{-1}(w)$ is $1/f'(z)$. In this case $f'(z) = f(z) = w$, and so $df^{-1}/dw = 1/w$ at each point $w \in A$, and the theorem is proved. ∎

SECOND PROOF (using the Cauchy-Riemann equations in polar form). In polar form, $\log z = \log r + i\theta$, and so $u(r, \theta) = \log r$, $v(r, \theta) = \theta$, which are C^∞ functions of r, θ. Also, the Cauchy-Riemann equations expressed in polar coordinates are a valid tool on the region A, as explained in Sec. 1.5 and Exercises 11 and 12 at the end of that section. But

$$\frac{\partial u}{\partial r} = \frac{1}{r} = \frac{1}{r} \frac{\partial v}{\partial \theta}$$

and

$$-\frac{1}{r} \frac{\partial u}{\partial \theta} = 0 = \frac{\partial v}{\partial r}$$

and so the Cauchy-Riemann equations hold. We also have

$$\frac{d}{dz} \log z = \frac{\partial}{\partial x} \log r + i \frac{\partial \theta}{\partial x}$$

$$= \frac{1}{r} \frac{\partial r}{\partial x} + i \frac{\partial \theta}{\partial x}$$

It is obvious that on A, $\partial r/\partial x = x/r$ and $\partial\theta/\partial x = -y/r^2$, using, for example, $\theta = \tan^{-1}(y/x)$, and so

$$\frac{d}{dz}\log z = \frac{x}{r^2} - \frac{iy}{r^2} = \frac{\bar{z}}{|z|^2} = \frac{1}{z} \quad \blacksquare$$

The domain on which the principal branch of $\log z$ is analytic is given in Fig. 1.6.1. Here is an example of another branch: $\log z = \log |z| + i \arg z$, $0 < \arg z < 2\pi$, is analytic on $\mathbf{C}\backslash\{x + iy \mid x \geq 0, y = 0\}$. We will use the principal branch unless otherwise stated.

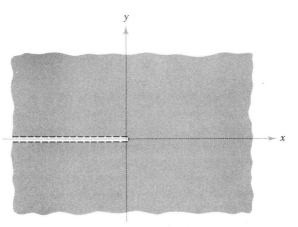

FIGURE 1.6.1   Domain of log z.

When using $\log z$ in compositions, we must be careful to stay in the domain of log. For example, let us consider $f(z) = \log z^2$ using the principal branch of log. This function is analytic on $A = \{z \mid z \neq 0 \text{ and } \arg z \neq \pm \pi/2\}$ by the following reasoning. Proposition 1.5.3 shows that z^2 is analytic on all of $\mathbf{C}$. The image of A under the map $z \mapsto z^2$ is precisely $\mathbf{C}\backslash\{x + iy \mid x \leq 0, y = 0\}$ (Why?), which is the set on which the principal branch of log is defined and analytic. By the chain rule, $z \mapsto \log z^2$ is analytic on A (see Fig. 1.6.2).

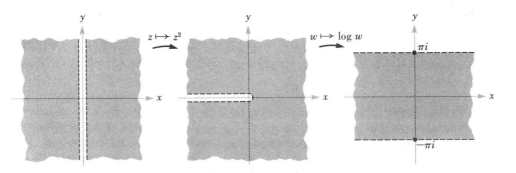

FIGURE 1.6.2   Domain of analyticity of log z^2.

The function $f(z) = \log z^2$ also illustrates the fact that a certain amount of caution must be exercised in manipulating logarithms. Consider the two functions $\log z^2$ and $2 \log z$. If we consider all possible values of the logarithm, then the two collections are the same. If, however, we pick a particular branch, for example, the principal branch, and use it for both, then the two are *not* always the same. For example, if $z = 1 + i$, then arg $z = 3\pi/4$ and $\log z^2 = \log 2 - \pi i/2$, while $2 \log z = \log 2 + 3\pi i/2$. The function $\log z^2$ is analytic on the plane with the imaginary axis deleted, while $2 \log z$ is analytic on the plane with the negative real axis deleted.

The Trigonometric Functions

Now that we have established properties of e^z, the differentiation of the sine and cosine functions follows readily.

1.6.3 PROPOSITION $\sin z$ *and* $\cos z$ *are entire functions with derivatives given by*

$$\frac{d}{dz} \sin z = \cos z \qquad and \qquad \frac{d}{dz} \cos z = -\sin z$$

PROOF $\sin z = (e^{iz} - e^{-iz})/2i$; using the sum rule and the chain rule and the fact that the exponential function is entire, we can conclude that $\sin z$ is entire and that $d(\sin z)/dz = \{ie^{iz} - (-ie^{-iz})\}/2i = (e^{iz} + e^{-iz})/2 = \cos z$. Similarly,

$$\frac{d}{dz} \cos z = \frac{d}{dz} \frac{1}{2} (e^{iz} + e^{-iz}) = \frac{i}{2} (e^{iz} - e^{-iz})$$

$$= -\frac{1}{2i} (e^{iz} - e^{-iz}) = -\sin z \qquad \blacksquare$$

We can also discuss $\sin^{-1} z$ and $\cos^{-1} z$ in somewhat the same way that we discussed $\log z$ (which is $\exp^{-1} z$ with appropriate domains and ranges). These functions are analyzed in Exercise 6 at the end of this section.

The Power Function

Let a and b be complex numbers. We recall that $a^b = e^{b \log a}$, which is, in general, multivalued according to the different branches for log that we choose. We now wish to consider the functions $z \mapsto z^b$ and $z \mapsto a^z$. Although these functions appear to be similar, their properties of analyticity are quite different. The case in which b is an integer was covered in Proposition 1.5.3. The situation for general b is slightly more complicated.

1.6.4 PROPOSITION

(i) *For any choice of branch for the log function, the function $z \mapsto a^z$ is entire and has derivative $z \mapsto (\log a)a^z$.*

(ii) *Fix a branch of the* log *function—for example, the principal branch. Then the function $z \mapsto z^b$ is analytic on the domain of the branch of* log *chosen and the derivative is $z \mapsto bz^{b-1}$.*

PROOF

(i) $a^z = e^{z \log a}$. By the chain rule this function is analytic on $\mathbf{C}$ with derivative $(\log a)$ $e^{z \log a} = (\log a)a^z$ ($\log a$ is merely a constant).

(ii) $z^b = e^{b \log z}$. This function is analytic on the domain of $\log z$, since $w \mapsto e^{bw}$ is entire. Also, by the chain rule,

$$\frac{d}{dz} z^b = \frac{b}{z} e^{b \log z} = \frac{b}{z} z^b$$

(That this equals bz^{b-1} follows from Exercise 20 at the end of Sec. 1.3.) ∎

Of course, in (ii), if b is an integer ≥ 0 we know that z^b is entire (with derivative bz^{b-1}). But in general, z^b is analytic only on the domain of $\log z$.

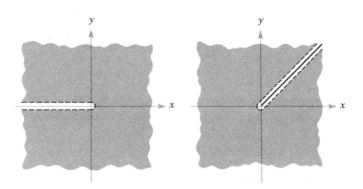

FIGURE 1.6.3   Regions of analyticity for $z \mapsto z^a$.

Let us emphasize what is stated in (ii). If we choose the principal branch of log z, which has domain $\mathbb{C}\backslash\{x + iy \mid y = 0 \text{ and } x \leq 0\}$ and range $\mathbf{R} \times] - \pi, \pi [$ (Why?), then $z \mapsto z^a$ is analytic on $\mathbb{C}\backslash\{x + iy \mid y = 0, x \leq 0\}$ (see Fig. 1.6.3). We could also choose the branch of log that has domain $\mathbb{C}\backslash\{x + ix \mid x \geq 0\}$ and range $\mathbf{R} \times] - 7\pi/4, \pi/4[$; then $z \mapsto z^a$ would be analytic on $\mathbb{C}\backslash\{x + ix \mid x \geq 0\}$.

The nth Root Function

One of the nth roots of z is given by $z^{1/n}$, for a choice of branch of log z. The other roots are given by the other choices of branches as in Sec. 1.3. The principal branch is the one that is usually used. Thus, from Proposition 1.6.4(ii), we get, as a special case.

1.6.5 PROPOSITION. *The function* $z \mapsto z^{1/n} = \sqrt[n]{z}$ *is analytic on the domain of* log z *(for example, the principal branch) and has derivative*

$$z \mapsto \frac{1}{n} z^{(1/n)-1}$$

As with log z, we must exercise care with the functions $z \mapsto z^b$, $z \mapsto \sqrt[n]{z}$ when composing with other functions to be sure we stay in the domain of analyticity. The procedure is illustrated for the square root function in Worked Example 1.6.8.

Worked Examples

1.6.6   *Differentiate the following functions, giving the appropriate region on which the functions are analytic:*

(a) e^{e^z} (b) $\sin(e^z)$

(c) $e^z/(z^2 + 3)$ (d) $\sqrt{e^z + 1}$

(e) $\cos \bar{z}$ (f) $1/(e^z - 1)$

(g) $\log(e^z + 1)$

Solution.

(a) $z \mapsto e^{e^z}$. e^z is entire; by the chain rule $z \mapsto e^{e^z}$ is entire. Also, by the chain rule the derivative at z is $e^z e^{e^z}$.

(b) $z \mapsto \sin e^z$. Both $z \mapsto e^z$ and $w \mapsto \sin w$ are entire, and so by the chain rule $z \mapsto \sin e^z$ is entire and the derivative at z is $(\cos e^z)e^z$.

(c) $z \mapsto e^z/(z^2 + 3)$. The map $z \mapsto e^z$ is entire and the map $z \mapsto 1/(z^2 + 3)$ is analytic on $\mathbb{C}\backslash\{\pm \sqrt{3}i\}$. Hence $z \mapsto e^z/(z^2 + 3)$ is analytic on $\mathbb{C}\backslash 3i\}$ and has derivative at $z \neq \pm\sqrt{3}i$ given by

$$\frac{e^z}{z^2 + 3} - \frac{e^z \cdot 2z}{(z^2 + 3)^2} = \frac{e^z(z^2 - 2z + 3)}{(z^2 + 3)^2}$$

(d) $z \mapsto \sqrt{e^z + 1}$. Choose the branch of the function $w \mapsto \sqrt{w}$ that is analytic on $\mathbb{C}\backslash\{x + iy \mid y = 0, x \leq 0\}$. Then we must choose the region A such that if $z \in A$, then $e^z + 1$ is not both real and ≤ 0.

Notice that e^z is real iff $y = \text{Im } z = n\pi$ for some integer n (Why?). When n is even, e^z is positive; when n is odd, e^z is negative. Here $|e^z| = e^x$, where $x = \text{Re } z$ and $e^x \geq 1$ iff $x \geq 0$. Therefore if we define $A = \mathbb{C}\backslash\{x + iy \mid x \geq 0, y = (2n + 1)\pi$, n running through positive and negative integers and zero$\}$ (as in Fig. 1.6.4), then $e^z + 1$ is real and ≤ 0 iff $z \notin A$. Since $e^z + 1$ is entire, it is certainly analytic on A. By the chain rule, $\sqrt{e^z + 1}$ is analytic on A with derivative at z given by $(e^z + 1)^{-1/2}(e^z)/2$.

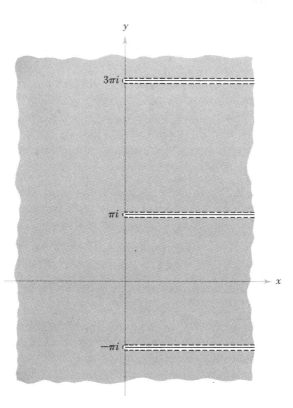

FIGURE 1.6.4 Region of analyticity of $z \to \sqrt{e^z + 1}$.

(e) $z \mapsto \cos \bar{z}$. Since $z = x + iy$, by Proposition 1.3.4, $\cos \bar{z} = \cos (x - iy) = \cos x \cos (-iy) - \sin x \sin (-iy) = \cos x \cosh y + i \sin x \sinh y$, so $u(x, y) = \cos x \cosh y$ and $v(x, y) = \sin x \sinh y$. Thus $\partial u/\partial x = -\sin x \cosh y$, $\partial v/\partial y = \sin x \cosh y$. If $\cos \bar{z}$ were analytic, $\partial u/\partial x$ would equal $\partial v/\partial y$, which would occur iff $\sin x = 0$ (that is, if $x = 0$, or if $x = \pi n$, $n = \pm 1, \pm 2, \ldots$). Thus there is not an open (nonempty) set A on which $z \mapsto \cos \bar{z}$ is analytic.

(f) $z \mapsto 1/(e^z - 1)$. By Proposition 1.5.3(iii) and the fact that $z \mapsto e^z - 1$ is entire, we conclude that $z \mapsto 1/(e^z - 1)$ is analytic on the set on which $e^z - 1 \neq 0$; namely, the set $A = \mathbb{C}\backslash\{z = 2\pi ni \mid n = 0, \pm 1, \pm 2, \ldots\}$. The derivative at z is $-e^z/(e^z - 1)^2$.

(g) $z \mapsto \log(e^z + 1)$. Since (the principal branch of) the log is defined and analytic on the same region as the square root, namely, $A = \mathbb{C}\setminus\{x + iy \mid y = 0, x \leq 0\}$, we can use the results of (d). By the chain rule and the results of (d), $z \mapsto \log(e^z + 1)$ is analytic on the region depicted in Fig. 1.6.4.

1.6.7 *Verify directly that the function e^z preserves the angles between lines parallel to the coordinate axes.*

Solution. The line determined by $y = y_0$ is mapped to the ray $\{x + ix \tan y_0 \mid x > 0\}$, and the line determined by $x = x_0$ is mapped to the circle $\{z$ such that $|z| = e^{x_0}\}$; see Fig. 1.3.7. The angle between any such ray and the tangent vector to the circle at the point of contact of the ray and the circle is $\pi/2$. Thus angles are preserved. This is consistent with the conformal mapping theorem (1.5.7).

1.6.8 *Show that a branch of the function $w \mapsto \sqrt{w}$ can be defined in such a way that $z \mapsto \sqrt{z^2 - 1}$ is analytic in the region shaded in Fig. 1.6.5, and using the notations of that figure, show that $\sqrt{z^2 - 1} = \sqrt{r_1 r_2} e^{i(\theta_1 + \theta_2)/2}$, where $0 < \theta_1 < 2\pi, -\pi < \theta_2 < \pi$.*

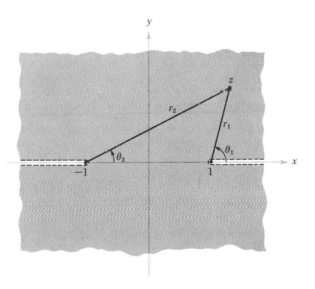

FIGURE 1.6.5 Domain of analyticity of $\sqrt{z^2 - 1}$.

Solution. If $\sqrt{z - 1}$ is a square root of $z - 1$ and $\sqrt{z + 1}$ is a square root of $z + 1$, then $\sqrt{z - 1} \cdot \sqrt{z + 1}$ is a square root of $z^2 - 1$ (Why?). For $z \mapsto \sqrt{z - 1}$ we may choose $\sqrt{}$ defined and analytic on $\mathbb{C}\setminus\{x + iy \mid y = 0$ and $x \geq 0\}$; thus $z \mapsto \sqrt{z - 1}$ is analytic on $\mathbb{C}\setminus\{x + iy \mid y = 0, x \geq 1\}$. For $z \mapsto \sqrt{z + 1}$ we may choose $\sqrt{}$ defined and analytic on $\mathbb{C}\setminus\{x + iy \mid y = 0$ and $x \leq 0\}$; therefore, $z \mapsto \sqrt{z + 1}$ is analytic on $\mathbb{C}\setminus\{x + iy \mid y = 0, x \leq -1\}$. Thus $z \mapsto \sqrt{z - 1}\sqrt{z + 1}$ is analytic on $\mathbb{C}\setminus\{x + iy \mid y = 0, |x| \geq 1\}$ with the appropriate branches of $\sqrt{}$ as indicated. With these branches we have $\sqrt{z - 1} = \sqrt{r_1} e^{i\theta_1/2}$ and $\sqrt{z + 1} = \sqrt{r_2} e^{i\theta_2/2}$, and so $\sqrt{z - 1}\sqrt{z + 1} = \sqrt{r_1 r_2}\, e^{i(\theta_1 + \theta_2)/2}$. Since $z \mapsto \sqrt{z - 1}\sqrt{z + 1}$ is analytic on $\mathbb{C}\setminus\{x + iy \mid y = 0, |x| \geq 1\}$, it should correspond to some branch of the square root function $\sqrt{\ }$ in the map $z \mapsto \sqrt{z^2 - 1}$. To see which one, note that $z \in \mathbf{C}\backslash\{x + iy \mid y = 0, |x| \geq 1\}$ iff $z^2 - 1 \in \mathbf{C}\backslash\{x + iy \mid y = 0, x \geq 0\}$ (Why?), and so if we take the $\sqrt{\ }$ defined and analytic on $\mathbf{C}\backslash\{x + iy \mid y = 0, x \geq 0\}$, then $z \mapsto \sqrt{z^2 - 1}$ is analytic on $\mathbf{C}\backslash\{x + iy \mid y = 0, |x| \geq 1\}$.

Note. The symbol $\sqrt{\ }$ has been used here to mean different branches of the square root function, the particular branch being clear from the context. The student should get in the habit of thinking of $\sqrt{\ }$ together with a choice of branch.

Exercises

1. Differentiate and give the appropriate region of analyticity for each of the following:
(a) $z^2 + z$
(b) $1/z$
(c) $\sin z/\cos z$
(d) $\exp\left(\dfrac{z^3 + 1}{z - 1}\right)$

2. Differentiate and give the appropriate region of analyticity for each of the following:
(a) 3^z
(b) $\log(z + 1)$
(c) $z^{(1+i)}$
(d) $\sqrt{z}$
(e) $\sqrt[3]{z}$

3. Determine whether the following complex limits exist and find their value if they do:
(a) $\displaystyle\lim_{z \to 0} \frac{e^z - 1}{z}$
(b) $\displaystyle\lim_{z \to 0} \frac{\sin |z|}{z}$

4. Determine whether the following complex limits exist and find their value if they do:
(a) $\displaystyle\lim_{z \to 1} \frac{\log z}{z - 1}$
(b) $\displaystyle\lim_{z \to 1} \frac{\bar{z} - 1}{z - 1}$

5. Is it true that $|\sin z| \leq 1$ for all $z \in \mathbf{C}$?

6. Solve $\sin z = w$ and show how to choose a domain and thus how to pick a particular branch of $\sin^{-1} z$ so that it is analytic on the domain. Give the derivative of this branch of $\sin^{-1} z$; see Exercise 35 at the end of Sec. 1.3.

7. Let $f(z) = 1/(1 - z)$; is it continuous on the interior of the unit circle?

8. Let $u(x, y)$ and $v(x, y)$ be real-valued functions defined on an open set $A \subset \mathbf{R}^2 = \mathbf{C}$ and suppose that they satisfy the Cauchy-Riemann equations on A. Show that $u_1(x, y) = [u(x, y)]^2 - [v(x, y)]^2$ and $v_1(x, y) = 2u(x, y)v(x, y)$ satisfy the Cauchy-Riemann equations on A and that the functions $u_2(x, y) = e^{u(x, y)} \cos v(x, y)$ and $v_2(x, y) = e^{u(x, y)} \sin v(x, y)$ also satisfy the Cauchy-Riemann equations on A. Can you do this without performing any computations?

9. Find the region of analyticity and the derivative of each of the following functions:
(a) $z/(z^2 - 1)$
(b) $e^{z + (1/z)}$

10. Find the region of analyticity and the derivative of each of the following functions.
(a) $\sqrt{z^3 - 1}$ (b) $\sin \sqrt{z}$

11. Find the minimum of $|e^{z^2}|$ for those z with $|z| \leq 1$.

12. Prove Proposition 1.6.5 using the method of the first proof of Proposition 1.6.2.

13. Where is $z \mapsto 2^{z^i}$ analytic? $z \mapsto z^{2z}$?

14. Define a branch of $\sqrt{1 + \sqrt{z}}$ and show that it is analytic.

REVIEW EXERCISES FOR CHAPTER 1

1. Compute e^i; $\log(1 + i)$; $\sin i$; 3^i; $e^{2\log(-1)}$.

2. For what values of z is $\log z^2 = 2 \log z$ if the principal branch of the logarithm is used on both sides of the equation?

3. Find the eighth roots of i.

4. Find all numbers z such that $z^2 - 1 + i$.

5. Solve $\cos z = \sqrt{3}$ for z.

6. Solve $\sin z = \sqrt{3}$ for z.

7. Describe geometrically the set of points $z \in \mathbb{C}$ satisfying
(a) $|z + i| = |z - i|$ (b) $|z - 1| = 3|z - 2|$

8. Describe geometrically the set of points $z \in \mathbb{C}$ satisfying
(a) $|z - 1| = |z + 1|$ (b) $|z - 1| = 2|z|$

9. Differentiate the following expressions on appropriate regions:

(a) $z^3 + 8$ (b) $\dfrac{1}{z^3 + 1}$

(c) $\exp(z^4 - 1)$ (d) $\sin(\log z^2)$

10. On what set is $\sqrt{z^2 - 2}$ analytic? Compute the derivative.

11. Describe the sets on which the following functions are analytic and compute their derivatives:

(a) $e^{1/z}$ (b) $\dfrac{1}{(1 - \sin z)^2}$

(c) $\dfrac{e^{az}}{a^2 + z^2}$ for a real

12. Repeat Review Exercise 11 for the following functions:

(a) $\exp \dfrac{1}{1 - az}$ for $a \in \mathbf{C}$

(b) $\dfrac{\sin z}{z}$

13. Can a single-valued (analytic) branch of $\log z$ be defined on the following sets?

(a) $\{z \mid 1 < |z| < 2\}$

(b) $\{z \mid \operatorname{Re} z > 0\}$

(c) $\{z \mid \operatorname{Re} z > \operatorname{Im} z\}$

14. Show that the map $z \mapsto z + 1/z$ maps the circle $\{z \text{ such that } |z| = c\}$ onto the ellipse described by $\{w = u + iv \mid u = (c + 1/c) \cos \theta, v - (c - 1/c) \sin \theta, 0 \leq \theta \leq 2\pi\}$. Can we allow $c = 1$?

15. Let f be analytic on A. Define $g: A \to \mathbf{C}$ by $g(z) = \overline{f(z)}$. When is g analytic?

16. Find the real and imaginary parts of $f(z) = z^3$ and verify directly that they satisfy the Cauchy-Riemann equations.

17. Let $f(x + iy) = (x^2 + 2y) + i(x^2 + y^2)$. At what points does $f'(z_0)$ exist?

18. Let $f: A \subset \mathbf{C} \to \mathbf{C}$ be analytic on an open set A. Let $\bar{A} = \{\bar{z} \mid z \in A\}$.

(a) Describe $\bar{A}$ geometrically.

(b) Define $g: \bar{A} \to \mathbf{C}$ by $g(z) = \overline{[f(\bar{z})]^2}$. Show that g is analytic.

19. Suppose that $f: A \subset \mathbf{C} \to \mathbf{C}$ is analytic on the open connected set A and that $f(z)$ is real for all $z \in A$. Show that f is constant.

20. Prove the Cauchy-Riemann equations as follows. Let $f: A \subset \mathbf{C} \to \mathbf{C}$ be differentiable at $z_0 = x_0 + iy_0$. Let $g_1(t) = t + iy_0$ and $g_2(t) = x_0 + it$. Apply the chain rule to $f \circ g_1$ and $f \circ g_2$ to prove the result.

21. Let $f(z)$ be analytic in the disk $|z - 1| < 1$. Suppose that $f'(z) = 1/z$, $f(1) = 0$. Prove that $f(z) = \log z$.

22. Use the inverse function theorem to prove the following result. Let $f: A \subset \mathbf{C} \to \mathbf{C}$ be analytic (where A is open and connected) and suppose that $f(A) \subset \{z \text{ such that } |z| = 3\}$. Then f is constant.

23. Prove that

$$\lim_{h \to 0} \frac{(z_0 + h)^n - z_0^n}{h} = nz_0^{n-1}$$

for any $z_0 \in \mathbf{C}$.

24. (a) If a polynomial $p(z) = a_0 + a_1 z + \cdots + a_n z^n$ has a root c, then show that we can write $p(z) = (z - c)h(z)$ where $h(z)$ is a polynomial of degree $n - 1$. (Use division of polynomials to show that $z - c$ divides $p(z)$.)

(b) Use part (a) to show that p can have no more than n roots.

(c) When is c a root of both $p(z)$ and $p'(z)$?

25. On what set is the function $z \mapsto z^z$ analytic? Compute its derivative.

26. Let g be analytic on the open set A. Let $B = \{z \in A \mid g(z) \neq 0\}$. Show that B is open and that $1/g$ is analytic on B.

27. Find and plot all solutions of $z^3 = -8i$.

28. Let $f: A \subset \mathbf{C} \to \mathbf{C}$ be analytic on the open set A and let $f'(z_0) \neq 0$ for every $z_0 \in A$. Show that $\{\operatorname{Re} f(z) \mid z \subset A\} \subset \mathbf{R}$ is open.

29. Show that $u(x, y) = x^3 - 3xy^2$, $v(x, y) = 3x^2y - y^3$ satisfy the Cauchy-Riemann equations. Comment on the result.

30. Prove that the following functions are continuous at $z = 0$:

(a) $f(z) = \begin{cases} (\operatorname{Re} z^2)^2/|z|^2 & z \neq 0 \\ 0 & z = 0 \end{cases}$ (b) $f(z) = |z|$

31. At what points z are the following functions differentiable?

(a) $f(z) = |z|^2$ (b) $f(z) = y - ix$

32. Use de Moivre's theorem to find the sum of $\sin x + \sin 2x + \cdots + \sin nx$.

33. For the function $u(x, y) = y^3 - 3x^2y$,

(a) Show that u is harmonic (see Proposition 1.5.11).

(b) Determine a conjugate function $v(x, y)$ such that $u + iv$ is analytic.

34. Consider the function $w(z) = 1/z$. Draw the level curves $u = $ constant. Discuss.

35. Determine the four different values of z that are mapped to unity by the function $w(z) = z^4$.

36. Suppose that $f(z)$ is analytic and satisfies the condition $|f(z)^2 - 1| < 1$ in a region Ω. Show that either $\operatorname{Re} f(z) > 0$ or that $\operatorname{Re} f(z) < 0$ throughout Ω.

37. Suppose that $f: \mathbf{C} \to \mathbf{C}$ is continuous and that $f(z) = f(2z)$ for all $z \in \mathbf{C}$. Show that f is constant on $\mathbf{C}$.

38. Suppose that $f: \mathbf{C} \to \mathbf{C}$ is entire and that $f(2z) = 2f(z)$ for all $z \in \mathbf{C}$. Show that there is a constant c such that $f(z) = cz$ for all z. (You might want to use Exercise 37).

CHAPTER 2

CAUCHY'S THEOREM

A convenient feature of complex analysis is that it is based on a few simple, yet powerful theorems from which most of the results of the subject follow. Foremost among these theorems is Cauchy's theorem, which enables us to prove, for example, that if f is analytic, then all the derivatives of f exist. Cauchy's theorem is the key to the development of the rest of the subject and its applications.

2.1 CONTOUR INTEGRALS

Definitions and Basic Properties

To be able to study Cauchy's theorem we first need to define contour integrals and to study their basic properties.

Let $h: [a, b] \subset \mathbf{R} \to \mathbf{C}$ be a function and set $h(t) = u(t) + iv(t)$. Suppose, for the sake of simplicity, that u and v are continuous. Define

$$\int_a^b h(t)\, dt = \int_a^b u(t)\, dt + i \int_a^b v(t)\, dt \in \mathbf{C}$$

where the integrals of u and v have the usual meaning from single-variable calculus. We want to extend this definition to integrals of functions along curves in $\mathbf{C}$. A continuous *curve* or *contour* in $\mathbf{C}$ is, by definition, a continuous map $\gamma: [a, b] \to \mathbf{C}$. The curve is called *piecewise C^1* if we can divide up the interval $[a, b]$ into finitely many subintervals $a = a_0 < a_1 < \cdots < a_n = b$ such that $\gamma'(t)$ exists on each open subinterval $]\, a_i, a_{i+1}\, [$ and is continuous on $[a_i, a_{i+1}]$; continuity on $[a_i, a_{i+1}]$ means that the limits $\lim_{t \to a_i +} \gamma'(t)$ and $\lim_{t \to a_{i+1} -} \gamma'(t)$ exist (see, for example, Fig. 2.1.1). Unless otherwise specified, curves will always be assumed to be continuous and piecewise C^1.

2.1.1 DEFINITION *Suppose that f is continuous and defined on an open set $A \subset \mathbf{C}$ and that $\gamma: [a, b] \to \mathbf{C}$ is a piecewise smooth curve satisfying $\gamma([a, b]) \subset A$. The expression*

$$\int_\gamma f = \int_\gamma f(z)\, dz = \sum_{i=0}^{n-1} \int_{a_i}^{a_{i+1}} f(\gamma(t))\gamma'(t)\, dt$$

*is called the **integral** of f along γ.*

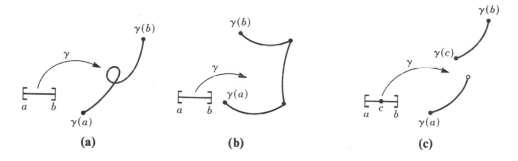

FIGURE 2.1.1 Curves in C. (a) smooth curve; (b) piecewise C^1 curve; (c) discontinuous curve.

The definition is analogous to the following definition of a line integral from calculus: Let $P(x, y)$ and $Q(x, y)$ be real-valued functions of x and y and let γ be a curve. Define

$$\int_\gamma P(x, y)\, dx + Q(x, y)\, dy = \sum_{i=0}^{n-1} \int_{a_i}^{a_{i+1}} \left[P(x(t), y(t)) \frac{dx}{dt} + Q(x(t), y(t)) \frac{dy}{dt} \right] dt$$

where $\gamma(t) = (x(t), y(t))$. The two definitions are related as follows.

2.1.2. PROPOSITION *If $f(z) = u(x, y) + iv(x, y)$, then*

$$\int_\gamma f = \int_\gamma [u(x, y)\, dx - v(x, y)\, dy] + i \int_\gamma [u(x, y)\, dy + v(x, y)\, dx]$$

PROOF

$$f(\gamma(t)) \cdot \gamma'(t) = [u(x(t), y(t)) + iv(x(t), y(t))] \cdot [x'(t) + iy'(t)]$$
$$= [u(x(t), y(t))x'(t) - v(x(t), y(t))y'(t)]$$
$$+ i[v(x(t), y(t))x'(t) + u(x(t), y(t))y'(t)]$$

Integrating both sides over $[a_i, a_{i+1}]$ with respect to t and using Definition 2.1.1 gives the desired result. ∎

This result can easily be remembered by formally writing

$$f(z)\, dz = (u + iv)(dx + i\, dy) = u\, dx - v\, dy + i(v\, dx + u\, dy)$$

For a curve $\gamma: [a, b] \to \mathbf{C}$, we define the *opposite curve*, $-\gamma: [a, b] \to \mathbf{C}$, by $(-\gamma)(t) = \gamma(a + b - t)$. This is merely γ traversed in the opposite sense (see Fig. 2.1.2).

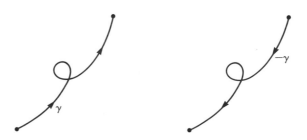

FIGURE 2.1.2 Opposite curve.

We also want to define the *join* or *sum* or *union* $\gamma_1 + \gamma_2$ of two curves γ_1 and γ_2. Intuitively, we want to join them at their endpoints to make a single curve (see Fig. 2.1.3). Precisely, let us suppose that $\gamma_1: [a, b] \to \mathbf{C}$ and that $\gamma_2: [b, c] \to \mathbf{C}$, with $\gamma_1(b) = \gamma_2(b)$. Let us define $\gamma_1 + \gamma_2: [a, c] \to \mathbf{C}$ by

$$(\gamma_1 + \gamma_2)(t) = \begin{cases} \gamma_1(t) & \text{if } t \in [a, b] \\ \gamma_2(t) & \text{if } t \in [b, c] \end{cases}$$

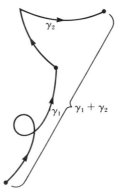

FIGURE 2.1.3 Join of two curves.

Clearly, if γ_1 and γ_2 are piecewise smooth, then so is $\gamma_1 + \gamma_2$. If the intervals $[a, b]$ and $[b, c]$ for γ_1 and γ_2 are not of this special form (the first interval ends where the second begins), then the formula is a little more complicated, but the special form will suffice for this text. The general sum $\gamma_1 + \cdots + \gamma_n$ is defined similarly.

Proposition 2.1.3 gives some properties of the integral that follow from the definitions given in this section. The student is asked to prove them in Exercise 6 at the end of this section.

2.1.3 PROPOSITION   *For (continuous) functions f, g, complex constants c_1, c_2, and piecewise C^1 curves γ, γ_1, γ_2, we have*

(i)
$$\int_\gamma (c_1 f + c_2 g) = c_1 \int_\gamma f + c_2 \int_\gamma g$$

(ii)
$$\int_{-\gamma} f = -\int_\gamma f$$

and

(iii)
$$\int_{\gamma_1 + \gamma_2} f = \int_{\gamma_1} f + \int_{\gamma_2} f$$

Of course, more general statements (that actually follow from the preceding) could be made; namely:

(i)
$$\int_\gamma \sum_{i=1}^n c_i f_i = \sum_{i=1}^n \left(c_i \int_\gamma f_i \right)$$

(iii)
$$\int_{\gamma_1 + \cdots + \gamma_n} f = \sum_{i=1}^n \int_{\gamma_i} f$$

To compute specific examples it is sometimes convenient to use the formula

$$\int_\gamma f = \int_\gamma (u\, dx - v\, dy) + i \int_\gamma (u\, dy + v\, dx)$$

However, it may be that we are not given γ as a map but are told only that it is, for example, "the straight line joining 0 to $i + 1$" or "the unit circle traversed counterclockwise." In such cases, we need to choose some explicit map $\gamma(t)$ that describes this geometrically given curve. Obviously, the same geometric curve can be described in different ways, and so the question arises whether the integral $\int_\gamma f$ is independent of that description.

To answer this question, we need the following definition.

2.1.4 DEFINITION *A (piecewise smooth) curve $\tilde{\gamma}: [\tilde{a}, \tilde{b}] \to \mathbb{C}$ is called a **reparametrization** of γ if there is a C^1 function $\alpha: [a, b] \to [\tilde{a}, \tilde{b}]$ with $\alpha'(t) > 0$, $\alpha(a) = \tilde{a}$, and $\alpha(b) = \tilde{b}$ such that $\gamma(t) = \tilde{\gamma}(\alpha(t))$.*

The conditions $\alpha'(t) > 0$ (hence α is increasing), $\alpha(a) = \tilde{a}$, and $\alpha(b) = \tilde{b}$ imply that $\tilde{\gamma}$ traverses the curve in the same sense as γ does. This is the precise meaning of the statement that γ and $\tilde{\gamma}$ represent the same (oriented) geometric curve (see Fig. 2.1.4). Also, the points in $[\tilde{a}, \tilde{b}]$ at which $\tilde{\gamma}'$ does not exist correspond under α to the points of $[a, b]$ at which γ' does not exist. (This is because α has a strictly increasing C^1 inverse.)

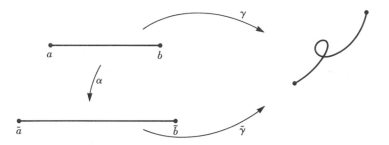

FIGURE 2.1.4   Reparametrization.

2.1.5 PROPOSITION   *If $\tilde{\gamma}$ is a reparametrization of γ, then*

$$\int_{\gamma} f = \int_{\tilde{\gamma}} f$$

for any continuous f defined on an open set containing the image of γ = image of $\tilde{\gamma}$.

PROOF   We can, by breaking up $[a, b]$ into subintervals, assume that γ is C^1. By definition.

$$\int_{\gamma} f = \int_a^b f(\gamma(t)) \cdot \gamma'(t) \, dt$$

By the chain rule, $\gamma'(t) = d\gamma(t)/dt = d\tilde{\gamma}(\alpha(t))/dt = \tilde{\gamma}'(\alpha(t)) \cdot \alpha'(t)$. Let $s = \alpha(t)$ be a new variable, so that $s = \tilde{a}$ when $t = a$ and $s = \tilde{b}$ when $t = b$. Then

$$\int_a^b f(\gamma(t))\gamma'(t) \, dt = \int_a^b f(\tilde{\gamma}(\alpha(t)))\tilde{\gamma}'(\alpha(t)) \frac{d\alpha}{dt} \, dt$$

$$= \int_{\tilde{a}}^{\tilde{b}} f(\tilde{\gamma}(s))\tilde{\gamma}'(s) \, ds$$

The changing of variables in a complex integral (here, from t to s) is justified by applying the usual real-variables rule to its real and imaginary parts. ■

This proposition "justifies" the use of any curve γ that describes a given oriented geometric curve to evaluate an integral.* This statement can be illustrated with an example. Let us evaluate $\int_\gamma x\,dz$, where γ is the straight line from $z = 0$ to $z = 1 + i$ (see Fig. 2.1.5). We choose the curve $\gamma \colon [0, 1] \to \mathbf{C}$, defined by $\gamma(t) = t + it$. Of course, when we write x in $\int_\gamma x\,dz$, we mean the function that gives the real part of any complex number (that is, $f(z) = x = \operatorname{Re} z$). Thus,

$$\int_\gamma x\,dz = \int_0^1 x \circ \gamma(t)\gamma'(t)\,dt$$

Hence

$$\int_\gamma x\,dz = \int_0^1 t(1 + i)\,dt = \frac{1 + i}{2}$$

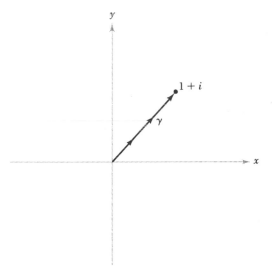

FIGURE 2.1.5 Curve from 0 to $1 + i$.

---

* Strictly speaking, this statement is not correct, since two maps with the same image need not be reparametrizations of one another. However, they are reparametrizations if we ignore points where $\gamma'(t) = 0$. The proposition can be generalized to cover this situation as well, but the complications that result from generalizing it to cover this case have been omitted to simplify the exposition.

An orientation is often described by saying that "γ goes from z_1 to z_2." However, if γ is a closed curve, on which $z_1 = z_2$, we need a different prescription. When solving examples, where the curves are always easy to visualize, the student should assume that a closed curve γ is traversed in the counterclockwise direction unless advised to the contrary.

From calculus, the *arc length* of a curve $\gamma: [a, b] \to \mathbf{C}$ is defined by

$$l(\gamma) = \int_a^b |\gamma'(t)| \, dt = \int_a^b \sqrt{x'(t)^2 + y'(t)^2} \, dt$$

Arc length, too, is independent of the parametrization, by a similar proof to that of Proposition 2.1.5. The student is familiar with the fact that the arc length of the unit circle is 2π, the perimeter of the unit square is 4, and so on.

The next result gives an important way to estimate integrals.

2.1.6 PROPOSITION   *Let f be continuous on an open set A and let γ be a piecewise C^1 curve in A. If there is a constant $M \geq 0$ such that $|f(z)| \leq M$ for all points z on γ (that is, for all z of the form $\gamma(t)$ for some t), then* $-M \leq f(z) \leq M$

$$\left| \int_\gamma f \right| \leq Ml(\gamma)$$

More generally, we have

$$\left| \int_\gamma f \right| \leq \int_\gamma |f| \, |dz|$$

where the latter integral is defined by

$$\int_\gamma |f| \, |dz| = \int_a^b |f(\gamma(t))| \, |\gamma'(t)| \, dt$$

PROOF   For a complex-valued function $g(t)$ on $[a, b]$, we have

$$\mathrm{Re} \int_a^b g(t) \, dt = \int_a^b \mathrm{Re} \, g(t) \, dt$$

since $\int_a^b g(t) \, dt = \int_a^b u(t) \, dt + i \int_a^b v(t) \, dt$ if $g(t) = u(t) + iv(t)$. Let us use this fact to prove that

$$\left| \int_a^b g(t) \, dt \right| \leq \int_a^b |g(t)| \, dt$$

(We would know, from calculus, how to prove this if g were real-valued, but here it is complex-valued.) For our proof, we let $\int_a^b g(t)\,dt = re^{i\theta}$ for fixed r and θ, where $r \geq 0$, so that $r = e^{-i\theta} \int_a^b g(t)\,dt = \int_a^b e^{-i\theta} g(t)\,dt$. Thus

$$r = \operatorname{Re} r = \operatorname{Re} \int_a^b e^{-i\theta} g(t)\,dt = \int_a^b \operatorname{Re}\,(e^{-i\theta} g(t))\,dt$$

But by Proposition 1.2.3(iii), $\operatorname{Re} e^{-i\theta} g(t) \leq |e^{-i\theta} g(t)| = |g(t)|$, since $|e^{-i\theta}| = 1$. Thus $\int_a^b \operatorname{Re}\,(e^{-i\theta} g(t))\,dt \leq \int_a^b |g(t)|\,dt$, and so

$$\left| \int_a^b g(t)\,dt \right| = r \leq \int_a^b |g(t)|\,dt \tag{1}$$

Therefore,

$$\left| \int_\gamma f \right| = \left| \int_a^b f(\gamma(t))\gamma'(t)\,dt \right| .$$
$$\leq \int_a^b |f(\gamma(t))\gamma'(t)|\,dt = \int_a^b |f(\gamma(t))|\,|\gamma'(t)|\,dt \tag{2}$$

by inequality (1) and the fact that $|zz'| = |z|\,|z'|$. Expression (2) is an ordinary real integral, and therefore since $|f(\gamma(t))| \leq M$, the expression (2) is bounded by $M \int_a^b |\gamma'(t)|\,dt = Ml(\gamma)$. ∎

This proposition provides a basic tool that we shall use in subsequent proofs to estimate the size of integrals. The student might try to prove this result directly in terms of the expression

$$\int_\gamma f = \int_\gamma u\,dx - v\,dy + i \int_\gamma (u\,dy + v\,dx)$$

to be convinced that the result is not altogether trivial. Another approach is given in the supplement to this section.

Fundamental Theorem of Calculus for Contour Integrals

The fundamental theorem of calculus is a basic fact in the calculus of real-valued functions which says essentially that the integral of the derivative of a function is just the difference of the values of the function at the endpoints of the interval of integration and that the indefinite integral of a function is an antiderivative for the function. Both of these assertions have important analogues for complex path integrals.

2.1.7 FUNDAMENTAL THEOREM OF CALCULUS FOR CONTOUR INTE-GRALS *Suppose that $\gamma : [0, 1] \rightarrow \mathbf{C}$ is a piecewise smooth curve and that F is a function defined and analytic on an open set G containing γ. Then*

$$\int_\gamma F'(z) \, dz = F(\gamma(1)) - F(\gamma(0))$$

In particular, if $\gamma(0) = \gamma(1)$, then

$$\int_\gamma F'(z) \, dz = 0$$

PROOF   The chain rule and the definition of the path integral will be used to reduce the problem to the standard fundamental theorem for calculus of real-valued functions of a real variable. Let g, u, and v be defined by

$$F(\gamma(t)) = g(t) = u(t) + iv(t)$$

Then

$$F'(\gamma(t))\gamma'(t) = g'(t) = u'(t) + iv'(t)$$

and so

$$\int_\gamma F'(z) \, dz = \int_0^1 F'(\gamma(t))\gamma'(t) \, dt = \int_0^1 g'(t) \, dt$$
$$= \int_0^1 u'(t) \, dt + i \int_0^1 v'(t) \, dt = [u(1) - u(0)] + i[v(1) - v(0)]$$
$$= [u(1) + iv(1)] - [u(0) + iv(0)] = g(1) - g(0)$$
$$= F(\gamma(1)) - F(\gamma(0)) \quad \blacksquare$$

Using this result can save a lot of effort in working examples. For instance, consider $\int_\gamma z^3 \, dz$ where γ is the portion of the ellipse $x^2 + 4y^2 = 1$ that joins $z = 1$ to $z = i/2$. To evaluate the integral we merely note that $z^3 = \frac{1}{4} (dz^4/dz)$ and so

$$\int_\gamma z^3 \, dz = \frac{z^4}{4}\bigg|_1^{i/2} = \left(\frac{1}{4}\right)\left(\frac{i}{2}\right)^4 - \left(\frac{1}{4}\right)(1)^4 = -\frac{15}{64}$$

Notice that we did not even need to parametrize the curve! By applying the fundamental theorem, we would have obtained the same answer for any curve joining these two points. We will investigate in a general context the independence of the value of an integral from the particular path used in the next subsection.

The fundamental theorem has many applications and ramifications, one of which is the following proof of a property of open connected sets which first appeared as Proposition 1.5.5. It is the analogue in the complex domain of the following principle so useful in calculus: A function whose derivative is identically 0 is constant.

2.1.8 COROLLARY *If f is a function defined and analytic on an open connected set $G \subset \mathbf{C}$, and if $f'(z) = 0$ for every point z in G, then f is constant on G.*

PROOF Fix a point z_0 in G and suppose that z is any other point in G. By Proposition 1.4.15 there is a smooth path γ from z_0 to z in G. By the last theorem, $f(z) - f(z_0) = \int_\gamma f'(\xi)\, d\xi = 0$. Therefore $f(z) = f(z_0)$. The value of f at any point of G is thus the same as its value at z_0. That is, f is constant on G. ∎

Path Independence of Integrals

The idea that an indefinite integral is an antiderivative does not carry over directly to the complex domain. What should we mean by the integral between two points? There are many possible paths. The connection comes up in the study of one of the central questions we will study in this chapter. Under what conditions is the value of an integral independent of which particular path is selected between the two points? Consider the following two examples.

EXAMPLE Let $z_0 = 1$ and $z_1 = -1$ and let $f(z) = 3z^2$. Then $F(z) = z^3$ is an antiderivative for f everywhere in the complex plane. Therefore, by the fundamental theorem, no matter what path γ we take from z_0 to z_1 we will have $\int_\gamma f(z)\, dz = F(z_1) - F(z_0) = 1^3 - (-1)^3 = 2$. The value of the integral does not depend on the particular path selected, but only on the function and the two endpoints. ▲

EXAMPLE Again let $z_0 = 1$ and $z_1 = -1$, but now take $f(z) = 1/z$. Let γ_1 be the upper half of the unit circle from 1 to -1. Then γ_1 is parametrized by $\gamma(t) = e^{it}$ for $0 \le t \le \pi$. Thus,

$$\int_{\gamma_1} f(z)\, dz = \int_0^\pi f(\gamma_1(t))\gamma_1'(t)\, dt = \int_0^\pi e^{-it}\, ie^{it}\, dt = \pi i$$

Now let γ_2 be the lower half of the unit circle from 1 to -1. Then γ_2 is parametrized by $\gamma_2(t) = e^{-it}$ for $0 \le t \le \pi$ and

$$\int_{\gamma_2} f(z)\, dz = \int_0^\pi f(\gamma_2(t))\gamma_2'(t)\, dt = \int_0^\pi e^{it}\,(-ie^{-it})\, dt = -\pi i$$

The values of the integral between z_0 and z_1 now are different for the two different paths. ▲

The dependence on the path in the second example is related to the problem of antiderivatives. "The" antiderivative of $f(z)$ ought to be log z. As we saw in Chap. 1, it is perfectly possible to define a branch of the logarithm function which is analytic along either one of the two curves, but it is *not* possible to define consistently a single branch of the logarithm on an open set containing both these curves at once. This way of looking at the difficulty is made precise in the next theorem.

2.1.9 PATH INDEPENDENCE THEOREM   *Suppose f is a continuous function on an open connected set G. Then the following are equivalent:*

(i) *Integrals are path-independent: If z_0 and z_1 are any two points in G and γ_0 and γ_1 are paths in G from z_0 to z_1, then $\int_{\gamma_0} f(z)\, dz = \int_{\gamma_1} f(z)\, dz$.*
(ii) *Integrals around closed curves are 0: If Γ is a closed curve (loop) lying in G, then $\int_\Gamma f(z)\, dz = 0$.*
(iii) *There is a (global) antiderivative for f on G: There is a function F defined and analytic on all of G such that $F'(z) = f(z)$ for all z in G.*

PROOF

(i) ⟺ (ii).   The equivalence of (i) and (ii) is obtained in the direction (ii) ⟹ (i) by joining the curves γ_0 and $-\gamma_1$ to form a closed curve Γ and in the direction (i) ⟹ (ii) by picking two points z_0 and z_1 along a closed curve and thinking of it as made up of two curves from one point to the other and then back to the first. The construction is illustrated in Fig. 2.1.6, and the computation runs as follows:

$$\int_\Gamma f(z)\, dz = \int_{\gamma_0} f(z)\, dz + \int_{-\gamma_1} f(z)\, dz = \int_{\gamma_0} f(z)\, dz - \int_{\gamma_1} f(z)\, dz$$

Thus, the integral along the closed loop Γ is 0 if and only if the integrals along the paths γ_0 and γ_1 are equal.

(iii) ⟹ (i).   This implication follows from the fundamental theorem. The value of the integral is $F(z_1) - F(z_0)$ regardless of which path is selected.
(i) ⟹ (iii).   We will attempt to use an integral ending at z to define the value of the antiderivative at z. Let z_0 be any point fixed in G, and let z be any other point in G. Since G is open and connected, it is path-connected and by Proposition 1.4.15 there is at least one smooth path in G from z_0 to z. Let γ be any such path and set $F(z) = \int_\gamma f(\xi)\, d\xi$. This defines a function F on G in a nonambiguous way, since (i) says that the value $F(z)$ depends only on z and *not* on

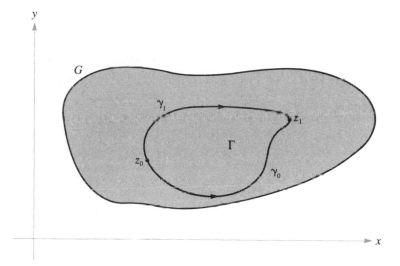

FIGURE 2.1.6 $\gamma_0 - \gamma_1 = \Gamma$.

the particular path selected so long as it stays in G. (Of course, it also depends on z_0, but that is fixed for the entire discussion.) We say that F is *well defined*. Our remaining task is to check that F is differentiable and that $F' = f$. This computation is illustrated in Fig. 2.1.7.

Let $\epsilon > 0$. Since G is open and f is continuous at z, there is a number $\delta > 0$ such that the disk $D(z; \delta) \subset G$ and $|f(\xi) - f(z)| < \epsilon$ whenever $|\xi - z| < \delta$. Suppose $|w - z| < \delta$. Connect z to w by a straight line segment ρ. Then all of ρ lies in $D(z; \delta)$ and

$$F(w) - F(z) = \int_{\gamma + \rho} f(\xi)\, d\xi - \int_\gamma f(\xi)\, d\xi = \int_\rho f(\xi)\, d\xi$$

Thus

$$\left| \frac{F(w) - F(z)}{w - z} - f(z) \right| = \frac{|F(w) - F(z) - (w - z)f(z)|}{|w - z|}$$

$$= \frac{|\int_\rho f(\xi)\, d\xi - f(z) \int_\rho 1\, d\xi|}{|w - z|}$$

$$= \frac{|\int_\rho [f(\xi) - f(z)]\, d\xi|}{|w - z|}$$

$$\leq \frac{\epsilon \,\text{length}\,(\rho)}{|w - z|} = \frac{\epsilon |w - z|}{|w - z|} = \epsilon$$

Thus the limit of the difference quotient is $f(z)$, and so F is differentiable and $F' = f$, as desired. ■

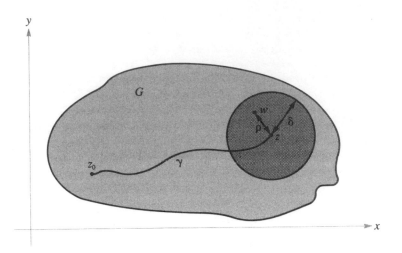

FIGURE 2.1.7 Defining an antiderivative by an integral.

The reader who is familiar with conservative force fields from calculus may recognize the constructions in the last proof. The integral of a force field along a path defines the work done by it (or in moving against it) along that path. The field is called *conservative* if the net work done along a closed path is always 0 or equivalently if the work done between two points is independent of the path taken between those points. If it is, then such an integral defines a quantity, called the *potential energy,* whose gradient is the original force field.

Worked Examples

2.1.10   *Evaluate the following integrals:*
(a) $\int_\gamma x \, dz$ *(γ is the circumference of the unit square).*
(b) $\int_\gamma e^z \, dz$ *(γ is the part of the unit circle joining 1 to i in a counterclockwise direction).*

Solution.
(a) Define $\gamma: [0, 4] \to \mathbf{C}$ as follows. $\gamma = \gamma_1 + \gamma_2 + \gamma_3 + \gamma_4$ where the four sides of the unit square are:

$$\gamma_1(t) = t + 0i; \, 0 \le t \le 1 \qquad \gamma_3(t) = (3 - t) + i; \, 2 \le t \le 3$$
$$\gamma_2(t) = 1 + (t - 1)i; \, 1 \le t \le 2 \qquad \gamma_4(t) = 0 + (4 - t)i; \, 3 \le t \le 4$$

We compute as follows:

$$\int_{\gamma_1} x \, dz = \int_0^1 [\mathrm{Re} \, (\gamma_1(t))]\gamma_1'(t) \, dt = \int_0^1 t \, dt = \frac{1}{2}$$

$$\int_{\gamma_2} x \, dz = \int_1^2 [\text{Re} \, (\gamma_2(t))] \gamma_2'(t) \, dt = \int_1^2 i \, dt = i$$

$$\int_{\gamma_3} x \, dz = \int_2^3 [\text{Re} \, (\gamma_3(t))] \gamma_3'(t) \, dt = \int_2^3 -(3-t) \, dt = -\frac{1}{2}$$

$$\int_{\gamma_4} x \, dz = \int_3^4 [\text{Re} \, (\gamma_4(t))] \gamma_4'(t) \, dt = \int_3^4 0 \, dt = 0$$

Hence

$$\int_{\gamma} x \, dz = \int_{\gamma_1} x \, dz + \int_{\gamma_2} x \, dz + \int_{\gamma_3} x \, dz + \int_{\gamma_4} x \, dz = \tfrac{1}{2} + i - \tfrac{1}{2} + 0 = i$$

(b) e^z is the derivative of the function e^z, and e^z is analytic on all of $\mathbb{C}$. Thus whatever parametrization we use for the part of the unit circle joining 1 to i in a counterclockwise direction, we will have $\int_{\gamma} e^z \, dz = e^i - e^1$ by the fundamental theorem. We can also use the original definition to evaluate the integral directly. Define $\gamma(t) = \cos t + i \sin t, 0 \le t \le \pi/2$. Hence

$$\int_{\gamma} e^z \, dz = \int_0^{\pi/2} \overset{f(\gamma(t))}{\overbrace{e^{\cos t + i \sin t}}} \overset{\gamma'(t)}{\overbrace{(-\sin t + i \cos t)}} \, dt$$

$$= \int_0^{\pi/2} [-e^{\cos t} \cos (\sin t) \cdot \sin t - e^{\cos t} \sin (\sin t) \cdot \cos t] \, dt$$

$$+ i \int_0^{\pi/2} [-e^{\cos t} \sin (\sin t) \cdot \sin t + e^{\cos t} \cos (\sin t) \cdot \cos t] \, dt$$

$$= e^{\cos t} \cos (\sin t) \Big|_0^{\pi/2} + i e^{\cos t} \sin (\sin t) \Big|_0^{\pi/2}$$

$$= e^{\cos t + i \sin t} \Big|_0^{\pi/2} = e^i - e^1$$

Clearly, the first method is easier and should be used whenever possible.

2.1.11   *Let γ be the upper half of the unit circle described counterclockwise. Show that*

$$\left| \int_{\gamma} \frac{e^z}{z} \, dz \right| \le \pi e$$

Solution. We use Proposition 2.1.6. The arc length of γ is

$$l(\gamma) = \int_0^{\pi} |\gamma'(t)| \, dt = \pi$$

since we can take $\gamma(t) = e^{it}, 0 \le t \le \pi$, and $\gamma'(t) = i e^{it}$ and thus $|\gamma'(t)| = 1$. Of course, this is what we would expect. The absolute value of e^z/z, with $z = e^{it} = \cos t + i \sin t$, is estimated by

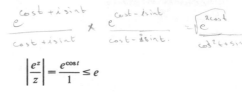

$$\left|\frac{e^z}{z}\right| = \frac{e^{\cos t}}{1} \le e$$

since $\cos t \le 1$. Thus $e = M$ is a bound for $|e^z/z|$ along γ, and therefore,

$$\left|\int_\gamma \frac{e^z}{z}\, dz\right| \le Ml(\gamma) = e\pi$$

2.1.12   *Let γ be the circle of radius r around $a \in \mathbf{C}$. Evaluate $\int_\gamma (z - a)^n\, dz$ for all integers $n = 0,$*
$\pm 1, \pm 2, \ldots$

Solution. First, let $n \ge 0$. Then

$$(z - a)^n = \frac{d}{dz}\frac{1}{n+1}(z - a)^{n+1}$$

which is the derivative of an analytic function, and so by the fundamental theorem 2.1.7,

$$\int_\gamma (z - a)^n \cdot dz = 0$$

Second, let $n \le -2$. Then again

$$(z - a)^n = \frac{d}{dz}\frac{1}{n+1}(z - a)^{n+1}.$$

which is analytic on $A = \mathbf{C}\backslash\{a\}$. (Note that this formula fails if $n = -1$.) Since γ lies in A, the fundamental theorem again shows that $\int_\gamma (z - a)^n\, dz = 0$.

Finally, let $n = -1$. It is easiest to proceed directly. We parametrize γ by $\gamma(\theta) = re^{i\theta} + a, 0 \le \theta \le 2\pi$ (see Fig. 2.1.8). By the chain rule, $\gamma'(\theta) = rie^{i\theta}$, and so

$$\int_\gamma \frac{1}{z - a}\, dz = \int_0^{2\pi} \frac{1}{(re^{i\theta} + a) - a}\, ire^{i\theta}\, d\theta = \int_0^{2\pi} i\, d\theta = 2\pi i$$

In summary

$$\int_\gamma (z - a)^n\, dz = \begin{cases} 0 & n \ne -1 \\ 2\pi i & n = -1 \end{cases}$$

This is a useful formula and we shall have occasion to use it later.

2.1.13   *Prove that there does not exist an analytic function f defined on $\mathbf{C}\backslash\{0\}$ such that $f'(z) = 1/z$.*

Solution. If such an f existed, then, using the fundamental theorem, we would conclude $\int_\gamma (1/z)\, dz = 0$, where γ is the unit circle. But by Worked Example 2.1.12 $\int_\gamma (1/z)\, dz = 2\pi i$, so no such f can exist.

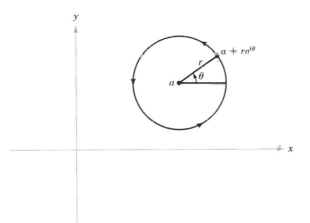

FIGURE 2.1.8 Parametrization of the circle of radius r and center a.

Note. Although $d(\log z)/dz = 1/z$, it does not contradict this example, because $\log z$ is not analytic on $\mathbb{C}\backslash\{0\}$; it is analytic only on $\mathbb{C}$ minus the negative x axis including zero.

SUPPLEMENT TO SECTION 2.1: RIEMANN SUMS

The theory of complex contour integrals can be based directly on a definition in terms of approximation by Riemann sums, as in calculus. If γ is a curve from a to b in the complex plane and f is a function defined along γ, we can choose intermediate points $a = z_0, z_1, z_2, \ldots, z_{n-1}, z_n = b$ on γ and form the sum

$$\sum_{k=1}^{n} f(z_k)(z_k - z_{k-1})$$

(see Fig. 2.1.9). As in calculus, if these sums approach a limit as the maximum of $|z_k - z_{k-1}|$ tends toward 0, we take that limit to be the value of the integral $\int_\gamma f(z)\, dz$.

The properties of the integral given in Proposition 2.1.3 follow from this approach much as the corresponding properties in real-variable calculus. To see that this leads to the same result as Definition 2.1.1 when γ is a C^1 curve, suppose that $z(t) = u(t) + iv(t)$ is a continuously differentiable parametrization of γ with $z(t_k) = z_k$. The mean value theorem guarantees numbers t_k' and t_k'' between t_{k-1} and t_k such that $z_k - z_{k-1} = [u'(t_k') + iv'(t_k'')](t_k \quad t_{k-1})$. Thus, the Riemann sums $\sum f(z_k)(z_k - z_{k-1})$ correspond to Riemann sums for $\int f(\gamma(t))\gamma'(t)\, dt$ after sorting out real and imaginary parts.

This approach to the integral allows the use of more general curves and is sometimes useful in writing approximations to the integral. For example, Proposition 2.1.6 may be

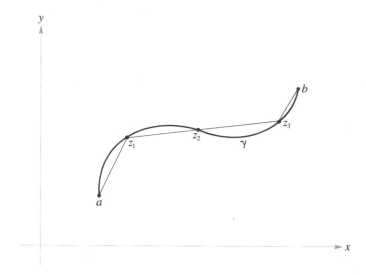

FIGURE 2.1.9 A polygonal approximation of γ.

established by using the triangle inequality: for any approximating Riemann sum we have

$$
\left| \sum f(z_k)(z_k - z_{k-1}) \right| \le \sum |f(z_k)||z_k - z_{k-1}|
$$
$$
\le M \sum |z_k - z_{k-1}|
$$
$$
\le Ml(\gamma)
$$

The last step uses the fact that $|z_k - z_{k-1}|$ is the length of the line segment from z_{k-1} to z_k, which is no greater than the distance between them along γ. Since the estimate holds for each approximating sum, it must hold for the integral, which is their limit.

Exercises

1. Evaluate the following:
(a) $\int_\gamma y \, dz$, where γ is the union of the line segments joining 0 to i and then to $i + 2$.
(b $\int_\gamma \sin 2z \, dz$, where γ is the line segment joining $i + 1$ to $-i$.
(c) $\int_\gamma z e^{z^2} \, dz$ where γ is the unit circle.

2. Evaluate the following:
(a) $\int_\gamma x \, dz$, where γ is the union of the line segments joining 0 to i and then to $i + 2$.
(b) $\int_\gamma (z^2 + 2z + 3) \, dz$, where γ is the straight line segment joining 1 to $2 + i$.

(c) $\int_\gamma \dfrac{1}{z - 1} \, dz$, where γ is the circle of radius 2 centered at 1 traveled once counterclockwise.

3. Evaluate $\int_\gamma (1/z)\, dz$, where γ is the circle of radius 1 centered at 2 traveled once counterclockwise.

4. Evaluate $\int_\gamma \dfrac{1}{z^2 - 2z}\, dz$, where γ is the curve in Exercise 3.

5. Does Re $\{\int_\gamma f\, dz\} = \int_\gamma \operatorname{Re} f\, dz$?

6. Prove Proposition 2.1.3.

7. Evaluate the following:
(a) $\int_\gamma \bar{z}\, dz$, where γ is the unit circle traversed once in a counterclockwise direction.
(b) $\int_\gamma (x^2 - y^2)\, dz$, where γ is the straight line from 0 to i.

8. Evaluate $\int_\gamma \bar{z}^2\, dz$ along two paths joining $(0, 0)$ to $(1, 1)$ as follows:
(a) γ is the straight line joining $(0, 0)$ to $(1, 1)$.
(b) γ is the broken line joining $(0, 0)$ to $(1, 0)$, then joining $(1, 0)$ to $(1, 1)$.
In view of your answers to 8(a) and (b) and the fundamental theorem, could $\bar{z}^2$ be the derivative of any analytic function $F(z)$?

9. Estimate the absolute value of

$$\int_\gamma \frac{dz}{2 + z^2}$$

where γ is the upper half of the unit circle.

10. Let C be the arc of the circle $|z| = 2$ that lies in the first quadrant. Show that

$$\left| \int_C \frac{dz}{z^2 + 1} \right| \le \frac{\pi}{3}$$

11. Evaluate the following:
(a) $\displaystyle\int_{|z|=1} \frac{dz}{z}$; $\displaystyle\int_{|z|=1} \frac{dz}{|z|}$; $\displaystyle\int_{|z|=1} \frac{|dz|}{z}$; $\displaystyle\int_{|z|=1} \left|\frac{dz}{z}\right|$
(b) $\int_\gamma z^2\, dz$, where γ is the curve given by $\gamma(t) = e^{it} \sin^3 t,\ 0 \le t \le \pi/2$.

12. Let γ be a closed curve lying entirely in $\mathbb{C}\backslash\{z \mid \operatorname{Re} z \le 0\}$. Show that $\int_\gamma (1/z)\, dz = 0$.

13. Evaluate $\int_\gamma z \sin z^2\, dz$ where γ is the unit circle.

14. Give some conditions on a closed curve γ that will guarantee that $\int_\gamma (1/z)\, dz = 0$.

15. Let γ be the unit circle. Prove that

$$\left| \int_\gamma \frac{\sin z}{z^2}\, dz \right| \le 2\pi e$$

16. Show that the arc length $l(\gamma)$ of a curve γ is unchanged if γ is reparametrized.

2.2 CAUCHY'S THEOREM: INTUITIVE VERSION

In mathematics it is important to have an intuitive understanding and to be able to express that intuition precisely. In this section Cauchy's theorem will be discussed rather informally; the proofs will be simple and the definitions somewhat casual (although entirely adequate for most applications). Section 2.3 provides more precise definitions, sharper theorems, and more complete proofs. That section can be omitted, but such omission is recommended only if there is a desire to proceed quickly to subsequent sections on applications of the theorem.

One form of Cauchy's theorem states that *if γ is a simple closed curve* (the word "simple" meaning that γ intersects itself only at its endpoints)* *and if f is analytic on and inside γ, then*

$$\int_\gamma f = 0$$

(see Fig. 2.2.1).

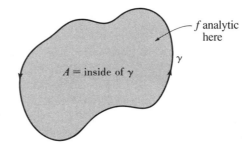

FIGURE 2.2.1   Cauchy's theorem: $\int_\gamma f = 0$.

---

* In appropriate situations γ does not actually have to be simple. This more general case is discussed in Sec. 2.3.

If the function f is not analytic on the whole region inside γ, then the integral may or may not be 0. For example, let γ be the unit circle and $f(z) = 1/z$. Then f is analytic at all points except $z = 0$, and indeed the integral is not zero. In fact,

$$\int_\gamma f = 2\pi i$$

by Worked Example 2.1.12. On the other hand, if $f(z) = 1/z^2$, then f is still analytic at all points except $z = 0$, but now the integral *is* 0. This value of 0 results *not* from Cauchy's theorem — f is not analytic everywhere inside γ — but rather from the fact that f has an antiderivative on $\mathbb{C}\backslash\{0\}$ namely, f is the derivative of $-1/z$. The path independence theorem (2.1.9) shows that the conclusion of Cauchy's theorem is closely tied to the existence of an antiderivative for f. This is made explicit in Proposition 2.2.5.

Our proof of Cauchy's theorem in this section utilizes a theorem from advanced calculus called *Green's theorem*. (The proof of Cauchy's theorem given in Sec. 2.3 is not based on this theorem.) *Green's theorem states that, given functions $P(x, y)$ and $Q(x, y)$,*

$$\int_\gamma P(x, y)\, dx + Q(x, y)\, dy = \iint_A \left[\frac{\partial Q}{\partial x}(x, y) - \frac{\partial P}{\partial y}(x, y) \right] dx\, dy \qquad (1)$$

Recall that if $\gamma: [a, b] \to \mathbb{C}$, $\gamma(t) = (x(t), y(t))$, then we define

$$\int_\gamma P(x, y)\, dx = \int_a^b P(x(t), y(t)) x'(t)\, dt$$

and

$$\int_\gamma Q(x, y)\, dy = \int_a^b Q(x(t), y(t)) y'(t)\, dt$$

In Eq. (1), A represents the "inside" of γ, γ is traversed in a counterclockwise direction, and P and Q are sufficiently smooth — class C^1 is sufficient. (See a calculus text such as J. Marsden and A. Weinstein, *Calculus III* (New York: Springer-Verlag, 1985), pp. 908–911, or J. Marsden and A. Tromba, *Vector Calculus,* 2d ed. (New York: W. H. Freeman and Company, 1981), pp. 404–413, for a proof of Eq. (1).

At this point the reader may ask, "What, precisely, is the inside of γ?" Intuitively the meaning of "inside" should be clear. The precise expression of the concept is based on the Jordan curve theorem ("Any simple closed curve has an 'inside' and an 'outside' "), which will be stated formally in the supplement to the next section.

Preliminary Version of Cauchy's Theorem

2.2.1 THEOREM *Suppose that f is analytic, with the derivative f' continuous on and inside a simple closed curve γ. Then*

$$\int_{\gamma} f = 0 \qquad\qquad (2)$$

PROOF Setting $f = u + iv$, we have

$$\int_{\gamma} f = \int_{\gamma} f(z)\, dz = \int_{\gamma} (u + iv)(dx + i\, dy)$$

$$= \int_{\gamma} (u\, dx - v\, dy) + i \int_{\gamma} (u\, dy + v\, dx)$$

By applying Green's theorem Eq. (1)) to each integral, we get

$$\int_{\gamma} f = \iint_{A} \left[-\frac{\partial v}{\partial x} - \frac{\partial u}{\partial y} \right] dx\, dy + i \iint_{A} \left[\frac{\partial u}{\partial x} - \frac{\partial v}{\partial y} \right] dx\, dy$$

Both terms are zero by the Cauchy-Riemann equations. ∎

A proof of a more precise version of Cauchy's theorem due to Édouard Goursat given in Sec. 2.3 does not assume f' to be continuous. The continuity of f' follows automatically, but this is not obvious. It also eliminates the assumption that the curve is simple. In many cases, the simplicity assumption can be avoided by viewing the path as being made up of two or more simple pieces. In Fig. 2.2.2, the "figure eight" can be treated as two simple loops.

As an example of the use of Eq. (2), let γ be the unit square and let $f(z) = \sin(e^{z^2})$. Then f is analytic on and inside γ (in fact, f is entire) and so $\int_{\gamma} f = 0$.

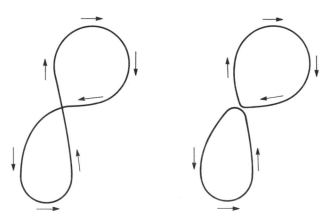

FIGURE 2.2.2   Treating a nonsimple curve as made up of several simple loops.

Deformation Theorem

It is important to be able to study functions that are not analytic on the entire inside of γ and whose integral therefore might not be zero. For example, $f(z) = 1/z$ fails to be analytic at $z = 0$, and $\int_\gamma f = 2\pi i$ where γ is the unit circle. (The point $z = 0$ is called a *singularity* of f.) To study such functions it is important to be able to replace $\int_\gamma f$ by $\int_{\tilde\gamma} f$, where $\tilde\gamma$ is a simpler curve (say, a circle). Then $\int_{\tilde\gamma} f$ can often be evaluated. The procedure that allows us to pass from γ to $\tilde\gamma$ is based on Cauchy's theorem and is as follows.

2.2.2 PRELIMINARY VERSION OF THE DEFORMATION THEOREM *Let f be analytic on a region A and let γ be a simple closed curve in A. Suppose that γ can be continuously deformed to another simple closed curve $\tilde\gamma$ without passing outside the region A. (We say that γ is **homotopic** to $\tilde\gamma$ in A.) Then*

$$\int_\gamma f = \int_{\tilde\gamma} f \tag{3}$$

NOTE The precise definition of "homotopic" is given in Sec. 2.3, and the assumption the curves are simple will be eliminated.

The deformation theorem is illustrated in Fig. 2.2.3. Note that f need not be analytic inside γ, so Cauchy's theorem does *not* imply that the integrals in Eq. (3) are zero. It is implicit in the statement of the deformation theorem that both γ and $\tilde\gamma$ are traversed in a counterclockwise direction.

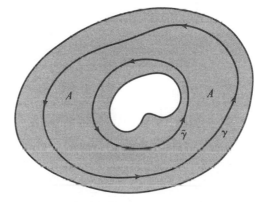

FIGURE 2.2.3 Deformation theorem.

PROOF   Consider Fig. 2.2.4, in which a curve γ_0 is drawn joining γ to $\tilde{\gamma}$; we assume such a curve can in fact be drawn (it can in all practical examples). We get a new curve consisting of γ, then γ_0, then $-\tilde{\gamma}$, and then $-\gamma_0$, in that order.

The inside of this curve is the shaded region in Fig. 2.2.4. In this region, f is analytic, so Cauchy's theorem (Eq. (2)) gives

$$\int_{\gamma+\gamma_0-\tilde{\gamma}-\gamma_0} f = 0$$

Strictly speaking, this new curve is not a simple closed curve, but such an objection can be taken care of by drawing two parallel copies of γ_0 and taking the limit as these copies converge together. From

$$\int_{\gamma+\gamma_0-\tilde{\gamma}-\gamma_0} f = 0$$

we get

$$\int_{\gamma} f + \int_{\gamma_0} f - \int_{\tilde{\gamma}} f - \int_{\gamma_0} f = 0$$

that is,

$$\int_{\gamma} f = \int_{\tilde{\gamma}} f$$

as required. ∎

The basic idea of this proof is also the basis of the more technical proof of the precise

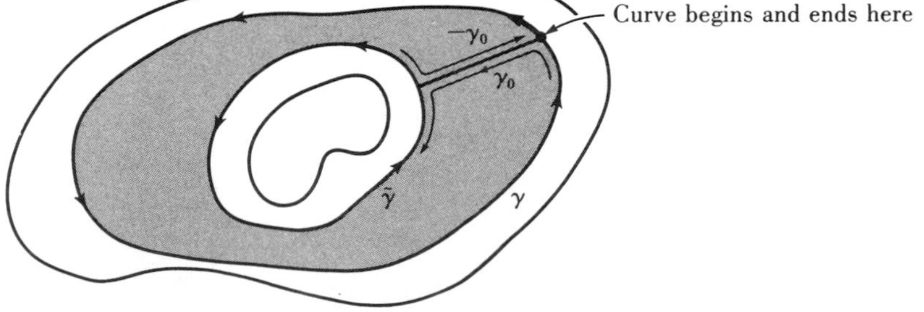

FIGURE 2.2.4   Curve used to prove the deformation theorem.

version of the deformation theorem considered in Sec. 2.3, but in these more technical proofs the simplicity of the key ideas sometimes gets lost.

Simply Connected Regions

A region $A \subset \mathbf{C}$ is called *simply connected* if A is connected and every closed curve γ in A can be deformed in A to some constant curve $\tilde{\gamma}(t) = z_0 \in A$: we also say that γ is *homotopic to a point* or is *contractible to a point*. Intuitively, a region is simply connected when it has no holes; this is because a curve that loops around a hole cannot be shrunk down to a point in A without leaving A (see Fig. 2.2.5). Therefore, the domain on which a function like $f(z) = 1/z$, which has a singularity, is analytic is *not* simply connected. Such regions are important because we shall want to study singularities in detail in Chap. 4.

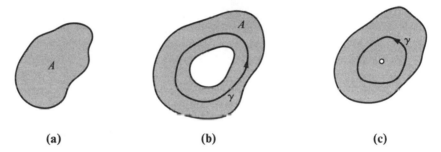
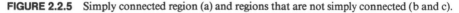

(a)   **(b)**   **(c)**

FIGURE 2.2.5   Simply connected region (a) and regions that are not simply connected (b and c).

By applying the Jordan curve theorem (see the supplement to Sec. 2.3), we can prove that a region is simply connected iff, for every simple closed curve γ in A, the inside of γ also lies in A. This conclusion is, intuitively, quite obvious and the student should try to be convinced that it is. We can also apply the theorem to prove that the inside of a simple closed curve is simply connected.

We can rewrite Cauchy's theorem in terms of simply connected regions as follows.

2.2.3 COROLLARY   *Let f be analytic on a simply connected region A and let γ be a simple closed (piecewise C^1) curve in A. Then*

$$\int_\gamma f = 0$$

Independence of Path and Antiderivatives

In the path independence theorem (2.1.9) we saw how to relate the vanishing of integrals along closed curves to path independence of integrals between points and to the existence of antiderivatives on regions. We can exploit these ideas in the present context.

2.2.4 PROPOSITION   *Suppose that f is analytic on a simply connected region A. Then for any two curves γ_1 and γ_2 joining two points z_0 and z_1 in A (as in Fig. 2.2.6), we have $\int_{\gamma_1} f = \int_{\gamma_2} f$.*

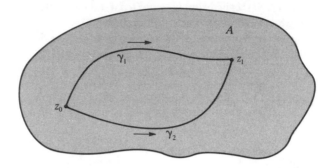

FIGURE 2.2.6   Independence of path.

PROOF   Consider the closed curve $\gamma = \gamma_2 - \gamma_1$. (In this proof we assume, because we have assumed simplicity above, that γ is a *simple* closed curve. As noted, we will show in the next section that this is not necessary.) By Cauchy's theorem,

$$0 = \int_{\gamma} f = \int_{\gamma_2} f - \int_{\gamma_1} f$$

and so

$$\int_{\gamma_2} f = \int_{\gamma_1} f$$

as required. ∎

Just as in Theorem 2.1.9, we also get the existence of an antiderivative for f on the region.

2.2.5 ANTIDERIVATIVE THEOREM   *Let f be defined and analytic on a simply connected region A. Then there is an analytic function F defined on A which is unique up to an additive constant, such that $F'(z) = f(z)$ for all z in A. We call F the **antiderivative** of f on A.*

PROOF The existence of the antiderivative follows from the path independence theorem. (Strictly speaking, we should get rid of the assumption of simple curves first.) The uniqueness assertion means that if F_0 is any other such function, then $F_0(z) = F(z) + C$ for some constant C. This follows because the region A is connected and $(F_0 - F)'(z) = F_0'(z) - F'(z) = f(z) - f(z) = 0$ for all z in A. Thus $F_0 - F$ is constant on A by Corollary 2.1.8. ∎

If A is not simply connected, this result does not hold. For example, if $A = \mathbb{C}\backslash\{0\}$ and $f(z) = 1/z$, there is no F defined on all of A with $F' = f$. (See Worked Example 2.1.13.) In some sense, F ought to be the logarithm, but we cannot define this in a consistent way on all of A. However, on any simply connected region not containing 0 we can find such an F. This is the basis of the following discussion.

The Logarithm Again

2.2.6 PROPOSITION *Let A be a simply connected region and let $0 \notin A$. Then there is an analytic function $F(z)$, unique up to the addition of multiples of $2\pi i$, such that $e^{F(z)} = z$.*

PROOF By the antiderivative theorem, there is an analytic function F with $F'(z) = 1/z$ on A. Let us fix a point $z_0 \in A$. Then z_0 lies in the domain of some branch of the log function defined in Sec. 1.6. If we adjust F by adding a constant so that $F(z_0) = \log z_0$, then at z_0, $e^{F(z_0)} = z_0$. We now want to show that $e^{F(z)} = z$ is true on all of A. To do this, we let $g(z) = e^{F(z)}/z$. Then, since $0 \notin A$, g is analytic on A, and since $F'(z) = 1/z$,

$$g'(z) = \frac{z \cdot \dfrac{1}{z} \cdot e^{F(z)} - 1 \cdot e^{F(z)}}{z^2} = 0$$

Thus g is constant on A. But $g = 1$ at z_0, and so g is 1 on all of A. Therefore, $e^{F(z)} = z$ on all of A.

For uniqueness, let F and G be functions analytic on A and let $e^{F(z)} = z$ and $e^{G(z)} = z$. Then $e^{F(z) - G(z)} = 1$, and so at a fixed z_0, $F(z_0) - G(z_0) = 2\pi n i$ for some integer n. But $F'(z) = 1/z = G'(z)$, and so we have $d(F - G)/dz = 0$, from which we conclude that $F - G = 2\pi n i$ on all of A. ∎

We write $F(z) = \log z$ and call such a choice of F a *branch* of log on A. Clearly, this procedure generalizes the procedure described in Sec. 1.6 and we get the usual log as defined in that section if A is $\mathbb{C}$ minus 0 and the negative real axis. Note that this A is simply connected. However, the A in this proposition can be more complicated, as depicted in Fig. 2.2.7.

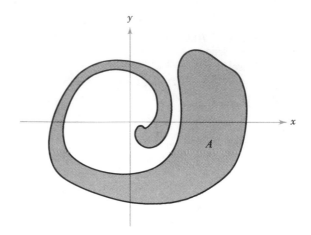

FIGURE 2.2.7   A possible domain for the log function.

Worked Examples

2.2.7   *Evaluate the following integrals:*
(a) $\int_\gamma e^z \, dz$, *where γ is the perimeter of the unit square.*
(b) $\int_\gamma 1/z^2 \, dz$, *where γ is the unit circle.*
(c) $\int_\gamma 1/z \, dz$, *where γ is the circle $3 + e^{i\theta}, 0 \le \theta \le 2\pi$.*
(d) $\int_\gamma z^2 \, dz$, *where γ is the segment joining $1 + i$ to 2.*

Solution.   analytic on the whole plane

(a) e^z is entire; thus by Cauchy's theorem, $\int_\gamma e^z \, dz = 0$, since γ is a simple closed curve. Alternatively, e^z is the derivative of e^z, and since γ is closed we can apply the path independence theorem (2.1.9).

(b) $1/z^2$ is defined and analytic on $\mathbf{C}\backslash\{0\}$ and is the derivative of $-1/z$, which is defined and analytic on $\mathbf{C}\backslash\{0\}$. By Theorem (2.1.9) and the fact that the unit circle lies in $\mathbf{C}\backslash\{0\}$, we have $\int_\gamma 1/z^2 \, dz = 0$. Alternatively, we can use Worked Example 2.1.12, for our solution.

(c) The circle $\gamma = 3 + e^{i\theta}, 0 \le \theta \le 2\pi$, does not pass through 0 or include 0 in its interior. Hence $1/z$ is analytic on γ and the interior of γ, so by Cauchy's theorem, $\int_\gamma 1/z \, dz = 0$. An alternative but less direct solution is the following. The region $\{x + iy \mid x > 0\}$ is simply connected and $1/z$ is analytic on it. Therefore, by Proposition 2.2.6, $1/z$ is the derivative of some analytic function $F(z)$ (one of the branches of log z) and thus, since γ is closed, we have by the path independence theorem that $\int_\gamma 1/z \, dz = 0$.

(d) z^2 is entire and is the derivative of $z^3/3$, which is also entire. By Thoerem 2.1.9,

$$\int_\gamma z^2 \, dz = \frac{1}{3} z^3 \Big|_{1+i}^2 = \frac{(2)^3}{3} - \frac{(1+i)^3}{3} = \frac{10}{3} - \frac{2i}{3}$$

Remark. In (a) and (c) the first method uses Cauchy's theorem; the alternative method is based on the more elementary fact that if $f(z)$ is the derivative of another analytic function, then the integral of f around a closed contour is zero (see the path independence theorem). On the other hand, we have the antiderivative theorem, stating that an analytic function defined on a simply connected region is the derivative of some other analytic function. The student is reminded that we required Cauchy's theorem to be able to prove the antiderivative theorem.

2.2.8 *Evaluate* $\int_{\gamma} (z + 1/z)^2 \, dz$ *where γ is the straight-line path from 1 to i.*

First solution. A direct approach is to parametrize the path by $z = \gamma(t) = (1 - t) + it$ for $0 \le t \le 1$, insert this into the integral, and compute it. We will not carry this out.

Second solution. Another approach is to notice that the integrand is analytic everywhere between γ and the arc γ_0 of the unit circle leading from 1 to i; see Fig. 2.2.8. Thus, the integrals are the same by Proposition 2.2.4. The second path is parametrized by $z = \gamma_0(t) = e^{it}$ for $0 \le t \le \pi/2$. There the integral becomes, with $f(z) = (z + 1/z)^2$,

$$\cos t = \frac{e^{it} + e^{-it}}{2}$$

$$\int_{\gamma} f = \int_{\gamma_0} f = \int_0^{\pi/2} (e^{it} + e^{-it})^2 i e^{it} \, dt = 4i \int_0^{\pi/2} (\cos^2 t)(\cos t + i \sin t) \, dt$$

$$= -4 \int_0^{\pi/2} \cos^2 t \sin t \, dt + 4i \int_0^{\pi/2} (1 - \sin^2 t)(\cos t) \, dt$$

$$= \left[\tfrac{4}{3} \cos^3 t + 4i (\sin t - \tfrac{1}{3} \sin^3 t) \right]_0^{\pi/2}$$

$$= \frac{-4 + 8i}{3}$$

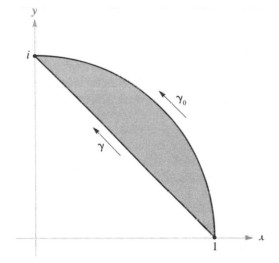

FIGURE 2.2.8 Two paths from 1 to i.

Third solution. The integrand has an antiderivative: $(z + 1/z)^2 = z^2 + 2 + 1/z^2 = (d/dz)\,(z^3/3 + 2z - z^{-1})$, which is valid everywhere along the path, so that

$$\int_1^i \left(z + \frac{1}{z} \right)^2 dz = \left(\frac{z^3}{3} + 2z - \frac{1}{z} \right) \bigg|_1^i$$

$$= \frac{-4 + 8i}{3}$$

2.2.9 *Use the deformation theorem to argue informally that if γ is a simple closed curve (not necessarily a circle) containing 0, then*

$$\int_\gamma \frac{1}{z}\, dz = 2\pi i$$

Solution. The inside of γ contains 0. Thus we can find an $r > 0$ such that the circle $\tilde{\gamma}$ of radius r and centered at 0 lies entirely on the inside of γ. Our intuition should tell us that we can deform γ to $\tilde{\gamma}$ without passing through 0 (that is, by staying in the region $A = \mathbb{C}\backslash\{0\}$ of analyticity of $1/z$; see Fig. 2.2.9). Therefore the deformation theorem and the calculation in Worked Example 2.1.12 show that

$$\int_\gamma \frac{1}{z}\, dz = \int_{\tilde{\gamma}} \frac{1}{z}\, dz = 2\pi i$$

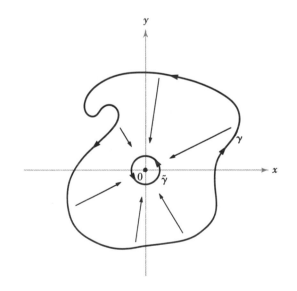

FIGURE 2.2.9 Deformation of γ to the circle $\tilde{\gamma}$.

2.2.10 *Outline a proof of this extension of the deformation theorem: Suppose that $\gamma_1, \ldots, \gamma_n$ are simple closed curves and that γ is a simple closed curve with f analytic on the region between γ and $\gamma_1, \ldots, \gamma_n$ (see Fig. 2.2.10). Then*

$$\int_\gamma f = \sum_{k=1}^n \int_{\gamma_k} f$$

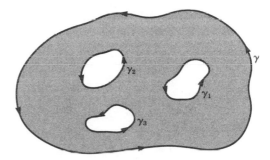

FIGURE 2.2.10 Generalized deformation theorem.

Solution. Draw curves $\tilde{\gamma}_1, \tilde{\gamma}_2, \ldots, \tilde{\gamma}_n$ joining γ to $\gamma_1, \ldots, \gamma_n$, respectively, as shown in Figure 2.2.11(a). Let ρ denote the curve drawn in Fig. 2.2.11(b). The inside of ρ is a region of analyticity of f, and so $\int_\rho f = 0$. But ρ consists of $\gamma, -\gamma_1, -\gamma_2, \ldots, -\gamma_n$, and each $\tilde{\gamma}_i$ traversed twice in opposite directions, and so the contributions from these last portions cancel. Thus

$$0 = \int_\gamma f + \int_{-\gamma_1} f + \cdots + \int_{-\gamma_n} f = \int_\gamma f - \sum_{i=1}^n \int_{\gamma_i} f$$

as required.

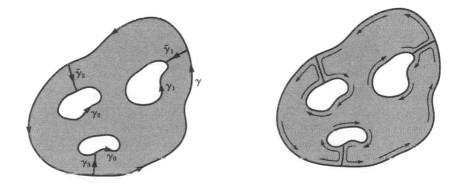

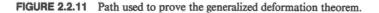

FIGURE 2.2.11 Path used to prove the generalized deformation theorem.

2.2.11   *Let $f(z)$ be analytic on a simply connected region A, except possibly not analytic at $z_0 \in A$. Suppose, however, that f is bounded in absolute value near z_0. Show that, for any simple closed curve γ containing z_0, $\int_\gamma f = 0$.*

Solution. Let $\epsilon > 0$ and let γ_ϵ be the circle of radius ϵ and with center z_0. By the deformation theorem, $\int_\gamma f = \int_{\gamma_\epsilon} f$. Let $|f(z)| \le M$ near z_0. Thus

$$\left| \int_{\gamma_\epsilon} f(z)\, dz \right| \le 2\pi \epsilon M$$

Hence for any $\epsilon > 0$,

$$\left| \int_\gamma f(z)\, dz \right| \le 2\pi \epsilon M$$

Letting $\epsilon \to 0$, we conclude $\int_\gamma f = 0$.

Exercises

1. Evaluate the following integrals:
   (a) $\int_\gamma (z^3 + 3)\, dz$, where γ is the upper half of the unit circle.
   (b) $\int_\gamma (z^3 + 3)\, dz$, where γ is the unit circle.
   (c) $\int_\gamma e^{1/z}\, dz$, where γ is a circle of radius 3 centered at $5i + 1$.
   (d) $\int_\gamma \cos [3 + 1/(z - 3)]\, dz$, where γ is a unit square with corners at $0, 1, 1 + i$, and i.

2. Let γ be a simple closed curve containing 0. Argue informally that

$$\int_\gamma \frac{1}{z^2}\, dz = 0$$

3. Let f be entire. Evaluate

$$\int_0^{2\pi} f(z_0 + re^{i\theta})\, e^{ki\theta}\, d\theta$$

for k an integer, $k \ge 1$.

4. Discuss the validity of the formula $\log z = \log r + i\theta$ for log on the region A shown in Fig. 2.2.7.

5. For what simple closed curves γ does the following hold:

$$\int_\gamma \frac{dz}{z^2 + z + 1} = 0$$

6. Evaluate $\int_\gamma (z - (1/z))\, dz$ where γ is the straight-line path from 1 to i.

7. Does Cauchy's theorem hold separately for the real and imaginary parts of f? If so, prove that it does; if not, give a counterexample.

8. Let γ_1 be the circle of radius 1 and let γ_2 be the circle of radius 2 (traversed counterclockwise and centered at the origin). Show that

$$\int_{\gamma_1} \frac{dz}{z^3 (z^2 + 10)} = \int_{\gamma_2} \frac{dz}{z^3 (z^2 + 10)}$$

9. Evaluate $\int_\gamma \sqrt{z}\, dz$ where γ is the upper half of the unit circle: first, directly; then, using the fundamental theorem (2.1.7).

10. Evaluate $\int_\gamma \sqrt{z^2 - 1}\, dz$ where γ is a circle of radius $\frac{1}{2}$ centered at 0.

11. Evaluate

$$\int_\gamma \frac{2z^2 - 15z + 30}{z^3 - 10z^2 + 32z - 32}\, dz$$

where γ is the circle $|z| = 3$. (*Hint.* Use partial fractions; one root of the denominator is $z = 2$.)

2.3 CAUCHY'S THEOREM: PRECISE VERSION

Section 2.2 developed some familiarity with theorems of the Cauchy type, the basic theme being that if a function is analytic everywhere inside a closed contour, then its integral around that contour must be 0. The principal goal of this section is to give a precise proof of a form of the theorem known as a *homotopy version of Cauchy's theorem*. This approach is taken because it makes precise the intuitive notion presented in the last section of the continuous deformation of a curve. The primary objective will be the precise formulation and proof of deformation theorems which say, roughly, that if a curve is continuously deformed through a region in which a function is analytic, then the integral along the curve does not change. In the course of proving the results, Green's theorem will not be used. Instead a different approach will be utilized. The reader will also notice that in this section references are made not to "simple closed curves" (except at the end of the section, when the link between Cauchy's theorem and the Jordan curve theorem is made) but rather merely to "closed curves." This is another technical advantage of the approach used here.

Local Version of Cauchy's Theorem

We begin by looking at the important special case in which the curve is contained in a disk on which the function is analytic. The method is the elegant and classical bisection procedure introduced by Édouard Goursat in 1883 (see *Acta Mathematica*, Vol. 14

(1884), and *Transactions of the American Mathematical Society,* Vol. 1 (1900), pp. 14–16).

For most of this section, "curve" means "piecewise C^1 curve." However, at one point in the development it will become important that this can be dropped and we can consider continuous curves if we are concerned only with integrating functions analytic on an open set containing the curve. The process of accomplishing this is somewhat indirect and is treated in supplement A to this section.

2.3.1 CAUCHY-GOURSAT THEOREM FOR A DISK   *Suppose that $f: D \to C$ is analytic on a disk $D = D(z_0; \rho) \subset C$; then*

(i) *f has an antiderivative on D; that is, there is a function $F: D \to C$ which is analytic on D and which satisfies $F'(z) = f(z)$ for all z in D.*

and

(ii) *If Γ is any closed curve in D, then $\int_\Gamma f = 0$.*

From the discussion in Sec. 2.1 on the path independence of integrals (see Theorem 2.1.9), we know that (i) and (ii) are equivalent in the sense that whichever we establish first, the other will follow readily from it. Our problem is how to obtain either one of them. In the proof of the path independence theorem (2.1.9), it was shown that (ii) follows easily from (i), and the construction of an antiderivative to get (i) was facilitated by the path independence of integrals. The plan here is curiously indirect.

First: Prove (ii) directly for the very special case in which Γ is the boundary of a rectangle.

Second: Show that this limited version of path independence is enough to carry out a construction of an antiderivative similar to that in the proof of the path independence theorem.

Third: With (i) thus established, part (ii) in its full generality follows as in the path independence theorem.

The first step is embodied in the following:

2.3.2 CAUCHY-GOURSAT THEOREM FOR A RECTANGLE   *Suppose R is a rectangular path with sides parallel to the axes and that f is a function defined and analytic on an open set G containing R and its interior. Then $\int_R f = 0$.*

A proof of this can almost be based on Green's theorem, as was outlined in Sec. 2.2. If the integral is interpreted in real-variable calculus terms as a pair of line integrals of vector-valued functions around the curve R, Green's theorem will convert it to double integrals of a certain quantity over the interior of R. The Cauchy-Riemann equations for f say that that quantity is identically 0, so the integral must be 0. The difficulty with this is that to apply Green's theorem we must know not only that f is differentiable but also that the derivative is continuous. We would rather not have to assume that. It turns out to be true, but we will use Cauchy's theorem to prove it, so we had best not use such an assumption in the proof of Cauchy's theorem or we will be caught in a very short logical circle. In 1883, Edouard Goursat noticed a clever way to establish the theorem directly without recourse to Green's theorem. It is essentially that method which is presented here. This has the logical advantage just mentioned in addition to *not* requiring the reader to be familiar with Green's theorem.*

PROOF Let P be the perimeter of R and Δ the length of its diagonal. Divide the rectangle R into four congruent smaller rectangles $R^{(1)}$, $R^{(2)}$, $R^{(3)}$, and $R^{(4)}$. If each is oriented in the counterclockwise direction, then cancellation along common edges leaves

$$\int_R f = \int_{R^{(1)}} f + \int_{R^{(2)}} f + \int_{R^{(3)}} f + \int_{R^{(4)}} f$$

Since

$$\left| \int_R f \right| \leq \left| \int_{R^{(1)}} f \right| + \left| \int_{R^{(2)}} f \right| + \left| \int_{R^{(3)}} f \right| + \left| \int_{R^{(4)}} f \right|$$

for at least one of the rectangles we must have $|\int_{R^{(k)}} f| \geq \frac{1}{4} |\int_R f|$. Call this subrectangle R_1. Notice that the perimeter and diagonal of R_1 are half those of R (Fig. 2.3.1). Now repeat this bisection process, obtaining a sequence $R_1, R_2, R_3, \ldots$ of smaller and smaller rectangles such that

(i) $\left| \int_{R_n} f \right| \geq \frac{1}{4} \left| \int_{R_{n-1}} f \right| \geq \cdots \geq \frac{1}{4^n} \left| \int_R f \right|$

(ii) Perimeter $(R_n) = \dfrac{1}{2^n}$ perimeter $(R) = \dfrac{P}{2^n}$

---

* There have been many other proofs given as well; for example, Pringsheim (*Transactions of the American Mathematical Society,* Vol. 2 (1902)) uses triangles rather than rectangles, which has some advantages. Cauchy's original proof is more like the one given in the previous section (he had the content of Green's theorem implicitly in his proof—in fact Green did not formulate Green's theorem as such until about 1830, whereas Cauchy's theorem is given in *Mémoire sur les intégrales définies prises éntre des limites imaginaires,* which appeared in 1825). A proof of Dixon is given in S. Lang's *Complex Variables* (New York: Spring-Verlag, 2d ed., 1985), and a proof based on homology is given in L. Ahlfors' *Complex Analysis* (New York: McGraw-Hill, 2d ed. 1966).

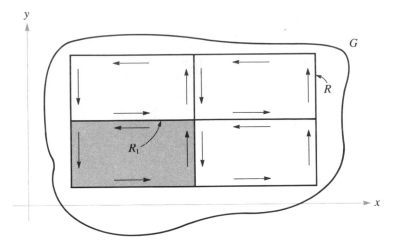

FIGURE 2.3.1   Bisection procedure.

and

(iii) Diagonal $(R_n) = \dfrac{1}{2^n}$ diagonal $(R) = \dfrac{\Delta}{2^n}$

(Fig. 2.3.2).

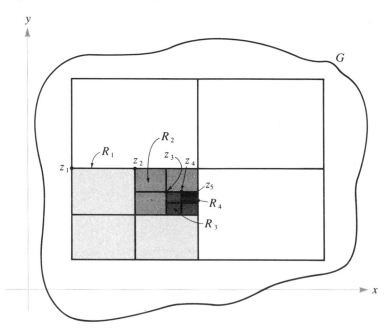

FIGURE 2.3.2   Goursat's repeated bisection process for the proof of Cauchy's theorem for a rectangle.

Since these rectangles are nested one within another and have diagonals tending to 0, they must shrink down to a single point w_0. To be precise, let z_n be the upper left-hand corner of R_n. If $m > n$ then $|z_n - z_m| \leq$ diagonal $(R_n) = \Delta/2^n$, and thus $\{z_n\}$ forms a Cauchy sequence which must converge to some point w_0. If z is any point on the rectangle R_n, then since all z_k with $k \geq n$ are within R_n, z can be no farther from w_0 than the length of the diagonal of R_n. That is, $|z - w_0| \leq \Delta/2^n$ for z in R_n.

From (i) we see that $|\int_R f| \leq 4^n |\int_{R_n} f|$. To obtain a sufficiently good estimate on the right side of this inequality, we use the differentiability of f at the point w_0.

For $\epsilon > 0$, there is a number $\delta > 0$ such that

$$\left| \frac{f(z) - f(w_0)}{z - w_0} - f'(w_0) \right| < \epsilon$$

whenever $|z - w_0| < \delta$. If we choose n large enough that $\Delta/2^n$ is less than δ, then

$$|f(z) - f(w_0) - (z - w_0)f'(w_0)| < \epsilon|z - w_0| \leq \epsilon\, \Delta/2^n$$

for all points z on the rectangle R_n. Furthermore, by the path independence theorem (2.1.9),

$$\int_{R_n} 1\, dz = 0 \qquad \text{and} \qquad \int_{R_n} (z - w_0)\, dz = 0$$

Since z is an antiderivative for 1, $(z - w_0)^2/2$ is an antiderivative for $(z - w_0)$, and the path R_n is closed. Thus,

$$\left| \int_R f \right| \leq 4^n \left| \int_{R_n} f \right|$$
$$= 4^n \left| \int_{R_n} f(z)\, dz - f(w_0) \int_{R_n} 1\, dz - f'(w_0) \int_{R_n} (z - w_0)\, dz \right|$$
$$\leq 4^n \left| \int_{R_n} [f(z) - f(w_0) - (z - w_0)f'(w_0)]\, dz \right|$$
$$\leq 4^n \int_{R_n} \left| f(z) - f(w_0) - (z - w_0)f'(w_0) \right| |dz|$$
$$\leq 4^n \left(\frac{\epsilon\Delta}{2^n} \right) \cdot \text{perimeter } (R_n)$$
$$= \epsilon\Delta P$$

Since this is true for every $\epsilon > 0$, we must have $|\int_R f| = 0$ and so $\int_R f = 0$, as desired. ▼

We can now carry out the second step of the proof of the Gauchy-Goursat theorem for a disk (2.3.1). Since the function f is analytic on the disk $D = D(z_0; \rho)$, the result for a rectangle just proved shows that the integral of f is 0 around any rectangle in D. This is enough to carry out a construction of an antiderivative for f very much like that done in the proof of the path independence theorem (2.1.9) and thus to establish part (i) of the theorem.

We will again define the antiderivative $F(z)$ as an integral from z to z_0. However, we do not yet know that such an integral is path-independent. Instead we will specify a particular choice of path and use the new information available—the analyticity of f and the geometry of the situation together with the rectangular case of Cauchy's theorem—to show that we get an antiderivative. For the duration of this proof we will use the notation $\langle\langle a, b \rangle\rangle$ to denote the polygonal path proceeding from a point a to a point b in two segments, first parallel to the x axis, then parallel to the y axis, as in Fig. 2.3.3.

If the point b is in a disk $D(a; \delta)$ centered at a, then the path $\langle\langle a, b \rangle\rangle$ is contained in that disk. Thus, for $z \in D$, we may define a function $F(z)$ by

$$F(z) = \int_{\langle\langle z_0, z \rangle\rangle} f(\xi)\, d\xi$$

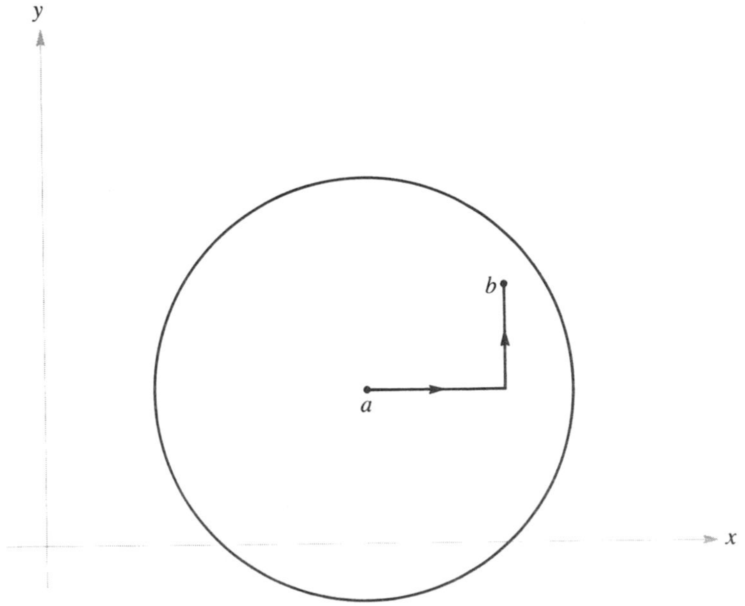

FIGURE 2.3.3   The path $\langle\langle a, b \rangle\rangle$.

We want to show that $F'(z) = f(z)$. To do this we need to show that

$$\operatorname*{limit}_{w \to z} = \frac{F(w) - F(z)}{w - z} = f(z)$$

Fixing $z \in D$ and $\epsilon > 0$, we use the fact that D is open and f is continuous on D to choose $\delta > 0$ small enough that $D(z; \delta) \subset D$ and $|f(z) - f(\xi)| < \epsilon$ for $\xi \in D(z; \delta)$. If $w \in D(z; \delta)$, then the path $\langle\langle z, w\rangle\rangle$ is contained in $D(z; \delta)$ and hence in D. The paths $\langle\langle z_0, z\rangle\rangle$ and $\langle\langle z_0, w\rangle\rangle$ are also contained in D, and these three paths fit together in a nice way with a rectangular path R also contained in D and having one corner at z; see Fig. 2.3.4. We can write, for the two cases in Fig. 2.3.4,

$$\int_{\langle\langle z_0, z\rangle\rangle} f(\xi)\, d\xi \pm \int_R f(\xi)\, d\xi + \int_{\langle\langle z, w\rangle\rangle} f(\xi)\, d\xi = \int_{\langle\langle z_0, w\rangle\rangle} f(\xi)\, d\xi$$

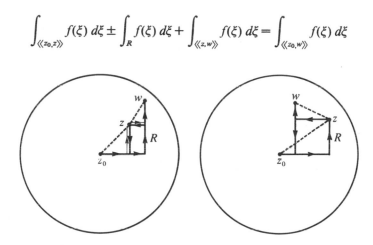

FIGURE 2.3.4 Two possible configurations for R, z_0, z, and w.

By the Cauchy-Goursat theorem for a rectangle (2.3.2), $\int_R f(\xi)\, d\xi = 0$, and so the preceding equation becomes

$$F(z) + \int_{\langle\langle z, w\rangle\rangle} f(\xi)\, d\xi = F(w)$$

Since neither side of the right triangle defined by $\langle\langle z, w\rangle\rangle$ can be any longer than its hypotenuse, which has length $|z - w|$, we conclude that $\text{length}(\langle\langle z, w\rangle\rangle) \leq 2|z - w|$, and so

$$\left| \frac{F(w) - F(z)}{w - z} - f(z) \right| = \frac{1}{|w - z|} \left| \int_{\langle\langle z, w\rangle\rangle} f(\xi)\, d\xi - f(z)(w - z) \right|$$

$$= \frac{1}{|w - z|} \left| \int_{\langle\langle z, w\rangle\rangle} f(\xi)\, d\xi - f(z) \int_{\langle\langle z, w\rangle\rangle} 1\, d\xi \right|$$

$$= \frac{1}{|w-z|} \left| \int_{\langle\langle z,w\rangle\rangle} [f(\xi) - f(z)] \, d\xi \right|$$

$$\leq \frac{1}{|w-z|} \int_{\langle\langle z,w\rangle\rangle} |f(\xi) - f(z)| |d\xi|$$

$$\leq \frac{1}{|w-z|} \epsilon \text{ length } (\langle\langle z, w\rangle\rangle) \leq \frac{1}{|w-z|} \epsilon \cdot 2|w-z| = 2\epsilon$$

Thus,

$$\operatorname*{limit}_{w\to z} = \frac{F(w) - F(z)}{w - z} = f(z)$$

and so $F'(z) = f(z)$, as desired. This establishes part (i) of the theorem. Since f has an antiderivative defined everywhere on D and γ is a closed curve in D, we have $\int_\gamma f = 0$ by the path independence theorem (2.1.9). This establishes part (ii) of the theorem and so the proof is complete. ∎

Deleted Neighborhoods

For technical reasons that will be apparent in Sec. 2.4, it will be useful to have the following varient of (2.3.2).

2.3.3 LEMMA   *Suppose that R is a rectangular path with sides parallel to the axes, and that f is a function defined on an open set G containing R and its interior, and that f is analytic on G except at some fixed point z_1 in G which is not on the path R. Suppose that at z_1, the function f satisfies $\operatorname*{limit}_{z\to z_1} (z - z_1) f(z) = 0$. Then $\int_R f = 0$.*

Notice that the condition in this lemma holds under any of the following three situations:

   (i) If f is bounded in a deleted neighborhood of z_1
   (ii) If f is continuous on G

or

   (iii) If $\operatorname*{limit}_{z\to z_1} f(z)$ exists

PROOF   If z_1 is outside R, then the situation is really just that of the Cauchy-Goursat theorem for a rectangle (2.3.2), so we may assume that z_1 is in the interior of R. For $\epsilon > 0$, there is a number $\delta > 0$ such that $|z - z_1| |f(z)| < \epsilon$ whenever $|z - z_1| < \delta$. Choose

δ small enough to do this and so that the square S of side length δ centered at z_1 lies entirely within R. Then everywhere along S we have $|f(z)| < \epsilon/|z - z_1|$. Now divide R into nine subrectangles by extending the sides of S, as shown in Fig. 2.3.5.

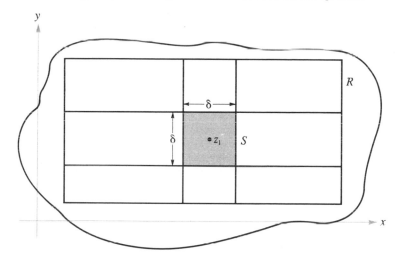

FIGURE 2.3.5 Construction of S and subdivision of R for the proof of Lemma 2.3.3.

By the Cauchy-Goursat theorem for a rectangle (2.3.2), the integrals of f around all eight of the subrectangles other than S are 0, so $\int_S f = \int_R f$. But along S we have

$$|f(z)| < \frac{\epsilon}{|z - z_1|} \leq \frac{\epsilon}{\delta/2} = \frac{2\epsilon}{\delta}$$

since $|z - z_1| \geq \delta/2$ along S. Thus

$$\left| \int_S f \right| \leq \text{length } (S) \frac{2\epsilon}{\delta} = 4\delta \frac{2\epsilon}{\delta} = 8\epsilon$$

Therefore $|\int_R f| \leq |\int_S f| \leq 8\epsilon$ for every $\epsilon > 0$. Thus we must have $|\int_R f| = 0$ and so $\int_R f = 0$, as desired. ▼

If we strengthen the assumption on f and assume that it is continuous at z_1, then we can drop the stipulation that z_1 not be on the path R.

2.3.4 LEMMA *Suppose that R is a rectangular path with sides parallel to the axes, and that f is a function defined and continuous on an open set G containing R and its interior, and that f is analytic on G except at some fixed point z_1 in G. Then $\int_R f = 0$.*

The only real problem is to make sure that the integral is well behaved if the rectangle R happens to pass through z_1. In that case, the subdivision is a little different, but the estimates are simpler.

PROOF   Again let $\epsilon > 0$. We may choose δ so that $|f(z) - f(z_1)| < \epsilon$ whenever $|z - z_1| < \delta$. If z_1 is not on R, then Lemma 2.3.3 applies. If it is on R, let S be half a square of side δ, and subdivide R as shown in Fig. 2.3.6.

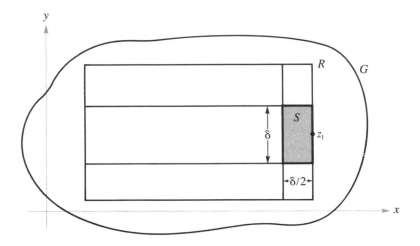

FIGURE 2.3.6   What happens if z_1 lies on R.

By the Cauchy-Goursat theorem for rectangles (2.3.2), the integrals of f around all five of the subrectangles other than S are 0, and so $\int_S f = \int_R f$. But along S we have $|f(z)| < \epsilon$. Thus, if we also require $\delta < \frac{1}{2}$, we get

$$\left| \int_S f \right| \leq \text{length } (S) \, \epsilon = 2\delta\epsilon < \epsilon$$

Therefore $|\int_R f| \leq |\int_S f| \leq \epsilon$ for every $\epsilon > 0$, and so we must have $|\int_R f| = 0$. Thus $\int_R f = 0$, as desired. ▼

If we use Lemma 2.3.4 instead of the Cauchy-Goursat theorem for a rectangle (2.3.2) in the proof of Theorem 2.3.1, we obtain the corresponding conclusions:

2.3.5 STRENGTHENED CAUCHY-GOURSAT THEOREM FOR A DISK   *The same conclusions as in the Cauchy-Goursat theorem for a disk (2.3.1) hold if we assume only that the function f is continuous on D and analytic on $G\backslash\{z_1\}$ for some fixed z_1 in D.*

Notice that continuity at z_1 is assumed. Again this is needed to apply Lemma 2.3.4 and to make sure that the integral $\int_\gamma f$ is defined even if γ passes through z_1. Notice also that a more complicated but parallel version of the same argument will produce the same conclusion if there are a finite number of "bad" points in G instead of just one.

Homotopy and Simply Connected Regions

To extend Cauchy's theorem to more general regions than disks or rectangles and to prove the deformation theorems, we must clarify the concept of deforming curves or homotopy that was discussed informally in Sec. 2.2. There are two situations to be treated: two different curves between the same two endpoints, and two closed curves which might not cross at all. For convenience we will assume that all curves are parametrized by the interval $[0, 1]$ unless specified otherwise. (This can always be done by reparametrizing if necessary.)

2.3.6 DEFINITION   *Suppose γ_0: $[0, 1] \to G$ and γ_1: $[0, 1] \to G$ are two continuous curves from z_0 to z_1 in a set G. We say that γ_0 is **homotopic with fixed endpoints** to γ_1 in G if there is a continuous function H: $[0, 1] \times [0, 1] \to G$ from the unit square $[0, 1] \times [0, 1]$ into G such that*

(i) $H(0, t) = \gamma_0(t)$ *for $0 \leq t \leq 1$*
(ii) $H(1, t) = \gamma_1(t)$ *for $0 \leq t \leq 1$*
(iii) $H(s, 0) = z_0$ *for $0 \leq s \leq 1$*

and

(iv) $H(s, 1) = z_1$ *for $0 \leq s \leq 1$*

The idea behind this definition is simple. As s ranges from 0 to 1, we have a family of curves that continuously change, or deform, from γ_0 to γ_1, as in Fig. 2.3.7. The reader should be aware that the picture need not be as simple in appearance as this illustration. The curves may twist and turn and cross over themselves or each other. No assumption is made that the curves are simple, but usually this does not matter. A little more notation may make the matter clearer. If we put $\gamma_s(t) = H(s, t)$, then each γ_s is a continuous curve from z_0 to z_1 in G. The initial curve is γ_0, and it corresponds to the left edge of the unit square. The final curve is γ_1, and it corresponds to the right edge of the square. The entire bottom edge goes to z_0 and the entire top edge to z_1. The curves γ_s are a continuously changing family of intermediate curves.

For example, the straight line segment from 0 to $1 + i$, which is parametrized by $\gamma_0(t) = t + ti$, is homotopic with fixed endpoints to the parabolic path from 0 to $1 + i$ parametrized by $\gamma_1(t) = t + t^2 i$; see Fig. 2.3.8.

Once possible homotopy from one curve to the other is

$$H(s, t) = t + t^{1+s} i$$

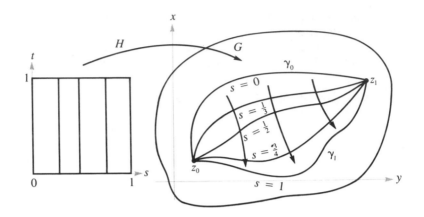

FIGURE 2.3.7 Fixed-endpoint homotopy.

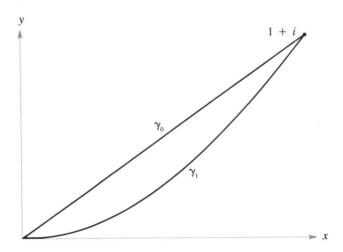

FIGURE 2.3.8 A straight-line path and a parabolic path from 0 to $1 + i$.

Of course, there is more than one way to get a homotopy between these curves. Another way makes $H(s, t)$ follow the straight line between $t + ti$ and $t + t^2i$:

$$H(s, t) = s(t + ti) + (1 - s)(t + t^2i) = t + [st + (1 - s)t^2]i$$

A slightly different definition is called for in the deformation of one *closed* curve to another.

2.3.7 DEFINITION *Suppose $\gamma_0: [0, 1] \to G$ and $\gamma_1: [0, 1] \to G$ are two continuous closed curves in a set G. We say that γ_0 and γ_1 are **homotopic as closed curves in** G if there is a continuous function $H: [0, 1] \times [0, 1] \to G$ from the unit square $[0, 1] \times [0, 1]$ into G such that*

(i) $H(0, t) = \gamma_0(t)$ *for $0 \le t \le 1$*
(ii) $H(1, t) = \gamma_1(t)$ *for $0 \le t \le 1$*

and

(iii) $H(s, 0) = H(s, 1)$ *for $0 \le s \le 1$*

Again, if we put $\gamma_s(t) = H(s, t)$, then each γ_s is a continuous curve in G. The third condition says that each of them is a closed curve; see Fig. 2.3.9.

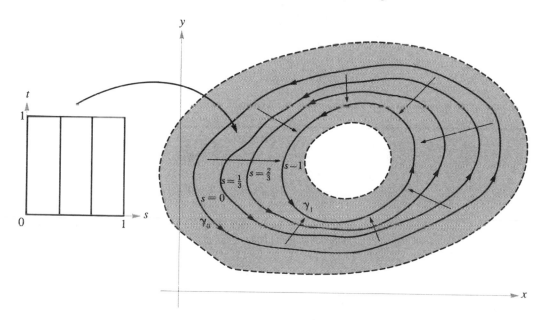

FIGURE 2.3.9 Closed-curve homotopy.

For example, the unit circle can be parametrized by $\gamma_0(t) = \cos t + i \sin t$, and the ellipse $x^2/4 + y^2 = 1$ by $\gamma_1(t) = 2 \cos t + i \sin t$. These curves are homotopic as closed curves in the annulus $G = \{z \mid \frac{1}{2} < |z| < 3\}$. One possibility for the homotopy is $H(s, t) = (1 + s) \cos t + i \sin t$. (See Fig. 2.3.10.)

If the hole were not in the middle of G in Fig. 2.3.10 but we had instead the solid disk $D = \{z \text{ such that } |z| < 3\}$, then either of the two curves could be continuously deformed down to a point. For example, $H(s, t) = (1 - s)\gamma_0(t)$ is a homotopy which shrinks the circle γ_0 down to a constant curve at the point 0. The intermediate curves γ_s are circles of

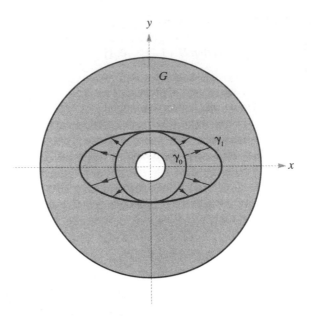

FIGURE 2.3.10   A circle homotopic to an ellipse.

radius $(1 - s)$ centered at 0. If γ_0 were any other curve in D, then the same definition of H would give a homotopy which continuously changed the scale of the curve until it shrank down to a point. Thus any curve in D is homotopic to a point in D. If there were a hole in the set as there is in the annulus in Fig. 2.3.10 then this could not be done if the curve surrounded the hole. This leads us to a more precise definition of the notion of simply connected regions that was introduced informally in Sec. 2.2.

2.3.8 DEFINITION   *A set G is called **simply connected** if every closed curve γ in G is homotopic (as a closed curve) to a point in G, that is, to some constant curve.*

The second homotopy between the straight line and the parabola in Fig. 2.3.8, which followed along straight-line segments, and the homotopy of a circle down to a point in the disk, are suggestive and lead to the definition of two important classes of simply connected sets. Recall that if z_0 and z_1 are any two points and $0 \leq s \leq 1$, then the point $sz_1 + (1 - s)z_0$ lies on the straight line segment between the two.

2.3.9 DEFINITION   *A set A is called **convex** if it contains the straight line segment between every pair of its points. That is, if z_0 and z_1 are in A, then so is $sz_1 + (1 - s)z_0$ for every number s between 0 and 1 (Fig. 2.3.11).*

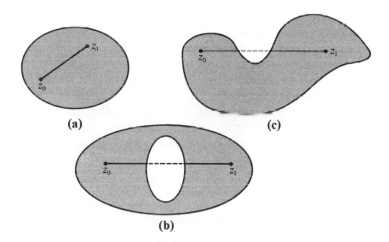

FIGURE 2.3.11   A set which is convex (a) and two which are not (b and c).

2.3.10 PROPOSITION   *If A is a convex region, then any two closed curves in A are homotopic as closed curves in A, and any two curves with the same endpoints are homotopic with fixed endpoints.*

PROOF   Let $\gamma_0: [0, 1] \to G$ and $\gamma_1: [0, 1] \to G$ be the two curves and define $H(s, t)$ by $H(s, t) = s\gamma_1(t) + (1 - s)\gamma_0(t)$. Then $H(s, t)$ lies on the straight line segment between $\gamma_0(t)$ and $\gamma_1(t)$ and so is in the set A. It is a continuous function, since γ_0 and γ_1 are continuous. At $s = 0$ we get $\gamma_0(t)$, and at $s = 1$ we get $\gamma_1(t)$. If they are closed curves, then

$$H(s, 0) = s\gamma_1(0) + (1 - s)\gamma_0(0) = s\gamma_1(1) + (1 - s)\gamma_0(1) = H(s, 1)$$

and so it is a closed-curve homotopy between the two. If they both go from z_0 to z_1, then $H(s, 0) = s\gamma_1(0) + (1 - s)\gamma_0(0) = sz_0 + (1 - s)z_0 = z_0$ and $H(s, 1) = s\gamma_1(1) + (1 - s)\gamma_0(1) = sz_1 + (1 - s)z_1 = z_1$, and so H is a fixed endpoint homotopy between the two. ∎

2.3.11 COROLLARY   *A convex region is simply connected.*

PROOF   Let z_0 be any point in the convex region A, and let γ be any closed curve in A. The constant curve at z_0, $\gamma_1(t) = z_0$ for all t, is certainly closed, and the two are homotopic by Proposition 2.3.10. ∎

A slightly more general type of simply connected region called a *starlike* (or *star-shaped*) *region* will be considered in the exercises. For more complicated regions, we often rely on our geometric intuition to determine when two curves are homotopic. In other words, we try to decide whether we can continuously deform one curve to the other without leaving our region. One reason is that we rarely use the homotopies H explicitly in practice; they are usually theoretical tools whose existence allows us to claim something else. Frequently this is the equality of two integrals. Also in many situations they might be quite complicated actually to write down. However, we must be prepared to substantiate our geometric intuition either with an explicit H or a proof of its existence in any particular situation.

Deformation Theorem

2.3.12 DEFORMATION THEOREM   *Suppose that f is an analytic function on an open set G and that γ_0 and γ_1 are piecewise C^1 curves in G.*
(i) *If γ_0 and γ_1 are paths from z_0 to z_1 and are homotopic in G with fixed endpoints, then*

$$\int_{\gamma_0} f = \int_{\gamma_1} f$$

(ii) *If γ_0 and γ_1 are closed curves which are homotopic as closed curves in G, then*

$$\int_{\gamma_0} f = \int_{\gamma_1} f$$

PROOF   The homotopy assumption means that there is a continuous function H: $[0, 1] \times [0, 1] \to G$ from the unit square into G which implements a continuous deformation from γ_0 to γ_1 in G. For each value of s, the function $\gamma_s(t) = H(s, t)$ is an intermediate curve taken on during the deformation. Similarly, for each fixed value of t, the function $\gamma_t(s) = H(s, t)$ traces out a curve crossing from $H(0, t) = \gamma_0(t)$ to $H(1, t) = \gamma_1(t)$. Thus a grid of horizontal and vertical lines in the square defines a corresponding grid of curves in G with the left edge of the square corresponding to γ_0 and the right edge to γ_1. In the fixed-endpoint case, $\gamma_0(s)$ is a constant curve at z_0 and $\gamma_1(s)$ is a constant curve at z_1. In the closed-curve case, they are the same curve, from $\gamma_0(0) (= \gamma_0(1))$ to $\gamma_1(0) (= \gamma_1(1))$. (See Figs. 2.3.12 and 2.3.13.) The reader is cautioned that the grid of curves in G need not look as nice as this illustration, since it may twist and cross over itself, becoming somewhat entangled in appearance like a fishnet thrown on the beach. However, this does not matter for the proof.

The idea of the proof is to use uniform continuity to restrict the problem to small disks, use Cauchy's theorem for a disk, and then put the pieces back together to obtain the desired result. We wish to partition the square $[0, 1] \times [0, 1]$ into smaller squares by choosing intermediate points $0 = s_0 < s_1 < s_2 < \cdots < s_n = 1$ and $0 = t_0 < t_1 < t_2 < \cdots < t_n = 1$ close enough together that each small square of the resulting grid is

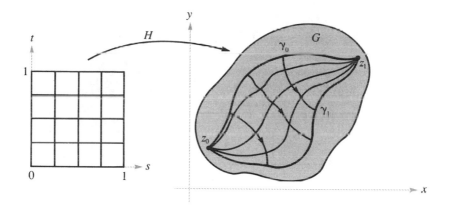

FIGURE 2.3.12   Fixed-endpoint homotopy.

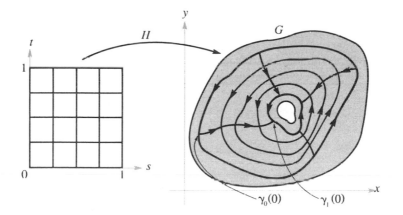

FIGURE 2.3.13   Closed-curve homotopy.

mapped into a disk which is wholly contained in G, as shown in Fig. 2.3.14. We will then be able to apply Cauchy's theorem for a disk to the integral around each of these small paths. Making the subdivision is no problem. The function H is continuous on the compact set $[0, 1] \times [0, 1]$, and so its image is a compact subset of G, by Proposition 1.4.19. By the distance lemma (1.4.21), it stays a positive distance ρ away from the closed set $\mathbb{C}\backslash G$. That is, $|H(s, t) - z| < \rho$ implies that $z \in G$. But we know (by Proposition 1.4.23) that H is actually uniformly continuous on the square. Therefore there is a number δ such that $|H(s, t) - H(s', t')| < \rho$ whenever distance$((s, t), (s', t')) = \sqrt{(s - s')^2 + (t - t')^2} < \delta$. If we choose the intermediate points equally spaced to break $[0, 1] \times [0, 1]$ into small squares with edge length $1/n$, the diagonal of each subsquare

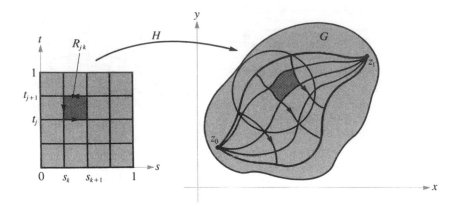

FIGURE 2.3.14 Subdivision for proof of the deformation theorem.

will have length less than δ if $n > \sqrt{2}/\delta$. If R_{kj} is the rectangle with corners at (s_{k-1}, t_{j-1}), (s_k, t_{j-1}), (s_k, t_j), (s_{k-1}, t_j), then the whole rectangle is mapped into the disk $D_{kj} = D(H(s_{k-1}, t_{j-1}); \rho)$ which is contained in G. Let Γ_{kj} be the closed curve described by $H(R_{kj})$ oriented by taking R_{kj} to be oriented in the counterclockwise direction. The image of each edge of each of the subsquares R_{kj} enters as part of two of the closed curves Γ_{kj}, and with opposite orientation, except those subsquares along the outer edge where t or s is 0 or 1. Notice that these edges actually piece together to make up the curves $\gamma_s(t)$ and $\lambda_t(s)$ discussed above. If we sum the integrals around all the loops Γ_{kj}, all the edges used twice will cancel out and leave only

$$\sum_{j=1}^{n} \sum_{k=1}^{n} \int_{\Gamma_{kj}} f = \int_{\lambda_0} f + \int_{\gamma_1} f - \int_{\lambda_1} f - \int_{\gamma_0} f$$

(see Fig. 2.3.15). Since Γ_{kj} is a closed curve lying entirely within the disk D_{kj} on which the function f is analytic, Cauchy's theorem for a disk implies that each integral in the sum on the left is 0, and so the right side is also 0. Thus,

$$0 = \int_{\lambda_0} f + \int_{\gamma_1} f - \int_{\lambda_1} f - \int_{\gamma_0} f$$

That is,

$$\int_{\lambda_1} f + \int_{\gamma_0} f = \int_{\lambda_0} f + \int_{\gamma_1} f \tag{1}$$

Up to this point the proofs for the fixed-endpoint case and the closed-curve case have been the same. Now they diverge a bit.

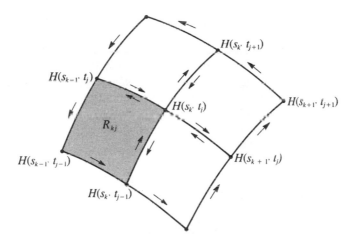

FIGURE 2.3.15 Cancellation of edges of the subsquares in the proof of the deformation theorems.

For the fixed-endpoint case, $\lambda_0(s) = H(s, 0) = z_0$ for all s, and $\lambda_1(s) = H(s, 1) = z_1$ for all s. Both are constant curves, that is, single points, and so $\int_{\lambda_0} f = \int_{\lambda_1} f = 0$.

For the closed-curve case, λ_0 and λ_1 are the same curve:

$$\lambda_0(s) = H(s, 0) = H(s, 1) = \lambda_1(s) \qquad \text{for all } s$$

and so $\int_{\lambda_0} f = \int_{\lambda_1} f$.

In either case, Eq. (1) becomes $\int_{\gamma_0} f = \int_{\gamma_1} f$, which is exactly what we want.

The proof just given is actually not quite valid. The difficulty may appear to be an uninteresting subtlety, but it is crucial. The function H has been assumed to be continuous, but no assumption was made about differentiability. Thus the curves $\gamma_s(t)$ and $\lambda_t(s)$ are continuous, but need not be piecewise C^1. Unfortunately, all our theory about contour integrals is based on piecewise C^1 curves. Thus, the integrals appearing above do not necessarily make sense. They would, and everything would be all right, if all the curves in question were piecewise C^1. Therefore we make one more provisional definition and assumption.

2.3.13 DEFINITION *A homotopy $H : [0, 1] \times [0, 1] \to G$ is called **smooth** if the intermediate curves $\gamma_s(t)$ are piecewise C^1 functions of t for each s and the cross curves $\lambda_t(s)$ are piecewise C^1 functions of s for each t.*

Provisional assumption. Assume that the homotopies in the deformation theorem are smooth.

With this extra assumption, all the curves in the proof above are piecewise C^1, the integrals all make sense; and the proof is valid. With this additional assumption in place, we will refer to the theorem as the *smooth deformation theorem.* Supplement A to this section shows how to use the smooth deformation theorem to make a reasonable definition for the integral of an analytic function along a continuous curve and to obtain the deformation theorem without the smoothness assumption by a use of the smooth deformation theorem. ∎

Cauchy's Theorem

With the power of the deformation theorem we are now able to state a fairly general form of Cauchy's theorem.

2.3.14 CAUCHY'S THEOREM   *Let f be analytic on a region G. Let γ be a closed curve in G which is homotopic to a point in G. Then*

$$\int_\gamma f = 0$$

PROOF   The curve γ is homotopic in G to a constant curve $\lambda(t) = z_0$ for all t. Therefore, $\int_\gamma f = \int_\lambda f = 0$. ∎

2.3.15 CAUCHY'S THEOREM FOR A SIMPLY-CONNECTED REGION   *If f is analytic on a simply connected region G and γ: [0, 1] → G is a closed curve in G, then*

$$\int_\gamma f = 0$$

PROOF   Every closed curve γ in G is homotopic to a point in G, and so the result follows from Cauchy's theorem. ∎

It is seldom needed in practice, but it is worth noting that with the material of the supplement to this section the curve γ and the homotopy taking it to a point need only to be assumed continuous and not necessarily piecewise C^1.

Antiderivatives and Logarithms Again

In the path independence theorem (2.1.9) we saw that the existence of antiderivatives is closely tied to path independence of integrals and to the vanishing of integrals around closed curves. Since we have now established Cauchy's theorem, we have, as a byproduct, also established the existence of antiderivatives.

2.3.16 ANTIDERIVATIVE THEOREM *Let f be analytic on a simply connected region G. Then there is an analytic function F defined on G such that $F'(z) = f(z)$ for every z in G. The function F is unique up to an additive constant in the sense that if F_0 is any other such function, then $F - F_0$ is constant on G.*

In Sec. 2.2 we showed that the existence of logarithms follows from the antiderivative theorem. Thus, we now have established the following.

2.3.17 PROPOSITION *Suppose G is a simply connected region and that 0 is not in G. Then there is an analytic function g defined on G which is uniquely defined up to the addition of multiples of $2\pi i$ and is such that $e^{g(z)} = z$ for all z in G.*

So long as we are careful about the domain of definition we can refer to $g(z)$ as log z. We say that it defines a branch of the logarithm function on the region G.

REMARK The deformation theorem, Cauchy's theorem, and all these consequences were proved from the conclusions of the local version of Cauchy's theorem (2.3.1). Thus, from Proposition 2.3.5 we see that all these conclusions remain valid if we assume merely that f is continuous on G and analytic on $G\backslash\{z_1\}$ for some fixed z_1 in G. In Sec. 2.4 it will be shown that this assumption actually implies that f is analytic on G, so that such a weakening of the hypotheses of the theorems is only apparent. But it is necessary for the logical development of the theory.

SUPPLEMENT A TO SECTION 2.3

The material of this supplement is separated from the body of the section because it is not essential to an understanding of Cauchy's theorem or of the material in subsequent chapters. The first portion of the supplement supplies the material promised earlier in the discussion of the deformation theorem. The smooth deformation theorem is used to show how the integral of an analytic function may be defined along a curve which is continuous but not necessarily piecewise C^1. Then this and the smooth deformation theorem itself are used to finish the proof of the deformation theorem. Supplement B will explore without proof, the relationship of Cauchy's theorem to a geometric result known as the Jordan curve theorem, which discusses what we mean by the inside and outside of a simple continuous closed curve.

Integrals along Continuous Curves

In the proof of the deformation or homotopy version of Cauchy's theorem we made the provisional assumption that the deformation was smooth in the sense that each intermediate curve $\gamma_s(t) = H(s, t)$ and each cross curve $\lambda_t(s) = H(s, t)$, thought of as curves traced out by the point $H(s, t)$ as either s or t, respectively, is held constant, are piecewise C^1. It was stated at that time that this is not really necessary. We really need only to assume that $H(s, t)$ is a continuous function of s and t so that each $\gamma_s(t)$ is a continuous curve. For the time being we will refer to the theorem with the C^1 assumption as the "smooth deformation theorem." The main reason for the assumption was that our whole definition of contour integrals was based on piecewise C^1 curves—after all, the derivative of the curve appears explicitly in the definition! In general we do not really know what the integral of a function along a curve which is continuous but not piecewise C^1 really is. In fact, such a general theory is not within our grasp. However, the situation is saved by the fact that we are not really interested in general functions, but only in analytic functions. This extra assumption about the function to be integrated makes up for the weaker information about the curve along which it is to be integrated. The approach taken here to overcome this difficulty may not be quite the most direct route to the deformation theorem, but it has the advantage of showing how we can make sense of the integral of an analytic function along a continuous curve. It also has the interesting feature of using the smooth deformation theorem in the process of showing that the smoothness assumption is not really needed. [Many of the ideas here are presented, more completely and a bit differently, in the paper by R. Redheffer, "The Homotopy Theorems of Function Theory," *American Mathematical Monthly,* Vol. 76 (1969), pp. 778–787, and are used there to do several other interesting things.]

Suppose f is an analytic function on an open set G and that $\gamma: [0, 1] \to G$ is a continuous (but not necessarily piecewise C^1) curve from z_0 to z_1 in G. We want to find a reasonable way to define $\int_\gamma f$. The outline of the program is this:

(i) We know what $\int_\lambda f$ means if λ is a piecewise C^1 curve in G from z_0 to z_1.

(ii) We show that there is at least one such λ which is "close to" γ by using the path covering lemma (1.4.24).

(iii) We show that if λ_0 and λ_1 are two such curves which are "close to" γ, then they are "close to" each other, and we use the smooth deformation theorem to show that $\int_{\lambda_0} f = \int_{\lambda_1} f$.

(iv) Because of (iii), $\int_\lambda f$ is the same for all the piecewise C^1 curves λ which are "close to" γ with the same endpoints, and we can take that common value as a reasonable definition for $\int_\gamma$.

To carry out this program, we must first define "close to." To do this, we define a type of distance between two parametrized curves with the same parameter interval, by moving out along both curves, recording at each parameter value t the distance between the corresponding points on the curves, and then taking the largest of these distances. This is illustrated in Fig. 2.3.16.

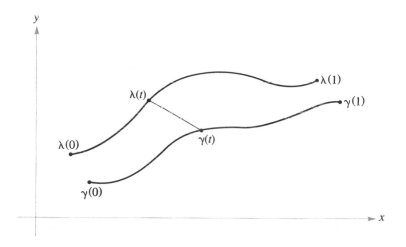

FIGURE 2.3.16   A "distance" between parametrized curves.

2.3.18 DEFINITION   *If λ: $[0, 1] \to C$ and γ: $[0, 1] \to C$ are parametrized curves in C, let* dist(λ, γ) = max $\{|\lambda(t) - \gamma(t)|$ *such that* $0 \le t \le 1\}$.

Now suppose G is an open set in C and γ: $[0, 1] \to G$ is a continuous curve from z_0 to z_1 in G. By the distance lemma (1.4.21), there is a positive distance ρ between the compact image of γ and the closed complement of G, that is, $|\gamma(t) - w| \ge \rho$ for $w \in C \backslash G$, so that $|\gamma(t) - z| < \rho$ implies that z is in G. The path covering lemma (1.4.24) provides a covering of the curve γ by a finite number of disks centered at points $\gamma(t_k)$ along the curve in such a way that each disk is contained in G and each contains the centers of the succeeding and preceding disks. The radius of these disks may be taken to be ρ for purposes of this proof.

We construct a piecewise C^1 curve λ in G by putting $\lambda(t_k) = \gamma(t_k)$ for $k = 0, 1, 2, \ldots$, n and then connecting these points by straight line segments. More precisely, for $t_{k-1} \le t \le t_k$, we put

$$\lambda(t) = \frac{(t - t_{k-1})\lambda(t_k) + (t_k - t)\lambda(t_{k-1})}{t_k - t_{k-1}}$$

Since the numbers $(t - t_{k-1})/(t_k - t_{k-1})$ and $(t_k - t)/(t_k - t_{k-1})$ are positive and add up to 1, the point $\lambda(t)$ traces out the straight line segment from $\gamma(t_{k-1})$ to $\gamma(t_k)$ as t goes from t_{k-1} to t_k, as in Fig. 2.3.17.

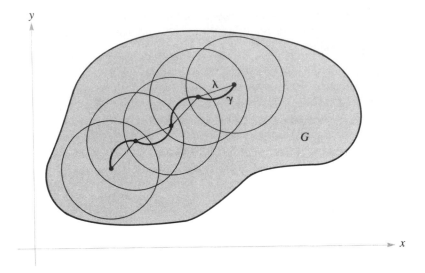

FIGURE 2.3.17 A piecewise smooth, in fact linear, approximation to a continuous curve.

The function $\lambda(t)$ is linear, and therefore is a differentiable function of t between t_{k-1} and t_k, and so λ is a piecewise C^1 path from z_0 to z_1. Furthermore, for each t, the points $\lambda(t)$ and $\gamma(t)$ both lie in the disk $D(\gamma(t_{k-1}); \rho)$, and so the curve λ lies in the set G and dist$(\lambda, \gamma) \leq 2\rho$. In fact, since $\lambda(t)$ is on the line between the centers and $\gamma(t)$ is in both disks $D(\gamma(t_{k-1}); \rho)$ and $D(\gamma(t_k); \rho)$, we have dist$(\lambda, \gamma) \leq \rho$. Since all three sides of the triangle

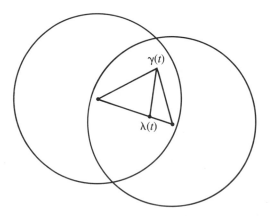

FIGURE 2.3.18   dist $(\lambda, \gamma) < \rho$.

shown have length less than ρ, the distance from $\lambda(t)$ to $\gamma(t)$ is also less than ρ. (See Fig. 2.3.18.)

This gives us the existence of at least one piecewise C^1 path which is "close to" γ. Step (iii) of the program is to show that the integrals along all such paths are the same. Suppose λ_0 and λ_1 are piecewise C^1 paths from z_0 to z_1 such that dist $(\lambda_0, \gamma) < \rho$ and dist $(\lambda_1, \gamma) < \rho$. Then both λ_0 and λ_1 lie in G. The smooth deformation theorem can be used to show that $\int_{\lambda_0} f = \int_{\lambda_1} f$. The required homotopy between the two curves can be accomplished by following the straight line from $\lambda_0(t)$ to $\lambda_1(t)$. (See Fig. 2.3.19.) For s and t between 0 and 1, define

$$H(s, t) = s\lambda_1(t) + (1 - s)\lambda_0(t)$$

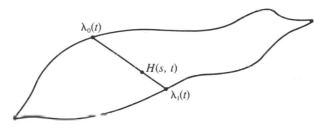

$\lambda_0(t)$

$H(s, t)$

$\lambda_1(t)$

FIGURE 2.3.19   Smooth homotopy from λ_0 to λ_1.

The function $H(s, t)$ is a piecewise C^1 function of s and of t. Trouble can occur only when $t = t_k$, $k = 0, 1, 2, \ldots n$, and so we need only check that the image always lies in G. But

$$
\begin{aligned}
|H(s, t) - \gamma(t)| &= |s\lambda_1(t) + (1 - s)\lambda_0(t) - \gamma(t)| \\
&= |s[\lambda_1(t) - \gamma(t)] + (1 - s)[\lambda_0(t) - \gamma(t)]| \\
&\leq s|\lambda_1(t) - \gamma(t)| + (1 - s)|\lambda_0(t) - \gamma(t)| \\
&\leq s\rho + (1 - s)\rho = \rho
\end{aligned}
$$

Thus $H(s, t) \in D(\gamma(t); \rho) \subset G$, and so the smooth deformation theorem applies to λ_0 and λ_1 and shows that $\int_{\lambda_0} f = \int_{\lambda_1} f$. This completes part (iii) of the program and shows that it makes sense to define the integral of an analytic function along a continuous curve as follows.

2.3.19 DEFINITION   *Suppose f is analytic on an open set G and that γ: $[0, 1] \to G$ is a continuous curve in G. If the distance from γ to the complement of G is ρ, let $\int_\gamma f = \int_\lambda f$, where λ is any piecewise C^1 curve in G which has the same endpoints as γ and which is "close to" γ in the sense that* dist$(\lambda, \gamma) < \rho$

The Deformation Theorem

With a bit of care, essentially the same idea used in the proof of step (iii) above can be used to obtain the deformation theorem (both for fixed endpoints and for closed curves) from the smooth deformation theorem. If H is a continuous homotopy from γ_0 to γ_1, then for s^* close to s, $\gamma_{s^*}(t)$ is close to $\gamma_s(t)$ and so γ_{s^*} is "close to" γ_s. If we choose piecewise C^1 curves λ and μ sufficiently "close to" γ_s and γ_{s^*} respectively, then λ will be "close to" μ, and following along the short straight line segment between $\lambda(t)$ and $\mu(t)$ will provide a smooth deformation from λ to μ. (See Fig. 2.3.20.) The smooth deformation theorem says that $\int_\lambda f = \int_\mu f$, so that the integral along γ_s is the same as that along γ_{s^*}. Thus if we shift s from 0 to 1 in steps sufficiently small that this argument applies at every step, the integral will never change and the integral along γ_0 will be the same as that along γ_1. That this actually can be done in a finite number of sufficiently small steps follows because H is a continuous function from the compact square $[0, 1] \times [0, 1]$, so that its image is a compact subset of G and lies at a positive distance from the closed complement of G.

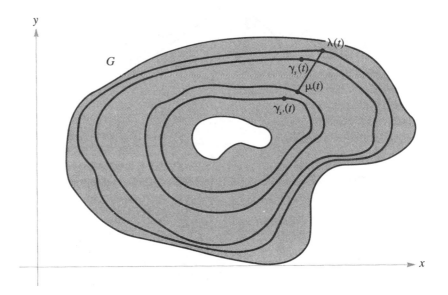

FIGURE 2.3.20   The deformation theorems can be obtained from the smooth deformation theorem.

SUPPLEMENT B TO SECTION 2.3

Relationship of Cauchy's Theorem to the Jordan Curve Theorem

An understanding of the Jordan curve theorem is not absolutely essential to an understanding of Cauchy's theorem or of the material in subsequent chapters. However, the

Jordan curve theorem is closely related to the hypotheses in Cauchy's theorem, and therefore it will be briefly considered here. In many practical examples the result of the Jordan curve theorem is geometrically obvious and can usually be proven directly. The general case of the theorem is quite difficult and will not be proven here.

2.3.20 JORDAN CURVE THEOREM   *Let $\gamma\colon [a, b] \to \mathbf{C}$ be a simple closed continuous curve in $\mathbf{C}$. Then $\mathbf{C}\backslash\gamma([a, b])$ can be written uniquely as the disjoint union of two regions I and O such that I is bounded (that is, lies in some large disk). The region I is called the **inside** of γ and O is called the **outside**. Region I is simply connected and γ is contractible to any point in $I \cup \gamma([a, b])$. The boundary of each of the two regions is $\gamma([a, b])$.*

The proof of this theorem uses more advanced mathematics and is beyond the scope of this book; see, for example, G. T. Whyburn, *Topological Analysis* (Princeton, N.J.: Princeton University Press, 1964).

Thus the Jordan curve theorem, combined with Cauchy's theorem (2.3.14), yields the following: *If f is analytic on a region A and γ is a simple closed curve in A and the inside of γ lies in A, then $\int_\gamma f = 0$.* This is one classical way of stating Cauchy's theorem. Although convenient in practice, it is theoretically awkward for two reasons: (1) it depends on the Jordan curve theorem for defining the concept of "inside"; and (2) γ is restricted to being a simple curve. The versions of the Cauchy theorem stated in Theorems 2.3.12 and 2.3.14 do not depend on the difficult Jordan curve theorem, are more general, and are just as easy to apply. On the other hand, the Jordan curve theorem reassures us that regions that we intuitively expect to be simply connected indeed are. (There is another way to describe the inside of a simple closed curve using the index, or winding number, of a curve; this method will be discussed in the next section.)

The general philosophy of this text is that we should use our geometric intuition to justify that a given region is simply connected or that two curves are homotopic, but with the realization that such knowledge is based on intuition and that to attempt to make it precise could be tedious. On the other hand, a precise argument should be used whenever possible and practical (see, for instance, the earlier argument that a convex region is simply connected).

Worked Examples

2.3.21   *Let A be the region bounded by the x axis and the curve $\sigma(\theta) = Re^{i\theta}$, $0 \le \theta \le \pi$, where $R > 0$ is fixed. Let $f(z) = e^{z^2}/(2R - z)^2$. Show that for any closed curve γ in A, $\int_\gamma f = 0$.*

Solution. First observe that f fails to be analytic only when $z = 2R$ and hence f is analytic on A, since $2R$ lies outside A (see Fig. 2.3.21). We claim that A is simply connected. That any two points in A can be joined by a straight line lying in A (that is, that A is convex) is obvious geometrically and also

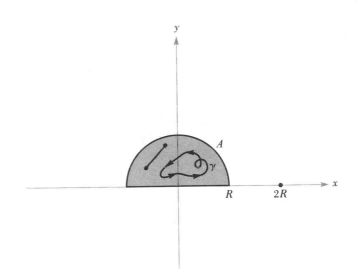

FIGURE 2.3.21   Convex region.

is a simple matter to check (which the student should do). Hence A is simply connected, by Corollary 2.3.11. By Cauchy's theorem, $\int_\gamma f = 0$ for any closed curve in A.

2.3.22   *Let $A = \{z \in \mathbf{C} \mid 1 < |z| < 4\}$. First intuitively, then precisely, show that A is not simply connected. Also show precisely that the circles $|z| = 2$ and $|z| = 3$ are homotopic in A.*

Solution. Intuitively, the circle $|z| = 2$ cannot be contracted continuously to a point without passing over the hole in A; that is, the set $\{z \in \mathbf{C}$ such that $|z| \le 1\}$.

Precisely, the function $1/z$ is analytic on A, and if A were simply connected, then we would have $\int_\gamma (1/z)\, dz = 0$ for any closed curve in A. But if we let $\gamma(t) = 2e^{it}$, $0 \le t \le 2\pi$, then we obtain

$$\int_\gamma \frac{dz}{z} = \int_0^{2\pi} \frac{1}{2e^{it}} \cdot 2ie^{it}\, dt = 2\pi i$$

Hence A is not simply connected.

Let $\gamma_1(t) = 2e^{it}$ and $\gamma_2(t) = 3e^{it}$, which represent the circles $|z| = 2$ and $|z| = 3$, respectively. Define $H(t, s) = 2e^{it} + se^{it}$; then H is a suitable homotopy between γ_1 and γ_2 in A. The effect of H is illustrated in Fig. 2.3.22.

2.3.23   *Let γ denote the unit circle $|z| = 1$. Let γ_1 and γ_2 be two circles of radius $\frac{1}{4}$ and centers $-\frac{1}{2}$ and $\frac{1}{2}$, respectively. Let A be a region containing γ, γ_1, and γ_2 and including the region between these curves (see Fig. 2.3.23). For f analytic on A, show that*

$$\int_\gamma f = \int_{\gamma_1} f + \int_{\gamma_2} f$$

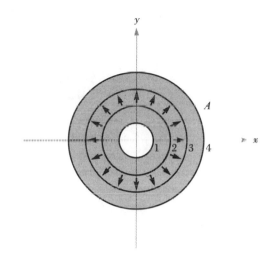

FIGURE 2.3.22 Region that is not simply connected.

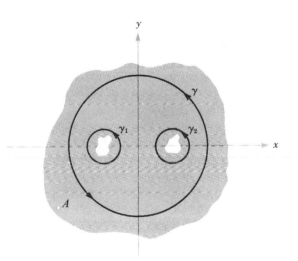

FIGURE 2.3.23 Region of analyticity of f.

Solution. We have

$$\gamma_1(t) = -\tfrac{1}{2} + \tfrac{1}{4}e^{it} \qquad 0 \le t \le 2\pi$$

and

$$\gamma_2(t) = \tfrac{1}{2} + \tfrac{1}{4}e^{it} \qquad 0 \le t \le 2\pi$$

Let $\tilde{\gamma}$ be the curve depicted in Fig. 2.3.24.

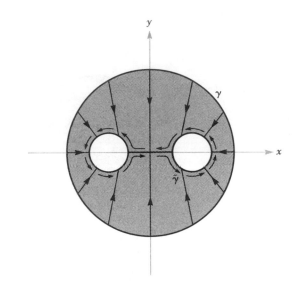

FIGURE 2.3.24   Construction of homotopy from γ to $\tilde{\gamma}$.

It is geometrically clear that γ is homotopic to $\tilde{\gamma}$ in A. The exact homotopy can be obtained easily by (1) reparametrizing $\tilde{\gamma}$ so that it has the same interval $[0, 2\pi]$ as γ, (2) defining $H(t, s) = s\gamma(t) + (1 - s)\tilde{\gamma}(t)$, and (3) checking that the straight line joining $\gamma(t)$ and $\tilde{\gamma}(t)$ remains in A (even though A is not convex). This method is illustrated in Fig. 2.3.24. By the deformation theorem, $\int_{\gamma} f = \int_{\tilde{\gamma}} f$. But $\int_{\tilde{\gamma}} f = \int_{\gamma_1} f + \int_{\gamma_2} f$, since $\tilde{\gamma} = \gamma_1 + \gamma_2 + \gamma_0 + (-\gamma_0)$, where γ_0 denotes the straight line joining $\frac{1}{2}$ to $-\frac{1}{2}$. Thus the assertion is proved. Compare this solution with that of Worked Example 2.2.10, Sec. 2.2, and note that here $\tilde{\gamma}$ does not have to be a simple closed curve, but merely a closed one.

Exercises

1.   Prove that $\mathbb{C}\backslash\{0\}$ is not simply connected.

2.   Show that every disk is convex.

3.   A region A is called *star-shaped with respect to z_0* if it contains the line segment between each of its points and z_0, that is, if $z \in A$ and $0 \le s \le 1$ imply that $sz_0 + (1 - s)z \in A$. The region is called *star-shaped* if there is at least one such point in A. Show that a star-shaped set is simply connected.

4.   Show that a set A is convex if and only if it is star-shaped with respect to each of its points (see Exercise 3).

5.   Let G be the region built as a union of two rectangular regions $G = \{z \text{ such that } |\text{Re } z| < 1 \text{ and } |\text{Im } z| < 3\} \cup \{z \text{ such that } |\text{Re } z| < 3 \text{ and } |\text{Im } z| < 1\}$. (This set is illustrated in Fig. 2.3.25.) Show that G is star-shaped. (See Exercise 3.)

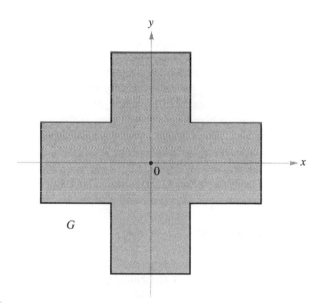

FIGURE 2.3.25 A star-shaped nonconvex region.

6. Complete the proof of Proposition 2.2.4.

7. Evaluate the following integrals without performing an explicit computation:

(a) $\displaystyle\int_\gamma \frac{dz}{z}$, where $\gamma(t) = \cos t + 2i \sin t$, $0 \le t \le 2\pi$

(b) $\displaystyle\int_\gamma \frac{dz}{z^2}$, where γ is defined as in (a)

(c) $\displaystyle\int_\gamma \frac{e^z\, dz}{z}$, where $\gamma(t) = 2 + e^{it}$, $0 \le t \le 2\pi$

(d) $\displaystyle\int_\gamma \frac{dz}{z^2 - 1}$, where γ is a circle of radius 1 centered at 1

8. Evaluate $\int_\gamma dz/z$ where γ is the line segment joining 1 to i.

9. (a) Let γ be a curve homotopic to the unit circle in $\mathbb{C}\backslash\{0\}$. Evaluate $\int_\gamma dz/z$.
(b) Evaluate $\int_\gamma dz/z$ where γ is the curve $\gamma(t) = 3\cos t + i\, 4 \sin t$, $0 \le t \le 2\pi$.

10. Evaluate the following:

(a) $\displaystyle\int_{|z|=\frac{1}{2}} \frac{dz}{(1 - z)^3}$

(b) $\displaystyle\int_{|z+1|=\frac{1}{2}} \frac{dz}{(1 - z)^3}$

(c) $\displaystyle\int_{|z-1|=\frac{1}{2}} \frac{dz}{(1 - z)^3}$

2.4 CAUCHY'S INTEGRAL FORMULA

One of the attractions of the theory of functions of one complex variable is that many powerful results can be made to flow rapidly from any one of several starting points. We have selected Cauchy's theorem as the starting point and are now in a position to draw some important consequences. For example, we shall see that a differentiable function must be infinitely differentiable and, in fact, analytic in the sense that the Taylor series converges to the function in some disk. The fundamental theorem of algebra, that every polynomial has a complex root, will be a side benefit. The connecting link to these results is a consequence of Cauchy's theorem which also bears his name. It says that the values of an analytic function are completely determined everywhere inside a closed curve by its values along the curve and gives an explicit formula linking these values.

Index of a Closed Path

There is a useful formula that expresses how many times a curve γ winds around a given point z_0 (see Fig. 2.4.1). This number of times is, informally, called the *index* of γ with respect to z_0. The term "index" will be formally defined in Definition 2.4.1.

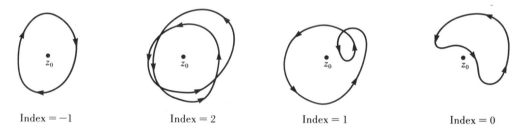

Index = -1 Index = 2   Index = 1   Index = 0

FIGURE 2.4.1   Index of a curve around a point.

The formula we shall use to compute the index is based on the computation that was done in Worked Example 2.1.12: If γ is the unit circle $\gamma(t) = e^{it}$, $0 \le t \le 2\pi$, then

$$2\pi i = \int_\gamma \frac{dz}{z}$$

If $\gamma(t) = e^{it}$, $0 \le t \le 2\pi n$, then γ encircles the origin n times, and we find in the same way that

$$n = \frac{1}{2\pi i} \int_\gamma \frac{dz}{z}$$

Now let us suppose that another closed curve $\tilde{\gamma}$ can be deformed to γ without passing through zero (that is, that $\tilde{\gamma}$ and γ are homotopic in the region $A = \mathbf{C}\backslash\{0\}$). Then again,

$$n = \frac{1}{2\pi i} \int_{\tilde{\gamma}} \frac{dz}{z} = \frac{1}{2\pi i} \int_{\gamma} \frac{dz}{z}$$

by the deformation theorem (see Sec. 2.2 or 2.3). Since $\tilde{\gamma}$ and γ are homotopic in $\mathbf{C}\backslash\{0\}$, it is intuitively reasonable that they wind around 0 the same number of times. Generally, for any point $z_0 \in \mathbf{C}$, the number of times a curve $\tilde{\gamma}$ winds around z_0 is seen to be

$$n = \frac{1}{2\pi i} \int_{\tilde{\gamma}} \frac{dz}{z - z_0}$$

by a similar argument. It is thus reasonably clear that

$$\frac{1}{2\pi i} \int_{\gamma} \frac{dz}{z - z_0} = \begin{cases} \pm 1 & \text{if } z_0 \text{ is inside } \gamma \\ 0 & \text{if } z_0 \text{ is outside } \gamma \end{cases}$$

for a *simple* closed curve γ (this can be established precisely using the Jordan curve theorem from Supplement B to Sec. 2.3). These ideas lead to the formulation in the following definition.

2.4.1 DEFINITION　*Let γ be a closed curve in $\mathbf{C}$ and $z_0 \in \mathbf{C}$ be a point not on γ. Then the **index** of γ with respect to z_0 (also called the **winding number** of γ with respect to z_0) is defined by*

$$I(\gamma, z_0) = \frac{1}{2\pi i} \int_{\gamma} \frac{dz}{z - z_0}$$

*We say that γ **winds around** z_0, $I(\gamma, z_0)$ times.*

The discussion that preceded this definition proves the following proposition, which is illustrated in Fig. 2.4.2.

2.4.2 PROPOSITION

(i) *The circle $\gamma(t) = z_0 + re^{it}$, $0 \le t \le 2\pi n$, $r > 0$, has index n with respect to z_0; the circle $-\gamma(t) = z_0 + re^{-it}$, $0 \le t \le 2\pi n$, has index $-n$.*

(ii) *If z_0 does not lie on either $\tilde{\gamma}$ or γ and if $\tilde{\gamma}$ and γ are homotopic in $\mathbf{C}\backslash\{z_0\}$, then*

$$I(\tilde{\gamma}, z_0) = I(\gamma, z_0)$$

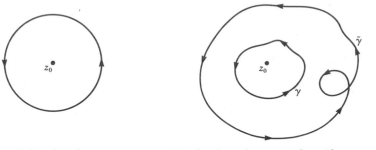

Index of circle $\tilde{\gamma}$ and γ have the same index with respect to z_0

FIGURE 2.4.2   Index = the number of times that z_0 is encircled.

Since homotopies can sometimes be awkward to deal with directly, it is customary merely to give an intuitive geometric argument that $I(\gamma, z_0)$ has a certain value, but again the student should be prepared to give a complete proof when called for (see Worked Example 2.4.12 at the end of this section).

The next proposition provides a check that the index $I(\gamma, z_0)$ is always an integer. This should be the case if Definition 2.4.1 actually represents the ideas illustrated in the figures.

2.4.3 INDEX THEOREM   *Let $\gamma : [a, b] \to \mathbf{C}$ be a (piecewise C^1) closed curve and z_0 a point not on γ; then $I(\gamma, z_0)$ is an integer.*

PROOF   Let

$$g(t) = \int_a^t \frac{\gamma'(s)}{\gamma(s) - z_0}\, ds$$

Then, at points where the integrand is continuous, the fundamental theorem of calculus gives

$$g'(t) = \frac{\gamma'(t)}{\gamma(t) - z_0}$$

Thus

$$\frac{d}{dt}\, e^{-g(t)}[\gamma(t) - z_0] = 0$$

at points where $g'(t)$ exists, and so $e^{-g(t)}[\gamma(t) - z_0]$ is piecewise constant on $[a, b]$. But $e^{-g(t)}[\gamma(t) - z_0]$ is continuous and therefore must be constant on $[a, b]$. This constant value is

$$e^{-g(a)}[\gamma(a) - z_0]$$

and so we get $e^{-g(b)}[\gamma(b) - z_0] = e^{-g(a)}[\gamma(a) - z_0]$. But $\gamma(b) = \gamma(a)$, so that $e^{-g(b)} = e^{-g(a)}$. On the other hand, $g(a) = 0$; hence $e^{-g(b)} = 1$. Thus $g(b) = 2\pi ni$ for an integer n, and the theorem follows. ■

The *inside* of a curve γ could be defined by $\{z \mid I(\gamma, z) \neq 0\}$, and this definition would agree with the "inside" defined by the Jordan curve theorem and also with the intuitive ideas illustrated in Fig. 2.4.1. Thus the inside of a closed curve can be defined purely analytically without applying the Jordan curve theorem.

Cauchy's Integral Formula

Cauchy's theorem will now be used to derive a very useful formula relating the value of an analytic function at z_0 to a certain integral.

2.4.4 CAUCHY'S INTEGRAL FORMULA *Let f be analytic on a region A, let γ be a closed curve in A that is homotopic to a point, and let $z_0 \in A$ be a point not on γ. Then*

$$f(z_0) \cdot I(\gamma, z_0) = \frac{1}{2\pi i} \int_\gamma \frac{f(z)}{z - z_0} \, dz \tag{1}$$

This formula* is often applied when γ is a simple closed curve and z_0 is inside γ. Then $I(\gamma, z_0) = 1$, so that formula (1) becomes

$$f(z_0) = \frac{1}{2\pi i} \int_\gamma \frac{f(z)}{z - z_0} \, dz \tag{1'}$$

Formula (1') is remarkable, for it says that the values of f on γ completely determine the values of f inside γ. In other words, the value of f is determined by its "boundary values."

---

* Cauchy's integral formula can be strengthened by requiring only that f be continuous on γ and analytic "inside" γ. This change makes little difference in solving most examples. For the proof of the strengthened theorem the methods of Supplement A to Sec. 2.3 and an approximation argument may be used. See also E. Hille, *Analytic Function Theory*, Vol. I (Boston: Ginn, 1959).

PROOF   The proof makes a clever use of the analyticity of f and the technical strength-ening of Cauchy's theorem for which we laid the groundwork in the strengthened Cauchy theorem for a disk (2.3.5), in which the function was allowed to be merely continuous and not necessarily analytic at one point. (See also Worked Example 2.2.11 and the remarks following Proposition 2.3.17.) Let

$$g(z) = \begin{cases} \dfrac{f(z) - f(z_0)}{z - z_0} & \text{if } z \neq z_0 \\[2mm] f'(z_0) & \text{if } z = z_0 \end{cases}$$

Then g is analytic except perhaps at z_0, and it is continuous at z_0 since f is differentiable there. Thus $\int_\gamma g = 0$, and so

$$0 = \int_\gamma g(z)\, dz = \int_\gamma \frac{f(z)}{z - z_0}\, dz - \int_\gamma \frac{f(z_0)}{z - z_0}\, dz$$

and

$$\int_\gamma \frac{f(z_0)}{z - z_0}\, dz = f(z_0) \int_\gamma \frac{1}{z - z_0}\, dz = 2\pi i f(z_0) I(\gamma, z_0)$$

and so the theorem follows. ∎

Formula (1) is extremely useful for computations. For example, we can immediately calculate

$$f(0) = \frac{1}{2\pi i} \int \frac{e^z}{z - 0}\, dz$$

$$\int_\gamma \frac{e^z}{z}\, dz = 2\pi i \cdot e^0 = 2\pi i$$

where γ is the unit circle. Here $f(z) = e^z$ and $z_0 = 0$.

Note that in formula (1) it is f and not the integrand $f(z)/(z - z_0)$ which is analytic on A; the integrand is analytic only on $A \backslash \{z_0\}$, and so we cannot use Cauchy's theorem to conclude that the integral is zero—in fact, the integral is usually nonzero.

Integrals of Cauchy Type

Cauchy's integral formula is a special and powerful formula for the value of f at z_0. We will now use it to show that all the higher derivatives of f also exist. The central trick in the proof is an idea that is often useful. We change point of view slightly. If we start assuming only that we know the values of a function along a curve, then we can consider integrals along the curve as defining new functions. The version of this idea used here is called an *integral of Cauchy type*.

2.4.5 DIFFERENTIABILITY OF CAUCHY-TYPE INTEGRALS *Let γ be a curve in C and let g be a continuous function defined along the curve (on the image γ ([a, b])). Set*

$$G(z) = \frac{1}{2\pi i} \int_\gamma \frac{g(\zeta)}{\zeta - z} \, d\zeta$$

Then G is analytic on C\γ([a, b]); in fact, G is infinitely differentiable, with the kth derivative given by

$$G^{(k)}(z) = \frac{k!}{2\pi i} \int_\gamma \frac{g(\zeta)}{(\zeta - z)^{k+1}} \, d\zeta \qquad k = 1, 2, 3, \ldots \tag{2}$$

The formula for the derivatives can be remembered by "differentiating with respect to z under the integral sign":

$$\frac{d}{dz} G(z) = \frac{1}{2\pi i} \frac{d}{dz} \int_\gamma \frac{g(\zeta)}{\zeta - z} \, d\zeta$$

$$= \frac{1}{2\pi i} \int_\gamma \frac{\partial}{\partial z} \left(\frac{g(\zeta)}{\zeta - z} \right) d\zeta$$

$$= \frac{1}{2\pi i} \int_\gamma \frac{g(\zeta)}{(\zeta - z)^2} \, d\zeta$$

The formal proof justifies this procedure and appears at the end of this section.

Existence of Higher Derivatives

Using integrals of Cauchy type we can show that a differentiable function of one complex variable is actually infinitely differentiable and at the same time give a formula for all the derivatives.

2.4.6 CAUCHY'S INTEGRAL FORMULA FOR DERIVATIVES *Let f be analytic on a region A. Then all the derivatives of f exist on A. Furthermore, for z_0 in A and γ any closed curve homotopic to a point in A with z_0 not on γ, we have*

$$f^{(k)}(z_0) \cdot I(\gamma, z_0) = \frac{k!}{2\pi i} \int_\gamma \frac{f(\zeta)}{(\zeta - z)^{k+1}} \, d\zeta \qquad k = 1, 2, 3, \ldots \tag{3}$$

where $f^{(k)}$ denotes the kth derivative of f.

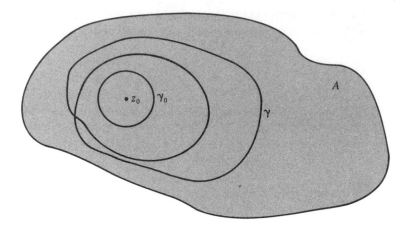

FIGURE 2.4.3 Let γ_0 be a circle centered at z_0 and small enough that it does not meet γ.

PROOF Since A is open and z_0 is not on γ, we can find a small circle γ_0 centered at z_0 with interior in A and such that γ does not cut across γ_0. (See Fig. 2.4.3.) For z in A and not on γ, define

$$G(z) = f(z) \cdot I(\gamma, z) = \frac{1}{2\pi i} \int_\gamma \frac{f(\zeta)}{\zeta - z} \, d\zeta$$

This is an integral of Cauchy type, and so it is infinitely differentiable on $A \backslash \gamma$, and

$$G^{(k)}(z) = \frac{k!}{2\pi i} \int_\gamma \frac{f(\zeta)}{(\zeta - z)^{k+1}} \, d\zeta \tag{4}$$

But

$$f(z) = \frac{1}{2\pi i} \int_{\gamma_0} \frac{f(\zeta)}{\zeta - z} \, d\zeta \quad \text{and} \quad I(\gamma, z) = \frac{1}{2\pi i} \int_\gamma \frac{1}{\zeta - z} \, d\zeta$$

are also integrals of Cauchy type, and so they are infinitely differentiable near z_0. In particular, the index is a continuous function of z so long as z does not cross γ—but it is also an integer. Thus it must stay constant except when z crosses the curve. In particular, it is constant inside γ_0. Thus $G^{(k)}(z_0) = f^{(k)}(z_0)I(\gamma, z_0)$. Combining this with Eq. (4) gives the desired result. ∎

Cauchy's Inequalities and Liouville's Theorem

2.4.7 CAUCHY'S INEQUALITIES *Let f be analytic on a region A and let γ be a circle with radius R and center z_0 that lies in A. Assume that the disk {z such that*

$|z - z_0| < R$} also lies in A. Suppose that $|f(z)| \le M$ for all z on γ. Then, for any $k = 0, 1, 2, \ldots$,

$$|f^{(k)}(z_0)| \le \frac{k!}{R^k} M \tag{5}$$

PROOF Since $I(\gamma, z_0) = 1$, from formula (3) we obtain

$$f^{(k)}(z_0) = \frac{k!}{2\pi i} \int_\gamma \frac{f(\zeta)}{(\zeta - z_0)^{k+1}} \, d\zeta$$

and hence

$$|f^{(k)}(z_0)| = \frac{k!}{2\pi} \left| \int_\gamma \frac{f(\zeta)}{(\zeta - z_0)^{k+1}} \, d\zeta \right|$$

Now

$$\left| \frac{f(\zeta)}{(\zeta - z_0)^{k+1}} \right| \le \frac{M}{R^{k+1}}$$

since $|\zeta - z_0| = R$ for ζ on γ, and so

$$|f^{(k)}(z_0)| \le \frac{k!}{2\pi} \cdot \frac{M}{R^{k+1}} \cdot l(\gamma)$$

But $l(\gamma) = 2\pi R$, so we get our result. ∎

This result states that although the kth derivatives of f can go to infinity as $k \to \infty$, they cannot grow too fast as $k \to \infty$; specifically, they can grow no faster than a constant times $k!/R^k$. We can use Cauchy's inequalities to derive the following surprising result: *the only bounded entire functions are constants.*

2.4.8 LIOUVILLE'S THEOREM* *If f is entire and there is a constant M such that $|f(z)| \le M$ for all $z \in \mathbf{C}$, then f is constant.*

---

* According to E. T. Whittaker and G. N. Watson, *A Course of Modern Analysis*, 4th ed. (London: Cambridge University Press, 1927), p. 105, Liouville's theorem is incorrectly attributed to Liouville by Borchardt (whom others copied), who heard it in Liouville's lectures in 1847. It is due to Cauchy, in *Comptes Rendus*, Vol. 19 (1844), pp. 1377–1378, although it may have been known to Gauss earlier (see the footnote on the next page).

PROOF   For any $z_0 \in \mathbf{C}$ we have, by formula (5), $|f'(z_0)| \leq M/R$. Let $R \to \infty$. Thus we conclude that $|f'(z_0)| = 0$ and therefore that $f'(z_0) = 0$, so f is constant. ∎

This is again a quite different property than any that could possibly hold for functions of a real variable. Certainly, there are many nonconstant bounded smooth functions of a real variable; for example, $f(x) = \sin x$.

Fundamental Theorem of Algebra

Next we shall prove a result that appears to be elementary and that the student has, in the past, probably taken for granted. Algebraically, the theorem is quite difficult.* However, there is a simple proof that uses the theorem of Liouville.

2.4.9 FUNDAMENTAL THEOREM OF ALGEBRA   *Let $a_0, a_1, \ldots, a_n$ be complex numbers and suppose that $n \geq 1$ and $a_n \neq 0$. Let $p(z) = a_0 + a_1 z + \cdots + a_n z^n$. Then there exists a point $z_0 \in \mathbf{C}$ such that $p(z_0) = 0$.*

NOTE   By Review Exercise 24 at the end of Chap. 1, the polynomial p can have no more than n roots. It follows by repeated factoring that p will have *exactly* n roots if they are counted according to their multiplicity.

PROOF   Suppose that $p(z_0) \neq 0$ for all $z_0 \in \mathbf{C}$. Then $f(z) = 1/p(z)$ is entire. Now $p(z)$ and hence $f(z)$ is not constant (because $a_n \neq 0$), so it suffices, by Liouville's theorem, to show that $f(z)$ is bounded.

To do so, we first show that $p(z) \to \infty$ as $z \to \infty$, or, equivalently, that $f(z) \to 0$ as $z \to \infty$. In other words, we prove that, given $M > 0$, there is a number $K > 0$ such that $|z| > K$ implies $|p(z)| > M$. From $p(z) = a_0 + a_1 z + \cdots + a_n z^n$ we have $|p(z)| \geq |a_n||z|^n - |a_0| - |a_1||z| - \cdots - |a_{n-1}||z|^{n-1}$. (We set $a_n z^n = p(z) - a_0 - a_1 z - \cdots - a_{n-1} z^{n-1}$ and apply the triangle inequality.) Let $a = |a_0| + |a_1| + \cdots + |a_{n-1}|$. If $|z| > 1$, then

$$|p(z)| \geq |z|^{n-1} \left(|a_n||z| - \frac{|a_0|}{|z|^{n-1}} - \frac{|a_1|}{|z|^{n-2}} - \cdots - \frac{|a_{n-1}|}{1} \right)$$
$$\geq |z|^{n-1}(|a_n||z| - a)$$

Let $K = \max\{1, (M + a)/|a_n|\}$; then, if $|z| > K$, we have $|p(z)| \geq M$.

---

* It was first proved by Karl Friedrich Gauss in his doctoral thesis in 1799. The present proof appears to be essentially due to Gauss as well (*Comm. Soc. Gott.*, Vol. 3 (1816), pp. 59–64).

Thus if $|z| > K$, we have $1/|p(z)| < 1/M$. But on the set of z for which $|z| \le K$, $1/p(z)$ is bounded in absolute value because it is continuous. If this bound for $1/p(z)$ is denoted by L, then on $\mathbf{C}$ we have $1/|p(z)| < \max(1/M, L)$, and so $|f(z)|$ is bounded on $\mathbf{C}$. ∎

Another argument for showing that $f(z) \to 0$ as $z \to \infty$ that is a little simpler but accepts the validity of various limit theorems is as follows:

$$f(z) = \frac{1}{a_n z^n + a_{n-1} z^{n-1} + \cdots + a_0}$$

$$= \frac{1/z^n}{a_n + a_{n-1}(1/z) + a_{n-2}(1/z^2) + \cdots + a_0(1/z^n)}$$

Letting $z \to \infty$, we get

$$\lim_{z \to \infty} f(z) = \frac{0}{a_n + 0 + \cdots + 0} = 0$$

since $a_n \ne 0$.

Morera's Theorem

The following theorem is a partial converse of Cauchy's theorem.

2.4.10 MORERA'S THEOREM *Let f be continuous on a region A, and suppose that $\int_\gamma f = 0$ for every closed curve in A. Then f is analytic on A, and $f = F'$ for some analytic function F on A.*

PROOF The existence of the antiderivative follows from the vanishing of integrals around closed curves and the path independence theorem (2.1.9). The antiderivative F is certainly analytic (its derivative is f). Therefore by Cauchy's integral formula for derivatives, it is infinitely differentiable. In particular, $F'' = f'$ exists. ∎

In applying Morera's theorem, one often wishes only to show that f is analytic on a region. If the region is not simply connected, f might not have an antiderivative on the whole region. But to show differentiability near a point one may restrict attention to a small neighborhood of the point and to special curves if convenient. This idea is illustrated in the following corollary and Worked Examples 2.4.16 and 2.4.17.

2.4.11 COROLLARY *Let f be continuous on a region A and analytic on $A \setminus \{z_0\}$ for a point $z_0 \in A$. Then f is analytic on A.*

PROOF To show analyticity at z_0, we may restrict attention to a small disk $D(z_0; \epsilon) \subset A$. If γ is any closed curve in this disk, then $\int_\gamma f = 0$, by the strengthened Cauchy theorem for a disk (2.3.5). Thus Morera's theorem implies f is analytic on this disk. We already know it is analytic on the rest of A. ∎

Technical Proof of Theorem 2.4.5

2.4.5 DIFFERENTIABILITY OF CAUCHY-TYPE INTEGRALS *Let γ be a curve in $\mathbf{C}$ and let g be a continuous function defined along the curve, on the image $\gamma([a, b])$. Set*

$$G(z) = \frac{1}{2\pi i} \int_\gamma \frac{g(\zeta)}{\zeta - z} \, d\zeta$$

Then G is analytic on $\mathbf{C} \setminus \gamma([a, b])$; in fact G is infinitely differentiable, with the kth derivative given by

$$G^{(k)}(z) = \frac{k!}{2\pi i} \int_\gamma \frac{g(\zeta)}{(\zeta - z)^{k+1}} \, d\zeta \qquad k = 1, 2, 3, \ldots \tag{2}$$

We will actually prove this with a somewhat weaker assumption on g than continuity. All we assume is that the function is bounded and integrable along γ. We call such functions *admissible*.

PROOF First we use several facts from the advanced calculus developed in Sec. 1.4. The image curve γ is a compact set since it is a continuous image of a closed bounded interval. If z_0 is not on γ, then by the distance lemma, it lies at a positive distance δ from it. If we let $\eta = \delta/2$ and U be the η disk around z_0, then $z \in U$ and ζ on γ implies that $|z - \zeta| \geq \eta$ since $|z - \zeta| \geq |\zeta - z_0| - |z_0 - z| \geq 2\eta - \eta$ (see Fig. 2.4.4).

We begin now with the case $k = 1$. We want to show that

$$\lim_{z \to z_0} \left[\frac{G(z) - G(z_0)}{z - z_0} - \frac{1}{2\pi i} \int_\gamma \frac{g(\zeta)}{(\zeta - z_0)^2} \, d\zeta \right] = 0$$

The expression in brackets may be written

$$\frac{G(z) - G(z_0)}{z - z_0} - \frac{1}{2\pi i} \int_\gamma \frac{g(\zeta)}{(\zeta - z_0)^2} \, d\zeta = \frac{(z - z_0)}{2\pi i} \int_\gamma \frac{g(\zeta)}{(\zeta - z_0)^2(\zeta - z)} \, d\zeta$$

which we get by using the identity

$$\frac{1}{z - z_0} \left(\frac{1}{\zeta - z} - \frac{1}{\zeta - z_0} \right) - \frac{1}{(\zeta - z_0)^2} = \frac{z - z_0}{(\zeta - z_0)^2(\zeta - z)}$$

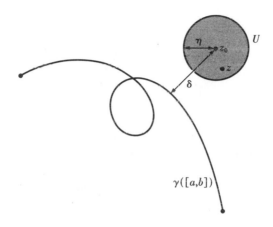

FIGURE 2.4.4 A point z_0 not on a curve γ is a positive distance from γ.

Let the η neighborhood U of z_0 be constructed as previously described and let M be the maximum of g on γ. Then $|(\zeta - z_0)^2(\zeta - z)| \geq \eta^2 \cdot \eta = \eta^3$, and so we have the estimate $|g(\zeta)/[(\zeta - z_0)^2(\zeta - z)]| \leq M\eta^{-3}$ (a fixed constant independent of ζ on γ and $z, z_0 \in U$). Thus

$$\left| \frac{z - z_0}{2\pi i} \int_\gamma \frac{g(\zeta)}{(\zeta - z_0)^2(\zeta - z)} \, d\zeta \right| \leq |z - z_0| \frac{M\eta^{-3}}{2\pi} l(\gamma)$$

This expression approaches 0 as $z \to z_0$, and so the limit is 0, as we wanted.

To prove the general case we proceed by induction on k. Suppose the theorem is known to hold for all admissible functions and all values of k from 1 to $n - 1$. We want to prove it works for $k = n$. We phrase the induction hypothesis this way since we will apply it not only to g, but also to $g(\zeta)/(\zeta - z_0)$, which is also bounded and integrable along γ. We know that G can be differentiated $n - 1$ times on $\mathbf{C} \backslash \gamma$ and that

$$G^{(n-1)}(z) = \frac{(n - 1)!}{2\pi i} \int_\gamma \frac{g(\zeta)}{(\zeta - z)^n} \, d\zeta$$

Let $z_0 \in \mathbf{C} \backslash \gamma([a, b])$. Then, using the identity

$$\frac{1}{(\zeta - z)^n} = \frac{1}{(\zeta - z)^{n-1}(\zeta - z_0)} + \frac{z - z_0}{(\zeta - z)^n(\zeta - z_0)}$$

we obtain

$$G^{(n-1)}(z) - G^{(n-1)}(z_0) = \frac{(n-1)!}{2\pi i} \left[\int_\gamma \frac{g(\zeta)}{(\zeta-z)^{n-1}(\zeta-z_0)} d\zeta - \int_\gamma \frac{g(\zeta)}{(\zeta-z_0)^n} d\zeta \right]$$
$$+ \frac{(n-1)!}{2\pi i} (z-z_0) \int_\gamma \frac{g(\zeta)}{(\zeta-z)^n(\zeta-z_0)} d\zeta \tag{6}$$

We can conclude from this equation that $G^{(n-1)}$ is continuous at z_0, for the following reason. By applying the induction hypothesis to $g(\zeta)/(\zeta-z_0)$, we see that

$$\int_\gamma \frac{g(\zeta)}{(\zeta-z)^{n-1}(\zeta-z_0)} d\zeta$$

is analytic as a function of z on the set $\mathbf{C}\backslash\gamma([a, b])$ and thus is continuous in z. Therefore,

$$\int_\gamma \frac{g(\zeta)}{(\zeta-z)^{n-1}(\zeta-z_0)} d\zeta \to \int_\gamma \frac{g(\zeta)}{(\zeta-z_0)^n} d\zeta$$

as $z \to z_0$. If the distance from z_0 to γ is 2η, if $|g(z)| < M$ on γ, and if $|z-z_0| < \eta$, we have

$$\left| \int_\gamma \frac{g(\zeta)}{(\zeta-z)^n(\zeta-z_0)} d\zeta \right| < \frac{M}{\eta^{n+1}} \cdot l(\gamma)$$

where $l(\gamma)$ is the length of γ. Hence

$$|z-z_0| \left| \int_\gamma \frac{g(\zeta)}{(\zeta-z)^n(\zeta-z_0)} d\zeta \right| \to 0$$

as $z \to z_0$, and therefore $G^{(n-1)}$ is continuous on $\mathbf{C}\backslash\gamma([a, b])$.

From Eq. (6) we obtain

$$\frac{G^{(n-1)}(z) - G^{(n-1)}(z_0)}{z-z_0}$$

$$= \frac{(n-1)!}{2\pi i} \frac{1}{(z-z_0)} \left[\int_\gamma \frac{g(\zeta)}{(\zeta-z)^{n-1}(\zeta-z_0)} d\zeta - \int_\gamma \frac{g(\zeta)}{(\zeta-z_0)^n} d\zeta \right]$$
$$+ \frac{(n-1)!}{2\pi i} \int_\gamma \frac{g(\zeta)}{(\zeta-z)^n(\zeta-z_0)} d\zeta \tag{7}$$

By applying the induction hypothesis to $g(\zeta)/(\zeta-z_0)$, we see that the first term on the right side of Eq. (7) converges to

$$\frac{(n-1)(n-1)!}{2\pi i} \int_\gamma \frac{g(\zeta)}{(\zeta-z_0)^{n+1}} d\zeta$$

as $z \to z_0$. In the paragraph following Eq. (6), it was shown that $G^{(n-1)}$ is continuous on $C\backslash\gamma([a, b])$, and this fact applied to $g(\zeta)/(\zeta - z_0)$, instead of to $g(\zeta)$, implies that

$$\int_\gamma \frac{g(\zeta)}{(\zeta - z)^n(\zeta - z_0)} d\zeta \to \int_\gamma \frac{g(\zeta)}{(\zeta - z_0)^{n+1}} d\zeta$$

as $z \to z_0$. Thus we have shown that as $z \to z_0$,

$$\frac{G^{(n-1)}(z) - G^{(n-1)}(z_0)}{z - z_0}$$

converges to

$$(n - 1) \frac{(n - 1)!}{2\pi i} \int_\gamma \frac{g(\zeta)}{(\zeta - z_0)^{n+1}} d\zeta + \frac{(n - 1)!}{2\pi i} \int_\gamma \frac{g(\zeta)}{(\zeta - z_0)^{n+1}} d\zeta$$

$$= \frac{n!}{2\pi i} \int_\gamma \frac{g(\zeta)}{(\zeta - z_0)^{n+1}} d\zeta$$

This concludes the induction and proves the theorem. ∎

Worked Examples

2.4.12   *Consider the curve γ defined by $\gamma(t) = (\cos t, 3 \sin t)$, $0 \le t \le 4\pi$. Show rigorously that $I(\gamma, 0) = 2$.*

Solution. Suppose that we can show that γ is homotopic in $C\backslash\{0\}$ to a circle $\tilde{\gamma}$ that is centered around the origin and that is traversed twice in the counterclockwise direction (that is, $\tilde{\gamma}(t) = e^{it}$, $0 \le t \le 4\pi$). Then, by Proposition 2.4.2, $I(\gamma, 0) = I(\tilde{\gamma}, 0) = 2$. A suitable homotopy is $H(t, s) = \cos t + i(3 - 2s) \sin t$, since H is continuous, $H(t, 0) = \gamma(t)$ and $H(t, 1) = \tilde{\gamma}(t)$, and H is never zero (see Fig. 2.4.5).

2.4.13   *Evaluate*

$$\int_\gamma \frac{\cos z}{z} dz \quad and \quad \int_\gamma \frac{\sin z}{z^2} dz$$

where γ is the unit circle.

Solution. The circle γ is contractible to a point in the region in which $\cos z$ is analytic, since in fact $\cos z$ is entire. Therefore, we can apply Cauchy's integral formula, observing that $I(\gamma, 0) = 1$, to obtain

$$1 = \cos 0 = \frac{1}{2\pi i} \int_\gamma \frac{\cos z}{z} dz$$

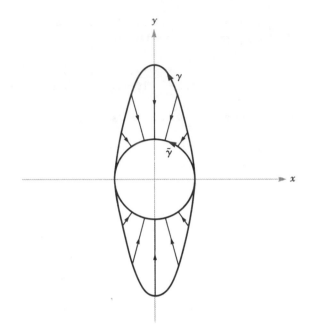

FIGURE 2.4.5   Homotopy of $\gamma(t) = (\cos t, 3 \sin t)$ to $\tilde{\gamma}(t) = (\cos t, \sin t)$.

so that

$$\int_\gamma \frac{\cos z}{z}\, dz = 2\pi i$$

By the Cauchy integral formula for derivatives, we have

$$\sin'(0) = \frac{1}{2\pi i} \int_\gamma \frac{\sin z}{z^2}\, dz$$

that is,

$$\int_\gamma \frac{\sin z}{z^2}\, dz = 2\pi i \cos 0 = 2\pi i$$

2.4.14   *Evaluate*

$$\int_\gamma \frac{e^z + z}{z - 2}\, dz$$

where γ is (i) the unit circle and (ii) a circle with radius 3 centered at 0.

Solution.

(i) γ is contractible to a point in $\mathbf{C}\backslash\{2\}$ (Why?), and $(e^z + z)/(z - 2)$ is analytic on $\mathbf{C}\backslash\{2\}$. Thus by Cauchy's theorem, $\int_\gamma (e^z + z)/(z - 2)\, dz = 0$.

(ii) Here γ is not contractible to a point in $\mathbf{C}\backslash\{2\}$. In fact, γ winds around $2 + 0i$ exactly once, so $I(\gamma, 2) = 1$. Precisely, γ is homotopic in $\mathbf{C}\backslash\{2\}$ to a circle γ centered at 2, and so $I(\gamma, 2) = 1$ by Proposition 2.4.2. Thus by Cauchy's integral formula, which is applicable since γ is contractible to a point and $e^z + z$ is analytic on all of $\mathbf{C}$, we have

$$\int_\gamma \frac{e^z + z}{z - 2}\, dz = 2\pi i(e^2 + 2)$$

2.4.15 *This example, which deals with analytic functions defined by integrals, generalizes Theorem 2.4.5. Let $f(z, w)$ be a continuous function of z, w for z in a region A and w on a curve γ. For each w on γ assume that f is analytic in z. Let*

$$F(z) = \int_\gamma f(z, w)\, dw$$

Then show that F is analytic and

$$F'(z) = \int_\gamma \frac{\partial f}{\partial z}(z, w)\, dw$$

where $\partial f/\partial z$ denotes the derivative of f with respect to z with w held fixed.

Solution. Let $z_0 \in A$. Let γ_0 be a circle in A around z_0 whose interior also lies in A. Then for z inside γ_0,

$$f(z, w) = \frac{1}{2\pi i} \int_\gamma \frac{f(\zeta, w)}{\zeta - z}\, d\zeta$$

by Cauchy's integral formula. Thus

$$F(z) = \frac{1}{2\pi i} \int_\gamma \left[\int_{\gamma_0} \frac{f(\zeta, w)}{\zeta - z}\, d\zeta \right] dw$$

Next we claim that we may invert the order of integration, thus obtaining

$$F(z) = \frac{1}{2\pi i} \int_{\gamma_0} \left[\int_\gamma \frac{f(\zeta, w)}{\zeta - z}\, dw \right] d\zeta = \frac{1}{2\pi i} \int_{\gamma_0} \frac{F(\zeta)}{\zeta - z}\, d\zeta$$

This procedure is justifiable because the integrand is continuous and when written out in terms of real integrals has the form

$$\int_a^b \int_\alpha^\beta h(x, y)\, dx\, dy + i \int_a^b \int_\alpha^\beta k(x, y)\, dx\, dy$$

We know from calculus that this order can be interchanged (Fubini's theorem); see, for instance, J. Marsden, *Elementary Classical Analysis* (San Francisco: W. H. Freeman and Company, 1974), Chap. 9.

Thus

$$F(z) = \frac{1}{2\pi i} \int_{\gamma_0} \frac{F(\zeta)}{\zeta - z} \, d\zeta$$

and so by Theorem 2.4.5, F is analytic inside γ_0 and

$$F'(z) = \frac{1}{2\pi i} \int_{\gamma_0} \frac{F(\zeta)}{(\zeta - z)^2} \, d\zeta = \frac{1}{2\pi i} \int_{\gamma_0} \int_{\gamma} \frac{f(\zeta, w)}{(\zeta - z)^2} \, dw \, d\zeta$$

$$= \frac{1}{2\pi i} \int_{\gamma} \int_{\gamma_0} \frac{f(\zeta, w)}{(\zeta - z)^2} \, d\zeta \, dw = \int_{\gamma} \frac{\partial f}{\partial z}(\zeta, w) \, dw$$

again by Cauchy's integral formula. Since z_0 is arbitrary we obtain the desired result.

Remark. f should be analytic in z but needs to be integrable only in the w variable, as is evident from the preceding proof; we merely need an adequate hypothesis to justify interchanging the order of integration.

2.4.16   *Prove the following assertion: Suppose f is continuous on a region A and that for each z_0 in A there is a disk $D = D(z_0; \rho)$ such that $\int_R f = 0$ for every rectangular path R in D with sides parallel to the axes. Then f is analytic on A (see Fig. 2.4.6).*

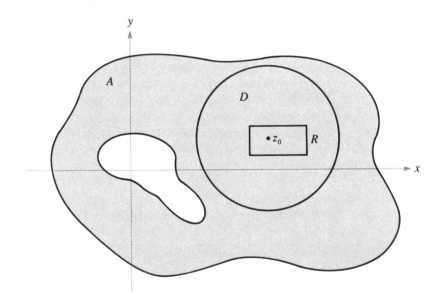

FIGURE 2.4.6   If $\int_R f = 0$, then f is analytic.

Solution. Let z_0 be in A. The vanishing of $\int_R f$ for rectangles in D was the conclusion of the Cauchy-Goursat theorem for a rectangle (2.3.2) and the tool used in the construction of the antiderivative for f in the proof of Cauchy's theorem for a disk. Thus the antiderivative exists on D (not necessarily on all of A at once). Analyticity on D follows as the proof of Morera's theorem, and so f is analytic near z_0. Since z_0 was an arbitrary point in A, f is analytic on A.

2.4.17 *Prove the following: Suppose A is a region which intersects the real axis and that f is a function continuous on A and analytic on $A \backslash \mathbf{R}$. Then f is analytic on A.*

Solution. We know f is analytic everywhere in A except on the real axis, so suppose $z_0 \in \mathbf{R}$. Since A is open there is a disk $D = D(z_0; \rho) \subseteq A$. Let R be a rectangular path in this disk with sides parallel to the axes. If R does not touch or cross the real axis, then $\int_R f = 0$ by Cauchy's theorem. If it does cross, as in Fig. 2.4.7, then $\int_R f = \int_{R_1} f + \int_{R_2} f$, where R_1 and R_2 are rectangles with one edge on the axis. (The edges on the axis are traversed in opposite directions and so cancel out.) Thus it is enough to show that $\int_R f = 0$ when R is a rectangle with one side on the real axis, as in Fig. 2.4.8. Let a and b be the ends of the edge on the axis and note that $b - a < \rho$.

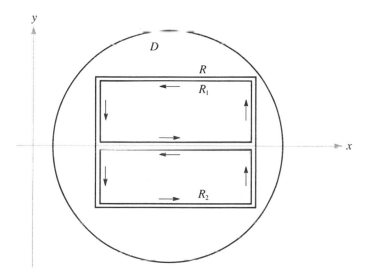

FIGURE 2.4.7   Construction used to show $\int_R f = 0$.

Let $c > 0$. Since f is continuous, it is uniformly continuous on the compact set composed of R and its interior, and so there is a $\delta > 0$ such that $|f(z_1) - f(z_2)| < \epsilon$ whenever $|z_1 - z_2| < \delta$ and z_1 and z_2 are in this set. We may also choose δ to be less than ϵ. Let M be the maximum of $|f(z)|$ on R and its interior, and let S be another rectangle the same as R except that the edge on the axis is moved a distance δ from the axis. Then with the notation of Fig. 2.4.8,

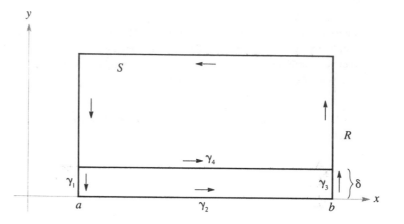

FIGURE 2.4.8   The rectangle pulled away slightly from the real axis.

$$\left|\int_R f - \int_S f\right| = \left|\int_{\gamma_1} f + \int_{\gamma_2} f + \int_{\gamma_3} f - \int_{\gamma_4} f\right|$$

$$\leq \left|\int_{\gamma_1} f\right| + \left|\int_{\gamma_2} f - \int_{\gamma_4} f\right| + \left|\int_{\gamma_3} f\right|$$

$$\leq \delta M + \left|\int_a^b [f(x) - f(x + \delta i)]\, dx\right| + \delta M$$

$$\leq 2\delta M + \int_a^b |f(x) - f(x + \delta i)|\, dx$$

$$\leq 2\delta M + \epsilon(b - a) \leq \epsilon(2M + \rho)$$

Since this holds for every $\epsilon > 0$, we must have $\int_R f - \int_S f = 0$. But $\int_S f = 0$ by Cauchy's theorem since S does not cross the axis and lies entirely within a region in which f is known to be analytic. Thus $\int_R f = 0$. We have shown that the conditions of Worked Example 2.4.16 apply, and so f is analytic on A.

Exercises

1. Evaluate the following integrals:

(a) $\displaystyle\int_\gamma \frac{z^2}{z - 1}\, dz$, where γ is a circle of radius 2, centered at 0

(b) $\displaystyle\int_\gamma \frac{e^z}{z^2}\, dz$, where γ is the unit circle

2. Evaluate the following integrals:

(a) $\displaystyle\int_\gamma \frac{z^2 - 1}{z^2 + 1}\, dz$, where γ is a circle of radius 2, centered at 0

(b) $\displaystyle\int_\gamma \frac{\sin e^z}{z}\, dz$, where γ is the unit circle

3. Let f be entire. If $|f(z)| \le M|z|^n$ for large $|z|$, for a constant M, and for an integer n, show that f is a polynomial of degree $\le n$.

4. Let f be analytic "inside and on" a simple closed curve γ. Suppose that $f = 0$ on γ. Show that $f = 0$ inside γ.

5. Evaluate the following integrals:

(a) $\displaystyle\int_\gamma \frac{dz}{z^3}$, where γ is the square with vertices $-1 - i,\ 1 - i,\ 1 + i,\ -1 + i$

(b) $\displaystyle\int_\gamma \frac{\sin z}{z^4}\, dz$, where γ is the unit circle

6. Let f be analytic on a region A and let γ be a closed curve in A. For any $z_0 \in A$ not on γ, show that

$$\int_\gamma \frac{f'(\zeta)}{\zeta - z_0}\, d\zeta = \int_\gamma \frac{f(\zeta)}{(\zeta - z_0)^2}\, d\zeta$$

Can you think of a way to generalize this result?

7. Suppose that we know $f(z)$ is analytic on $|z| < 1$ and that $|f(z)| \le 1$. What estimate can be made about $|f'(0)|$?

8. Suppose that f is entire and that $\displaystyle\lim_{z \to \infty} f(z)/z = 0$. Prove that f is constant.

9. Prove that if γ is a circle, $\gamma(t) = z_0 + re^{it}$, $0 \le t \le 2\pi$, then for every z inside γ (that is, $|z - z_0| < r$), $I(\gamma, z) = 1$.

10. Use Worked Example 2.4.15 to show that

$$F(z) = \int_0^1 e^{-z^2 x^2}\, dx$$

is analytic in z. What is $F'(z)$?

11. Show that if F is analytic on A, then so is f where

$$f(z) = \frac{F(z) - F(z_0)}{z - z_0}$$

if $z \ne z_0$ and $f(z_0) = F'(z_0)$ where z_0 is some point in A.

12. Prove that if the image of γ lies in a simply connected region A and if $z_0 \notin A$, then $I(\gamma, z_0) = 0$.

13. Use Worked Example 2.1.12 (where appropriate) and Cauchy's integral formula to evaluate the following integrals; γ is the circle $|z| = 2$ in each case.

(a) $\displaystyle\int_\gamma \frac{dz}{z^2 - 1}$

(b) $\displaystyle\int_\gamma \frac{dz}{z^2 + z + 1}$

(c) $\displaystyle\int_\gamma \frac{dz}{z^2 - 8}$

(d) $\displaystyle\int_\gamma \frac{dz}{z^2 + 2z - 3}$

14. Prove that $\int_0^\pi e^{\cos\theta} \cos(\sin\theta)\, d\theta = \pi$ by considering $\int_\gamma (e^z/z)\, dz$, where γ is the unit circle.

15. Evaluate

$$\int_C \frac{|z|e^z}{z^2}\, dz$$

where C is the circumference of the circle of radius 2 around the origin.

16. Consider the function $f(z) = 1/z^2$.
(a) It satisfies $\int_\gamma f(z)\, dz = 0$ for all closed contours γ (not passing through the origin) but is not analytic at $z = 0$. Does this statement contradict Morera's theorem?
(b) It is bounded as $z \to \infty$ but is not a constant. Does this statement contradict Liouville's theorem?

17. Let $f(z)$ be entire and let $|f(z)| \geq 1$ on the whole complex plane. Prove that f is constant.

18. Does $\int_{|z|=1} \dfrac{e^z}{z^2}\, dz = 0$? Does $\int_{|z|=1} \dfrac{\cos z}{z^2}\, dz = 0$?

19. Evaluate

(a) $\displaystyle\int_{|z-1|=2} \frac{dz}{z^2 - 2i}$ $\pi i/2 + i(1/2)$

(b) $\displaystyle\int_{|z|=2} \frac{dz}{z^2(z^2 + 16)}$ $\circ$

20. Prove that for closed curves γ_1, γ_2,

$$I(-\gamma_1, z_0) = -I(\gamma_1, z_0)$$

and

$$I(\gamma_1 + \gamma_2, z_0) = I(\gamma_1, z_0) + I(\gamma_2, z_0)$$

Interpret these results geometrically.

21. Let f be analytic inside and on the circle $\gamma: |z - z_0| = R$. Prove that

$$\frac{f(z_1) - f(z_2)}{z_1 - z_2} - f'(z_0) = \frac{1}{2\pi i} \int_\gamma \left[\frac{1}{(z - z_1)(z - z_2)} - \frac{1}{(z - z_0)^2} \right] f(z) \, dz$$

for z_1, z_2 inside γ.

2.5 MAXIMUM MODULUS THEOREM AND HARMONIC FUNCTIONS

One of the most striking and powerful consequences of the Cauchy integral formula is the maximum modulus theorem, also called the maximum modulus principle. It states that if f is a nonconstant analytic function on a region A, then $|f|$ cannot have a local maximum anywhere inside A — it can attain a maximum only on the boundary of A. This theorem and the Cauchy integral formula will be used to develop some of the important properties of harmonic functions.

Maximum Modulus Theorem

The central fact of the maximum modulus principle can perhaps best be stated as follows: if an analytic function has a local maximum (of its absolute value) at a point, then it must be constant near that point. A preliminary version of the theorem is thus:

2.5.1 MAXIMUM MODULUS PRINCIPLE — LOCAL VERSION *Let f be analytic on a region A and suppose that $|f|$ has a relative maximum at $z_0 \in A$. (That is, $|f(z)| \leq |f(z_0)|$ for all z in some neighborhood of z_0.) Then f is constant in some neighborhood of z_0.*

The proof rests on a striking consequence of the Cauchy integral formula: the value of an analytic function at the center of a circle is the average of its values around the circle. All this will be made precise shortly, but the local version of the principle follows essentially because an average cannot be greater than or equal to all the values unless they are all equal.

2.5.2 MEAN VALUE PROPERTY *Let f be analytic inside and on a circle of radius r and center z_0 (that is, analytic on a region containing the circle and its interior). Then*

$$f(z_0) = \frac{1}{2\pi} \int_0^{2\pi} f(z_0 + re^{i\theta}) \, d\theta \tag{1}$$

PROOF   By Cauchy's integral formula,

$$f(z_0) = \frac{1}{2\pi i} \int_\gamma \frac{f(z)}{z - z_0} \, dz$$

where $\gamma(\theta) = z_0 + re^{i\theta}$, $0 \leq \theta \leq 2\pi$. But by definition of the integral,

$$\frac{1}{2\pi i} \int_\gamma \frac{f(z)}{z - z_0} \, dz = \frac{1}{2\pi i} \int_0^{2\pi} \frac{f(z_0 + re^{i\theta})}{re^{i\theta}} \, rie^{i\theta} \, d\theta = \frac{1}{2\pi} \int_0^{2\pi} f(z_0 + re^{i\theta}) \, d\theta \qquad \blacksquare$$

It is worth noting that as long as we are integrating all the way around the circle, it does not matter through what range of 2π the angle goes. A simple change of variable shows that, for example, $\int_0^{2\pi} f(z_0 + re^{i\theta}) \, d\theta = \int_{-\pi}^{\pi} f(z_0 + re^{i\theta}) \, d\theta$.

The mean value property will now be used to establish 2.5.1. The idea is that if $f(z_0)$ is at least as great as all the other values of f near z_0 and also equal to the average of those values around small circles centered at z_0, then $|f(z)|$ must be constant near z_0. Once we know that $|f|$ is constant, it follows from the Cauchy-Riemann equations that f is itself constant. (See Worked Example 1.5.19.)

PROOF OF THEOREM 2.5.1.   Suppose that f is analytic and has a relative maximum at z_0, so that $|f(z)| \leq |f(z_0)|$ on some disk $D_0 = D(z_0; r_0)$. We want to show that $|f(z)| = |f(z_0)|$ on D_0, so suppose instead that there is a point z_1 in D_0 where strict inequality holds: $|f(z_1)| < |f(z_0)|$. Let $z_1 = z_0 + re^{ia}$ with $r < r_0$. Since f is continuous, there are positive numbers ϵ and δ such that $|f(z_0 + re^{i\theta})| < |f(z_0)| - \delta$ whenever $|\theta - a| < \epsilon$. Equivalently, $|f(z_0 + re^{i(a+\phi)})| < |f(z_0)| - \delta$ whenever $|\phi| < \epsilon$. We now obtain a contradiction by using the mean value property and considering separately that part of the integral over the circle where we know the function is smaller (Fig. 2.5.1); namely,

$$|f(z_0)| = \left| \frac{1}{2\pi} \int_{-\pi}^{\pi} f(z_0 + re^{i(a+\phi)}) \, d\phi \right|$$

$$= \left| \frac{1}{2\pi} \int_{-\pi}^{-\epsilon} f(z_0 + re^{i(a+\phi)}) \, d\phi + \frac{1}{2\pi} \int_{-\epsilon}^{\epsilon} f(z_0 + re^{i(a+\phi)}) \, d\phi \right.$$

$$\left. + \frac{1}{2\pi} \int_{\epsilon}^{\pi} f(z_0 + re^{i(a+\phi)}) \, d\phi \right|$$

$$\leq \frac{1}{2\pi} \int_{-\pi}^{-\epsilon} \left| f(z_0 + re^{i(a+\phi)}) \right| \, d\phi + \frac{1}{2\pi} \int_{-\epsilon}^{\epsilon} \left| f(z_0 + re^{i(a+\phi)}) \right| \, d\phi$$

$$+ \frac{1}{2\pi} \int_{\epsilon}^{\pi} \left| f(z_0 + re^{i(a+\phi)}) \right| \, d\phi$$

In the first and third integrals the integrand is no greater than $|f(z_0)|$ and the interval length is $\pi - \epsilon$. Thus each of these integrals is no more than $|f(z_0)|(\pi - \epsilon)$. In the middle

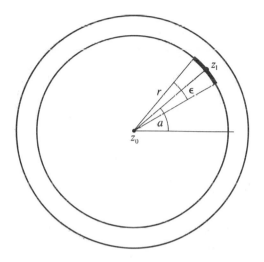

FIGURE 2.5.1 Construction for the proof of the maximum modulus principle—local version. $|\phi| < \epsilon$ gives a part of the circle where $|f|$ is known to be smaller.

integral, the interval length is 2ϵ and the integrand is less than $|f(z_0)| - \delta$. Hence this integral is no more than $(|f(z_0)| - \delta)2\epsilon$. Putting these together gives

$$|f(z_0)| < \frac{1}{2\pi} [|f(z_0)|(\pi - \epsilon) + (|f(z_0)| - \delta)2\epsilon + |f(z_0)|(\pi - \epsilon)]$$

or

$$|f(z_0)| < |f(z_0)| - \frac{\epsilon\delta}{\pi}$$

This obvious impossibility shows that there can be no such point z in D_0 with $|f(z)| < |f(z_0)|$. The only remaining possibility is that $|f(z)| = |f(z_0)|$ for all z in D_0.

Thus $|f|$ is constant on D_0. Use of the Cauchy-Riemann equations as in Worked Example 1.5.19 shows that the function f itself must be constant. This is exactly what we wanted. ∎

The local version of the maximum modulus principle says that an analytic function cannot have a local maximum point unless it is constant near that point. We will see in Chap. 6 that more is true. A function analytic on an open *connected* set cannot have a local maximum anywhere in that set unless it is constant on the whole set. We turn our attention here to a somewhat different global version of the principle. We investigate absolute maxima, that is, the largest value which $|f(z)|$ takes anywhere in the set. We shall show that this can be found only on the edge or boundary of the set. In Sec. 1.4, we saw that a real-valued function continuous on a closed bounded set actually attains a finite maximum but that it might not if the set fails to be closed or bounded.

Closure and Boundary

The intuition in Sec. 1.4 was that a set is closed if it contains all its boundary points and open if it contains none of them. Thus if we start with a set A and add to it any of its boundary points which happen to be missing we should obtain a closed set containing A. This is true, but there are some technical problems. One is that we really have no definition yet for "boundary."

2.5.3 DEFINITION   *Let A be a set. The **closure** of A, denoted by $\overline{A}$ or by $\mathrm{cl}\,(A)$, consists of A together with the limit points of all convergent sequences of points of A.*

This produces the desired result, a smallest closed set containing A.

2.5.4 PROPOSITION   *If $A \subset \mathbf{C}$, then*

(i) $A \subset \mathrm{cl}\,(A)$.
(ii) *A is closed if and only if $A = \mathrm{cl}\,(A)$.*
(iii) *If $A \subset C$ and C is closed, then $\mathrm{cl}\,(A) \subset C$.*
(iv) $\mathrm{cl}\,(A)$ *is closed.*

PROOF   The first assertion is immediate from the definition. The basic tool for the remainder is Proposition 1.4.8, which states that a set is closed if and only if it contains the limits of all convergent sequences of its points. If we let $\mathrm{limit}\,(A) = \{w \,|\, \text{there is a}$ sequence of points in A convergent to $w\}$, then $A \subset \mathrm{limit}\,(A)$, since constant sequences certainly converge. The closure was defined by $\mathrm{cl}\,(A) = A \cup \mathrm{limit}\,(A)$, and so we actually have $\mathrm{cl}\,(A) = \mathrm{limit}\,(A)$. But Proposition 1.4.8 says exactly that A is closed if and only if $\mathrm{limit}\,(A) \subset A$, and so (ii) is established. It also shows that if C is closed and $A \subset C$, then $\mathrm{limit}\,(A) \subset C$, and so we have (iii). The only remaining gap is to show that $\mathrm{cl}\,(A)$ is actually closed. To do this we need only show that $\mathrm{cl}\,(A) = \mathrm{cl}\,(\mathrm{cl}\,(A))$ that is limit $(A) = \mathrm{limit}\,(\mathrm{limit}\,(A))$. Since $\mathrm{limit}\,(A) \subset \mathrm{limit}\,(\mathrm{limit}\,(A))$ automatically, it remains to show that $\mathrm{limit}\,(\mathrm{limit}\,(A)) \subset \mathrm{limit}\,(A)$. Suppose $z_1, z_2, z_3, \ldots$ is a sequence of points in $\mathrm{limit}\,(A)$ such that $\lim\limits_{n \to \infty} z_n = w$. We want to show that w is in $\mathrm{limit}\,(A)$. Each z_n is in $\mathrm{limit}\,(A)$, so there are points w_n in A with $|w_n - z_n| < 1/n$. This forces $\lim\limits_{n \to \infty} w_n = w$, and so $w \in \mathrm{limit}$

(A), as desired. ∎

The boundary of a set A is the set of points on the "edge" of A. If w is in the boundary, we should be able to approach it through A and through the complement of A. This leads to the following definition.

2.5.5 DEFINITION *If A is a set, the **boundary** of A is defined by*

$$\text{bd}\,(A) = \text{cl}\,(A) \cap \text{cl}\,(\mathbf{C}\backslash A)$$

It is not hard to see that $\text{cl}\,(A) = A \cup \text{bd}\,(A)$. See Worked Example 2.5.17.

A Global Maximum Modulus Principle

Now we are ready for the promised global version of the maximum modulus principle.

2.5.6 MAXIMUM MODULUS PRINCIPLE *Let A be an open, connected, and bounded set and $f: \text{cl}\,(A) \to \mathbf{C}$ be a function analytic on A and continuous on $\text{cl}\,(A)$. Let M be the maximum of $|f(z)|$ on $\text{bd}\,(A)$; that is, M is the least upper bound or supremum of $|f(z)|$ as z ranges through $\text{bd}\,(A)$. In symbols, $M = \sup\{|f(z)| \text{ such that } z \in \text{bd}\,(A)\}$. Then*

(i) $|f(z)| \leq M$ *for all $z \in A$*

and

(ii) *If $|f(z)| = M$ for some $z \in A$, then f is constant on A.*

This theorem states that the maximum of f occurs on the boundary of A and that if that maximum is attained on A itself then f must be constant. This is a very striking result and is certainly a very special property of analytic functions. The values of $|f|$ inside a region A must be smaller than the largest value of $|f|$ on the boundary of A. One must exercise some care. For example, the maximum modulus principle in this form need *not* be true if A is not bounded. In such a case the function need not be bounded on A even if it is on $\text{bd}\,(A)$. (See Exercise 3.) In applications of this theorem, A will often be the inside of a simple closed curve γ, and so $\text{cl}\,(A)$ will be $A \cup \gamma$, and $\text{bd}\,(A)$ will be γ.

It is reasonably clear that if A is bounded, so is its closure. If $|z| \leq B$ for all z in A and z_1, $z_2, z_3, \ldots$ is a sequence in A converging to w, then $|z_n|$ converges to $|w|$, and so $|w| \leq B$. Thus $\text{cl}\,(A) = \text{limit}\,(A)$ is also bounded by B.

From the extreme value theorem (1.4.20), we know that a continuous real-valued function on a closed bounded set attains a maximum on that set. It follows that if $M' = \sup\{|f(z)| \text{ such that } z \in \text{cl}\,(A)\}$, then $M' = |f(a)|$ for some $a \in \text{cl}\,(A)$.

PROOF OF THE MAXIMUM MODULUS PRINCIPLE   The first step is to show that $M = M'$, where $M' = \sup \{|f(z)|$ such that $z \in \text{cl } (A)\}$ and $M = \sup \{|f(z)|$ such that $z \in \text{bd } (A)\}$.

In this first step, there are two cases. First, let us suppose that there is no $a \in A$ such that $|f(a)| = M'$. Hence there must be an $a \in \text{bd } (A)$ such that $|f(a)| = M'$ (because we know that there is some $a \in \text{cl } (A) = A \cup \text{bd } (A)$ such that $|f(a)| = M'$). But then we must have $|f(a)| = M' = M$. Second, let us suppose that there is an $a \in A$ such that $|f(a)| = M'$. By the local version (2.5.1), the set $B = \{z \in A \mid f(z) = f(a)\}$ is open, since every $z \in B$ has a neighborhood of which f is the constant value $f(z) = f(a)$. On the other hand, B is the inverse image of the closed set $\{f(a)\}$ by the continuous map f restricted to A; therefore, B is a closed set in A. Thus B is both open and closed in A and, of course, nonempty, so by the basic facts about connectedness we have $B = A$. Hence f is the constant value $f(a)$ on A and so, by the continuity of f on cl (A), the value of f equals the constant value $f(a)$ on cl (A) as well (Why?). Thus in this case we also have $M = M'$.

Since $M = M'$, we obviously have $|f(z)| \le M$ for all $z \in A$ and thus (i) is proved. If $|f(z)| = M = M'$ for some $z \in A$, corresponding to the second case, then, as shown there, f is constant on A, and thus (ii) is proved. ∎

Schwarz Lemma

The next theorem is an example of an application of the maximum modulus theorem. This result is not one of the most basic results of the theory, but it further indicates the type of severe restrictions that analyticity imposes.

2.5.7 SCHWARZ LEMMA   *Let f be analytic on the open unit disk $A = \{z \in \mathbf{C}$ such that $|z| < 1\}$ and suppose that $|f(z)| \le 1$ for $z \in A$ and $f(0) = 0$. Then $|f(z)| \le |z|$ for all $z \in A$ and $|f'(0)| \le 1$. If $|f(z_0)| = |z_0|$ for some $z_0 \in A$, $z_0 \ne 0$, then $f(z) = cz$ for all $z \in A$ for some constant c, $|c| = 1$.*

PROOF   Let

$$g(z) = \begin{cases} \dfrac{f(z)}{z} & \text{if } z \ne 0 \\[2mm] f'(0) & \text{if } z = 0 \end{cases}$$

The function g is analytic on A because it is continuous on A and analytic on $A \backslash \{0\}$ (see Corollary 2.4.11 to Morera's theorem. Let $A_r = \{z$ such that $|z| \le r\}$ for $0 < r < 1$ (see Fig. 2.5.2). Then g is analytic on A_r, and on $|z| = r$, $|g(z)| = |f(z)/z| \le 1/r$. By the maximum modulus principle $|g(z)| \le 1/r$ on all of A_r; that is, $|f(z)| \le |z|/r$ on A_r. But by holding $z \in A$ fixed, we can let $r \to 1$ to obtain $|f(z)| \le |z|$. Clearly, $|g(0)| \le 1$; that is, $|f'(0)| \le 1$.

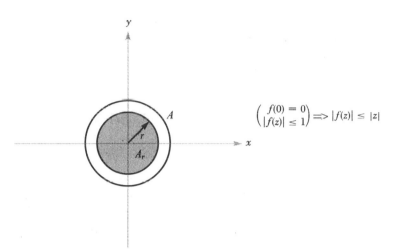

FIGURE 2.5.2 Schwarz lemma.

If $|f(z_0)| = |z_0|$, $z_0 \neq 0$, then $|g(z_0)| = 1$ is maximized in A_r, where $|z_0| < r < 1$, and so g is constant on A_r. The constant is independent of r (Why?), and the theorem is proved. ∎

The Schwarz lemma is a tool for many elegant and occasionally useful geometric results of complex analysis. A generalization that is useful for obtaining accurate estimates of bounds for functions is known as the *Lindelöf principle*, which is as follows: *Suppose that f and g are analytic on $|z| < 1$, that g maps $|z| < 1$ one to one onto a set G, that $f(0) = g(0)$, and that the range of f is contained in G. Then $|f'(0)| \le |g'(0)|$ and the image of $|z| < r$, for $r < 1$, under f is contained in its image under g.* This principle is made particularly useful by the convenient availability of linear fractional transformations for the role of g; these have the form $g(z) = (az + b)/(cz + d)$. As will be proved in Chap. 5, they take circles into circles, and so the g image of the disk $|z| < r$ is usually easy to find (see Exercise 4 for further details).

For a useful survey of some of the more geometric results and a bibliography, see T. H. MacGregor, "Geometric Problems in Complex Analysis," *American Mathematical Monthly* (May 1972), p. 447.

Harmonic Functions and Harmonic Conjugates

If f is analytic on A and $f = u + iv$, we know that u and v are infinitely differentiable and are harmonic (by Theorem 2.4.6 and Proposition 1.5.11). Let us now show that the converse is also true.

2.5.8 PROPOSITION *Let A be a region in* **C** *and let u be a twice continuously differentiable harmonic function on A. Then u is C^∞, and in a neighborhood of each point $z_0 \in A$, u is the real part of some analytic function. If A is simply connected, there is an analytic function f on A such that $u = \operatorname{Re} f$.*

Thus, a harmonic function is always the real part of an analytic function f (or the imaginary part of the analytic function if) at least locally, and on all of the domain of that function if the domain is simply connected.

PROOF   We prove the last statement of the theorem first. Let us consider the function $g = (\partial u/\partial x) - i(\partial u/\partial y)$. We claim that g is analytic. Setting $g = U + iV$ where $U = \partial u/\partial x$ and $V = -\partial u/\partial y$, we must check that U and V have continuous first partials and that they satisfy the Cauchy-Riemann equations. Indeed, the functions $\partial U/\partial x = \partial^2 u/\partial x^2$ and $\partial V/\partial y = -\partial^2 u/\partial y^2$ are continuous by assumption and are equal since $\nabla^2 u = 0$. Also, by the equality of mixed partials,

$$\frac{\partial U}{\partial y} = \frac{\partial^2 u}{\partial y\, \partial x} = \frac{\partial^2 u}{\partial x\, \partial y} = -\frac{\partial V}{\partial x}$$

Thus, we conclude that g is analytic. Furthermore, if A is simply connected, there is an analytic function f on A such that $f' = g$ (by the Antiderivative Theorem, 2.2.5 or 2.3.16). Let $f = \tilde{u} + i\tilde{v}$. Then $f' = (\partial \tilde{u}/\partial x) - i(\partial \tilde{u}/\partial y)$, and thus $\partial \tilde{u}/\partial x = \partial u/\partial x$ and $\partial \tilde{u}/\partial y = \partial u/\partial y$. Thus $\tilde{u}$ differs from u by a constant. Adjusting f by subtracting this constant, we get $u = \operatorname{Re} f$.

Now we prove the first statement. If D is a disk around z_0 in A, it is simply connected. Therefore, as a result of what we have just proven, we can write $u = \operatorname{Re} f$ for some analytic f on D. Thus since f is C^∞, u is also C^∞ on a neighborhood of each point in A, and so is C^∞ on A. ∎

Recall that when there is an analytic function f such that u and v are related by $f = u + iv$, we say that u and v are *harmonic conjugates*. Since if is analytic, $-v$ and u are also harmonic conjugates. Be careful! The order matters; if v is a harmonic conjugate of u, then u is probably not a harmonic conjugate of v. Instead, $-u$ is! The preceding proposition says that *on any simply connected region A, any harmonic function has a harmonic conjugate* $v = \operatorname{Im} f$. Since the Cauchy-Reimann equations ($\partial u/\partial x = \partial v/\partial y$ and $\partial u/\partial y = -\partial v/\partial x$) must hold, v is uniquely determined up to the addition of a constant. These equations may be used as a practical method for finding v when u is given (see Worked Example 1.5.20). Another way of obtaining the harmonic conjugate of u on a disk, by defining it directly in terms of an integral, was indicated in Exercise 32 of Sec. 1.5.

Mean Value Property and Maximum Principle for Harmonic Functions

One reason why Proposition 2.5.8 is important is that it enables us to deduce properties of harmonic functions from corresponding properties of analytic functions. This is done in the next theorem.

2.5.9 MEAN VALUE PROPERTY FOR HARMONIC FUNCTIONS *Let u be harmonic on a region containing a circle of radius r around $z_0 = x_0 + iy_0$ and its interior. Then*

$$u(x_0, y_0) = \frac{1}{2\pi} \int_0^{2\pi} u(z_0 + re^{i\theta}) \, d\theta \tag{3}$$

PROOF By Proposition 2.5.8, there is an analytic function f defined on a region containing this circle and its interior such that $u = \operatorname{Re} f$. This containing region may be chosen to be a slightly larger disk. The existence of a slightly larger circle in A is intuitively clear; the precise proof is given in Worked Example 1.4.28. By the mean value property for f,

$$f(z_0) = \frac{1}{2\pi} \int_0^{2\pi} f(z_0 + re^{i\theta}) \, d\theta$$

Taking the real part of both sides of the equation gives the desired result. ∎

From this result we can deduce, in a way similar to the way we deduced Theorem 2.5.1, the following fact.

2.5.10 MAXIMUM PRINCIPLE FOR HARMONIC FUNCTIONS — LOCAL VERSION *Let u be harmonic on a region A. Suppose that u has a relative maximum at $z_0 \in A$ (that is, $u(z) \le u(z_0)$ for z near z_0). Then u is constant in a neighborhood of z_0.*

In this theorem "maximum" can be replaced by "minimum" (see Exercise 6).
Instead of actually going through a proof for $u(z)$ similar to the proof of Theorem 2.5.1, we can use that result to give a quick proof.

PROOF On a disk around z_0, $u = \operatorname{Re} f$ for some analytic f. Then $e^{f(z)}$ is analytic and $|e^{f(z)}| = e^{u(z)}$. Thus since e^x is strictly increasing in x for all real x, the maxima of u are the same as those of $|e^f|$. By Theorem 2.5.1, e^f is constant in a neighborhood of z_0; therefore, e^u and hence u are also (again because e^x is strictly increasing for x real). ∎

From this result we deduce, exactly as the maximum modulus principle was deduced from its local version, the following more useful version.

2.5.11 MAXIMUM PRINCIPLE FOR HARMONIC FUNCTIONS *Let $A \subset \mathbb{C}$ be open, connected, and bounded. Let u: cl $(A) \rightarrow \mathbb{R}$ be continuous and harmonic on A. Let M be the maximum of u on* bd (A). *Then*   analytic.

(i) *$u(x, y) \leq M$ for all $(x, y) \in A$.*
(ii) *If $u(x, y) = M$ for some $(x, y) \in A$, then u is constant on A.*

There is a corresponding result for the minimum. Let m denote the minimum of u on bd (A). Then

(i) $u(x, y) \geq m$ for $(x, y) \in A$.
(ii) If $u(x, y) = m$ for some $(x, y) \in A$, then u is constant.

The minimum principle for harmonic functions may be deduced from the maximum principle for harmonic functions by applying the latter to $-u$.

Dirichlet Problem for the Disk and Poisson's Formula

There is a very important problem in mathematics and physics called the *Dirichlet problem*. It is this: Let A be an open bounded region and let u_0 be a given continuous function on bd (A). Find a real-valued function u on cl (A) that is continuous on cl (A) and harmonic on A and that equals u_0 on bd (A).

There do exist theorems stating that if the boundary bd (A) is "sufficiently smooth," then there always is a solution u. These theorems are quite difficult. However, we can easily show that the solution is always unique.

2.5.12 UNIQUENESS FOR THE DIRICHLET PROBLEM *The solution to the Dirichlet problem is unique (assuming that there is a solution).*

PROOF Let u and $\tilde{u}$ be two solutions. Let $\phi = u - \tilde{u}$. Then ϕ is harmonic and $\phi = 0$ on bd (A). We must show that $\phi = 0$.

By the maximum principle for harmonic functions, $\phi(x, y) \leq 0$ inside A. Similarly, from the corresponding minimum principle, $\phi(x, y) \geq 0$ on A. Thus $\phi = 0$. ∎

We want to find the solution to the Dirichlet problem for the case where the region is an open disk. To do so we derive a formula that explicitly expresses the values of the solution in terms of its values on the boundary of the disk.

2.5.13 POISSON'S FORMULA *If u is defined and continuous on the closed disk {z such that |z| ≤ r} and is harmonic on the open disk $D(0, r) = \{z$ such that $|z| < r\}$, then for $\rho < r$,*

$$u(\rho e^{i\phi}) = \frac{r^2 - \rho^2}{2\pi} \int_0^{2\pi} \frac{u(re^{i\theta})}{r^2 - 2r\rho \cos(\theta - \phi) + \rho^2} \, d\theta \tag{4}$$

or

$$u(z) = \frac{1}{2\pi} \int_0^{2\pi} u(re^{i\theta}) \frac{r^2 - |z|^2}{|re^{i\theta} - z|^2} \, d\theta \tag{4'}$$

NOTES

(i) The technical parts of the following proof require an acquaintance with the idea of *uniform convergence*. The student who has not studied uniform convergence in advanced calculus may wish to reread this proof after studying Sec. 3.1, where the relevant ideas are discussed.

(ii) If we set $z = 0$ in Eq. (4') we recover the mean value property of harmonic functions.

PROOF Since u is harmonic on $D(0; r)$ and $D(0; r)$ is simply connected, there is an analytic function f defined on $D(0, r)$ such that $u = \operatorname{Re} f$.

Let $0 < s < r$ and let γ_s be the circle $|z| = s$. Then, by Cauchy's integral formula, we have

$$f(z) = \frac{1}{2\pi i} \int_{\gamma_s} \frac{f(\zeta)}{\zeta - z} \, d\zeta$$

for all z such that $|z| < s$. We can manipulate this expression into a form suitable for taking real parts.

Let $\tilde{z} = s^2/\bar{z}$, which is called the *reflection* of z in the circle $|\zeta| = s$. Reflection is pictured geometrically in Fig. 2.5.3.

Thus if z lies inside the circle, then $\tilde{z}$ lies outside the circle, and therefore

$$\frac{1}{2\pi i} \int_{\gamma_s} \frac{f(\zeta)}{\zeta - \tilde{z}} \, d\zeta = 0$$

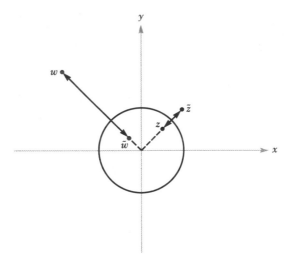

FIGURE 2.5.3   Reflection of a complex number in a circle.

for $|z| < s$. We may subtract this integral from

$$f(z) = \frac{1}{2\pi i} \int_{\gamma_s} \frac{f(\zeta)}{\zeta - z} \, d\zeta$$

to obtain

$$f(z) = \frac{1}{2\pi i} \int_{\gamma_s} f(\zeta) \left(\frac{1}{\zeta - z} - \frac{1}{\zeta - \tilde{z}} \right) d\zeta$$

Observing that $|\zeta| = s$, we can simplify as follows:

$$\frac{1}{\zeta - z} - \frac{1}{\zeta - \tilde{z}} = \frac{1}{\zeta - z} - \frac{1}{\zeta - |\zeta|^2/\bar{z}} = \frac{1}{\zeta - z} - \frac{\bar{z}}{\zeta(\bar{z} - \bar{\zeta})}$$

$$= \frac{-\zeta\bar{z} + |\zeta|^2 + \zeta\bar{z} - |z|^2}{\zeta|\zeta - z|^2} = \frac{|\zeta|^2 - |z|^2}{\zeta|\zeta - z|^2}$$

Hence we have

$$f(z) = \frac{1}{2\pi i} \int_{\gamma_s} \frac{f(\zeta)(|\zeta|^2 - |z|^2)}{\zeta|\zeta - z|^2} \, d\zeta$$

that is,

$$f(\rho e^{i\phi}) = \frac{1}{2\pi} \int_0^{2\pi} \frac{f(se^{i\theta})(s^2 - \rho^2)}{|se^{i\theta} - \rho e^{i\phi}|^2} \, d\theta$$

where $\rho < s$. Noting that $|se^{i\theta} - \rho e^{i\phi}|^2 = s^2 + \rho^2 - 2s\rho \cos(\theta - \phi)$ and taking the real parts on both sides of the equations, we obtain

$$u(\rho e^{i\phi}) = \frac{1}{2\pi} \int_0^{2\pi} \frac{u(se^{i\theta})(s^2 - \rho^2) \, d\theta}{s^2 + \rho^2 - 2s\rho \cos(\theta - \phi)}$$

Keeping ρ and ϕ fixed, we observe that this formula is true for any s such that $\rho < s < r$. Since u is continuous on the closure of $D(0; r)$ and since the function $s^2 + \rho^2 - 2s\rho \cos(\theta - \phi)$ is never zero whenever $s > \rho$, we conclude that for $s > \rho$, $[u(se^{i\theta})(s^2 - \rho^2)]/[s^2 + \rho^2 - 2s\rho \cos(\theta - \phi)]$ is a continuous function of s and θ and hence (with ρ, ϕ fixed) is uniformly continuous on the compact set, $0 \leq \theta \leq 2\pi$, $(r + \rho)/2 \leq s \leq r$. Consequently, as $s \to r$,

$$\frac{u(se^{i\theta})(s^2 - \rho^2)}{s^2 + \rho^2 - 2s\rho \cos(\theta - \phi)} \to \frac{u(re^{i\theta})(r^2 - \rho^2)}{r^2 + \rho^2 - 2r\rho \cos(\theta - \phi)}$$

uniformly in θ, which implies that as $s \to r$,

$$\frac{1}{2\pi} \int_0^{2\pi} \frac{u(se^{i\theta})(s^2 - r^2)}{s^2 + \rho^2 - 2s\rho \cos(\theta - \phi)} \, d\theta \to \frac{1}{2\pi} \int_0^{2\pi} \frac{u(re^{i\theta})(r^2 - \rho^2)}{r^2 + \rho^2 - 2r\rho \cos(\theta - \phi)} \, d\theta$$

(See Proposition 3.1.9, if you are not familiar with this result about the convergence of integrals.) Thus

$$u(\rho e^{i\phi}) = \frac{r^2 - \rho^2}{2\pi} \int_0^{2\pi} \frac{u(re^{i\theta})}{r^2 + \rho^2 - 2r\rho \cos(\theta - \phi)} \, d\theta \quad \blacksquare$$

Poisson's formula (Eq. (4)) also enables us to find a solution to the Dirichlet problem for the case where the region is a disk. Suppose that we are given the continuous function u_0 defined on the circle $|z| = r$. We define u by formula (4):

$$u(\rho e^{i\phi}) = \frac{r^2 - \rho^2}{2\pi} \int_0^{2\pi} \frac{u_0(re^{i\theta})}{r^2 - 2r\rho \cos(\theta - \phi) + \rho^2} \, d\theta$$

for $\rho < r$ and $u(re^{i\phi}) = u_0(re^{i\phi})$. While it is relatively simple to show that u is harmonic in $D(0; r)$ (see Review Exercise 18), it is more difficult to show that u is continuous on the closure of $D(0; r)$. The expression must be examined in the critical case in which $\rho \to r$ in which the integrand takes the indeterminate form $0/0$ near $\theta = \phi$. A proof will not be given here.

Worked Examples

2.5.14 *Let f be analytic and nonzero in a region A. Show that $|f|$ has no strict local minima in A. If f has zeros in A, show by example that this conclusion does not hold.*

Solution. Since f is analytic and nonzero in A, $1/f$ is analytic in A. By the maximum modulus theorem, $1/|f|$ can have no local maxima in A unless $1/|f|$ is constant. Thus $1/|f|$ has no strict local maxima in A. Hence $|f|$ can have no strict local minima in A. The identity function $I: z \mapsto z$ is analytic on the unit disk and $|I|$ has a strict minimum at the origin.

2.5.15 *Find the maximum of $|\sin z|$ on $[0, 2\pi] \times [0, 2\pi]$.*

Solution. Since $\sin z$ is entire we can apply the maximum modulus principle, which tells us that the maximum occurs on the boundary of this square. Now $|\sin z|^2 = \sinh^2 y + \sin^2 x$ because $\sin(x + iy) = \sin x \cosh y + i \cos x \cdot \sinh y$, $\sin^2 x + \cos^2 x = 1$, and $\cosh^2 y - \sinh^2 y = 1$. On the boundary $y = 0$, $|\sin z|^2$ has maximum 1; for $x = 0$ the maximum is $\sinh^2 2\pi$, since $\sinh y$ increases with y; for $x = 2\pi$ the maximum is again $\sinh^2 2\pi$; for $y = 2\pi$ the maximum is $\sinh^2 2\pi + 1$. Thus the maximum of $|\sin z|^2$ occurs at $x = \pi/2$, $y = 2\pi$, and is $\sinh^2 2\pi + 1 = \cosh^2 2\pi$. Therefore, the maximum of $|\sin z|$ on $[0, 2\pi] \times [0, 2\pi]$ is $\cosh 2\pi$.

2.5.16 *Find the maximum of $u(x, y) = \sin x \cosh y$ on the unit square $[0, 1] \times [0, 1]$.*

Solution. $u(x, y)$ is a harmonic function and is not constant, and so the maximum of u on the unit square $[0, 1] \times [0, 1]$ occurs on the boundary. The maximum of $\sin x \cosh y$ is $\sin(1) \cosh(1)$, since both sin and cosh are increasing on the interval $[0, 1]$.

2.5.17 *Let A be a set. Show that $\text{cl}(A) = A \cup \text{bd}(A)$.*

Solution. $A \subset \text{cl}(A)$ and $\text{bd}(A) \subset \text{cl}(A)$, and so $A \cup \text{bd}(A) \subset \text{cl}(A)$. In the other direction, if z is in $\text{cl}(A)$ and not in A, then

$$z \in (\text{limit}(A)) \cap (C\backslash A) \subset \text{cl}(A) \cap \text{cl}(C\backslash A) = \text{bd}(A)$$

Thus $\text{cl}(A) \subset A \cup \text{bd}(A)$.

2.5.18 *Prove the following: Under the conditions of the Schwarz lemma, if $|f'(0)| = 1$, then $f(z) = cz$ for all z in $D(0; 1)$ for some constant c with $|c| = 1$.*

Solution. Let C be the circle $\{z \text{ such that } |z| = r\}$ with $r < 1$, and let M be the maximum value of $|f(z)|$ on C. By the Schwarz lemma, if $f(z)$ is not a constant multiple of z, then $M < r$. Now use the Cauchy integral formula for derivatives:

$$|f'(0)| = \left| \frac{1}{2\pi i} \int_C \frac{f(z)}{z^2} \, dz \right| \le \frac{1}{2\pi} 2\pi r \frac{M}{r^2} < 1$$

Therefore the inequality $|f'(0)| \le 1$ given by the Schwarz lemma must be strict unless $f(z)$ is a constant multiple of z.

2.5.19 *Suppose f and g are one-to-one analytic functions from the unit disk D onto D which satisfy $f(0) = g(0)$ and $g'(0) = f'(0) \ne 0$. Show that $f(z) = g(z)$ for all z in D.*

Solution. The function $h(z) = g^{-1}(f(z))$ is analytic from D to D and $h(0) = g^{-1}(f(0)) = g^{-1}(g(0)) = 0$. Since $g(h(z)) = f(z)$, we have $g'(h(z))$. $h'(0) = f'(0)$, so $h'(0) = f'(0)/g'(0) = 1$. Example 2.5.18 shows that $h(z) = cz$ for a constant c; since $h'(0) = 1$, $c = 1$. Thus $f(z) = g(h(z)) = g(z)$.

Remark. We will see in Chap. 5 that the assumption that f and g are one-to-one *forces* the derivative to be nonzero, so this assumption is really superfluous.

Exercises

1. Find the maximum of $|e^z|$ on $|z| \le 1$.

2. Find the maximum of $|\cos z|$ on $[0, 2\pi] \times [0, 2\pi]$.

3. Give an example to show that the interpretation of the maximum modulus principle that reads "The absolute value of an analytic function on a region is always smaller than its maximum on the boundary of the region" is false if the region is not bounded. The region in your example should be something other than all of **C** so that the boundary is not empty.

4. (a) Show that for $|z_0| < R$, the mapping

$$T: z \mapsto \frac{R(z - z_0)}{R^2 - \bar{z}_0 z}$$

takes the open disk of radius R one to one onto the disk of radius 1 and takes z_0 to the origin. (*Hint.* Use the maximum modulus theorem and verify that $z_0 \mapsto 0$ and $|z| = R$ implies that $|Tz| = 1$.)
(b) Suppose that f is analytic on the open disk $|z| < R$ and that $|f(z)| < M$ for $|z| < R$. Suppose also that $f(z_0) = w_0$. Show that

$$\left| \frac{M[f(z) - w_0]}{M^2 - \bar{w}_0 f(z)} \right| \le \left| \frac{R(z - z_0)}{R^2 - \bar{z}_0 z} \right|$$

(This is a generalization of the Schwarz lemma.)

5. Let f and g be continuous on cl (A) and analytic on A, where A is an open, connected, bounded region. If $f = g$ on bd (A), show that $f = g$ on all of cl (A).

6. Let u be harmonic on the bounded region A and continuous on cl (A). Then show that u takes its minimum only on bd (A) unless u is constant. (Compare this exercise with Worked Example 2.5.14.) (*Hint.* Consider $-u$.)

7. Find the maximum of $|e^{z^2}|$ on the unit disk.

8. Find the maximum of $u = \text{Re } z^3$ on the unit square $[0, 1] \times [0, 1]$.

9. Find harmonic conjugates for each of the following functions (specify a region in each case):
(a) $u(x, y) = \sinh x \sin y$ (b) $u(x, y) = \log \sqrt{x^2 + y^2}$
(c) $u(x, y) = e^x \cos y$

10. Show that $u(x, y) = \log \sqrt{x^2 + y^2}$ is harmonic but that it has no harmonic conjugate on $\mathbb{C}\backslash\{0\}$.

11. Verify directly that the level curves of Re e^z and Im e^z intersect orthogonally.

12. (a) Prove that

$$\int_0^{2\pi} \frac{R^2 - r^2}{R^2 - 2rR \cos (\theta - \phi) + r^2} \, d\theta = 2\pi$$

for $R > r$ and any ϕ. (*Hint.* Use the uniqueness of the solution to the Dirichlet problem.)
(b) Solve the Dirichlet problem directly for the boundary condition $u_0(z) = x$. Deduce that for $r < 1$,

$$r \cos \phi = \frac{1}{2\pi} \int_0^{2\pi} \frac{(1 - r^2) \cos \theta}{1 - 2r \cos (\theta - \phi) + r^2} \, d\theta$$

13. Let f be analytic and let $f'(z) \neq 0$ on a region A. Let $z_0 \in A$ and assume that $f(z_0) \neq 0$. Given $\epsilon > 0$, show that there exist a $z \in A$ and a $\zeta \in A$ such that $|z - z_0| < \epsilon, |\zeta - z_0| < \epsilon$, and

$$|f(z)| > |f(z_0)| \qquad |f(\zeta)| < |f(z_0)|$$

(*Hint.* Use the maximum modulus theorem.)

14. Prove *Hadamard's three-circle theorem:* Let f be analytic on a region containing the set R in Fig. 2.5.4; let $R = \{z \mid r_1 \leq |z| \leq r_3\}$; and assume $0 < r_1 < r_2 < r_3$. Let M_1, M_2, M_3 be the maxima of $|f|$ on the circles $|z| = r_1, r_2, r_3$, respectively. Then $M_2^{\log (r_3/r_1)} \leq M_1^{\log (r_3/r_2)} M_3^{\log (r_2/r_1)}$. (*Hint.* Let $\lambda = -\log (M_3/M_1)/\log (r_3/r_1)$ and consider $g(z) = z^\lambda f(z)$. Apply the maximum principle to g; be careful about the domain of analyticity of g.)

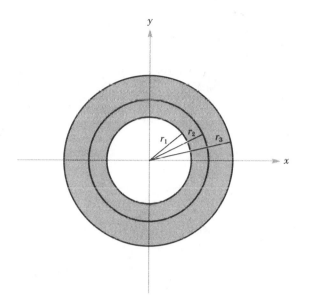

FIGURE 2.5.4   Hadamard's three-circle theorem.

15.   Let g be analytic on $\{z$ such that $|z| < 1\}$ and assume that $|g(z)| = |z|$ for all $|z| < 1$. Show that $g(z) = e^{i\theta}z$ for some constant $\theta \in [0, 2\pi]$. (*Hint.* Use the Schwarz lemma.)

16.   Prove: if u is continuous and satisfies the mean value property, then u is C^∞ and is harmonic. (*Hint.* Use Poisson's formula.)

17.   Evaluate $\int_\gamma dz/(z^2 - 1)$ where γ is the circle $|z| = 2$.

18.   The function $f(z)$ is analytic over the whole complex plane and Im $f \le 0$. Prove that f is a constant.

REVIEW EXERCISES FOR CHAPTER 2

1.   Evaluate the following integrals:

(a) $\displaystyle\int_\gamma \sin z \, dz$, where γ is the unit circle

(b) $\displaystyle\int_\gamma \frac{\sin z}{z} \, dz$, where γ is the unit circle

(c) $\displaystyle\int_\gamma \frac{\sin z}{z^2} \, dz$, where γ is the unit circle

(d) $\displaystyle\int_{|z|=1} \frac{\sin e^z}{z^2} \, dz$

2. Let f and g be analytic in a region A. Prove: If $|f| = |g|$ on A and $f \neq 0$ except at isolated points, then $f(z) = e^{i\theta}g(z)$ on A for some constant θ, $0 \le \theta < 2\pi$.

3. Let f be analytic on $\{z \text{ such that } |z| > 1\}$. Show that if γ, is the circle of radius $r > 1$ and center 0, then $\int_{\gamma} f$ is independent of r.

4. Let $f(z) = P(z)/Q(z)$ where P and Q are polynomials and Q has a degree of at least 2 more than that of P.

(a) Argue that if R is sufficiently large, there is a constant M such that

$$\left| \frac{P(z)}{Q(z)} \right| \le \frac{M}{|z|^2} \qquad \text{for } |z| \ge R$$

(b) If γ is a circle of radius r and center 0 with r large enough that f is analytic outside γ, prove that $\int_{\gamma} f(z)\, dz = 0$. (*Hint.* Use Exercise 4 and let $r \to \infty$.)

(c) Evaluate $\int_{\gamma} dz/(1 + z^2)$ where γ is a circle of radius 2 and center 0.

5. Evaluate $\int_{\gamma} f$ where $f(x + iy) = x^2 + iy^2$ and γ is the line joining 1 to i.

6. Let u be a bounded harmonic function on $\mathbf{C}$. Prove u is constant.

7. Let f be analytic on the open connected set A and suppose that there is a $z_0 \in A$ such that $|f(z)| \le |f(z_0)|$ for all $z \in A$. Then show that f is constant on A.

8. Let f be entire and let $|f(z)| \le M$ for z on the circle $|z| = R$; let R be fixed. Then prove that

$$|f^{(k)}(re^{i\theta})| \le \frac{k!M}{(R - r)^k} \qquad k = 0, 1, 2, \ldots$$

for all $0 \le r < R$.

9. Find a harmonic conjugate for

$$u(x, y) = \frac{x^2 + y^2 - x}{(x - 1)^2 + y^2}$$

on a suitable domain.

10. Let f be analytic on A and let $f'(z_0) \neq 0$. Show that if γ is a sufficiently small circle centered at z_0, then

$$\frac{2\pi i}{f'(z_0)} = \int_{\gamma} \frac{dz}{f(z) - f(z_0)}$$

(*Hint.* Use the inverse function theorem.)

11. Evaluate

$$\int_0^{2\pi} e^{-i\theta} e^{e^{i\theta}} \, d\theta.$$

12. Let f and g be analytic in a region A and let $g'(z) \neq 0$ for all $z \in A$; let g be one-to-one and let γ be a closed curve in A. Then for z not on γ, prove that

$$f(z)I(\gamma, z) = \frac{g'(z)}{2\pi i} \int_\gamma \frac{f(\zeta)}{g(\zeta) - g(z)} \, d\zeta$$

(*Hint.* Apply the Cauchy integral theorem to

$$h(\zeta) = \begin{cases} \dfrac{f(\zeta)(\zeta - z)}{g(\zeta) - g(z)} & z \neq \zeta \\ \dfrac{f(\zeta)}{g'(\zeta)} & z = \zeta \end{cases}$$

Apply this result to the case in which $g(z) = e^z$.)

13. Simplify: $e^{\log i}$; $\log i$; $\log (-i)$; $i^{\log(-1)}$.

14. Let $A = \mathbf{C}$ minus the negative real axis and zero. Show that $\log z = \int_{\gamma_z} d\zeta/\zeta$ where γ_z is any curve in A joining 1 to z. Is A simply connected?

15. Let f be analytic on a region A and let f be nonzero. Let γ be a curve homotopic to a point in A. Show that

$$\int_\gamma \frac{f'(z)}{f(z)} \, dz = 0$$

16. Let f be analytic on and inside the unit circle. Suppose that the image of the unit circle $|z| = 1$ lies in the disk $D = \{z \text{ such that } |z - z_0| < r\}$. Show that the image of the whole inside of the unit circle lies in D. Illustrate with e^z.

17. Is $\int_\gamma x \, dx + x \, dy$ always zero if γ is a closed curve?

18. Show that Poisson's formula may be written

$$u(z) = \operatorname{Re}\left[\frac{1}{2\pi i} \int_{\gamma_r} \frac{\zeta + z}{\zeta - z} u(\zeta) \frac{d\zeta}{\zeta} \right]$$

Then write a formula for the harmonic conjugate of u. Use this formula to show that if $u(\zeta)$ is continuous on the boundary, then $u(z)$ is harmonic inside it.

19. Let $f = u + iv$ be analytic on a region A. Indicate which of the following are analytic on A:

(a) $u - iv$ (b) $-u - iv$ Y_{es}

(c) $iu - v$

20. If f is analytic on and inside the unit disk, then show that

$$f(re^{i\phi}) = \frac{1}{2\pi} \int_0^{2\pi} \frac{f(e^{i\theta})}{1 - re^{i(\phi-\theta)}} \, d\theta \qquad r < 1$$

21. Compute the cube roots of $8i$.

22. Discuss the following sketch of a proof for Cauchy's theorem: Suppose f is analytic in a convex region G containing 0 and γ is a closed curve in G. Define $F(t) = t \int_\gamma f(tz) \, dz$ for $0 \le t \le 1$. Cauchy's theorem is that $F(1) = 0$. Compute that $F'(t) = \int_\gamma f(tz) \, dz + t \int_\gamma zf'(tz) \, dz$, and integrate the second integral by parts to obtain

$$F'(t) = \int_\gamma f(tz) \, dz + t \left\{ \left[\frac{zf(tz)}{t} \right]_\gamma - \frac{1}{t} \int_\gamma f(tz) \, dz \right\} = 0$$

so that $F(1) = F(0) = 0$. (See Philip M. Morse and Herman Feshbach, *Methods of Mathematical Physics,* Part I (New York: McGraw-Hill Book Co., 1953), pp. 364–365.)

23. Prove *Harnack's inequality:* If u is harmonic and nonnegative for $|z| \le R$, then

$$u(0) \frac{R - |z|}{R + |z|} \le u(z) \le u(0) \frac{R + |z|}{R - |z|}$$

24. Prove the deformation theorem by differentiating under the integral sign and integrating by parts. (Assume these operations are valid.)

CHAPTER 3

SERIES REPRESENTATION
OF ANALYTIC FUNCTIONS

There is an important alternative way to define an analytic function. In some developments of complex function theory, a function f is called *analytic* if it is locally representable as a convergent powers series.* If this can be done at all, that series must be the *Taylor series* of f. As in real-variable calculus, the Taylor series for f with center at a is the series

$$f(a) + f'(a)(z - a) + \frac{1}{2}f''(a)(z - a)^2 + \cdots - \sum_{n=0}^{\infty} \frac{f^{(n)}(a)}{n!}(z - a)^n$$

Thus an analytic function is one which is infinitely differentiable so that the Taylor series can be written down and such that the resulting series converges to the function. With real variables each of these tasks can present a problem. For example, the function

$$f(x) - \begin{cases} x^2 & \text{for } x \geq 0 \\ -x^2 & \text{for } x \leq 0 \end{cases}$$

is differentiable, but $f'(x) = 2|x|$. Thus, the second derivative does not exist at 0. Even if all the derivatives exist, the Taylor series might not converge to the function. The function

$$f(x) = \begin{cases} e^{-1/x^2} & \text{for } x \neq 0 \\ 0 & \text{for } x = 0 \end{cases}$$

is an example. Using induction one can check that $f^{(k)}(0)$ exists for all k (in the real-variable sense) and $f^{(k)}(0) = 0$. Here all coefficients of the Taylor series at 0 are 0 and so the resulting series is zero, which does not equal $f(x)$ in any nontrivial interval around 0.

---

* See, for example, H. Cartan, *Elementary Theory of Analytic Functions of One or Several Complex Variables* (Reading, Mass.: Addison-Wesley, 1963).

One of the nice things about complex analysis is that neither of these difficulties arises. Assuming the existence of a complex derivative is much stronger than assuming a real derivative exists. We discovered in Chap. 2 that as soon as the first derivative exists on a region all the higher ones must also. We will find in this chapter that the second difficulty also disappears. If f is analytic on a region A and z_0 is in A, then the Taylor series of f centered at z_0 must converge to f on the largest open disk centered at z_0 and contained in A.

The reader is probably familiar with the geometric series

$$\frac{1}{1-t} = 1 + t + t^2 + t^3 + \cdots = \sum_{n=0}^{\infty} t^n$$

which is valid provided $|t| < 1$. We will show in Sec. 3.1 that this works just as well for complex numbers as for real. In Sec. 3.2 we will use it to expand the integrand in the Cauchy integral formula as an infinite series, integrate this series term by term, and use the Cauchy integral formula for derivatives to recognize the resulting coefficients as the correct ones for a Taylor series. Some preparation in Sec. 3.1 is necessary for several reasons. The steps of the argument above require justification. We also want to investigate the series representation of a function analytic on a deleted neighborhood, that is, a function with an isolated singularity. This will be done in Sec. 3.3, also on the basis of the preparation in Sec. 3.1. The resulting series, called the *Laurent series,* yields valuable information about the behavior of functions near singularities, and this behavior is the key to the subject of residues and its subsequent applications.

CONVERGENT SERIES OF ANALYTIC FUNCTIONS

We shall use the Cauchy integral formula to determine when the limit of a convergent sequence or series of analytic functions is an analytic function and when the derivative (or integral) of the limit is the limit of the derivative (or integral) of the terms in the sequence or series. The basic type of convergence studied in this chapter is uniform convergence; the Weierstrass M test is the basic tool used to determine such convergence. We shall be especially interested in the special case of power series (studied in Sec. 3.2), but we should be aware that some important functions are convergent series that are not power series, such as the Riemann zeta function (see Worked Example 3.1.15).

The proofs of the first few results are slightly technical, and since they are analogous to the case of real series, they appear at the end of the section.

Convergence of Sequences and Series

3.1.1 DEFINITION   *A sequence z_n of complex numbers is said to **converge** to a complex number z_0 if, for all $\epsilon > 0$, there is an integer N such that $n \geq N$ implies that $|z_n - z_0| < \epsilon$. Convergence of z_n to z_0 is denoted by $z_n \to z_0$. An infinite series $\sum_{k=1}^{\infty} a_k$ is said to **converge** to S and we write $\sum_{k=1}^{\infty} a_k = S$ if the sequence of partial sums defined by $s_n = \sum_{k=1}^{n} a_k$ converges to S.*

The limit z_0 is unique; that is, a sequence can converge to only one point z_0. (This and other properties of limits were discussed in Sec. 1.4.) A sequence z_n converges iff it is a *Cauchy sequence*, in other words, if, for every $\epsilon > 0$, there is an N such that $n, m \geq N$ implies that $|z_n - z_m| < \epsilon$. (Equivalently, the definition of Cauchy sequence can read: For every $\epsilon > 0$ there is an N such that $n \geq N$ implies that $|z_n - z_{n+p}| < \epsilon$ for every integer $p = 0, 1, 2, \ldots$.) This property of $C = R^2$ follows from the corresponding property of R, and we shall accept it from advanced calculus.

Corresponding statements for the series $\Sigma_{k=1}^{\infty} a_k$ can be made if we consider the sequence of partial sums $s_n = \Sigma_{k=1}^{n} a_k$. Since $s_{n+p} - s_n = \Sigma_{k=n+1}^{n+p} a_k$, the Cauchy criterion becomes: $\Sigma_{k=1}^{\infty} a_k$ converges iff, for every $\epsilon > 0$, there is an N such that $n \geq N$ implies that $|\Sigma_{k=n+1}^{n+p} a_k| < \epsilon$ for all $p = 1, 2, 3, \ldots$. As a particular case of the Cauchy criterion, with $p = 1$ we see that *if* $\Sigma_{k=1}^{\infty} a_k$ *converges, then* $a_k \rightarrow 0$. The converse is not necessarily true, as the harmonic series $\Sigma_{k=1}^{\infty} 1/k$ from calculus demonstrates.

As with real series, a complex series $\Sigma_{k=1}^{\infty} a_k$ is said to *converge absolutely* if $\Sigma_{k=1}^{\infty} |a_k|$ converges. Using the Cauchy criterion, we get:

3.1.2 PROPOSITION   *If* $\Sigma_{k=1}^{\infty} a_k$ *converges absolutely, then it converges.*

The proof of this theorem is found at the end of the section. The example $\Sigma_{k=1}^{\infty} (-1)^k/k$ from calculus shows that the converse is not true; that is, this series converges, but not absolutely.

This proposition is important because $\Sigma_{k=1}^{\infty} |a_k|$ is a *real* series, and the usual tests for real series that we know from calculus can be applied. Some of those tests are included in the next proposition (again the proof appears at the end of the section).

3.1.3 PROPOSITION

(i) *Geometric series: If* $|r| < 1$, *then* $\Sigma_{n=0}^{\infty} r^n$ *converges to* $1/(1-r)$ *and diverges (does not converge) if* $|r| \geq 1$.

(ii) *Comparison test: If* $\Sigma_{k=1}^{\infty} b_k$ *converges and* $0 \leq a_k \leq b_k$, *then* $\Sigma_{k=1}^{\infty} a_k$ *converges; if* $\Sigma_{k=1}^{\infty} c_k$ *diverges and* $0 \leq c_k \leq d_k$, *then* $\Sigma_{k=1}^{\infty} d_k$ *diverges.*

(iii) *p-series test:* $\Sigma_{n=1}^{\infty} n^{-p}$ *converges if* $p > 1$ *and diverges to* ∞ *(that is, the partial sums increase without bound) if* $p \leq 1$.

(iv) *Ratio test: Suppose that* $\lim_{n \to \infty} \left| \dfrac{a_{n+1}}{a_n} \right|$ *exists and is strictly less than* 1. *Then* $\Sigma_{n=1}^{\infty} a_n$ *converges absolutely. If the limit is strictly greater than* 1, *the series diverges. If the limit equals* 1, *the test fails.*

(v) *Root test: Suppose that* $\lim_{n \to \infty} (|a_n|)^{1/n}$ *exists and is strictly less than* 1. *Then* $\Sigma_{n=1}^{\infty} a_n$ *converges absolutely. If the limit is strictly greater than* 1, *the series diverges; if the limit equals* 1, *the test fails.*

There are a few other tests that we shall occasionally call upon from calculus, such as the alternating series test and the integral test. We assume the reader will review these as the need arises.

Uniform Convergence

Suppose that $f_n: A \to \mathbf{C}$ is a sequence of functions all defined on the set A. The sequence is said to *converge pointwise* iff, for each $z \in A$, the sequence $f_n(z)$ converges. The limit defines a new function $f(z)$ on A. A more important kind of convergence is called uniform convergence and is defined as follows.

3.1.4 DEFINITION *A sequence $f_n: A \to \mathbf{C}$ of functions defined on a set A is said to* **converge uniformly** *to a function f if, for every $\epsilon > 0$, there is an N such that $n \geq N$ implies that $|f_n(z) - f(z)| < \epsilon$ for all $z \in A$. This is written "$f_n \to f$ uniformly on A."*

A series $\sum_{k=1}^{\infty} g_k(z)$ is said to **converge pointwise** *if the corresponding partial sums $s_n(z) = \sum_{k=1}^{n} g_k(z)$ converge pointwise. A series $\sum_{k=1}^{\infty} g_k(z)$ is said to* **converge uniformly** *iff $s_n(z)$ converges uniformly.*

Obviously, uniform convergence implies pointwise convergence. The difference between uniform and pointwise convergence is as follows. For pointwise convergence, given $\epsilon > 0$, the N required is allowed to vary from point to point, whereas for uniform convergence we must be able to find a *single N* that works for all z.

It is difficult to draw the graph of a complex-valued function of a complex variable, since it would require four real dimensions, but the corresponding notions for real-val-

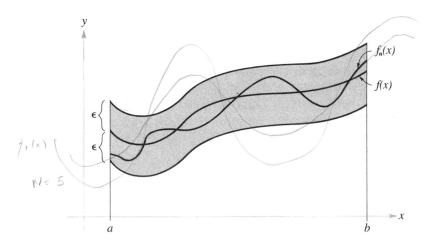

FIGURE 3.1.1 Uniform convergence on an interval $[a, b]$.

ued functions are instructive to illustrate. The geometric meaning of uniform convergence is shown in Fig. 3.1.1. If $\epsilon > 0$, then for large enough n, the graph $y = f_n(x)$ must stay inside the "ϵ tube" around the graph of f. It is important to notice that the concept of uniformity depends not only on the functions involved but also on the set on which we are working. Convergence might be uniform on one set but not on a larger set. The following example illustrates this point.

The sequence of functions $f_n(x) = x^n$ converges pointwise to the zero function $f(x) = 0$ for x in the half-open interval $[0, 1[$, but the convergence is not uniform. The function value x^n takes much longer to get close to 0 for x close to 1 than for x close to 0; by taking x close enough to 1, we need arbitrarily large values of n. The convergence is uniform on any closed subinterval $[0, r]$ with $r < 1$. Since the worst case is at $x = r$, whatever n works there also works for all smaller x. See Fig. 3.1.2.

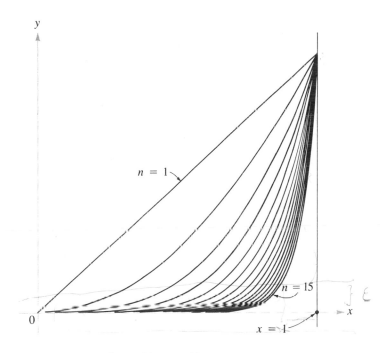

FIGURE 3.1.2 Convergence of x^n to 0 is not uniform on $\{x \mid 0 \le x < 1\}$.

3.1.5 CAUCHY CRITERION

(i) *A sequence $f_n(z)$ converges uniformly on A iff, for any $\epsilon > 0$, there is an N such that $n \ge N$ implies that $|f_n(z) - f_{n+p}(z)| < \epsilon$ for all $z \subset A$ and all $p = 1, 2, 3, \ldots$.*

(ii) *A series $\sum_{k=1}^{\infty} g_k(z)$ converges uniformly on A iff, for all $\epsilon > 0$, there is an N such that $n \ge N$ implies that*

$$\left| \sum_{k=n+1}^{n+p} g_k(z) \right| < \epsilon$$

for all $z \in A$ and all $p = 1, 2, \ldots$.

The next result states a basic property of uniform convergence.

3.1.6 PROPOSITION   *If the sequence f_n consists of continuous functions defined on A and if $f_n \to f$ uniformly, then f is continuous on A. Similarly, if the functions $g_k(z)$ are continuous and $g(z) = \sum_{k=1}^{\infty} g_k(z)$ converges uniformly on A, then g is continuous on A.*

(The results 3.1.5 and 3.1.6 are also proved at the end of the section.)

In other words, a uniform limit of continuous functions is continuous. If the convergence is not uniform, then the limit might be discontinuous. For example, let

$$f_n(x) = \begin{cases} -1 & \text{for } x \le -1/n \\ nx & \text{for } -1/n < x < 1/n \\ 1 & \text{for } 1/n \le x \end{cases}$$

and

$$f(x) = \begin{cases} -1 & \text{for } -\infty < x < 0 \\ 0 & \text{for } x = 0 \\ 1 & \text{for } 0 < x < \infty \end{cases}$$

as illustrated in Fig. 3.1.3. The functions f_n converge pointwise to f on the whole line, but the convergence is not uniform on any interval which contains 0, since for very small nonzero values of x, n may have to be quite large to bring $f_n(x)$ within a specified distance of $f(x)$. Each of the functions f_n is continuous, but the limit function is not.

The Weierstrass M Test

The Weierstrass M test is one of the most useful theoretical and practical tools for showing that a series converges uniformly. It does not always work, but it is effective in many cases.

3.1.7 WEIERSTRASS M TEST   *Let g_n be a sequence of functions defined on a set $A \subset \mathbf{C}$. Suppose that there is a sequence of real constants $M_n \ge 0$ such that*

(i) $|g_n(z)| \le M_n$ *for all $z \in A$*

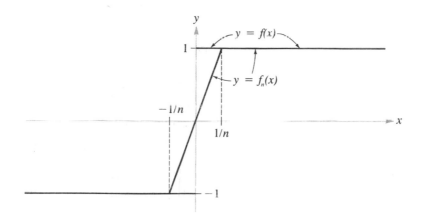

FIGURE 3.1.3 A nonuniform limit of continuous functions need not be continuous.

and

(ii) $\Sigma_{n=1}^{\infty} M_n$ *converges.*

  Then $\Sigma_{n=1}^{\infty} g_n$ *converges absolutely and uniformly on A.*

PROOF Since ΣM_n converges, for any $\epsilon > 0$ there is an N such that $n \geq N$ implies $\Sigma_{k=n+1}^{n+p} M_k < \epsilon$ for all $p = 1, 2, 3, \ldots$. (Absolute value bars are not needed on $\Sigma_{k=n+1}^{n+p} M_k$, because $M_n \geq 0$.) Thus $n \geq N$ implies

$$\left| \sum_{k=n+1}^{n+p} g_k(z) \right| \leq \sum_{k=n+1}^{n+p} |g_k(z)| \leq \sum_{k=n+1}^{n+p} M_k < \epsilon$$

and so by the Cauchy criterion we have the desired result. ∎

  For example, consider the series $g(z) = \Sigma_{n=1}^{\infty} z^n/n$. It will be shown that this series converges uniformly on the sets $A_r = \{z \text{ such that } |z| \leq r\}$ for each $0 \leq r < 1$. (We cannot let $r = 1$.) Here $g_n(z) = z^n/n$ and $|g_n(z)| = |z|^n/n \leq r^n/n$ since $|z| \leq r$. Therefore, let $M_n = r^n/n$. But $r^n/n \leq r^n$ and $\Sigma_{n=1}^{\infty} r^n$ converges for $0 \leq r < 1$. Thus ΣM_n converges, and so by the Weierstrass M test, the given series converges uniformly on A_r. It converges pointwise on $A = \{z \text{ such that } |z| < 1\}$ since each $z \in A$ lies in some A_r, for r close enough to 1 (see Fig. 3.1.4).

  This series does not, however, converge uniformly on A. Indeed, if it did, $\Sigma x^n/n$ would converge uniformly on $[0, 1[$. Suppose that this were true. Then for any $\epsilon > 0$ there would be an N such that $n \geq N$ would imply that

$$\frac{x^n}{n} + \frac{x^{n+1}}{n+1} + \cdots + \frac{x^{n+p}}{n+p} < \epsilon$$

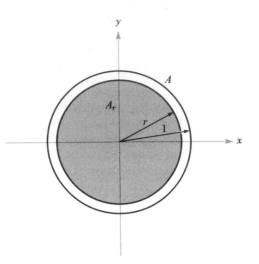

FIGURE 3.1.4   Region of convergence of $\Sigma \, (z^n/n)$: uniformly on A_r, pointwise on A.

for all $x \in [0, 1[$ and $p = 0, 1, 2, \ldots$. But the harmonic-type series

$$\frac{1}{N} + \frac{1}{N+1} + \cdots$$

diverges to infinity (that is, the partial sums $\to \infty$) and so we can choose p such that

$$\frac{1}{N} + \cdots + \frac{1}{N+p} > 2\epsilon$$

Next, we choose x so close to 1 that $x^{N+p} > \frac{1}{2}$. Then

$$\frac{x^N}{N} + \cdots + \frac{x^{N+p}}{N+p} > x^{N+p}\left(\frac{1}{N} + \cdots + \frac{1}{N+p}\right) > \epsilon$$

which is a contradiction. However, note that $g(z)$ is still continuous on A because it is continuous at each z, since each z lies in some A_r on which we do have uniform convergence.

Series of Analytic Functions

The next result is one of the main theorems concerning the convergence of analytic functions. The theorem was formulated by Karl Weierstrass in approximately 1860.

3.1.8 ANALYTIC CONVERGENCE THEOREM

(i) *Let A be a region in $\mathbf{C}$ and let f_n be a sequence of analytic functions defined on A. If $f_n \to f$ uniformly on every closed disk contained in A, then f is analytic. Furthermore, $f'_n \to f'$ pointwise on A and uniformly on every closed disk in A (see Fig. 3.1.5).*

(ii) *If g_k is a sequence of analytic functions defined on a region A in $\mathbf{C}$ and $g(z) = \sum_{k=1}^{\infty} g_k(z)$ converges uniformly on every closed disk in A, then g is analytic on A and $g'(z) = \sum_{k=1}^{\infty} g'_k(z)$ pointwise on A and also uniformly on every closed disk contained in A.*

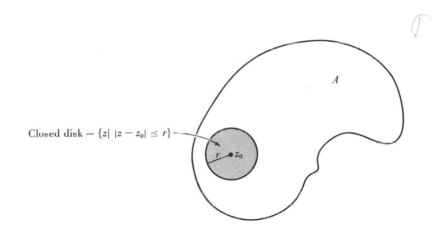

FIGURE 3.1.5 Convergence of analytic functions.

This theorem reveals yet another remarkable property of analytic functions that is not shared by functions of a real variable (compare Sec. 2.4). Uniform convergence usually is not sufficient to justify differentiation of a series term by term, but for analytic functions it is sufficient.

OPTIONAL REMARK Some sort of assumption about uniformity really is needed. The example illustrated in Fig. 3.1.3 shows, at least for functions of real variables, that a pointwise limit of continuous functions need not be continuous, much less differentiable. Worked Example 3.1.11 shows that this can still happen even if the functions of the sequence are all infinitely differentiable. With the experience in the last few sections of the good behavior of analytic functions, one might hope that pointwise convergence of analytic functions might be enough, but it is not. For example, there is a sequence of polynomials which converges pointwise on the unit disk to the branch of the square root function defined by $(re^{i\theta})^{1/2} = \sqrt{r}e^{i\theta/2}$ with $-\pi/2 < \theta \le \pi/2$. The limit function is not differentiable on the unit disk. In fact it has a jump discontinuity across the negative real axis. Existence of such a sequence is not obvious, but it can be proved with the help of an approximation theorem known as Runge's theorem, or Mergelyan's theorem; see W. Rudin, *Real and Complex Analysis* (New York: McGraw-Hill, 1966), pp. 255 and 386.

The proof of the analytic convergence theorem depends on Morera's theorem and Cauchy's integral formula, which were studied in Sec. 2.4. To prepare for this proof let us first analyze a result concerning integration of sequences and series.

3.1.9 PROPOSITION   *Let γ: $[a, b] \to A$ be a curve in a region A and let f_n be a sequence of continuous functions defined on $\gamma([a, b])$ which converge uniformly to f on $\gamma([a, b])$. Then*

$$\int_\gamma f_n \to \int_\gamma f$$

Similarly, if $\sum_{n=1}^\infty g_n(z)$ converges uniformly on γ, then

$$\int_\gamma \left(\sum_{n=1}^\infty g_n(z) \right) dz = \sum_{n=1}^\infty \int_\gamma g_n(z) \, dz$$

PROOF   The function f is continuous by Proposition 3.1.6, and so is integrable. Given $\epsilon > 0$, we can choose N such that $n \geq N$ implies that $|f_n(z) - f(z)| < \epsilon$ for all z on γ. Then, by Proposition 2.1.6,

$$\left| \int_\gamma f_n - \int_\gamma f \right| = \left| \int_\gamma f_n - f \right| \leq \int_\gamma |f_n(z) - f(z)| \, |dz| < \epsilon l(\gamma)$$

from which the first assertion follows. The second assertion is obtained by applying the first to the partial sums. (The student should write out the details.) ∎

PROOF OF THE ANALYTIC CONVERGENCE THEOREM (3.1.8)   As usual, it suffices to prove (i). Let $z_0 \in A$ and let $\{z$ such that $|z - z_0| \leq r\}$ be a closed disk around z_0 entirely contained in A. (Why does such a disk exist?) Consider $D(z_0; r) = \{z$ such that $|z - z_0| < r\}$, which is a simply connected region because it is convex. Since $f_n \to f$ uniformly on $\{z$ such that $|z - z_0| \leq r\}$, it is clear that $f_n \to f$ uniformly on $D(z_0; r)$. We wish to show that f is analytic on $D(z_0; r)$. To do this we use Morera's theorem (2.4.10). By Proposition 3.1.6, f is continuous on $D(z_0; r)$. Let γ be any closed curve in $D(z_0; r)$. Since f_n is analytic, $\int_\gamma f_n = 0$ by Cauchy's theorem and by the fact that $D(z_0; r)$ is simply connected. But by Proposition 3.1.9, $\int_\gamma f_n \to \int_\gamma f$ and so $\int_\gamma f = 0$. Thus by Morera's theorem f is analytic on $D(z_0; r)$.

We must still show that $f'_n \to f'$ uniformly on closed disks. To do this we use Cauchy's integral formula. Let $B = \{z$ such that $|z - z_0| \le r\}$ be a closed disk in A. We can draw a circle γ of radius $\rho > r$ centered at z_0 that contains B entirely in its interior (see Worked Example 1.4.28 and Fig. 3.1.6).

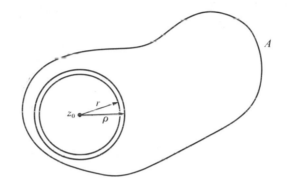

FIGURE 3.1.6   A closed disk in an open set can be slightly enlarged.

For any $z \in B$,

$$f'_n(z) = \frac{1}{2\pi i} \int_\gamma \frac{f_n(\zeta)}{(\zeta - z)^2} \, d\zeta \quad \text{and} \quad f'(z) = \frac{1}{2\pi i} \int_\gamma \frac{f(\zeta)}{(\zeta - z)^2} \, d\zeta$$

by the Cauchy integral formula. By hypothesis, $f_n \to f$ uniformly on the closed disk $\{z$ such that $|z - z_0| \le \rho\}$, which lies entirely in A. Then, given $\epsilon > 0$, we pick N such that $n \ge N$ implies that $|f_n(z) - f(z)| < \epsilon$ for all z in this disk (which we can do by hypothesis). Since γ is the boundary of this disk, $n \ge N$ implies that $|f_n(\zeta) - f(\zeta)| < \epsilon$ on γ. Let us note that

$$|f'_n(z) - f'(z)| = \left| \frac{1}{2\pi i} \int_\gamma \frac{f_n(\zeta) - f(\zeta)}{(\zeta - z)^2} \, d\zeta \right|$$

and observe that for ζ and γ and $z \in B$, $|\zeta - z| \ge \rho - r$. Hence $n \ge N$ implies that

$$|f'_n(z) - f'(z)| \le \frac{1}{2\pi} \cdot \frac{\epsilon}{(\rho - r)^2} \cdot l(\gamma) = \frac{\epsilon \rho}{(\rho - r)^2}$$

Since ρ and r are fixed constants that are independent of $z \in B$, we get the desired result. ∎

Applying the analytic convergence theorem repeatedly we see that the kth derivatives $f_n^{(k)}$ converge to $f^{(k)}$ uniformly on closed disks in A. Notice also that this theorem does not assume uniform convergence on all of A. For example, $\sum_{n=1}^{\infty} z^n/n$ on $A = \{z \text{ such that } |z| < 1\}$ converges uniformly on the sets $A_r = \{z \text{ such that } |z| \le r\}$ for $0 \le r < 1$ (as we saw in the preceding example) and hence converges uniformly on all closed disks in A. Thus we can conclude that $\sum z^n/n$ is analytic on A and that the derivative is $\sum z^{n-1}$, which also converges on A. However, as that example demonstrated, we do have pointwise but not uniform convergence on A; convergence is uniform only on each closed subdisk in A.

Technical Proofs

3.1.2 PROPOSITION   *If $\sum_{k=1}^{\infty} a_k$ converges absolutely, then it converges.*

PROOF   By the Cauchy criterion, given $\epsilon > 0$ there is an N such that $n \ge N$ implies

$$\sum_{k=n+1}^{n+p} |a_k| < \epsilon \qquad p = 1, 2, \ldots$$

But

$$\left| \sum_{k=n+1}^{n+p} a_k \right| \le \sum_{k=n+1}^{n+p} |a_k| < \epsilon$$

by the triangle inequality (see Sec. 1.2). Thus by the Cauchy criterion, $\sum_{k=1}^{\infty} a_k$ converges. ∎

3.1.3 PROPOSITION

(i) **Geometric series:** *If $|r| < 1$, then $\sum_{n=0}^{\infty} r^n$ converges to $1/(1-r)$ and diverges if $|r| \ge 1$.*

(ii) **Comparison test:** *If $\sum_{k=1}^{\infty} b_k$ converges and $0 \le a_k \le b_k$, then $\sum_{k=1}^{\infty} a_k$ converges; if $\sum_{k=1}^{\infty} c_k$ diverges and $0 \le c_k \le d_k$, then $\sum_{k=1}^{\infty} d_k$ diverges.*

(iii) **p-series test:** *$\sum_{n=1}^{\infty} n^{-p}$ converges if $p > 1$ and diverges to ∞ (that is, the partial sums increase without bound) if $p \le 1$.*

(iv) **Ratio test:** *Suppose that $\lim_{n \to \infty} \left| \dfrac{a_{n+1}}{a_n} \right|$ exists and is strictly less than 1. Then $\sum_{n=1}^{\infty} a_n$ converges absolutely. If the limit is strictly greater than 1, the series diverges. If the limit equals 1, the test fails.*

(v) **Root test:** *Suppose that $\lim_{n \to \infty} (|a_n|)^{1/n}$ exists and is strictly less than 1. Then $\sum_{n=1}^{\infty} a_n$ converges absolutely. If the limit is strictly greater than 1, the series diverges; if the limit equals 1, the test fails.*

PROOF

(i) By elementary algebra,

$$1 + r + r^2 + \cdots + r^n = \frac{1 - r^{n+1}}{1 - r}$$

if $r \neq 1$. Since $r^{n+1} \to 0$ as $n \to \infty$ if $|r| < 1$, and since $|r|^{n+1} \to \infty$ if $|r| > 1$, we have convergence if $|r| < 1$ and divergence if $|r| > 1$. Obviously, $\Sigma_{n=0}^{\infty} r^n$ diverges if $|r| = 1$, since $r^n \not\to 0$.

(ii) The partial sums of the series $\Sigma_{k=1}^{\infty} b_k$ form a Cauchy sequence and thus the partial sums of the series $\Sigma_{k=1}^{\infty} a_k$ also form a Cauchy sequence, since for any k and p we have $a_k + a_{k+1} + \cdots + a_{k+p} \leq b_k + b_{k+1} + \cdots + b_{k+p}$. Hence $\Sigma_{k=1}^{\infty} a_k$ converges. A positive series can diverge only to $+\infty$, and so given $M > 0$, we can find k_0 such that $k \geq k_0$ implies $c_1 + c_2 + \cdots + c_k \geq M$. Therefore, for $k \geq k_0$, $d_1 + d_2 + \cdots + d_k \geq M$, and so $\Sigma_{k=1}^{\infty} d_k$ also diverges to ∞.

(iii) First suppose that $p \leq 1$; in this case $1/n^p \geq 1/n$ for all $n = 1, 2, \ldots$. Therefore, by (ii), $\Sigma_{n=1}^{\infty} 1/n^p$ will diverge if $\Sigma_{n=1}^{\infty} 1/n$ diverges. We now recall the proof of this from calculus:[*] If $s_k = 1/1 + 1/2 + \cdots + 1/k$, then s_k is a strictly increasing sequence of positive real numbers. Write s_{2^k} as follows:

$$s_{2^k} = 1 + \frac{1}{2} + \left(\frac{1}{3} + \frac{1}{4}\right) + \left(\frac{1}{5} + \frac{1}{6} + \frac{1}{7} + \frac{1}{8}\right)$$

$$+ \cdots + \left(\frac{1}{2^{k-1} + 1} + \cdots + \frac{1}{2^k}\right)$$

$$\geq 1 + \frac{1}{2} + \left(\frac{1}{2}\right) + \left(\frac{1}{2}\right) + \cdots + \left(\frac{1}{2}\right) = 1 + \frac{k}{2}$$

Hence s_k can be made arbitrarily large if k is made sufficiently large; thus $\Sigma_{n=1}^{\infty} 1/n$ diverges.

Now suppose that $p > 1$. If we let

$$s_k = \frac{1}{1^p} + \frac{1}{2^p} + \frac{1}{3^p} + \cdots + \frac{1}{k^p}$$

then s_k is an increasing sequence of positive real numbers. On the other hand,

---

[*] We can also prove (iii) by using the integral test for positive series (see any calculus text). The demonstration given here also proves the *Cauchy condensation test:* Let $\Sigma\, a_n$ be a series of positive terms with $a_{n+1} \leq a_n$. Then $\Sigma\, a_n$ converges iff $\Sigma_{j=1}^{\infty} 2^j a_{2^j}$ converges (see G. J. Porter, "An Alternative to the Integral Test for Infinite Series," *American Mathematical Monthly,* Vol. 79 (1972), p. 634).

$$S_{2^k-1} = \frac{1}{1^p} + \left(\frac{1}{2^p} + \frac{1}{3^p}\right) + \left(\frac{1}{4^p} + \frac{1}{5^p} + \frac{1}{6^p} + \frac{1}{7^p}\right) + \cdots$$

$$+ \left(\frac{1}{(2^{k-1})^p} + \cdots + \frac{1}{(2^k - 1)^p}\right) \le \frac{1}{1^p} + \frac{2}{2^p} + \frac{4}{4^p} + \frac{2^{k-1}}{(2^{k-1})^p}$$

$$= \frac{1}{1^{p-1}} + \frac{1}{2^{p-1}} + \frac{1}{4^{p-1}} + \cdots + \frac{1}{(2^{k-1})^{p-1}} < \frac{1}{1 - 1/2^{p-1}}$$

(Why?). Thus the sequence $\{s_k\}$ is bounded from above by $1/(1 - 1/2^{p-1})$; hence $\sum_{n=1}^{\infty} 1/n^p$ converges.

(iv) Suppose that $\underset{n\to\infty}{\text{limit}} \left|\frac{a_{n+1}}{a_n}\right| = r < 1$. Choose r' such that $r < r' < 1$ and let N be such that $n \ge N$ implies

$$\left|\frac{a_{n+1}}{a_n}\right| < r'$$

Then $|a_{N+p}| < |a_N|(r')^p$. Consider the series $|a_1| + \cdots + |a_N| + |a_N|r' + |a_N|(r')^2 + |a_N|(r')^3 + \cdots$. This converges to

$$|a_1| + \cdots + |a_{N-1}| + \frac{|a_N|}{1 - r'}$$

By (ii) we can conclude that $\sum_{k=1}^{\infty} |a_k|$ converges. If $\underset{n\to\infty}{\text{limit}} \left|\frac{a_{n+1}}{a_n}\right| = r > 1$, choose r' such that $1 < r' < r$ and let N be such that $n \ge N$ implies that $\left|\frac{a_{n+1}}{a_n}\right| > r'$. Hence $|a_{N+p}| > (r')^p|a_N|$, and so $\underset{n\to\infty}{\text{limit}} |a_N| = \infty$, whereas the limit would have to be zero if the sum converged (see Exercise 10). Thus $\sum_{k=1}^{\infty} a_k$ diverges. To see that the test fails if $\underset{n\to\infty}{\text{limit}} \left|\frac{a_{n+1}}{a_n}\right| = 1$, consider the series $1 + 1 + 1 + \cdots$, and $\sum_{n=1}^{\infty} 1/n^p$ for $p > 1$.

In both cases $\underset{n\to\infty}{\text{limit}} \left|\frac{a_{n+1}}{a_n}\right| = 1$, but the first series diverges and the second converges.

(v) Suppose that $\underset{n\to\infty}{\text{limit}} |a_n|^{1/n} = r < 1$. Choose r' such that $r < r' < 1$ and N such that $n \ge N$ implies that $|a_n|^{1/n} < r'$; in other words, that $|a_n| < (r')^n$. The series $|a_1| + |a_2| + \cdots + |a_{N-1}| + (r')^N + (r')^{N+1} + \cdots$ converges to $|a_1| + |a_2| + \cdots + |a_{N-1}| + (r')^N/(1 - r')$, and so by (ii), $\sum_{k=1}^{\infty} a_k$ converges. If $\underset{n\to\infty}{\text{limit}} |a_n|^{1/n} = r > 1$, choose $1 < r' < r$ and N such that $n \ge N$ implies that $|a_n|^{1/n} > r'$ or, in other words, that $|a_n| > (r')^n$. Hence $\underset{n\to\infty}{\text{limit}} |a_n| = \infty$. Therefore, $\sum_{k=1}^{\infty} a_k$ diverges.

To show that the test fails when $\underset{n\to\infty}{\text{limit}} |a_n|^{1/n} = 1$, we use these limits from calculus:

$$\lim_{n\to\infty} \left(\frac{1}{n}\right)^{1/n} = 1 \quad \text{and} \quad \lim_{n\to\infty} \left(\frac{1}{n^2}\right)^{1/n} - 1$$

(take logarithms and use L'Hôpital's rule to show that $(\log x)/x \to 0$ as $x \to \infty$). But $\sum_{n=1}^{\infty} 1/n$ diverges and $\sum_{n=1}^{\infty} 1/n^2$ converges. ∎

3.1.5 CAUCHY CRITERION

(i) *A sequence $f_n(z)$ converges uniformly on A iff for any $\epsilon > 0$, there is an N such that $n \geq N$ implies that $|f_n(z) - f_{n+p}(z)| < \epsilon$ for all $z \in A$ and all $p = 1, 2, 3, \ldots$.*

(ii) *A series $\sum_{k=1}^{\infty} g_k(z)$ converges uniformly on A iff for all $\epsilon > 0$, there is an N such that $n \geq N$ implies*

$$\left| \sum_{k=n+1}^{n+p} g_k(z) \right| < \epsilon$$

for all $z \in A$ and $p = 1, 2, \ldots$.

PROOF

(i) First we prove the "if" part. Let $f(z) = \lim_{n\to\infty} f_n(z)$, which exists because for each z, $f_n(z)$ is a Cauchy sequence. We wish to show that $f_n \to f$ uniformly on A. Given $\epsilon > 0$, choose N such that $|f_n(z) - f_{n+p}(z)| < \epsilon/2$, for $n \geq N$ and all $p \geq 1$. The first step is to show that for any z and any $n \geq N$, $|f_n(z) - f(z)| < \epsilon$. For $z \in A$, choose p large enough so that $|f_{n+p}(z) - f(z)| < \epsilon/2$, which is possible by pointwise convergence. Then, by the triangle inequality, $|f_n(z) - f(z)| \leq |f_n(z) - f_{n+p}(z)| + |f_{n+p}(z) - f(z)| < \epsilon/2 + \epsilon/2 = \epsilon$. (Notice that although p depends on z, N does not.)

Conversely, if $f_n \to f$ uniformly, given $\epsilon > 0$ choose N such that $n \geq N$ implies $|f_n(z) - f(z)| < \epsilon/2$ for all z. Since $n + p \geq N$, $|f_n(z) - f_{n+p}(z)| \leq |f_n(z) - f(z)| + |f(z) - f_{n+p}(z)| < \epsilon/2 + \epsilon/2 = \epsilon$.

(ii) By applying (i) to the partial sums, we deduce (ii). ∎

3.1.6 PROPOSITION
If the functions f_n are continuous on A and $f_n \to f$ uniformly, then f is continuous. Similarly, if the functions $g_k(z)$ are continuous and $g(z) = \sum_{k=1}^{\infty} g_k(z)$ converges uniformly on A, g is continuous on A.

PROOF It suffices to prove the assertion for sequences (Why?). We wish to show that for $z_0 \in A$, given $\epsilon > 0$, there is a $\delta > 0$ such that $|z - z_0| < \delta$ implies that $|f(z) - f(z_0)| < \epsilon$. Choose N such that $|f_N(z) - f(z)| < \epsilon/3$ for all $z \in A$. Since f_N is continuous, there is a $\delta > 0$ such that $|f_N(z) - f_N(z_0)| < \epsilon/3$ if $|z - z_0| < \delta$. Thus $|f(z) - f(z_0)| \leq |f(z) - f_N(z)| + |f_N(z) - f_N(z_0)| + |f_N(z_0) - f(z_0)| < \epsilon/3 + \epsilon/3 + \epsilon/3 = \epsilon$. ∎

Note that in the last step we need an N that is independent of z to conclude that both $|f_N(z) - f(z)| < \epsilon/3$ and $|f_N(z_0) - f(z_0)| < \epsilon/3$.

Worked Examples

3.1.10 *Show that the sequence of functions $f_n(x) = \sin (x/n)$ converges uniformly to the constant function $f(x) = 0$ for x in the interval $[0, \pi]$.*

Solution. From calculus, $\sin \theta$ is increasing and $\sin \theta \leq \theta$ for $0 \leq \theta \leq \pi/2$. Thus, if $x \in [0, \pi]$ and $n \geq 2$, then $|f_n(x) - f(x)| = |\sin (x/n)| \leq \sin (\pi/n) \leq \pi/n$. (See Fig. 3.1.7.) Therefore $|f_n(x) - f(x)| < \epsilon$ provided $n > \max (2, 2/\epsilon)$. The same n works for all x in the interval, and so the convergence is uniform on $[0, \pi]$.

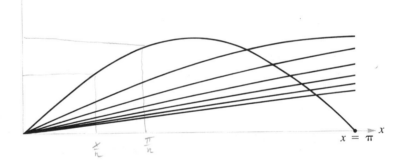

FIGURE 3.1.7 $y = \sin (x/n)$ for $n = 1$ through 7.

3.1.11 *Show that the sequence of functions $f_n(x) = \arctan (nx)$ converges for x in the interval $[-5, 5]$ to the function*

$$f(x) = \begin{cases} -\pi/2 & \text{for } x < 0 \\ 0 & \text{for } x = 0 \\ \pi/2 & \text{for } x > 0 \end{cases}$$

but that the convergence is not uniform. (See Fig. 3.1.8.)

Solution. If $x > 0$, then $|f_n(x) - f(x)| = |\arctan (nx) - \pi/2|$. We know that arctan (nx) is an increasing function of x with limit $\pi/2$ as $x \to \infty$. Therefore $|\arctan (nx) - \pi/2| < \epsilon$ if and only if $nx > \tan (\pi/2 - \epsilon)$. For any particular value of x, large enough values of n will work, but by taking x close to 0 we can force the required n to be quite large. Thus we have convergence but not uniform convergence. (Similar discussions apply for $x \leq 0$.) One can see indirectly that the convergence must not be uniform. If it were, then the limit function would be continuous, by Proposition 3.1.6. But it is not.

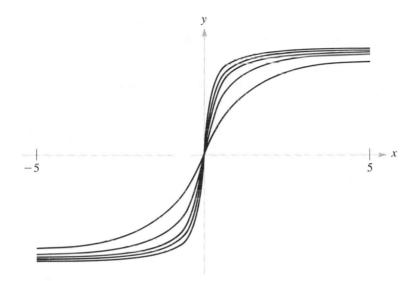

FIGURE 3.1.8 $y = \arctan(nx)$ for $n = 1$ through 5.

The next three examples develop the important special case of the geometric series and show how the tools of this section can be applied to it to obtain some interesting results. The behavior of these examples is typical of the more general power series studied in the next section.

3.1.12   *Show that the series $\sum_{n=0}^{\infty} z^n$ converges on the open unit disk $D = D(0; 1)$ to the analytic function $f(z) = 1/(1 - z)$. Prove the convergence is uniform and absolute on every closed disk $D_r = \{z$ such that $|z| \leq r\}$ with $r < 1$.*

Solution. If $z \in D$, then $z \in D_r$ whenever $|z| \leq r < 1$. Hence convergence at z follows from the second assertion. To prove it, suppose z is in D_r. Then $|z^n| < r^n$. Since $\sum r^n$ converges (Proposition 3.1.3(i)), the Weierstrass M test applies with $M_n = r^n$, and our series converges uniformly and absolutely on D_r. We have run into one of the shortcomings of a tool like the Weierstrass M test. We have shown that the series converges but have not identified the limit. To do this notice that

$$1 - z^{n+1} = (1 - z)(1 + z + z^2 + \cdots + z^n)$$

so that

$$\left| \frac{1}{1 - z} - \sum_{k=0}^{n} z^k \right| = \frac{|z|^{n+1}}{|1 - z|} \leq \frac{r^{n+1}}{1 - r}$$

Since $r < 1$, this goes to 0 as $n \to \infty$ and we have our result.

3.1.13   *Show that the series $\Sigma_{n=1}^{\infty} nz^{n-1} = \Sigma_{n=0}^{\infty} (n+1)z^n$ converges on the open unit disk D to $g(z) = 1/(1-z)^2$. The convergence is uniform and absolute on every closed disk contained in D.*

Solution. If B is any closed disk contained in D, then $B \subset D_r$ for some closed disk D_r, as in the last example. The series Σz^n converges uniformly and absolutely to $f(z) = 1/(1-z)$ on D_r, and so on B. By the analytic convergence theorem (3.1.8(ii)), the series of derivatives converges uniformly on every closed disk in D to $f'(z)$. That is, $\Sigma_{n=1}^{\infty} nz^{n-1} = f'(z) = 1/(1-z)^2$, as desired. The convergence is absolute by comparison. If $|z| \le r < 1$, then $|nz^{n-1}| < nr^{n-1}$, but Σnr^{n-1} converges by the argument just given.

3.1.14   *Show that the series $\Sigma_{n=1}^{\infty} (-1)^{n-1} z^n/n$ converges uniformly and absolutely to $\log(1+z)$ on the open unit disk, where $\log(\rho e^{i\theta}) = \log \rho + i\theta$ with $-\pi < \theta < \pi$.*

Solution. We know that the given formula for log defines a branch of logarithm on the disk $D(1; 1)$. In fact, it is the same as that described by the construction $\log w = \int_\gamma (1/\zeta)\, d\zeta$ where γ is a straight-line path from 1 to w. By the path independence guaranteed by Cauchy's theorem we can integrate first along a circular arc (constant $r = 1$) and then along a ray from the origin (constant θ) to get from 1 to $w = \rho e^{i\theta}$ (see Fig. 3.1.9). This gives

$$\int_\gamma \frac{1}{\zeta}\, d\zeta = \int_0^\theta e^{-i\phi} i e^{i\phi}\, d\phi + \int_1^\rho \frac{1}{re^{i\theta}} e^{i\theta}\, dr = i\theta + \log \rho$$

Changing variables to $\xi = \zeta - 1$ gives $\log w = \int_\mu 1/(\xi+1)\, d\xi = \int_\mu 1/[1-(-\xi)]\, d\xi$, the path μ being a straight line from 1 to $z = w - 1$ in the open unit disk $D = D(0; 1)$. By Worked Example

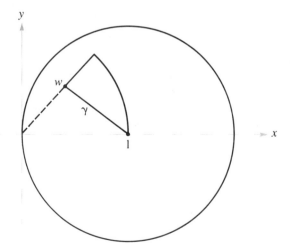

FIGURE 3.1.9   Paths for computing log w on $D(1; 1)$.

3.1.12, the integrand may be expanded in an infinite series $\Sigma_{n=0}^{\infty} (-\xi)^n$ which converges uniformly on μ. The analytic convergence theorem allows us to integrate term by term to obtain

$$\log w = \int_{\mu} \left[\sum (-\xi)^n \right] d\xi = \sum \left[\int_{\mu} (-\xi)^n d\xi \right] = \sum_{n=0}^{\infty} \frac{(-1)^n (w-1)^{n+1}}{n+1}$$

$$= \sum_{n=1}^{\infty} \frac{(-1)^{n-1}(w-1)^n}{n}$$

This works for every w in $D(1; 1)$. Setting $z = w - 1$ gives $\log (z + 1) = \Sigma_{n=1}^{\infty} (-1)^{n-1}z^n/n$ for all z in $D(0; 1)$. Again, the convergence is uniform and absolute on any D, with $r < 1$. Indeed, since $|z| \leq r$ implies $|(-1)^n z^n/n| \leq r^n/n \leq r^n$ and Σr^n converges, the Weierstrass M test applies with $M_n = r^n$.

3.1.15 *Show that the Riemann ζ function, defined by*

$$\zeta(z) = \sum_{n=1}^{\infty} n^{-z}$$

is analytic on the region $A = \{z \mid \mathrm{Re}\ z > 1\}$. Compute $\zeta'(z)$ on that set.

Solution. We use the analytic convergence theorem (3.1.8). We must be careful to try to prove uniform convergence only on closed disks in A and not on all of A. In this example we do not in fact have uniform convergence on all of A (see Exercise 8).

Let B be a closed disk in A and let δ be its distance from the line $\mathrm{Re}\ z = 1$ (Fig. 3.1.10). We shall show that $\Sigma_{n=1}^{\infty} n^{-z}$ converges uniformly on B. Here $n^{-z} = e^{-z \log n}$, where $\log n$ means the usual log of real numbers. Now $|n^{-z}| = |e^{-z \log n}| = e^{-x \log n} = n^{-x}$. But $x \geq 1 + \delta$ if $z \in B$, and so $|n^{-z}| \leq n^{-(1+\delta)}$ for all $z \in B$. Let us, therefore, choose $M_n = n^{-(1+\delta)}$.

By Proposition 3.1.3(iii), $\Sigma_{n=1}^{\infty} M_n$ converges. Thus by the Weierstrass M test, our series $\Sigma_{n=1}^{\infty} n^{-z}$ converges uniformly on B. Hence ζ is analytic on A. Also by the analytic convergence theorem, we can differentiate term by term to give

$$\zeta'(z) = - \sum_{n=1}^{\infty} (\log n)n^{-z}$$

which we know must also converge on A (and uniformly on closed disks in A).

3.1.16 *Show that*

$$f(z) = \sum_{n=1}^{\infty} \frac{z^n}{n^3}$$

is analytic on $A = \{z$ such that $|z| < 1\}$. Write a series for $f'(z)$.

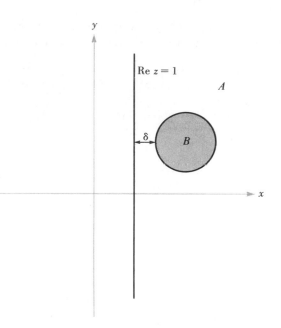

FIGURE 3.1.10 The domain of analyticity of the Riemann zeta function.

Solution. Again we use the analytic convergence theorem. (We note that this is a power series that can alternatively be dealt with after the student has read Sec. 3.2).

In this case we actually have uniform convergence on all of A. Let $M_n = 1/n^2$. Clearly, $\Sigma\, M_n$ converges and $|z^n/n^2| < 1/n^2 = M_n$ for all $z \in A$. Thus, by the Weierstrass M test, $\Sigma_{n=1}^{\infty}\, z^n/n^2$ converges uniformly on A; therefore, the series converges on closed disks in A. Thus the sum $\Sigma_{n=1}^{\infty}\, z^n/n^2$ is an analytic function on A. Furthermore,

$$f'(z) = \sum_{n=1}^{\infty} \frac{nz^{n-1}}{n^2} = \sum_{n=1}^{\infty} \frac{z^{n-1}}{n}$$

(This series for $f'(z)$ does not converge for $z = 1$, so f cannot be extended to be analytic on any region containing the closed unit disk.)

3.1.17 *Compute*

$$\int_{\gamma} \left(\sum_{n=-1}^{\infty} z^n \right) dz$$

where γ is a circle of radius $\tfrac{1}{2}$.

Solution. Let B be a closed disk in $A = \{z$ such that $|z| < 1\}$ a distance of δ from the circle $|z| = 1$. For $z \in B, |z^n| = |z|^n \le (1 - \delta)^n$ for $n \ge 0$. We choose $M_n = (1 - \delta)^n$ and note that $\Sigma\, M_n$ is convergent.

Therefore, $\sum_{n=0}^{\infty} z^n$ is uniformly convergent on B, so by the analytic convergence theorem, $\sum_{n=0}^{\infty} z^n$ is analytic on A. Therefore,

$$\int_{\gamma} \left(\sum_{n=-1}^{\infty} z^n \right) dz = \int_{\gamma} \frac{1}{z} dz + \int_{\gamma} \left(\sum_{n=0}^{\infty} z^n \right) dz = 2\pi i$$

by Example 2.1.12 and Cauchy's theorem.

3.1.18 *This example and the next illustrate how the Cauchy integral formula can often be used to obtain uniformity where it might not be expected.*

DEFINITION *A family of functions $\mathcal{S}$ defined on a set G is said to be **uniformly bounded on closed disks in** G if for each closed disk $B \subset G$ there is a number $M(B)$ such that $|f(z)| \le M(B)$ for all z in B and for all f in $\mathcal{S}$.*

Prove the following: If $f_1, f_2, f_3, \ldots$ is a sequence of functions analytic on a region G which is uniformly bounded on closed disks in G, then the sequence of derivatives $f'_1, f'_2, f'_3, \ldots$ is also uniformly bounded on closed disks in G.

Solution. Suppose $B = \{z \text{ such that } |z - z_0| \le r\}$ is a closed disk in G. Since B is closed and G is open, Worked Example 1.4.28 shows that there is a number ρ with $B \subset D(z_0; \rho) \subset G$. Let $R = (r + \rho)/2$ and $D = \{z \text{ such that } |z - z_0| \le R\}$. By hypothesis, there is a number $N(D)$ such that $|f_n(z)| \le N(D)$ for all n and all z in D. If Γ is the boundary circle of D, the Cauchy integral formula for derivatives gives, for any z in B,

$$|f'_n(z)| = \left| \frac{1}{2\pi i} \int_{\Gamma} \frac{f_n(\zeta)}{(\zeta - z)^2} d\zeta \right| \le \frac{1}{2\pi} \left[\frac{N(D)}{(R - r)^2} \right] 2\pi r$$

Thus if we put $M(B) = N(D)R/(R - r)^2$, we will have $|f'_n(z)| \le M(B)$ for all n and for all z in B, as desired.

3.1.19 **DEFINITION** *A family of functions $\mathcal{S}$ defined on a set B is said to be **uniformly equicontinuous on** B if for each $\epsilon > 0$ there is a number $\delta > 0$ such that $|f(\zeta) - f(\xi)| < \epsilon$ for all f in $\mathcal{S}$ whenever ζ and ξ are in B and $|\zeta - \xi| < \delta$.*

That is, for each ϵ, the same δ can be made to work for all functions in the family $\mathcal{S}$, and everywhere in the set B.

Prove: If $f_1, f_2, f_3, \ldots$ is a sequence of functions analytic on a region G which is uniformly bounded on closed disks in G, then this family of functions is uniformly equicontinuous on every closed disk in G.

Solution. Let B be a closed disk in G. By the last example, there is a number $M(B)$ such that $|f'_n(z)| \le M(R)$ for every n and for all z in B. Let γ be the straight line from ζ to ξ in B. Since that straight line is contained in B, we have $|f_n(\zeta) - f_n(\xi)| = |\int_{\gamma} f'_n(z) \, dz| \le \int_{\gamma} |f'_n(z)| \, |dz| \le M(B)|\zeta - \xi|$. Thus, given $\epsilon > 0$, we can satisfy the definition of uniform equicontinuity on B by setting $\delta = \epsilon/M(B)$.

Exercises

1. Do the following sequences converge, and if so, what is their limit?

(a) $z_n = (-1)^n + \dfrac{i}{n+1}$

(b) $z_n = \dfrac{n!}{n^n} i^n$

2. Let c be a complex constant. Let $z_0 = 0$ and $z_1 = c$, and define a sequence by putting $z_{n+1} = z_n^2 + c$.

(a) Show that if $|c| > 2$, then $\displaystyle\lim_{n \to \infty} z_n = \infty$. (*Hint.* Let $r = |c| - 1$ and use induction to show that $|z_n| > |c| r^{n-1}$ for all n.)

(b) Show that if $|c| \leq 2$ and there is a value of k with $|z_k| > 2$, then $\displaystyle\lim_{n \to \infty} z_n = \infty$. (*Hint.* Let $r = |z_k| - 1$, and show that $|z_{k+p}| \geq |z_k| r^p$ for all $p \geq 0$.)

Remark: Those values of c for which the sequence z_n defined in this problem stays bounded form a very interesting set with many pretty patterns called the *Mandelbrot set.* See A. K. Dewdney, "Computer Recreations," *Scientific American,* August 1985. See also Figure 6.S.B.1.

3. What is the limit of the sequence $f_n(x) = (1 + x)^{1/n}$, $x \geq 0$? Does it converge uniformly?

4. (a) Show that the series $\sum_{n=0}^{\infty} 1/(n^2 + z)$ converges on the set $\mathbf{C} \setminus \{z = ni \mid n \text{ is an integer}\}$.
(b) Show that the convergence is uniform and absolute on any closed disk contained in this region.

5. (a) Show that the sequence of functions $f_n(z) = z^n$ converges uniformly to the zero function $f(z) = 0$ on every closed disk $D_r = \{z \text{ such that } |z| \leq r\}$ with $r < 1$.
(b) Is the convergence uniform on the open unit disk $D(0; 1)$?

6. (a) Show that the sequence of functions $f_n(z) = \cos(x/n)$ converges uniformly to the constant function $f(x) = 1$ for $x \in [0, \pi]$.
(b) Show that it converges pointwise to 1 on all of $\mathbf{R}$.
(c) Is the convergence uniform on all of $\mathbf{R}$?

7. Test the following series for absolute convergence and convergence:

(a) $\displaystyle\sum_{2}^{\infty} \frac{i^n}{\log n}$

(b) $\displaystyle\sum_{1}^{\infty} \frac{i^n}{n}$

8. Prove that

$$\zeta(z) = \sum_{n=1}^{\infty} \frac{1}{n^z}$$

does not converge uniformly on $A = \{z \mid \operatorname{Re} z > 1\}$.

9. If $\sum_{k=1}^{\infty} g_k(z)$ is a uniformly convergent series of continuous functions and if $z_n \to z$, show that

$$\lim_{n \to \infty} \sum_{k=1}^{\infty} g_k(z_n) = \sum_{k=1}^{\infty} g_k(z)$$

10. If $\sum_{k=1}^{\infty} a_k$ converges, prove that $a_k \to 0$. If $\sum_{k=1}^{\infty} g_k(z)$ converges uniformly, show that $g_k \to 0$ uniformly.

11. Show that

$$\sum_{n=1}^{\infty} \frac{1}{z^n}$$

is analytic on $A = \{z \text{ such that } |z| > 1\}$.

12. Show that

$$\sum_{n=1}^{\infty} \frac{1}{n! z^n}$$

is analytic on $\mathbb{C}\setminus\{0\}$. Compute its integral around the unit circle.

13. Show that $\sum_{n=1}^{\infty} e^{-n} \sin nz$ is analytic in the region $A = \{z \mid -1 < \text{Im } z < 1\}$.

14. Prove that the series

$$\sum_{n=1}^{\infty} \frac{z^n}{1 + z^{2n}}$$

converges in both the interior and exterior of the unit circle and represents an analytic function in each region.

15. Show that $\sum_{n=1}^{\infty} (\log n)^k n^{-z}$ is analytic on $\{z \mid \text{Re } z > 1\}$. (*Hint.* Use the result of Worked Example 3.1.15.)

16. Let f be an analytic function on the disk $D(0; 2)$ such that $|f(z)| \leq 7$ for all $z \in D(0; 2)$. Prove that there exists a $\delta > 0$ such that if $z_1, z_2 \in D(0; 1)$, and if $|z_1 - z_2| < \delta$, then $|f(z_1) - f(z_2)| < \frac{1}{10}$. Find a numerical value of δ independent of f that has this property. (*Hint.* Use the Cauchy integral formula.)

17. If $f_n(z) \to f(z)$ uniformly on a region A, and if f_n is analytic on A, is it true that $f'_n(z) \to f'(z)$ uniformly on A? (*Hint.* See Worked Example 3.1.16.)

18. Prove: $f_n \to f$ uniformly on every closed disk in a region A iff $f_n \to f$ uniformly on every compact (closed and bounded) subset of A.

19. Find a suitable region in which

$$\sum_{n=1}^{\infty} \frac{(2z-1)^n}{n}$$

is analytic.

20. Let f_n be analytic on a bounded region A and continuous on cl (A), $n = 1, 2, 3, \ldots$. Suppose that the functions f_n converge uniformly on bd (A). Then prove that the functions f_n converge uniformly to an analytic function on A. (*Hint.* Use the maximum modulus theorem.)

3.2 POWER SERIES AND TAYLOR'S THEOREM

This section will consider a special kind of series called power series, which have the form $\sum_{n=0}^{\infty} a_n(z-z_0)^n$. We shall examine their convergence properties and show that a function is analytic iff it is locally representable as a convergent power series. To obtain this representation, we first need to establish Taylor's theorem, which asserts that if f is analytic on an open disk centered at z_0, then the *Taylor series* of f,

$$\sum_{n=0}^{\infty} \frac{f^{(n)}(z_0)}{n!} (z-z_0)^n$$

converges on the disk and equals $f(z)$ everywhere on that disk.

In proving the results of this section we shall use the techniques developed in Sec. 3.1 and Cauchy's integral formula.

Convergence of Power Series

A *power series* is a series of the form $\sum_{n=0}^{\infty} a_n(z-z_0)^n$. (Here a_n and $z_0 \in \mathbf{C}$ are fixed complex numbers.) Each term $a_n(z-z_0)^n$ is entire, and so in proving that the sum is analytic on a region, we can use the analytic convergence theorem (3.1.8). The basic fact to remember about power series is that the appropriate domain of analyticity is the interior of a circle centered at z_0. This is established in the first theorem.

3.2.1 POWER SERIES CONVERGENCE THEOREM *Let* $\sum_{n=0}^{\infty} a_n(z-z_0)^n$ *be a power series. There is a unique number* $R \geq 0$, *possibly* $+\infty$, *called the **radius of convergence,** such that if* $|z-z_0| < R$, *the series converges, and if* $|z-z_0| > R$, *the series diverges. Furthermore, the convergence is uniform and absolute on every closed disk in*

$A = \{z \in \mathbf{C} \text{ such that } |z - z_0| < R\}$. *No general statement about convergence can be made if* $|z - z_0| = R$. (See Fig. 3.2.1.)

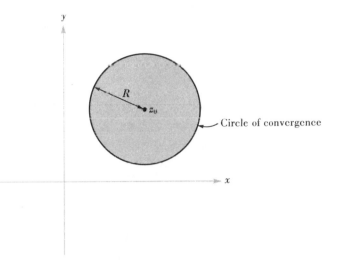

FIGURE 3.2.1   Convergence of power series. Series converges within circle; series diverges outside circle.

Thus on the region $A = \{z \in \mathbf{C} \text{ such that } |z - z_0| < R\}$, the series converges and we have divergence at z if $|z - z_0| > R$. The circle $|z - z_0| = R$ is called the *circle of convergence* of the given power series. Practical methods of calculating R use the ratio and root tests (3.2.5).

PROOF   Set $R = \sup \{r \geq 0 \mid \Sigma_{n=0}^{\infty} |a_n| r^n \text{ converges}\}$, where sup means the least upper bound of that set of real numbers. We shall show that R has the desired properties. The following lemma is useful in this regard.

3.2.2 ABEL-WEIERSTRASS LEMMA   *Suppose that* $r_0 \geq 0$ *and that* $|a_n| r_0^n \leq M$ *for all n, where M is some constant. Then for* $r < r_0$, $\Sigma_{n=0}^{\infty} a_n (z - z_0)^n$ *converges uniformly and absolutely on the closed disk* $A_r = \{z \text{ such that } |z - z_0| \leq r\}$.

PROOF   For $z \in A_r$, we have

$$|a_n(z - z_0)^n| \leq |a_n| r^n \leq M \left(\frac{r}{r_0}\right)^n$$

Let

$$M_n = M \left(\frac{r}{r_0}\right)^n$$

Since $r/r_0 < 1$, $\Sigma\, M_n$ converges. Thus by the Weierstrass M test (3.1.7), the series converges uniformly and absolutely on A_r. ▼

Now we can prove the first part of the power series convergence theorem. Let $r_0 < R$. By the definition of R there is an r_1, $r_0 < r_1 \le R$ such that $\Sigma\, |a_n| r_1^n$ converges. Therefore $\Sigma_{n=0}^{\infty} |a_n| r_0^n$ converges by the comparison test. The terms $|a_n| r_0^n$ are bounded (in fact $\to 0$), and so by the Abel-Weierstrass lemma the series converges uniformly and absolutely on A_r for any $r < r_0$. Since any z with $|z - z_0| < R$ lies in some A_r and since we can always choose r_0 such that $r < r_0 < R$, we have convergence at z.

Now suppose that $|z_1 - z_0| > R$ and $\Sigma\, a_n(z_1 - z_0)^n$ converges. We shall derive a contradiction. The terms $a_n(z_1 - z_0)^n$ are bounded in absolute value because they approach zero. Thus, by the Abel-Weierstrass lemma, if $R < r < |z_1 - z_0|$, then $\Sigma\, a_n(z_1 - z_0)^n$ converges absolutely if $z_1 \in A_r$. Therefore, $\Sigma\, |a_n| r^n$ converges. But this would mean, by definition of R, that $R < R$.

We have proved that the convergence is uniform and absolute on every *strictly smaller* closed disk A_r and hence on any closed disk in A. ■

Combining the analytic convergence and power series convergence theorems, we may deduce the following:

3.2.3 ANALYTICITY OF POWER SERIES   *A power series $\Sigma_{n=0}^{\infty} a_n(z - z_0)^n$ is an analytic function on the inside of its circle of convergence.*

We also know that we can differentiate term by term. Thus:

3.2.4 DIFFERENTIATION OF POWER SERIES   *Let $f(z) = \Sigma_{n=0}^{\infty} a_n(z - z_0)^n$ be the analytic function defined on the inside of the circle of convergence of the given power series. Then $f'(z) = \Sigma_{n=1}^{\infty} n a_n(z - z_0)^{n-1}$, and this series has the same circle of convergence as $\Sigma\, a_n(z - z_0)^n$. Furthermore, the coefficients a_n are given by $a_n = f^{(n)}(z_0)/n!$.*

PROOF   We know from the analytic convergence theorem (3.1.8) that the derivative $f'(z) = \Sigma_{n=1}^{\infty} a_n(z - z_0)^{n-1}$ converges on $A = D(z_0; R) = \{z \in \mathbb{C} \text{ such that } |z - z_0| < R\}$.

To show that the derived series has the same circle of convergence as the original series, we need only show that it diverges for $|z - z_0| > R$. If it did converge at some point z_1 with $|z_1 - z_0| = r_0 > R$, then $na_n r_0^{n-1}$ would be bounded. Thus $a_n r_0^n = (na_n r_0^{n-1})(r_0/n)$ would also be bounded and so $\Sigma\, a_n(z - z_0)^n$ would converge for $R \le |z - z_0| < r_0$ by the Abel-Weierstrass lemma. But this contradicts the maximal property of R from the power series convergence theorem (3.2.1). This establishes the assertion about the radius of convergence.

To identify the coefficients, set $z = z_0$ in the formula defining $f(z)$ to find $f(z_0) = a_0$. Proceeding inductively, we find

$$f^{(n)}(z) = n!a_n + \sum_{k=n+1}^{\infty} k(k-1)(k-2)\cdots(k-n+1)(z-z_0)^{k-n}$$

and setting $z = z_0$, we get $f^{(n)}(z_0) = n!a_n$. ∎

It is important to notice just what has been done in the last assertion of this theorem. The coefficients of a power series around a particular center are completely determined by the function which that series represents. Thus if two apparently different series have been obtained for the same function about the same center, they must in fact be the same.

3.2.5 UNIQUENESS OF POWER SERIES *Power series expansions around the same center are unique. If $\Sigma_{n=0}^{\infty} a_n(z - z_0)^n = f(z) = \Sigma_{n=0}^{\infty} b_n(z - z_0)^n$ for all z in some nontrivial disk $D(z_0; r)$ with $r > 0$, then $a_n = b_n$ for $n = 1, 2, 3, \ldots$.*

PROOF The last assertion of the differentiation of power series theorem says $a_n = f^{(n)}(z_0)/n! = b_n$. ∎

This observation may be used in a number of ways. In particular, it says that whatever tricks we can use to find a convergent power series representing a function, it must be the Taylor series. It can also help us use power series in the solution of differential equations and other problems. Several of these tricks and ideas for the manipulation and application of power series are demonstrated in the worked examples.

We will now obtain some practical methods of computing the radius of convergence R. (The method of defining R given in the proof of Theorem 3.2.1 is not useful for computing R in specific examples.)

3.2.6 PROPOSITION   *Consider a power series $\sum_{n=0}^{\infty} a_n(z - z_0)^n$.*

(i) **Ratio test:** *If*

$$\lim_{n\to\infty} \frac{|a_n|}{|a_{n+1}|}$$

exists, then it equals R, the radius of convergence of the series.

(ii) **Root test** *If $\rho = \lim_{n\to\infty} \sqrt[n]{|a_n|}$ exists, then $R = 1/\rho$ is the radius of convergence. (Set $R = \infty$ if $\rho = 0$; set $R = 0$ if $\rho = \infty$.)*

PROOF   To prove both cases we show that $R = \sup \{r \geq 0 \mid \sum_{n=0}^{\infty} |a_n| r^n < \infty\}$.

(i) By the ratio test (Proposition 3.1.3) we know that $\sum_{n=0}^{\infty} |a_n| r^n$ converges or diverges as

$$\lim_{n\to\infty} \frac{|a_{n+1} r^{n+1}|}{|a_n r^n|} < 1 \quad \text{or} \quad > 1$$

that is, according to whether

$$\lim_{n\to\infty} \frac{|a_n|}{|a_{n+1}|} > r \quad \text{or} \quad \lim_{n\to\infty} \frac{|a_n|}{|a_{n+1}|} < r$$

Thus, by the characterization of R in the power series convergence theorem (3.2.1), the limit equals R.

(ii) By the root test (Proposition 3.1.3) we know that $\sum_{n=0}^{\infty} |a_n| r^n$ converges or diverges as $\lim_{n\to\infty} (|a_n| r^n)^{1/n} < 1$ or > 1; that is, according to whether

$$r < 1/\lim_{n\to\infty} |a_n|^{1/n} \text{ or } r > 1/\lim_{n\to\infty} |a_n|^{1/n}$$

The result follows as in (i). ∎

For example:

The series $\sum_{n=0}^{\infty} z^n$ has radius of convergence 1 since $a_n = 1$, and thus we have $\lim_{n\to\infty} |a_n/a_{n+1}| = 1$.

The series $\sum_{n=0}^{\infty} z^n/n!$ has radius of convergence $R = +\infty$ (that is, the function is entire), since $a_n = 1/n!$, and so $|a_n/a_{n+1}| = n + 1 \to \infty$.

The series $\sum_{n=0}^{\infty} n! z^n$ has radius of convergence $R = 0$ as $|a_n/a_{n+1}| = 1/(n + 1) \to 0$. (This function thus does not have a nontrivial region of analyticity.)

REMARK By refining the root test, it is possible to show that $R = 1/\rho$ where $\rho = \lim_{n\to\infty} \sup \sqrt[n]{|a_n|}$, which always exists. ($\lim_{n\to\infty} \sup c_n = \lim_{n\to\infty} (\sup \{c_n, c_{n+1}, \ldots\})$, by definition. This is known as Hadamard's formula for the radius of convergence.) There is no analogous refinement for the ratio test (known to us).

Taylor's Theorem

It is obvious from the preceding computations that if $f: A \to \mathbf{C}$ equals, in a small disk around each $z_0 \in A$, a convergent power series, then f is analytic. The converse is also true: If f is analytic it equals, on every disk in its domain, a convergent power series. This is made explicit in the next theorem.

3.2.7 TAYLOR'S THEOREM *Let f be analytic on a region A. Let $z_0 \in A$ and let $A_r = \{z \text{ such that } |z - z_0| < r\}$ be contained in A (usually the largest open disk possible is used: if $r = \infty$, $A_r = A = \mathbf{C}$) (see Fig. 3.2.2). Then for every $z \in A_r$, the series*

$$\sum_{n=0}^{\infty} \frac{f^{(n)}(z_0)}{n!} (z - z_0)^n$$

converges on A_r (that is, has a radius of convergence $\geq r$), and we have

$$f(z) = \sum_{n=0}^{\infty} \frac{f^{(n)}(z_0)}{n!} (z - z_0)^n \tag{1}$$

*(We use the convention $0! = 1$.) The series of Eq. (1) is called the **Taylor series** of f around the point z_0.*

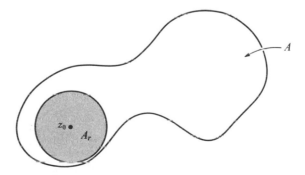

FIGURE 3.2.2 Taylor's theorem.

Before proving this result let us study an example that illustrates its usefulness. Consider $f(z) = e^z$. Here f is analytic, and $f^{(n)}(z) = e^z$ for all n, so that $f^{(n)}(0) = 1$ and thus

$$e^z = \sum_{n=0}^{\infty} \frac{z^n}{n!} \tag{2}$$

which is valid for all $z \in \mathbf{C}$, since e^z is entire. Table 3.2.1 lists the Taylor series of some common elementary functions. The Taylor series around the point $z_0 = 0$ is sometimes called the *Maclaurin series*.

All of the series in Table 3.2.1 are important and useful. They may be established by taking successive derivatives and using Taylor's theorem. The binomial series should be familiar from algebra. The Taylor series for many functions can be found by other means, using the special properties of power series which allow their manipulation. Some of these are presented in the worked examples. We have already found the geomet-

Table 3.2.1   Some Common Expansions

Function	Taylor series around 0	Where valid
$\dfrac{1}{1-z}$	$\displaystyle\sum_{n=0}^{\infty} z^n$ (geometric series)	$\lvert z \rvert < 1$
e^z	$\displaystyle\sum_{n=0}^{\infty} \frac{z^n}{n!}$	all z
$\sin z$	$z - \dfrac{z^3}{3!} + \dfrac{z^5}{5!} - \cdots = \displaystyle\sum_{n=1}^{\infty} (-1)^{n+1} \dfrac{z^{2n-1}}{(2n-1)!}$	all z
$\cos z$	$1 - \dfrac{z^2}{2} + \dfrac{z^4}{4!} - \dfrac{z^6}{6!} + \cdots = \displaystyle\sum_{n=0}^{\infty} (-1)^n \dfrac{z^{2n}}{(2n)!}$	all z
$\log(1+z)$ (principal branch)	$\displaystyle\sum_{n=1}^{\infty} \dfrac{(-1)^{n-1}}{n} z^n$	$\lvert z \rvert < 1$
$(1+z)^\alpha$ (principal branch) with $\alpha \in \mathbf{C}$ fixed	$\displaystyle\sum_{n=0}^{\infty} \binom{\alpha}{n} z^n$ (binomial series), where $\dbinom{\alpha}{n} = \dfrac{\alpha(\alpha-1)\cdots(\alpha-n+1)}{n!}$ $\left(\text{let } \dbinom{\alpha}{n} \text{ be zero if } \alpha \text{ is an integer} < n \text{ and let } \dbinom{\alpha}{0} = 1\right)$	

ric series for $1/(1 - z)$ and the series for $\log(1 + z)$ in Worked Examples 3.1.12 and 3.1.14. It is particularly important to notice that we have done this for the geometric series, since we will use it in the proof of Taylor's theorem which follows.

PROOF OF TAYLOR'S THEOREM Let $0 < \sigma < r$ and let γ be the circle $\gamma(t) = z_0 + \sigma e^{it}$, $0 \le t \le 2\pi$, of radius σ centered at z_0. If z is any point inside γ, Cauchy's integral formula gives

$$f(z) = \frac{1}{2\pi i} \int_\gamma \frac{f(\zeta)}{\zeta - z} \, d\zeta$$

The plan is to use the geometric series to expand the integrand as a power series in $z - z_0$ and then use Proposition 3.1.9 to integrate term by term. Finally the coefficients of the resulting integrated series are recognized to be those of the Taylor series by the Cauchy integral formula for derivatives. Since z is inside the circle γ and ζ is on its boundary, we have $|(z - z_0)/(\zeta - z_0)| < 1$. The geometric series of Worked Example 3.1.12 allows the following expansion:

$$\frac{1}{\zeta - z} = \frac{1}{\zeta - z_0} \cdot \frac{1}{1 - \dfrac{z - z_0}{\zeta - z_0}} = \frac{1}{\zeta - z_0} \sum_{n=0}^{\infty} \left(\frac{z - z_0}{\zeta - z_0} \right)^n$$

so that

$$f(z) = \frac{1}{2\pi i} \int_\gamma \left[\frac{f(\zeta)}{\zeta - z_0} \sum_{n=0}^{\infty} \left(\frac{z - z_0}{\zeta - z_0} \right)^n \right] d\zeta = \frac{1}{2\pi i} \int_\gamma \left[\sum_{n=0}^{\infty} \frac{f(\zeta)(z - z_0)^n}{(\zeta - z_0)^{n+1}} \right] d\zeta$$

Furthermore, since the curve γ stays away from the boundary of the disk of convergence, Worked Example 3.1.12 also shows that the convergence of the series

$$\sum_{n=0}^{\infty} \left(\frac{z - z_0}{\zeta - z_0} \right)^n$$

is uniform in ζ as ζ goes around the circle γ with z fixed. Also, $f(\zeta)/(\zeta - z_0)$ is a continuous function of ζ around the circle γ, so it is bounded there. It follows that the series

$$\sum_{n=0}^{\infty} \frac{f(\zeta)(z - z_0)^n}{(\zeta - z_0)^{n+1}}$$

converges uniformly on γ to $f(\zeta)/(\zeta - z)$. (The first series satisfies the Cauchy criterion uniformly in ζ, so it still satisfies it after being multiplied by something which remains bounded. The student is asked to supply the details in Exercise 21.). By Proposition 3.1.9, we have

$$f(z) = \sum_{n=0}^{\infty} \frac{1}{2\pi i} \int_{\gamma} \frac{f(\zeta)(z-z_0)^n}{(\zeta - z_0)^{n+1}} \, d\zeta$$

$$= \sum_{n=0}^{\infty} \left[(z-z_0)^n \frac{1}{2\pi i} \int_{\gamma} \frac{f(\zeta)}{(\zeta - z_0)^{n+1}} \, d\zeta \right]$$

$$= \sum_{n=0}^{\infty} \left[(z-z_0)^n \frac{f^{(n)}(z_0)}{n!} \right]$$

as desired. The last equality is just the Cauchy integral formula for derivatives. Since the radius of the circle γ was arbitrary, so long as it fit inside the region of analyticity, this representation of $f(z)$ is valid in the largest open disk centered at z_0 which is contained in the region A. ■

The following consequence of this theorem was mentioned informally at the beginning of this section.

3.2.8 COROLLARY *Let A be a region in $\mathbf{C}$ and let f be a complex-valued function defined on A. Then f is analytic on A if and only if for each z_0 in A there is a number $r > 0$ such that the disk $D(z_0; r) \subset A$ and f equals a convergent power series on $D(z_0; r)$.*

PROOF Taylor's theorem shows that every analytic function is equal to a power series, in fact to its Taylor series, on every disk in A. On the other hand, if $f(z)$ is equal to a convergent power series on $D(z_0; r)$, then $D(z_0; r)$ must be in the interior of the circle of convergence of the series and so f must be analytic on $D(z_0; r)$. Since there is such a disk and convergent power series for each z_0 in A, it follows from the analyticity of power series (3.2.3) that f is analytic on A. ■

The condition in this corollary may thus be taken as an alternative definition of "analytic." We have shown that the notions of differentiability on a region and analyticity on a region coincide for functions of a complex variable. (*Caution:* remember that they do *not* coincide for real variables.) Cauchy's theorem, Cauchy's integral formulas, and Taylor's theorem are among the most fundamental theorems of complex analysis.

In specific examples, the derivatives of f may be complicated and finding the Taylor series may be made easier by searching directly for a convergent series which represents f rather than computing the derivatives. By Corollary 3.2.5, if $f(z) = \sum_{n=0}^{\infty} a_n(z - z_0)^n$ and the series converges, then it must be the Taylor series. In fact we can sometimes then use Taylor's theorem to tell us formulas for the derivatives, having found the series by other means. Some of these tricks for manipulating series and applications are found in the worked examples.

Worked Examples

3.2.9 *Use the series expansions given in Table 3.2.1 to confirm the identity $e^{iz} = \cos z + i \sin z$ for all z.*

Solution.

$$\cos z + i \sin z = \sum_{n=0}^{\infty} \frac{(-1)^n z^{2n}}{(2n)!} + i \sum_{n=1}^{\infty} \frac{(-1)^{n+1} z^{2n-1}}{(2n-1)!}$$

$$= \sum_{n=0}^{\infty} \frac{(iz)^{2n}}{(2n)!} + \sum_{n=1}^{\infty} \frac{(-i)i^{2n}z^{2n-1}}{(2n-1)!}$$

$$= \sum_{n=0}^{\infty} \frac{(iz)^{2n}}{(2n)!} + \sum_{n=1}^{\infty} \frac{(iz)^{2n-1}}{(2n-1)!}$$

$$= \sum_{k=0}^{\infty} \frac{(iz)^k}{k!} = e^{iz}$$

as desired.

3.2.10 *Can a power series $\Sigma\, a_n (z-2)^n$ converge at $z = 0$ but diverge at $z = 3$?*

Solution. No. If it converges at $z = 0$ this implies, by the power series convergence theorem (3.2.1), that the radius of convergence R satisfies $R \geq 2$. But $z = 3$ lies inside that circle, so the series would converge there (see Fig. 3.2.3).

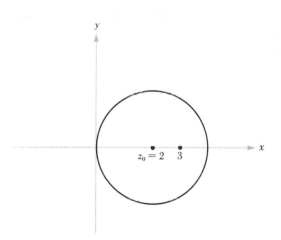

FIGURE 3.2.3 The circle of convergence for the power series in Worked Example 3.2.10 must be at least this big.

3.2.11 *Find the Taylor series around $z_0 = 0$ for $f(z) = 1/(4 + z^2)$ and calculate the radius of convergence.*

Solution. Write $f(z) = \frac{1}{4}\{1/[1 - (-z^2/4)]\}$. We know that so long as $|w| < 1$, then $1/(1 - w) = \sum_{n=0}^{\infty} w^n$. Replacing w by $-z^2/4$ gives

$$f(z) = \frac{1}{4} \sum_{n=0}^{\infty} \left(-\frac{z^2}{4}\right)^n = \sum_{n=0}^{\infty} (-1)^n 4^{-(n+1)} z^{2n}$$

so long as $|-z^2/4| < 1$; that is, so long as $|z| < 2$. Therefore the radius of convergence is 2. Notice that this is the largest disk around $z_0 = 0$ on which f is analytic, since analyticity fails at $z = \pm 2i$.

3.2.12 *Find the Taylor series of* $\log(1 + z)$ *around $z = 0$ and give its radius of convergence* (see Table 3.2.1).

First Solution. We have already done this problem as Worked Example 3.1.14 using the geometric series and term-by-term integration.

Second Solution. We use the principal branch of log so that the function $f(z) = \log(1 + z)$ is defined at $z = 0$. Since f is analytic on the region $A = \mathbb{C}\backslash\{x + iy \mid y = 0, x \le -1\}$ shown in Fig. 3.2.4, the radius of convergence of the Taylor series will be ≥ 1 by Taylor's theorem (3.2.7). That it is *exactly* 1 can be shown as follows. We know that

$$f(0) = \log 1 = 0$$

$$f'(z) = \frac{1}{z + 1} \quad \text{and so} \quad f'(0) = 1$$

$$f''(z) = -\frac{1}{(z + 1)^2} \quad \text{and so} \quad f''(0) = -1$$

and

$$f'''(z) = \frac{2}{(z + 1)^3} \quad \text{and so} \quad f'''(0) = 2$$

Inductively, we see that

$$f^{(n)}(z) = \frac{(n - 1)!(-1)^{n-1}}{(z + 1)^n}$$

so that $f^{(n)}(0) = (n - 1)!(-1)^{n-1}$. Thus the Taylor series is

$$\sum_{n=0}^{\infty} \frac{f^{(n)}(0)}{n!} z^n = \sum_{n=1}^{\infty} \frac{(-1)^{n-1}}{n} z^n$$

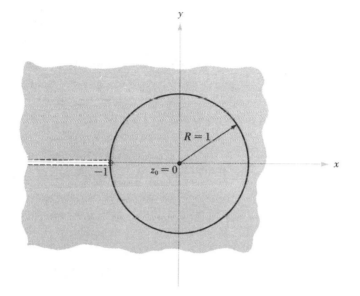

FIGURE 3.2.4 Taylor series of log $(1 + z)$.

(in agreement with Table 3.2.1). When $z = -1$, it is the harmonic series which diverges, so the radius of convergence is ≤ 1 and thus is exactly 1. (A general procedure to follow for determining the exact radius of convergence without computing the series is found in Exercise 19.)

3.2.13 *Suppose that $\Sigma\, a_n z^n$ and $\Sigma\, b_n z^n$ have radii of convergence $\geq r_0$. Define $c_n = \Sigma_{k=0}^{n} a_k b_{n-k}$. Prove that $\Sigma\, c_n z^n$ has radius of convergence $\geq r_0$ and that inside this circle of radius r_0 we have*

$$\sum_{n=0}^{\infty} c_n z^n = \left(\sum_{n=0}^{\infty} a_n z^n \right)\left(\sum_{n=0}^{\infty} b_n z^n \right)$$

Solution. This way of multiplying out two power series is a generalization of the manner in which polynomials are multiplied. A direct proof can be given but would be somewhat lengthy. If we use Taylor's theorem, the proof is fairly simple. Let $f(z) = \Sigma_{n=0}^{\infty} a_n z^n$, $g(z) = \Sigma_{n=0}^{\infty} b_n z^n$, and let $A = \{z$ such that $|z| < r_0\}$. Then f and g are analytic on A, so fg is also analytic on A. By Taylor's theorem we can write

$$(f \cdot g)(z) = \sum_{n=0}^{\infty} \frac{(f \cdot g)^{(n)}(0)}{n!} z^n$$

for all z in A. It is a simple exercise (as in calculus) to show by induction that the nth derivative of the product $f(z)g(z)$ is given by

$$(f \cdot g)^{(n)}(z) = \sum_{k=0}^{n} \binom{n}{k} f^{(k)}(z) g^{(n-k)}(z)$$

where

$$\binom{n}{k} = \frac{n!}{k!(n-k)!}$$

Hence

$$\frac{(f \cdot g)^{(n)}(0)}{n!} = \sum_{k=0}^{n} \frac{1}{k!(n-k)!} f^{(k)}(0) g^{(n-k)}(0) = \sum_{k=0}^{n} a_k b_{n-k}$$

Thus $\sum_{n=0}^{\infty} c_n z^n$ converges on A (and therefore, by Taylor's theorem, the radius of convergence is $\geq r_0$) and on A

$$\sum_{n=0}^{\infty} c_n z^n = (f \cdot g)(z) = \left(\sum_{n=0}^{\infty} a_n z^n \right) \left(\sum_{n=0}^{\infty} b_n z^n \right)$$

3.2.14 *Compute the Taylor series around $z = 0$ and give the radii of convergence for*
(a) $z/(z-1)$
and
(b) $e^z/(1-z)$ *(Compute the first few terms only.)*

Solution.
(a) From the binomial expansion (Table 3.2.1), we have $(1-z)^{-1} = 1 + z + z^2 + \cdots$ valid for $|z| < 1$. Hence for $|z| < 1$, $z/(z-1) = -z(1 + z + z^2 + \cdots) = -z - z^2 - z^3 - z^4 - \cdots$. By the uniqueness of representation by power series, this is the Taylor series of $z/(z-1)$ around 0. By observing that $z/(z-1)$ is analytic on the open disk $|z| < 1$, we know by the Taylor theorem that the Taylor series must have a radius of convergence ≥ 1. Of course, a close analysis of the series $-z - z^2 - z^3 - z^4 - \cdots$, using the ratio test or the root test, shows that the radius of convergence is exactly 1.
(b) $1/(1-z) = 1 + z + z^2 + \cdots$ for $|z| < 1$ and $e^z = 1 + z + z^2/2 + \cdots$ for all z. Thus by Worked Example 3.2.13 we get the series for the product by formally multiplying the two series out as if they were polynomials; the result must still converge for $|z| < 1$. We get

$$\frac{e^z}{1-z} = (1 + z + z^2 + z^3 + \cdots) \left(1 + z + \frac{z^2}{2} + \frac{z^3}{3!} + \cdots \right)$$

$$= 1 + (z + z) + \left(\frac{z^2}{2} + z^2 + z^2 \right) + \left(\frac{z^3}{6} + \frac{z^3}{2} + z^3 + z^3 \right) + \cdots$$

$$= 1 + 2z + \frac{5z^2}{2} + \frac{8z^3}{3} + \cdots$$

In this last series the general term has no simple form. Note that this method is faster than computing $f^{(k)}(0)$ for moderately large k.

3.2.15 *Find the Taylor series around $z_0 = 0$ for $f(z) = 1/(z^2 - 5z + 6)$ and determine the radius of convergence.*

First Solution. Factor the denominator, use the geometric series twice, and multiply the resulting power series:

$$f(z) = \frac{1}{(z-3)(z-2)} = \frac{1}{6}\left(\frac{1}{1-z/3}\right)\left(\frac{1}{1-z/2}\right)$$

If $|z| < 2$, then each of the last two terms may be expanded by the geometric series:

$$f(z) = \frac{1}{6}\left[\sum_{n=0}^{\infty}\left(\frac{z}{3}\right)^n\right]\left[\sum_{n=0}^{\infty}\left(\frac{z}{2}\right)^n\right] = \frac{1}{6}\sum_{n=0}^{\infty}\left[\sum_{k=0}^{n}(\tfrac{1}{3})^k(\tfrac{1}{2})^{n-k}\right]z^n$$

In general, $a^{n+1} - b^{n+1} = (a - b)\sum_{k=0}^{n}a^k b^{n-k}$, and so the last equation becomes

$$f(z) = \frac{1}{6}\sum_{n=0}^{\infty}\left[\frac{(\tfrac{1}{3})^{n+1} - (\tfrac{1}{2})^{n+1}}{\tfrac{1}{3} - \tfrac{1}{2}}\right]z^n = \sum_{n=0}^{\infty}[(\tfrac{1}{2})^{n+1} - (\tfrac{1}{3})^{n+1}]z^n$$

All these expansions and operations are valid if $|z| < 2$, and so the radius of convergence should be 2. Indeed, analyticity fails only at $z = 2$ and $z = 3$. The nearest of these to $z_0 = 0$ is 2. Therefore the radius of convergence is 2.

Second Solution. Instead of using the algebraic identity employed above and the multiplication of series, we can use partial fractions and addition of series:

$$f(z) = \frac{1}{(z-3)(z-2)} = \frac{1}{z-3} - \frac{1}{z-2}$$

$$= \frac{1}{2}\left(\frac{1}{1-z/2}\right) - \frac{1}{3}\left(\frac{1}{1-z/3}\right)$$

$$= \frac{1}{2}\sum_{n=0}^{\infty}\left(\frac{z}{2}\right)^n - \frac{1}{3}\sum_{n=0}^{\infty}\left(\frac{z}{3}\right)^n$$

$$= \sum_{n=0}^{\infty}[(\tfrac{1}{2})^{n+1} - (\tfrac{1}{3})^{n+1}]z^n$$

as before.

3.2.16 *Compute the first few terms of the Taylor series for $f(z) = \sec z$ around $z_0 = 0$. What is the radius of convergence?*

Solution. Suppose $f(z) = \sec z = 1/\cos z = a_0 + a_1 z + a_2 z^2 + \cdots$. Multiplying by $\cos z$,

$$1 = \left(1 - \frac{z^2}{2} + \frac{z^4}{24} - \cdots\right)(a_0 + a_1 z + a_2 z^2 + a_3 z^3 + a_4 z^4 + \cdots)$$

$$= a_0 + a_1 z + \left(a_2 - \frac{a_0}{2}\right)z^2 + \left(a_3 - \frac{a_1}{2}\right)z^3 + \left(a_4 - \frac{a_2}{2} + \frac{a_0}{24}\right)z^4 + \cdots$$

Therefore (since there can be only one series expansion for the constant function $g(z) = 1$) we have

$$a_0 = 1 \qquad a_1 = 0 \qquad a_2 = \frac{a_0}{2} = \frac{1}{2} \qquad a_3 = \frac{a_1}{2} = 0 \qquad a_4 = \frac{a_2}{2} - \frac{a_0}{24} = \frac{5}{24}$$

The radius of convergence must be $\pi/2$, since the nearest points to 0 at which analyticity fails are $z = \pm\pi/2$, the points where $\cos z = 0$.

3.2.17 *Application to Differential Equations: Find a function $f(z)$ such that $f(0) = 0$ and $f'(x) = 3f(x) + 2$ for all x.*

Solution. Suppose that there is a solution f which is the restriction to the real axis of a function which is analytic in $\mathbf{C}$. Therefore, it will have a power series expansion $f(z) = \sum_{n=0}^{\infty} a_n z^n$. Since $f'(z) = \sum_{n=1}^{\infty} n a_n z^{n-1}$, we must have

$$\sum_{n=1}^{\infty} n a_n z^{n-1} = 3\left(\sum_{n=0}^{\infty} a_n z^n\right) + 2$$

or

$$\sum_{n=0}^{\infty} (n+1)a_{n+1} z^n = (2 + 3a_0) + \sum_{n=1}^{\infty} 3a_n z^n$$

Thus

$$0 = (2 + 3a_0 - a_1) + \sum_{n=1}^{\infty} [3a_n - (n+1)a_{n+1}]z^n$$

We know $a_0 = f(0) = 0$. Therefore $a_1 = 2$. For $n \geq 1$, $a_{n+1} = 3a_n/(n+1)$, and so $a_2 = 3a_1/2$, $a_3 = 3^2 a_1/(3)(2)$, $a_4 = 3^3 a_1/(4)(3)(2)$, $\ldots$, $a_n = 3^{n-1}a_1/n! = (\frac{2}{3})3^n/n!$. (Notice that this formula also gives $a_1 = 2$.) Thus if there is a power series which represents a solution it must be

$$f(z) = \frac{2}{3}\sum_{n=1}^{\infty} \frac{3^n}{n!} z^n$$

Taking derivatives term by term confirms that this is a solution. In this case we can even recognize the function which the series represents:

$$f(z) = \frac{2}{3} \sum_{n=1}^{\infty} \frac{(3z)^n}{n!} = \frac{2}{3} \left[\left(\sum_{n=0}^{\infty} \frac{(3z)^n}{n!} \right) - 1 \right] = \tfrac{2}{3}(e^{3z} - 1)$$

The reader should check that this does solve the original problem.*

3.2.18 *Generating Function for the Hermite Polynomials: The function* $f(z) = e^{2xz - z^2}$ *is analytic everywhere and so has a power series expansion in powers of* z *whose coefficients depend on* x. *If we put* $f(x) = \sum_{n=0}^{\infty} H_n(x) z^n / n!$, *then the functions* $H_n(x)$ *are called the* **Hermite polynomials.** *(One needs to check that they are in fact polynomials in* x.) *The function* f *is called a* **generating function.** *Compute* $H_0(x)$, $H_1(x)$, *and* $H_2(x)$.

Solution.

$$H_0(x) = f(0) = 1$$
$$H_1(x) = f'(0) = (2x - 2z)e^{2xz - z^2}|_{z=0} = 2x$$
$$H_2(x) = f''(0) = [-2e^{2xz - z^2} + (2x - 2z)^2 e^{2xz - z^2}]_{z=0} = 4x^2 - 2$$

Proceeding inductively, we see that $f^{(k)}(z)$ always is a polynomial in x and z multiplied by $e^{2xz - z^2}$, and so evaluation at $z = 0$ will always produce a polynomial in x.

Exercises

1. Find the radius of convergence of each of the following power series:

(a) $\sum_{n=0}^{\infty} n z^n$

(b) $\sum_{n=0}^{\infty} \frac{z^n}{e^n}$

(c) $\sum_{n=1}^{\infty} n! \frac{z^n}{n^n}$

(d) $\sum_{n=1}^{\infty} \frac{z^n}{n}$

2. Find the radius of convergence of each of the following power series:

(a) $\sum_{n=0}^{\infty} n^2 z^n$

(b) $\sum_{n=0}^{\infty} \frac{z^{2n}}{4^n}$

(c) $\sum_{n=0}^{\infty} n! z^n$

(d) $\sum_{n=0}^{\infty} \frac{z^n}{1 + 2^n}$

3. Compute the Taylor series of the following functions around the indicated points and determine the set on which the series converges:

(a) e^z, $z_0 = 1$

(b) $1/z$, $z_0 = 1$

---

* For additional applications of power series to differential equations, see, for example, J. Marsden and A. Weinstein, *Calculus II* (New York: Springer-Verlag, 1985), Sec. 12.6, or virtually any text on differential equations.

4. Establish the Taylor series for $\sin z$, $\cos z$, and $(1 + z)^\alpha$ in Table 3.2.1.

5. Compute the Taylor series of the following. (Give only the first few terms where appropriate.)
(a) $(\sin z)/z$, $z_0 = 1$ (b) $z^2 e^z$; $z_0 = 0$
(c) $e^z \sin z$, $z_0 = 0$

6. Compute the first four terms of the Taylor series of $1/(1 + e^z)$ around $z_0 = 0$. What is the radius of convergence?

7. Compute the Taylor series of the following around the indicated points:
(a) e^{z^2}, $z_0 = 0$ (b) $1/(z - 1)(z - 2)$, $z_0 = 0$

8. Compute the Taylor series of the following around the indicated point:
(a) $\sin z^2$, $z_0 = 0$ (b) e^{2z}, $z_0 = 0$

9. Compute the first few terms in the Taylor expansion of $\sqrt{z^2 - 1}$ around 0.

10. Let $f(z) = \sum_{n=0}^\infty a_n z^n$ and $g(z) = \sum_{n=0}^\infty b_n z^n$ converge for $|z| < R$. Let γ be a circle of radius $r < R$ and define

$$F(z) = \frac{1}{2\pi i} \int_\gamma \frac{f(\zeta)}{\zeta} g\left(\frac{z}{\zeta}\right) d\zeta$$

Show that $F(z) = \sum_{n=0}^\infty a_n b_n z^n$. (*Hint.* Use Worked Example 2.4.15.)

11. Establish the following:

$$\sinh z = \sum_{n=1}^\infty \frac{z^{2n-1}}{(2n - 1)!} \quad \text{and} \quad \cosh z = \sum_{n=0}^\infty \frac{z^{2n}}{(2n)!}$$

12. What is the flaw in the following reasoning? Since $e^z = \sum_{n=0}^\infty z^n/n!$, we get $e^{1/z} = \sum_{n=0}^\infty 1/(n! z^n)$. Since this converges (because e^z is entire) and since the Taylor expansion is unique, the Taylor expansion of $f(z) = e^{1/z}$ around $z = 0$ is $\sum_{n=0}^\infty z^{-n}/n!$.

13. Differentiate the series for $1/(1 - z)$ to obtain expansions for

$$\frac{1}{(1 - z)^2} \quad \text{and} \quad \frac{1}{(1 - z)^3}$$

Give the radius of convergence.

14. Let $f(z) = \sum a_n z^n$ have radius of convergence R and let $A = \{z$ such that $|z| < R\}$. Let $z_0 \in A$ and $\tilde{R}$ be the radius of convergence of the Taylor series of f around z_0. Prove that $R - |z_0| \le \tilde{R} \le R + |z_0|$.

15. If $\sum_{n=0}^{\infty} a_n z^n$ has radius of convergence R, show that $\sum_{n=0}^{\infty} (\mathrm{Re}\, a_n) z^n$ has radius of convergence $\geq R$.

16. Let $f(z) = \sum a_n z^n$ be a power series with radius of convergence $R > 0$. For any closed curve γ in $A = \{z$ such that $|z| < R\}$ show that $\int_\gamma f = 0$ by
(a) using Cauchy's theorem.
(b) justifying term-by-term integration

17. In what region does

$$\sum_{n=1}^{\infty} \frac{\sin nz}{2^n}$$

represent an analytic function? What about

$$\sum_{n=1}^{\infty} \frac{\sin nz}{n^2} \quad ?$$

18. Find the first few terms of the Taylor expansion of $\tan z = (\sin z)/(\cos z)$ around $z = 0$. (*Hint.* We know that such an expansion exists. Write

$$\frac{\sin z}{\cos z} = a_0 + a_1 z + a_2 z^2 + \cdots$$

Multiply by

$$\cos z = 1 - \frac{z^2}{2!} + \frac{z^4}{4!} - \cdots$$

and use Worked Example 3.2.13 to solve for a_0, a_1, a_2.)

19. Let f be analytic on the region A, let $z_0 \in A$, and let D be the largest open disk centered at z_0 and contained in A.
(a) If f is unbounded on D, then show that the radius of D equals the radius of convergence of the Taylor series for f at z_0.
(b) If there exists no analytic extension of f (that is, if there are no $\tilde{f}$ and A' such that $\tilde{f}$ is analytic on A', $A' \supset A$, $A' \neq A$, and $f = \tilde{f}|A$), then show by an example that the radius of convergence of the Taylor series of f at z_0 can still be greater than the radius of D. (*Hint.* Use the principal branch of $\log (1 + z)$ with $z_0 = -2 + i$.)

20. Prove: A power series converges absolutely everywhere or nowhere on its circle of convergence. Give an example to show that each case can occur.

21. If $\sum g_n(z)$ converges uniformly on a set $B \subset \mathbb{C}$ and $h(z)$ is a bounded function on B, prove that $\sum [h(z) g_n(z)]$ converges uniformly on B to $h(z)[\sum g_n(z)]$.

22. Let $f(z) = \sum_{n=0}^{\infty} a_n z^n$ converge for $|z| < R$. If $0 < r < R$, show that $f(z) = \sum_{n=0}^{\infty} a_n r^n e^{in\theta}$ where $z = re^{i\theta}$ and

$$a_n = \frac{1}{2\pi r^n} \int_0^{2\pi} f(re^{i\theta}) e^{-in\theta} \, d\theta \qquad (3)$$

and that

$$\frac{1}{2\pi} \int_0^{2\pi} |f(re^{i\theta})|^2 \, d\theta = \sum_{n=0}^{\infty} |a_n|^2 r^{2n} \qquad (4)$$

Eq. (4) is referred to as *Parseval's theorem*, and we say that Eq. (3) expresses the Taylor series as a *Fourier series*. (*Hint. Use the Cauchy integral formula for a_n and expand $f\bar{f}$ in a series, and then integrate term by term.*)

23. Let $H_n(x)$ be the Hermite polynomials introduced in Worked Example 3.2.18. Show that $H_1(x) = 2xH_0(x)$ and that for $n \geq 1$

$$H_{n+1}(x) = 2xH_n(x) - 2nH_{n-1}(x)$$

24. Compute the Taylor expansion of $\zeta(z) = \sum_{n=1}^{\infty} n^{-z}$ around $z = 2$ (see Worked Example 3.1.15).

25. Find a function such that $f(0) = 1$ and $f'(x) = x + 2f(x)$ for all x (see Worked Example 3.2.17).

26. Find a function f such that $f(0) = 1$ and $f'(x) = xf(x)$ for all x.

3.3 LAURENT'S SERIES AND CLASSIFICATION OF SINGULARITIES

The Taylor series enables us to find a convergent power series expansion around z_0 for $f(z)$ when f is analytic in a whole disk around z_0. Thus the Taylor expansion does not apply to functions like $f(z) = 1/z$ or e^z/z^2, around $z_0 = 0$ that fail to be analytic at $z = 0$. For such functions there is another expansion, called the *Laurent expansion* (formulated in approximately 1840), that uses inverse powers of z rather than powers of z. This expansion is particularly important in the study of singular points of functions and leads to another fundamental result of complex analysis, the residue theorem, which is studied in the next chapter.

Laurent's Theorem

3.3.1 LAURENT EXPANSION THEOREM *Let $r_1 \geq 0$, $r_2 > r_1$, and $z_0 \in \mathbf{C}$, and consider the region $A = \{z \in \mathbf{C} \mid r_1 < |z - z_0| < r_2\}$ (see Fig. 3.3.1). We allow $r_1 = 0$ or $r_2 = \infty$ (or both). Let f be analytic on the region A. Then we can write*

$$f(z) = \sum_{n=0}^{\infty} a_n(z - z_0)^n + \sum_{n=1}^{\infty} \frac{b_n}{(z - z_0)^n} \tag{1}$$

where both series on the right side of the equation converge absolutely on A and uniformly on any set of the form $B_{\rho_1,\rho_2} = \{z \mid \rho_1 \leq |z - z_0| \leq \rho_2\}$ *where* $r_1 < \rho_1 < \rho_2 < r_2$. *If* γ *is a circle around* z_0 *with radius* r, $r_1 < r < r_2$, *then the coefficients are given by*

$$a_n = \frac{1}{2\pi i} \int_{\gamma} \frac{f(\zeta)}{(\zeta - z_0)^{n+1}} \, d\zeta \qquad n = 0, 1, 2, \ldots$$

and (2)

$$b_n = \frac{1}{2\pi i} \int_{\gamma} f(\zeta)(\zeta - z_0)^{n-1} \, d\zeta \qquad n = 1, 2, \ldots$$

(If we set $b_n = a_{-n}$, *then the first formula covers both cases.) The series in Eq. (1) for f is called the **Laurent series** or **Laurent expansion** around* z_0 *in the annulus A. Any pointwise convergent expansion of f of this form equals the Laurent expansion; in other words, the Laurent expansion is unique.*

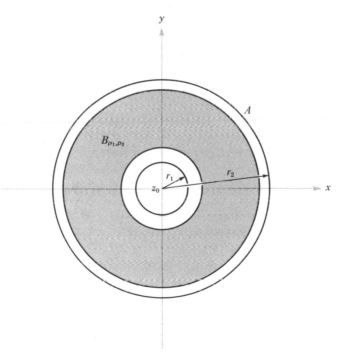

FIGURE 3.3.1   Laurent series, with $z_0 = 0$.

NOTE   We *cannot* set $a_n = f^{(n)}(z_0)/n!$ as we did with the Taylor expansion. Indeed, $f^{(n)}(z_0)$ is not even defined, since $z_0 \notin A$.

The equations (2) for a_n and b_n are not very practical for computing the Laurent series of a given function f. Tricks can be used to obtain some expansions of the desired form, and the uniqueness of the expansion will guarantee it is the desired one. A few techniques are given in the following text and in the worked examples.

In the following proof we shall see that the power series part of f, that is,

$$\sum_{n=0}^{\infty} a_n (z - z_0)^n$$

converges, and so is analytic *inside* the circle $|z - z_0| = r_2$, whereas the singular part,

$$\sum_{n=1}^{\infty} \frac{b_n}{(z - z_0)^n}$$

converges *outside* $|z - z_0| = r_1$. The sum therefore converges *between* these circles.

The student is cautioned that uniqueness is dependent on the choice of A. For example, if $A = \{z \text{ such that } |z| > 1\}$, then $f(z) = 1/[z(z - 1)]$ has the Laurent expansion

$$f(z) = \frac{1}{z(z - 1)} = \frac{1}{z}\left[\frac{1}{z\left(1 - \frac{1}{z}\right)} \right] = \frac{1}{z^2}\left(1 + \frac{1}{z} + \frac{1}{z^2} + \cdots \right) = \frac{1}{z^2} + \frac{1}{z^3} + \cdots$$

(valid if $|z| > 1$), whereas on $A = \{z \text{ such that } 0 < |z| < 1\}$, it has the expansion

$$f(z) = \frac{1}{z(z - 1)} = -\frac{1}{z}(1 + z + z^2 + \cdots) = -\left(\frac{1}{z} + 1 + z + z^2 + \cdots \right)$$

(valid for $0 < |z| < 1$). By uniqueness these are *the* Laurent expansions for the appropriate regions.

PROOF OF THE LAURENT EXPANSION THEOREM   As with the proof of Taylor's theorem, we begin with Cauchy's integral formula. We will first show uniform convergence of the stated series on B_{ρ_1, ρ_2}, where a_n and b_n are defined by equation pair (2). Since all the circles γ of radius r are homotopic to each other in A as long as $r_1 < r < r_2$ (Why?), the numbers a_n and b_n are independent of r, and so

$$a_n = \frac{1}{2\pi i} \int_{\gamma_1} \frac{f(\zeta)}{(\zeta - z_0)^{n+1}} \, d\zeta$$

and

$$b_n = \frac{1}{2\pi i} \int_{\gamma_2} f(\zeta)(\zeta - z_0)^{n-1} \, d\zeta$$

where γ_1 is a circle of radius $\tilde{\rho}_1$ and γ_2 is a circle of radius $\tilde{\rho}_2$, and where $r_1 < \tilde{\rho}_1 < \rho_1 < \rho_2 < \tilde{\rho}_2 < r_2$ (see Fig. 3.3.2).

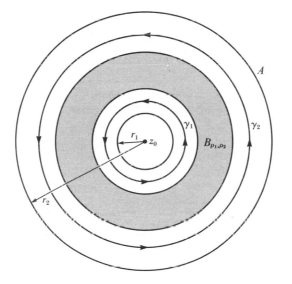

FIGURE 3.3.2   Construction of the curves γ_1 and γ_2.

For $z \in B_{\rho_1, \rho_2}$ we have

$$f(z) = \frac{1}{2\pi i} \int_{\gamma_2} \frac{f(\zeta)}{\zeta - z} \, d\zeta - \frac{1}{2\pi i} \int_{\gamma_1} \frac{f(\zeta)}{\zeta - z} \, d\zeta$$

by Cauchy's integral formula (see Exercise 5).

As in Taylor's theorem, for ζ on γ_2 (and z fixed inside γ_2),

$$\frac{1}{\zeta - z} = \frac{1}{\zeta - z_0} + \frac{z - z_0}{(\zeta - z_0)^2} + \frac{(z - z_0)^2}{(\zeta - z_0)^3} + \cdots$$

which converges uniformly in ζ on γ_2.

We may integrate term by term (by Proposition 3.1.9 and the fact that $f(\zeta)$ is bounded — see Exercise 21, Sec. 3.2) and thus obtain

$$\frac{1}{2\pi i} \int_{\gamma_2} \frac{f(\zeta)}{\zeta - z} \, d\zeta = \sum_{n=0}^{\infty} \frac{1}{2\pi i} \left[\int_{\gamma_2} \frac{f(\zeta)}{(\zeta - z_0)^{n+1}} \, d\zeta \right] (z - z_0)^n = \sum_{n=0}^{\infty} a_n (z - z_0)^n$$

Since this power series converges for z inside γ_2, it converges uniformly on strictly smaller disks (in particular, on B_{ρ_1,ρ_2}).

$$\frac{-1}{\zeta - z} = \frac{1}{(z - z_0)\left(1 - \dfrac{\zeta - z_0}{z - z_0}\right)} = \frac{1}{z - z_0} + \frac{\zeta - z_0}{(z - z_0)^2} + \frac{(\zeta - z_0)^2}{(z - z_0)^3} + \cdots$$

converges uniformly with respect to ζ on γ_1. Thus

$$\frac{-1}{2\pi i}\int_{\gamma_1}\frac{f(\zeta)}{\zeta - z}\,d\zeta = \sum_{n=1}^{\infty}\frac{1}{2\pi i}\left[\int_{\gamma_1}f(\zeta)\cdot(\zeta - z_0)^{n-1}\,d\zeta\right]\cdot\frac{1}{(z - z_0)^n} = \sum_{n=1}^{\infty}\frac{b_n}{(z - z_0)^n}$$

This series converges for z outside γ_1, so the convergence is uniform outside any strictly larger circle. This fact can be proved in the same way as the analogous fact for power series by using the Abel-Weierstrass lemma in Sec. 3.2. (Another method is to make the transformation $w = 1/(z - z_0)$ and apply the power series result to $\sum_{n=1}^{\infty} b_n w^n$.) The student is asked to do this in Exercise 15.

We have now proved the existence of the Laurent expansion. To show uniqueness, let us suppose that we have an expansion for f:

$$f(z) = \sum_{n=0}^{\infty} a_n(z - z_0)^n + \sum_{n=1}^{\infty}\frac{b_n}{(z - z_0)^n}$$

If this converges in A it will, by the preceding remarks, do so uniformly on the circle γ, so we can form

$$\frac{f(z)}{(z - z_0)^{k+1}} = \sum_{n=0}^{\infty} a_n(z - z_0)^{n-k-1} + \sum_{n=1}^{\infty}\frac{b_n}{(z - z_0)^{n+k+1}}$$

which also converges uniformly. We then integrate term by term. By Worked Example 2.1.12, we have

$$\int_{\gamma}(z - z_0)^m\,dz = \begin{cases} 0 & m \neq -1 \\ 2\pi i & m = -1 \end{cases}$$

Thus if $k \geq 0$, each term of the second series and all those of the first except that with $n = k$ integrate to 0 around γ. Hence

$$\int_{\gamma}\frac{f(z)}{(z - z_0)^{k+1}}\,dz = 2\pi i a_k$$

Similarly, if $k \leq -1$, all terms integrate to 0 except that in the second series with $n = -k$, and so

$$\int_{\gamma} \frac{f(z)}{(z-z_0)^{k+1}} \, dz = 2\pi i b_{-k}$$

That is,

$$b_n = \frac{1}{2\pi i} \int_{\gamma} f(z)(z-z_0)^{n-1} \, dz \qquad \text{for } n \geq 1$$

Thus, the coefficients a_n, b_n are uniquely determined by f and so the proof is complete. ∎

Isolated Singularities: Classification of Singular Points

We want to look in more detail at the special case of the Laurent expansion theorem when $r_1 = 0$. In this case, f is analytic on $\{z \mid 0 < |z-z_0| < r_2\}$, which is the deleted r_2 neighborhood of z_0 (see Fig. 3.3.3), and we say that z_0 is an *isolated singularity* of f. Thus we can expand f in a Laurent series as follows:

$$f(z) = \cdots + \frac{b_n}{(z-z_0)^n} + \cdots + \frac{b_1}{z-z_0} + a_0 + a_1(z-z_0) + a_2(z-z_0)^2 + \cdots$$

$$(3)$$

(valid for $0 < |z-z_0| < r_2$).

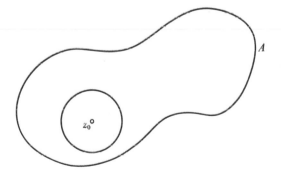

FIGURE 3.3.3 Isolated singularity.

3.3.2 DEFINITION *If f is analytic on a region A that contains some deleted ϵ neighborhood of z_0, then z_0 is called an* **isolated singularity.** *(Thus the preceding Laurent expansion (Eq. (3)) is valid in such a deleted ϵ neighborhood.)*

If z_0 is an isolated singularity of f and if all but a finite number of the b_n in Eq. (3) are zero, then z_0 is called a **pole** *of f. If k is the highest integer such that $b_k \neq 0$, z_0 is called a* **pole of order** *k. (To emphasize that $k \neq \infty$, we sometimes say "a pole of finite order k.") If z_0 is a first-order pole, we also say it is a* **simple pole.** *If an infinite number of b_k's are nonzero, z_0 is called an* **essential singularity.** *(Sometimes this z_0 is called a pole of infinite order.) "Pole" shall always mean a pole of finite order.*

We call b_1 the **residue** *of f at z_0.*

If all the b_k's are zero, we say that z_0 is a **removable singularity.**

A function that is analytic in a region A, except for poles in A, is called **meromorphic in** *A. The phrase "f is a meromorphic function" means that f is meromorphic in* **C.**

Thus f has a pole of order k iff its Laurent expansion has the form

$$\frac{b_k}{(z - z_0)^k} + \cdots + \frac{b_1}{z - z_0} + a_0 + a_1(z - z_0) + \cdots$$

The part

$$\frac{b_k}{(z - z_0)^k} + \cdots + \frac{b_1}{z - z_0}$$

which is often called the *principal part* of f at z_0, tells just "how singular" f is at z_0.

If f has a removable singularity, then

$$f(z) = \sum_{n=0}^{\infty} a_n(z - z_0)^n$$

is a convergent power series. Thus if we set $f(z_0) = a_0$, f will be analytic at z_0. In other words, f has a removable singularity at z_0 iff f can be defined at z_0 in such a way that f becomes analytic at z_0.

As we shall see in Chap. 4, finding the Laurent expansion is not as important as being able to compute the residue b_1, and this computation can often be done without computing the Laurent series. Techniques for doing so will be studied in Sec. 4.1. The important property of b_1 not shared by other coefficients is stated in the next result.

3.3.3 PROPOSITION   *Let f be analytic on a region A and have an isolated singularity at z_0 with residue b_1 at z_0. If γ is any circle around z_0 in A whose interior, except for the point z_0, lies in A, then*

$$\int_\gamma f(z) \, dz = b_1 \cdot 2\pi i \tag{4}$$

This conclusion follows from the formula for b_1 in the Laurent expansion theorem. The point is that we can compute b_1 by methods other than Eq. (4) and therefore we can use Eq. (4) to compute $\int_\gamma f$. For example, if $z \neq 0$, then

$$e^{1/z} = 1 + \frac{1}{z} + \frac{1}{2!z^2} + \frac{1}{3!z^3} + \cdots$$

(Why?), and so $e^{1/z}$ has an essential singularity at $z = 0$ and $b_1 = 1$. Thus $\int_\gamma e^{1/z}\,dz = 2\pi i$ for any circle γ around 0.

The following proposition characterizes the various types of singularities.

3.3.4 PROPOSITION *Let f be analytic on a region A and have an isolated singularity at z_0.*

(i) *z_0 is a removable singularity iff any one of the following conditions holds:*
  (1) *f is bounded in a deleted neighborhood of z_0*
  (2) *$\lim\limits_{z \to z_0} f(z)$ exists*

  or

  (3) *$\lim\limits_{z \to z_0} (z - z_0)f(z) = 0$*

(ii) *z_0 is a simple pole iff $\lim\limits_{z \to z_0} (z - z_0)f(z)$ exists and is unequal to zero. This limit equals the residue of f at z_0.*

(iii) *z_0 is a pole of order $\leq k$ (or possibly a removable singularity) iff any one of the following conditions holds:*
  (1) *There are a constant $M > 0$ and an integer $k \geq 1$ such that*

  $$|f(z)| \leq \frac{M}{|z - z_0|^k}$$

  in a deleted neighborhood of z_0
  (2) *$\lim\limits_{z \to z_0} (z - z_0)^{k+1}f(z) = 0$*

  or

  (3) *$\lim\limits_{z \to z_0} (z - z_0)^k f(z)$ exists*

(iv) *z_0 is a pole of order $k \geq 1$ iff there is an analytic function ϕ defined on a neighborhood U of z_0 such that $U \backslash \{z_0\} \subset A$, $\phi(z_0) \neq 0$, and $f(z) = \phi(z)/(z - z_0)^k$ for all $z \in U, z \neq z_0$.*

PROOF

(i) If z_0 is a removable singularity, then in a deleted neighborhood of z_0 we have $f(z) = \sum_{n=0}^\infty a_n(z - z_0)^n$. Since this series represents an analytic function in an undeleted neighborhood of z_0, obviously conditions 1, 2, and 3 hold. Conditions 1 and 2 each obviously imply condition 3, and so it remains to be shown that condition 3 implies that z_0 is a removable singularity for f. We must prove that each b_k in the Laurent expansion of f around z_0 is 0. Now

$$b_k = \frac{1}{2\pi i} \int_{\gamma_r} f(\zeta)(\zeta - z_0)^{k-1} \, d\zeta$$

where γ, is a circle in A whose interior (except for z_0) lies in A. Let $\epsilon > 0$ be given. By condition 3 we can choose $r > 0$ with $r < 1$ such that on γ_r we have the estimate $|f(\zeta)| < \epsilon/|\zeta - z_0| = \epsilon/r$. Then

$$|b_k| \le \frac{1}{2\pi} \int_{\gamma_r} |f(\zeta)||\zeta - z_0|^{k-1} \, |d\zeta|$$

$$\le \frac{1}{2\pi} \frac{\epsilon}{r} r^{k-1} \int_{\gamma_r} |d\zeta| = \frac{1}{2\pi} \frac{\epsilon}{r} r^{k-1} 2\pi r$$

$$= \epsilon r^{k-1} \le \epsilon$$

Thus $|b_k| \le \epsilon$. Since ϵ was arbitrary, $b_k = 0$. We shall use (iii) to prove (ii), so (iii) is proved next.

(iii) This statement follows by applying (i) to the function $(z - z_0)^k f(z)$, which is analytic on A. (The student should write out the details).

(ii) If z_0 is a simple pole, then in a deleted neighborhod of z_0,

$$f(z) = \frac{b_1}{z - z_0} + \sum_{n=0}^{\infty} a_n (z - z_0)^n = \frac{b_1}{z - z_0} + h(z)$$

where h is analytic at z_0 and where $b_1 \ne 0$ by the Laurent expansion. Hence $\lim_{z \to z_0} (z - z_0) f(z) = \lim_{z \to z_0} [b_1 + (z - z_0)h(z)] = b_1$. On the other hand, suppose that $\lim_{z \to z_0} (z - z_0) f(z)$ exists and is unequal to zero. Thus $\lim_{z \to z_0} (z - z_0)^2 f(z) = 0$. By the result obtained in (iii), this shows that

$$f(z) = \frac{b_1}{(z - z_0)} + \sum_{n=0}^{\infty} a_n (z - z_0)^n = \frac{b_1}{z - z_0} + h(z)$$

for some constant b_1 and analytic function h where b_1 may or may not be zero. But then $(z - z_0) f(z) = b_1 + (z - z_0)h(z)$, and so $\lim_{z \to z_0} (z - z_0) f(z) = b_1$. Thus, in fact, $b_1 \ne 0$, and therefore f has a simple pole at z_0.

(iv) By definition, z_0 is a pole of order $k \ge 1$ iff

$$f(z) = \frac{b_k}{(z-z_0)^k} + \frac{b_{k-1}}{(z-z_0)^{k-1}} + \cdots + \frac{b_1}{z-z_0} + \sum_{n=0}^{\infty} a_n(z-z_0)^n$$

$$= \frac{1}{(z-z_0)^k} \left[b_k + b_{k-1}(z-z_0) + \cdots + b_1(z-z_0)^{k-1} + \sum_{n=0}^{\infty} a_n(z-z_0)^{n+k} \right]$$

(where $b_k \neq 0$). This expansion is valid in a deleted neighborhood of z_0. Let $\phi(z) = b_k + b_{k-1}(z-z_0) + \cdots + b(z-z_0)^{k-1} + \sum_{n=0}^{\infty} a_n(z-z_0)^{n+k}$. Then $\phi(z)$ is analytic in the corresponding undeleted neighborhood (since it is a convergent power series) and $\phi(z_0) = b_k \neq 0$. Conversely, given such a ϕ, we can retrace these steps to show that z_0 is a pole of order $k \geq 1$. ∎

Zeros of Order k

Let f be analytic on a region A and let $z_0 \in A$. We say that f has a zero *of order k* at z_0 iff $f(z_0) = 0, \ldots, f^{(k-1)}(z_0) = 0, f^{(k)}(z_0) \neq 0$.

From the Taylor expansion

$$f(z) = \sum_{n=0}^{\infty} \frac{f^{(n)}(z_0)}{n!} (z-z_0)^n$$

we see that f has a zero of order k iff, in a neighborhood of z_0, we can write $f(z) = (z-z_0)^k g(z)$ where $g(z)$ is analytic at z_0 and $g(z_0) = f^{(k)}(z_0)/k! \neq 0$. From Proposition 3.3.4 (iv), letting $\phi(z) = g(z)^{-1}$, we get the following.

3.3.5 PROPOSITION *If f is analytic in a neighborhood of z_0, then f has a zero of order k at z_0 iff $1/f(z)$ has a pole of order k at z_0. If h is analytic and $h(z_0) \neq 0$, then $h(z)/f(z)$ also has a pole of order k at z_0.*

Obviously, if z_0 is a zero of f and f is not identically equal to zero in a neighborhood of z_0, then z_0 has some finite order k. (Otherwise the Taylor series would be identically zero.)

Essential Singularities

In practical examples, many singularities are poles. It is not hard to show that if $f(z)$ has a pole (of finite order k) at z_0, then $|f(z)| \to \infty$ as $z \to z_0$ (see Exercise 7). However, in case of an essential singularity, $|f|$ will not, in general, approach ∞ as $z \to z_0$. In fact there is the following result, proved by C. E. Picard in 1879.

3.3.6 PICARD THEOREM *Let f have an essential singularity at z_0 and let U be any (arbitrarily small) deleted neighborhood of z_0. Then for all $w \in \mathbf{C}$, except perhaps one value, the equation $f(z) = w$ has infinitely many solutions z in U.*

This theorem actually belongs in a more advanced course.* However, we can easily prove a simpler version.

3.3.7 CASORATI-WEIERSTRASS THEOREM *Let f have an essential singularity at z_0 and let $w \in \mathbf{C}$. Then there exist $z_1, z_2, z_3, \ldots$ such that $z_n \to z_0$ and $f(z_n) \to w$.*

PROOF If the assertion were false, there would be a deleted neighborhood U of z_0 and an $\epsilon > 0$ such that $|f(z) - w| \geq \epsilon$ for all $z \in U$ (Why?). Let $g(z) = 1/[f(z) - w]$. Thus on U, g is analytic and is bounded, so z_0 is a removable singularity by Proposition 3.3.4(i). Let k be the order of the zero of g at z_0 (set $k = 0$ if $g(z_0) \neq 0$). (The order must be finite because otherwise, as mentioned previously, by the Taylor theorem, g would be zero in a neighborhood of z_0, whereas g is 0 nowhere on U.) Thus $f(z) = w + 1/g(z)$ is either analytic (if $k = 0$) or has a pole of order k by Proposition 3.3.5. This conclusion contradicts our assumption that f has an essential singularity. ■

For another interpretation of this result see Exercise 20.

Worked Examples

3.3.8   *Find the Laurent expansions of the following functions (with z_0, r_1, r_2 as indicated):*
(a) $(z + 1)/z$; $z_0 = 0, r_1 = 0, r_2 = \infty$
(b) $z/(z^2 + 1)$; $z_0 = i, r_1 = 0, r_2 = 2$

Solution.

(a) $\dfrac{z + 1}{z} = \dfrac{1}{z} + 1$

This equation is in the form of the Laurent expansion and so, by uniqueness, it equals it; that is, $b_k = 0$ for $k > 1, b_1 = 1, a_0 = 1, a_k = 0$ for $k \geq 1$.
(b) A partial-fraction expansion gives

---

* See E. C. Titchmarsh, *The Theory of Functions* (New York: Oxford University Press, 1939), p. 283.

$$\frac{z}{z^2+1} = \frac{z}{(z+i)(z-i)} = \frac{1}{2}\frac{1}{z-i} + \frac{1}{2}\frac{1}{z+i}$$

Because $1/(z+i)$ is analytic near $z=i$, it can be expanded as a power series in $z-i$ by using the geometric series (see Fig. 3.3.4):

$$\frac{1}{z+i} = \frac{1}{2i+(z-i)} = \frac{1}{2i}\frac{1}{1-\left(-\dfrac{z-i}{2i}\right)}$$

$$= \frac{1}{2i}\sum_{n=0}^{\infty}\left(-\frac{z-i}{2i}\right)^n = \sum_{n=0}^{\infty} i^{n-1}2^{-n-1}(z-i)^n$$

Thus, the Laurent expansion is

$$\frac{z}{z^2+1} = \frac{1}{2}(z-i)^{-1} + \sum_{n=0}^{\infty} i^{n-1}2^{-n-2}(z-i)^n$$

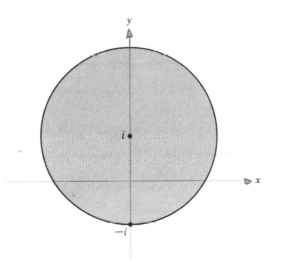

FIGURE 3.3.4 Region of convergence for the expansion of $1/(z+i)$.

3.3.9 *Determine the order of the pole of each of the following functions at the indicated singularity:*
(a) $(\cos z)/z^2$, $z_0 = 0$ (b) $(e^z - 1)/z^2$, $z_0 = 0$
(c) $(z+1)/(z-1)$, $z_0 = 0$

Solution.
(a) z^2 has a zero of order 2 and $\cos 0 = 1$, so $(\cos z)/z^2$ has a pole of order 2 by Proposition 3.3.5.

Alternatively,

$$\frac{\cos z}{z^2} = \frac{1}{z^2}\left(1 - \frac{z^2}{2!} + \frac{z^4}{4!} - \cdots\right) = \frac{1}{z^2} - \frac{1}{2!} + \frac{z^2}{4!} - \cdots$$

and so again the pole is of order 2.

(b) The numerator vanishes at 0, and so Proposition 3.3.5 does not apply. But

$$\frac{e^z - 1}{z^2} = \frac{1}{z^2}\left[\left(1 + z + \frac{z^2}{2!} + \cdots\right) - 1\right] = \frac{1}{z} + \frac{1}{2!} + \frac{z}{3!} + \frac{z^2}{4!} + \cdots$$

and so the pole is simple.

(c) There is no pole since the function is analytic at 0.

3.3.10 *Determine which of the following functions have removable singularities at $z_0 = 0$:*
(a) $(\sin z)/z$ (b) e^z/z
(c) $(e^z - 1)^2/z^2$ (d) $z/(e^z - 1)$

Solution.

(a) $\lim\limits_{z \to 0} z \cdot (\sin z)/z = \lim\limits_{z \to 0} \sin z = 0$, and so the singularity is removable (by Proposition 3.3.4(i)).

Alternatively,

$$\frac{\sin z}{z} = \frac{1}{z}\left(z - \frac{z^3}{3!} + \cdots\right) = 1 - \frac{z^2}{3!} + \frac{z^4}{5!} - \cdots$$

(b) $\lim\limits_{z \to 0} z \cdot e^z/z = 1$, and so the pole is simple (the singularity is not removable).

(c) $(e^z - 1)/z$ has a removable singularity, since $\lim\limits_{z \to 0} z \cdot (e^z - 1)/z = 0$, and so $[(e^z - 1)/z]^2$ also has a removable singularity.

(d) $\lim\limits_{z \to 0} z/(e^z - 1) = 1$, because $(e^z - 1)/z = 1 + z/2 + z^2/3! + \cdots \to 1$ as $z \to 0$. Thus $z/(e^z - 1)$ has a removable singularity.

3.3.11 *Show that if f and g are analytic at z_0 and g has a zero of order n and f a zero of order k with $k \geq n$, then g/f has a pole of order $k - n$. What if $k < n$?*

Solution. We can factor the Taylor series for f and g centered at z_0 to obtain analytic functions F and G such that $f(z) = (z - z_0)^k F(z)$ and $g(z) = (z - z_0)^n G(z)$ for points z near z_0 and for $F(z_0) \neq 0$ and $G(z_0) \neq 0$. Thus $f(z)/g(z) = (z - z_0)^{k-n} F(z)/G(z)$. The first derivative of f/g which does not have a factor of $(z - z_0)$ is $(f/g)^{(k-n)}$, and

$$\left(\frac{f}{g}\right)^{(k-n)}(z_0) = (k - n)!\,\frac{F(z_0)}{G(z_0)} \neq 0$$

Thus f/g has a zero of order $k - n$ at z_0 by 3.3.5. If $k < n$ then a similar argument shows that g/f has a zero of order $n - k$.

Exercises

1. Find the Laurent series expansions of the following functions around $z_0 = 0$ in the regions indicated:

(a) $\sin(1/z)$, $0 < |z| < \infty$

(b) $1/z(z + 1)$, $0 < |z| < 1$

(c) $z/(z + 1)$, $0 < |z| < 1$

(d) e^z/z^2, $0 < |z| < \infty$

2. Find the Laurent series expansion of $1/z(z + 1)$ around $z_0 = 0$ valid in the region $1 < |z| < \infty$.

3. Find the Laurent series expansion of $z/(z + 1)$ around $z_0 = 0$ valid in the region $1 < |z| < \infty$.

4. Expand $\dfrac{1}{z(z - 1)(z - 2)}$ in a Laurent series in the following regions:

(a) $0 < |z| < 1$

(b) $1 < |z| < 2$

5. Let γ_1 and γ_2 be two concentric circles around z_0 of radii R_1 and R_2, $R_1 < R_2$. If z lies between the circles and f is analytic on a region containing γ_1, γ_2, and the region between them, show that

$$f(z) = \frac{1}{2\pi i} \int_{\gamma_2} \frac{f(\zeta)}{\zeta - z} \, d\zeta - \frac{1}{2\pi i} \int_{\gamma_1} \frac{f(\zeta)}{\zeta - z} \, d\zeta$$

6. Suppose the Laurent series of $f(z) = e^{1/z}/(1 - z)$ valid for $0 < |z| < 1$ is $\sum_{n=-\infty}^{\infty} c_n z^n$. Compute c_{-2}, c_{-1}, c_0, c_1, and c_2.

7. Let f have a pole at z_0 of order $k \geq 1$. Prove that $f(z) \to \infty$ as $z \to z_0$. (Hint: Use part (iv) of Proposition 3.3.4.)

8. Prove, using the Taylor series, the following complex version of *l'Hôpital's rule:* Let $f(z)$ and $g(z)$ be analytic, both having zeros of order k at z_0. Then $f(z)/g(z)$ has a removable singularity and

$$\lim_{z \to z_0} \frac{f(z)}{g(z)} = \frac{f^{(k)}(z_0)}{g^{(k)}(z_0)}$$

9. Which of the following functions have removable singularities at the indicated points:

(a) $\dfrac{\cos(z - 1)}{z^2}$, $z_0 = 0$

(b) $z/(z - 1)$, $z_0 = 1$

(c) $f(z)/(z - z_0)^k$ if f has a zero at z_0 of order k

10. If f is analytic on a region containing a circle γ and its interior and has a zero of order 1 only at z_0 inside or on γ, show that

$$z_0 = \frac{1}{2\pi i} \int_\gamma \frac{z f'(z)}{f(z)} \, dz$$

(*Hint.* Let $f(z) = (z - z_0)\phi(z)$ and apply the Cauchy integral formula.)

11. Find the first few terms in the Laurent expansion of $1/(e^z - 1)$ around $z = 0$. (*Hint.* Show that since $1/(e^z - 1)$ has a simple pole,

$$\frac{1}{e^z - 1} = \frac{b_1}{z} + a_0 + a_1 z + a_2 z^2 + \cdots$$

Then cross-multiply (using Worked Example 3.2.13) and solve for b_1, a_0, a_1.)

12. For f as in the Laurent expansion theorem 3.3.1, show that if $r_1 < r < r_2$, then

$$\int_0^{2\pi} |f(z_0 + re^{i\theta})|^2 \, d\theta = 2\pi \sum_{n=0}^{\infty} |a_n|^2 r^{2n} + 2\pi \sum_{n=1}^{\infty} |b_n|^2 r^{-2n}$$

13. Use the Hint of Exercise 11 to find the first few terms in the Laurent expansion of $\cot z = (\cos z)/(\sin z)$ around $z = 0$.

14. Define a branch of $\sqrt{z^2 - 1}$ that is analytic except for the segment $[-1, 1]$ on the real axis. Determine the first few terms in the Laurent expansion that is valid for $|z| > 1$.

15. If

$$\sum_{n=1}^{\infty} \frac{b_n}{(z - z_0)^n}$$

converges for $|z - z_0| > R$, prove that it necessarily converges *uniformly* on the set $F_r = \{z \text{ such that } |z - z_0| > r\}$ for $r > R$. (*Hint.* Adapt the Abel-Weierstrass lemma and the Weierstrass M test to this case.)

16. Let f have a zero at z_0 of multiplicity k. Show that the residue of f'/f at z_0 is k.

17. Discuss the singularities of $1/\cos(1/z)$.

18. Evaluate $\int_\gamma z^n e^{1/z} \, dz$ where γ is the circle of radius 1 centered at 0 and traveled once in the counterclockwise direction.

19. Find the residues of the following functions at the indicated points:
(a) $1/(z^2 - 1)$, $z = 1$
(b) $z/(z^2 - 1)$, $z = 1$
(c) $(e^z - 1)/z^2$, $z = 0$
(d) $(e^z - 1)/z$, $z = 0$

20. (a) Let z_0 be an essential singularity of f and let U be any deleted neighborhood of z_0. Prove that the closure of $f(U)$ is $\mathbf{C}$.

(b) Assuming the Picard theorem (3.3.6), derive the "little Picard theorem": *the image of an entire nonconstant function misses at most one point of* $\mathbf{C}$.

REVIEW EXERCISES FOR CHAPTER 3

1. Find the Taylor expansion of $\log z$ (principal branch of the logarithm) around $z = 1$.

2. Where are the poles of $1/\cos z$ and what are their orders?

3. Find the Laurent expansion of $1/(z^2 + z^3)$ around $z = 0$.

4. The $2n$th derivative of $f(z) = e^{z^2}$ at $z = 0$ is given by $(2n)!/n!$. Prove this without actually computing the $2n$th derivative.

5. Expand $z^2 \sin z^2$ in a Taylor series around $z = 0$.

6. If f is analytic and nonconstant on any open set in a region A, prove that the zeros of f are isolated. (In other words, prove that if z_0 is a zero, there is a neighborhood of z_0 in which there are no other zeros.)

7. Verify the Picard theorem (3.3.6) for the function $e^{1/z}$.

8. Let $\exp[t(z - 1/z)/2] = \sum_{n=-\infty}^{\infty} J_n(t)z^n$ be the Laurent expansion for each fixed $t \in \mathbf{R}$. $J_n(t)$ is called the *Bessel function* of order n. Show that

(a) $J_n(t) = \dfrac{1}{\pi} \displaystyle\int_0^\pi \cos(t \sin \theta - n\theta)\, d\theta$ (b) $J_{-n}(t) = (-1)^n J_n(t)$

9. Find the radii of convergence of

(a) $\displaystyle\sum_{n=0}^{\infty} \frac{2^n}{n^2} z^n$ (b) $\displaystyle\sum_{n=0}^{\infty} z^{n!}$

10. Let $\sum_{n=0}^{\infty} a_n(z - z_0)^n$ be a power series with radius of convergence $R > 0$. If $0 < r < R$, show that there is a constant M such that $|a_n| \le Mr^{-n}$, $n = 0, 1, 2, \ldots$.

11. Let f be analytic on $\mathbf{C}\backslash\{0\}$. Show that the Laurent expansions of f valid in the regions $\{z$ such that $|z| > 0\}$ and $\{z$ such that $|z| > 1\}$ are the same.

12. Suppose that f is analytic on the open unit disk $|z| < 1$ and that there is a constant M such that $|f^{(k)}(0)| \le M^k$ for all k. Show that f can be extended to an entire function.

13. Suppose that f is analytic in a region containing the unit disk $|z| \le 1$, that $f(0) = 0$, and that $|f(z)| < 1$ if $|z| = 1$. Show that there are no $z \ne 0$ with $|z| < 1$ and $f(z) = z$. (*Hint.* Use the Schwarz lemma.)

14. What is the radius of convergence of the Taylor expansion of

$$f(z) = \frac{e^z}{(z - 1)(z + 1)(z - 2)(z - 3)}$$

when expanded around $z = i$?

15. Evaluate

$$\int_\gamma \frac{z^2 + e^z}{z(z - 3)} \, dz$$

where γ is the unit circle.

16. Suppose that $\sum_{n=0}^\infty a_n$ converges but that $\sum_{n=0}^\infty |a_n|$ diverges. Show that $\sum_{n=0}^\infty a_n z^n$ has a radius of convergence equal to 1. Answer the same question but assume that the series $\sum_{n=0}^\infty a_n$ converges and that $\sum_{n=0}^\infty n|a_n|$ diverges.

17. Find the Laurent expansion of

$$f(z) = \frac{1}{z(z^2 + 1)}$$

that is valid for
(a) $0 < |z| < 1$ (b) $1 < |z|$

18. Find the Laurent series expansion of $f(z) = 1/(1 + z^2) + 1/(3 - z)$ valid in each of the following regions:
(a) $\{z \text{ such that } |z| < 1\}$ (b) $\{z \text{ such that } 1 < |z| < 3\}$
(c) $\{z \text{ such that } |z| > 3\}$

19. Let f be entire and let $g(z) = \sum_{n=0}^\infty a_n z^n$ have radius of convergence R. Can you find another power series $\sum b_n z^n$ with radius of convergence $\ge R$ such that

$$\sum_{n=0}^\infty b_n z^n = f\left(\sum_{n=0}^\infty a_n z^n\right)$$

20. Let f be entire and suppose that $f(z) \to \infty$ as $z \to \infty$. Show that f is a polynomial. (*Hint.* Show that $f(1/z)$ has a pole of finite order at $z = 0$.)

21. Let f have an isolated singularity at z_0. Show that if $f(z)$ is bounded in a deleted neighborhood of z_0, then $\lim\limits_{z \to z_0} f(z)$ exists.

22. Let f be analytic on $|z| < 1$. Show that the inequality $|f^{(k)}(0)| \geq k! 5^k$ cannot hold for all k.

23. Evaluate

$$\int_0^{2\pi} e^{e^{i\theta}} \, d\theta$$

24. Let $f(x)$ be entire and satisfy these two conditions:
(a) $f'(z) = f(z)$ (b) $f(0) = 1$
Show that $f(z) = e^z$. If you replace (a) by $f(z_1 + z_2) = f(z_1)f(z_2)$, show that $f(z) = e^{az}$ for some constant a.

25. Determine the order of the poles of the following functions at their singularities:
(a) $\dfrac{e^z(z - 3)}{(z - 1)(z - 5)}$ (b) $(e^z - 1)/z$
(c) $(e^z - 2)/z$ (d) $(\cos z)/(1 - z)$

26. Identify the singularities of $f(z) = z/(e^z - 1)(e^z - 2)$ and classify each as removable, essential, or a pole of specified order.

27. Evaluate $\int_\gamma e^z/z^2 \, dz$ where γ is the unit circle.

28. (a) Show by example that the mean value theorem for analytic functions is not true. In other words, let f be defined on the region A and let z_1, $z_2 \subset A$ be such that the straight line joining z_1 to z_2 lies in A. Show that there need not be a z_0 on this straight line such that

$$f'(z_0) = \frac{f(z_1) - f(z_2)}{z_1 - z_2}$$

(b) If, however, $|f'(z_0)| \leq M$ on this line, prove that $|f(z_1) - f(z_2)| \leq M|z_1 - z_2|$ and generally that if $|f'(z_0)| \leq M$ on a curve γ joining z_1 to z_2, then $|f(z_1) - f(z_2)| \leq Ml(\gamma)$.

29. Let $f(z) = (z^2 - 1)/[\cos(\pi z) + 1]$ have the series expansion $\sum_{n=0}^{\infty} a_n z^n$ near $z = 0$.
(a) Compute a_0, a_1, and a_2.
(b) Identify the singularities of f and classify each as essential or a pole of specified order.
(c) What is the radius of convergence of the series?

30. If $f(z) = f(-z)$ and $f(z) = \sum_{n=0}^{\infty} a_n z^n$ is convergent on a disk $|z| < R$, $R > 0$, show that $a_n = 0$ for $n = 1, 3, 5, 7, \ldots$.

31. If f is entire and is bounded on the real axis, then f is constant. Prove or give a counter example.

32. Let f be analytic on a region A containing $\{z$ such that $|z - z_0| < R\}$ so that

$$f(z) = \sum_{n=0}^{\infty} \frac{f^{(n)}(z_0)(z - z_0)^n}{n!}$$

Let $R_n(z)$ equal $f(z)$ minus the nth partial sum. (R_n is thus the remainder.) Let $\rho < R$ and let M be the maximum of f on $\{z$ such that $|z - z_0| = R\}$. Show that $|z - z_0| \le \rho$ implies that

$$|R_n(z)| \le M \left(\frac{\rho}{R}\right)^{n+1} \frac{1}{1 - \rho/R}$$

33. The *Bernoulli numbers* B_n are related to the coefficients of the power series of $z/(e^z - 1)$ by the formula

$$\frac{z}{e^z - 1} = \sum_{n=0}^{\infty} \frac{B_n}{n!} z^n$$

(a) Determine the radius of convergence of the series

$$\sum_{n=0}^{\infty} \frac{B_n}{n!} z^n$$

(b) Using the Cauchy integral formulas and the contour $|z| = 1$, find an integral expression for B_n of the form

$$B_n = \int_0^{2\pi} g_n(\theta)\, d\theta$$

(for suitable functions $g_n(\theta)$, where $0 \le \theta \le 2\pi$).

34. The *Legendre polynomials* $P_n(\alpha)$ are defined to be the coefficients of z^n in the Taylor development

$$(1 - 2\alpha z + z^2)^{-1/2} = \sum_{n=0}^{\infty} P_n(\alpha) z^n$$

Prove that $P_n(\alpha)$ is a polynomial of degree n and find P_1, P_2, P_3, P_4.

35. Find the radius of convergence of the power series $\sum_{n=0}^{\infty} 2^n z^{n^2}$.

36. Prove:

(a) $\left(\dfrac{z^n}{n!}\right)^2 = \dfrac{1}{2\pi i}\displaystyle\int_\gamma \dfrac{z^n e^{zt}}{n! t^n}\dfrac{dt}{t}$ where γ is the unit circle

(b) $\displaystyle\sum_{n=0}^{\infty}\left(\dfrac{z^n}{n!}\right)^2 = \dfrac{1}{2\pi}\int_0^{2\pi} e^{2z\cos\theta}\, d\theta$

37. Find a power series which solves the functional equation $f(z) = z + f(z^2)$ and show that there is only one power series which solves the equation with $f(0) = 0$.

38. What is wrong with the following argument? Consider

$$f(z) = \cdots + \frac{1}{z^3} + \frac{1}{z^2} + \frac{1}{z} + 1 + z + z^2 + \cdots$$

Note that

$$z + z^2 + \cdots = \frac{z}{1-z}$$

whereas

$$1 + \frac{1}{z} + \frac{1}{z^2} + \cdots = \frac{1}{1 - 1/z} = \frac{-z}{1-z}$$

Hence $f(z) = 0$. Is f in fact the zero function?

39. Suppose f is an entire function and that $|f^{(k)}(0)| \leq 1$ for all $k \geq 0$. Show that $|f(z)| \leq e^{|z|}$ for all $z \in \mathbf{C}$.

40. Let

$$f(z) = \frac{(z-1)^2(z+3)}{1 - \sin(\pi z/2)}$$

(a) Find all the singularities of f and identify each as a removable singularity, a pole (give the order) or an essential singularity.

(b) If $f(z) = a_0 + a_1 z + a_2 z^2 + \cdots$ is the Taylor expansion of f centered at 0, find a_0, a_1, and a_2.

(c) What is the radius of convergence of the series in (b)?

CHAPTER 4

CALCULUS OF RESIDUES

This chapter focuses on the residue theorem, which states that the integral of an analytic function f around a closed contour equals $2\pi i$ times the sum of the residues of f inside the contour. We shall use this theorem in our first main application of complex analysis, the evaluation of definite integrals. So that the student will gain ample facility in handling residues, the chapter begins with the techniques for computing residues of functions at isolated singularities.

4.1 CALCULATION OF RESIDUES

We recall from Sec. 3.3 that if f has an isolated singularity at z_0, then f admits a Laurent expansion that is valid in a deleted neighborhood of z_0:

$$f(z) = \cdots + \frac{b_2}{(z - z_0)^2} + \frac{b_1}{(z - z_0)} + a_0 + a_1(z - z_0) + \cdots$$

where b_1 is called the *residue* of f at z_0. This is written

$$b_1 = \text{Res}\,(f, z_0)$$

We want to develop techniques for computing the residue without having to find the Laurent expansion. Of course, if the Laurent expansion is known there is no problem. For example, since

$$e^{1/z} = 1 + \frac{1}{z} + \frac{1}{2z^2} + \cdots + \frac{1}{n!z^n} + \cdots$$

$f(z) = e^{1/z}$ has residue 1 at $z_0 = 0$.

The reader should keep the following in mind when dealing with residues: The residue at z_0 is simply the coefficient of $(z - z_0)^{-1}$ in the Laurent series for the function for an annular region $\{z \mid 0 < |z - z_0| < R\}$ near z_0. Any of the tricks learned in the last chapter for computing this may help. This section concentrates on obtaining mechanical formulas which will always work, but a little thought will sometimes produce a trick which is easier. (This is illustrated in some of the worked examples.) Particularly important is the case of a pole (in contrast to an essential singularity). For this case we have fairly straightforward techniques that are easy to apply if the order of the pole is not too large.

If we are given an f defined on a region A with an isolated singularity at z_0, then we proceed in the following way to find the residue. First we decide whether we can easily find the first few terms in the Laurent expansion. If so, the residue of f at z_0 will be the coefficient of $1/(z - z_0)$ in the expansion. If not, then we guess the order of singularity, verify it according to the rules that will be developed in this section (some rules were already developed in Proposition 3.3.4), and calculate the residue according to these rules. (The rules are summarized in Table 4.1.1.) If we have any doubt as to what order to guess, we should work systematically by first guessing removable singularity, then simple pole, and so on, checking against Table 4.1.1 until we obtain a verified answer.

Removable Singularities

Let f be analytic in a deleted neighborhood $U\backslash\{z_0\}$ of z_0. It was shown in Sec. 3.3 that f has a removable singularity at z_0 iff $\lim_{z \to z_0} (z - z_0)f(z) = 0$. The following theorem covers many important cases and is sometimes the easiest to use.

4.1.1 PROPOSITION   *If $g(z)$ and $h(z)$ are analytic and have zeros at z_0 of the same order, then $f(z) = g(z)/h(z)$ has a removable singularity at z_0.*

PROOF   By Proposition 3.3.5, we can write $g(z) = (z - z_0)^k \tilde{g}(z)$ where $\tilde{g}(z_0) \neq 0$ and $h(z) = (z - z_0)^k \tilde{h}(z)$ where $\tilde{h}(z_0) \neq 0$ and $\tilde{g}$ and $\tilde{h}$ are analytic and nonzero at z_0. Thus $f(z) = \tilde{g}(z)/\tilde{h}(z)$ is analytic at z_0. ■

Likewise, if g has a zero at z_0 of order greater than h, then g/h has a removable singularity at z_0.

EXAMPLES

(i) $e^z/(z - 1)$ has no singularity at $z_0 = 0$.
(ii) $(e^z - 1)/z$ has a removable singularity at 0 because $e^z - 1$ and z have zeros of order 1. (They vanish but their derivatives do not.)
(iii) $z^2/\sin^2 z$ has a removable singularity at $z_0 = 0$ because both the numerator and the denominator have zeros of order 2. ■

The preceding discussion is summarized in lines 1 and 2 of Table 4.1.1.

Simple Poles

By Proposition 3.3.4, if $\lim_{z \to z_0} (z - z_0) f(z)$ exists and is nonzero, then f has a simple pole at z_0 and this limit equals the residue. Let us apply this result to obtain a useful method for computing residues.

4.1.2 PROPOSITION   *Let g and h be analytic at z_0 and assume that $g(z_0) \neq 0$, $h(z_0) = 0$, and $h'(z_0) \neq 0$. Then $f(z) = g(z)/h(z)$ has a simple pole at z_0 and*

$$\operatorname{Res}(f, z_0) = \frac{g(z_0)}{h'(z_0)} \tag{1}$$

PROOF   We know that

$$\lim_{z \to z_0} \frac{h(z) - h(z_0)}{z - z_0} = \lim_{z \to z_0} \frac{h(z)}{z - z_0} = h'(z_0) \neq 0$$

so that

$$\lim_{z \to z_0} \frac{z - z_0}{h(z)} = \frac{1}{h'(z_0)}$$

Thus

$$\lim_{z \to z_0} (z - z_0) \frac{g(z)}{h(z)} = \frac{g(z_0)}{h'(z_0)}$$

exists and therefore equals the residue.   ∎

ALTERNATIVE PROOF $h(z_0) = 0$ and $h'(z_0) \neq 0$ imply that $h(z) = \tilde{h}(z)(z - z_0)$ where $\tilde{h}(z)$ is analytic at z_0, and that $\tilde{h}(z_0) = h'(z_0) \neq 0$. Thus we can write $g(z)/h(z) = g(z)/\tilde{h}(z)(z - z_0)$ and $g(z)/\tilde{h}(z)$ is analytic at z_0. Hence there is a Taylor series $g(z)/\tilde{h}(z) = \sum_{n=0}^{\infty} a_n (z - z_0)^n$, where $a_0 = g(z_0)/\tilde{h}(z_0)$. Therefore, $g(z)/\tilde{h}(z)(z - z_0) = \sum_{n=0}^{\infty} a_n (z - z_0)^{n-1}$ is the Laurent expansion and $a_0 = g(z_0)/\tilde{h}(z_0) = g(z_0)/h'(z_0)$ is the residue.   ∎

Generally, if $g(z)$ has a zero of order k, and h has a zero of order l with $l > k$, then $g(z)/h(z)$ has a pole of order $l - k$. To see this, write $g(z) = (z - z_0)^k \tilde{g}(z)$ and $h(z) = (z - z_0)^l \tilde{h}(z)$ where $\tilde{g}(z_0) \neq 0$ and $\tilde{h}(z_0) \neq 0$. Thus

$$\frac{g(z)}{h(z)} = \frac{\phi(z)}{(z - z_0)^{l-k}}$$

where $\phi(z) = \tilde{g}(z)/\tilde{h}(z)$, which is analytic at z_0 because $\tilde{h}(z_0) \neq 0$ (and hence $\tilde{h}(z) \neq 0$ in a neighborhood of z_0). Thus our assertion follows from Proposition 3.3.4. If $l = k + 1$, we have a simple pole and can obtain the residue from the next proposition. This generalizes the preceding result.

4.1.3 PROPOSITION *Suppose that $g(z)$ has a zero of order k at z_0 and that $h(z)$ has a zero of order $k + 1$. Then $g(z)/h(z)$ has a simple pole with residue given by*

$$\text{Res}\left(\frac{g}{h}, z_0\right) = (k + 1)\frac{g^{(k)}(z_0)}{h^{(k+1)}(z_0)} \tag{2}$$

PROOF By Taylor's theorem and the fact that $g(z_0) = 0, \ldots, g^{(k-1)}(z_0) = 0$, we can write

$$g(z) = \frac{(z - z_0)^k}{k!} g^{(k)}(z_0) + (z - z_0)^{k+1}\tilde{g}(z)$$

where $\tilde{g}$ is analytic. Similarly,

$$h(z) = \frac{(z - z_0)^{k+1}}{(k + 1)!} h^{(k+1)}(z_0) + (z - z_0)^{k+2}\tilde{h}(z)$$

Thus

$$(z - z_0)\frac{g(z)}{h(z)} = \frac{g^{(k)}(z_0)/k! + (z - z_0)\tilde{g}(z)}{h^{(k+1)}(z_0)/(k + 1)! + (z - z_0)\tilde{h}(z)}$$

As $z \to z_0$, this converges (by the quotient theorem for limits) to

$$(k + 1)\frac{g^{(k)}(z_0)}{h^{(k+1)}(z_0)}$$

which proves our assertion. ∎

We could also have proved that

$$\lim_{z \to z_0} (z - z_0)\frac{g(z)}{h(z)} = (k + 1)\frac{g^{(k)}(z_0)}{h^{(k+1)}(z_0)}$$

by using l'Hôpital's rule (see Exercise 8, Sec. 3.3) and observing that both $(z - z_0)g(z)$ and $h(z)$ are analytic at z_0 with z_0 a zero of order $(k + 1)$.

EXAMPLES

(i) e^z/z at $z = 0$. In this case 0 is not a zero of e^z but is a first-order zero of z, so the residue at 0 is $1 \cdot e^0/1 = 1$. Clearly, Proposition 4.1.2 also applies.

(ii) $e^z/\sin z$ at 0. e^z is not zero at the point $z = 0$, and, since $\cos 0 = 1$, 0 is a first-order zero of $\sin z$. Thus the residue is $e^0/\cos 0 = 1$.

(iii) $z/(z^2 + 1)$ at $z = i$. Here $g(z) = z$, $h(z) = z^2 + 1$. Therefore, $g(i) = i \neq 0$ and $h(i) = 0$, $h'(i) = 2i \neq 0$. Thus the residue at i is $g(i)/h'(i) = \frac{1}{2}$.

(iv) $z/(z^4 - 1)$ at $z = 1$. Here $g(z) = z$, and $h(z) = z^4 - 1$. Thus $g(1) = 1 \neq 0$ and $h(1) = 0$, $h'(1) = 4 \neq 0$, and so the residue is $\frac{1}{4}$.

(v) $z/(1 - \cos z)$ at $z = 0$. Here $g(0) = 0$ and $g'(z) = 1 \neq 0$, so 0 is a simple zero of g. Also $h(0) = 0$, $h'(0) = \sin 0 = 0$, and $h''(0) = \cos 0 = 1 \neq 0$, so 0 is a double zero of h. Thus, by Eq. (2) (see also line 5 of Table 4.1.1), the residue at 0 is

$$2 \cdot \frac{g'(0)}{h''(0)} = 2 \cdot \frac{1}{1} = 2 \quad \blacktriangle$$

Double Poles

As the order of poles increases, the formulas become more complicated and the residues become more laborious to obtain. However, for second-order poles the situation is still relatively simple. Probably the most useful formula for finding the residue in this case is Eq. (3) in the following result.

4.1.4 PROPOSITION   *Let g and h be analytic at z_0 and let $g(z_0) \neq 0$, $h(z_0) = 0$, $h'(z_0) = 0$, and $h''(z_0) \neq 0$. Then $g(z)/h(z)$ has a second-order pole at z_0 and the residue is*

$$\text{Res}\left(\frac{g}{h}, z_0\right) = 2\frac{g'(z_0)}{h''(z_0)} - \frac{2}{3}\frac{g(z_0)h'''(z_0)}{[h''(z_0)]^2} \tag{3}$$

PROOF   Since g has no zero and h has a second-order zero, we know that the pole is of second order (see the remark preceding Proposition 4.1.3). Thus we may write the Laurent series in the form

$$\frac{g(z)}{h(z)} = \frac{b_2}{(z - z_0)^2} + \frac{b_1}{z - z_0} + a_0 + a_1(z - z_0) + a_2(z - z_0)^2 + \cdots$$

and we want to compute b_1. We can write

$$g(z) = g(z_0) + g'(z_0)(z - z_0) + \frac{g''(z_0)}{2}(z - z_0)^2 + \cdots$$

and

$$h(z) = \frac{h''(z_0)}{2}(z - z_0)^2 + \frac{h'''(z_0)}{6}(z - z_0)^3 + \cdots$$

Therefore,

$$g(z) = h(z)\left[\frac{b_2}{(z - z_0)^2} + \frac{b_1}{z - z_0} + a_0 + a_1(z - z_0) + \cdots\right]$$

$$= \left[\frac{h''(z_0)}{2} + \frac{h'''(z_0)}{6}(z - z_0) + \cdots\right] \cdot [b_2 + b_1(z - z_0) + a_0(z - z_0)^2 + \cdots]$$

We can multiply out these two convergent power series as if they were polynomials (see Worked Example 3.2.13). The result is

$$g(z) = \frac{b_2 h''(z_0)}{2} + \left[\frac{b_2 h'''(z_0)}{6} + \frac{b_1 h''(z_0)}{2}\right](z - z_0) + \cdots$$

Since these two power series are equal we can conclude that the coefficients are equal. Therefore,

$$g(z_0) = \frac{b_2 h''(z_0)}{2} \quad \text{and} \quad g'(z_0) = \frac{b_2 h'''(z_0)}{6} + \frac{b_1 h''(z_0)}{2}$$

Solving for b_1 yields the theorem. ∎

NOTE For a second-order pole of the form $g(z)/(z - z_0)^2$ where $g(z_0) \neq 0$, Eq. (3) simplifies to $g'(z_0)$.

The following result may be proved in an analogous manner.

4.1.5 PROPOSITION *Let g and h be analytic at z_0 and let $g(z_0) = 0$, $g'(z_0) \neq 0$, $h(z_0) = 0$, $h'(z_0) = 0$, $h''(z_0) = 0$, and $h'''(z_0) \neq 0$. Then g/h has a second-order pole at z_0 with residue*

$$\frac{3g''(z_0)}{h'''(z_0)} - \frac{3}{2}\frac{g'(z_0)h^{(iv)}(z_0)}{[h'''(z_0)]^2} \tag{4}$$

The proof is left for the student as Exercise 4.

EXAMPLES

(i) $e^z/(z-1)^2$ has a second-order pole at $z_0 = 1$; here we choose $g(z) = e^z$, $h(z) = (z-1)^2$ and note that $g(1) = e \neq 0$, $h(z_0) = 0$, $h'(z_0) = 2(z_0-1) = 0$, and $h''(z_0) = 2 \neq 0$. Therefore, by Eq. (3), the residue at 1 is $(2 \cdot e)/2 - \frac{2}{3} \cdot (e \cdot 0)/2^2 = e$.

(ii) $2(e^z - 1)/\sin^3 z$ with $z_0 = 0$. Here we choose $g(z) = e^z - 1$, $h(z) = \sin^3 z$ and then note that $g(0) = 0$, $g'(0) \neq 0$, $h(0) = 0$, $h'(0) = 3 \sin^2 0 \cdot \cos 0 = 0$, $h''(z) = 6 \sin z \cdot \cos^2 z - 3 \sin^3 z$ (which is equal to zero at 0), and finally $h'''(z) = 6 \cos^3 z - 12 \sin^2 z \cdot \cos z - 9 \sin^2 z \cdot \cos z$ (which is 6 at $z = 0$). We also compute $h^{(iv)}(0) = 0$. Thus by Eq. (4) the residue is $3 \cdot \frac{1}{6} = \frac{1}{2}$. ▲

Higher-Order Poles

For poles of order greater than 2 we could develop formulas in the same manner in which we developed the preceding ones, but they would be quite complicated. Instead two general methods can be used. The first is described in the next proposition.

4.1.6 PROPOSITION *Let f have an isolated singularity at z_0 and let k be the smallest integer ≥ 0 such that $\lim\limits_{z \to z_0} (z - z_0)^k f(z)$ exists. Then $f(z)$ has a pole of order k at z_0 and, if we let $\phi(z) = (z - z_0)^k f(z)$, then ϕ can be defined uniquely at z_0 so that ϕ is analytic at z_0 and*

$$\text{Res}\,(f, z_0) = \frac{\phi^{(k-1)}(z_0)}{(k-1)!} \tag{5}$$

PROOF Since $\lim\limits_{z \to z_0} (z - z_0)^k f(z)$ exists, $\phi(z) = (z - z_0)^k f(z)$ has a removable singularity at z_0, by Proposition 3.3.4. Thus in a neighborhood of z_0, $\phi(z) = (z - z_0)^k f(z) = b_k + b_{k-1}(z - z_0) + \cdots + b_1(z - z_0)^{k-1} + a_0(z - z_0)^k + \cdots$ and so

$$f(z) = \frac{b_k}{(z - z_0)^k} + \frac{b_{k-1}}{(z - z_0)^{k-1}} + \cdots + \frac{b_1}{(z - z_0)} + a_0 + a_1(z - z_0) + \cdots$$

If $b_k = 0$, then $\lim\limits_{z \to z_0} (z - z_0)^{k-1} f(z)$ exists, which contradicts the hypothesis about k. Thus z_0 is a pole of order k. Finally, consider the expansion for $\phi(z)$, and differentiate it $k - 1$ times at z_0, to obtain $\phi^{(k-1)}(z_0) = (k - 1)! b_1$. ■

In this theorem it is Eq. (5) that is important rather than the test for the order. It usually is easier to test the order by writing (if possible) $f = g/h$ and showing that h has a zero of order k greater than g. Then we have a pole of order k (as was explained in the text preceding Proposition 4.1.3).

Let us now suppose that the form of f makes application of Proposition 4.1.6 inconvenient. (For example, consider $e^z/\sin^4 z$ with $z_0 = 0$. Here $k = 4$, since the numerator has no zero and the denominator has a zero of order 4.) There is an alternative method that generalizes Proposition 4.1.4. Suppose that $f(z) = g(z)/h(z)$ and that $h(z)$ has a zero of order k more than g at z_0; therefore, f has a pole of order k. We may write

$$\frac{g(z)}{h(z)} = \frac{b_k}{(z - z_0)^k} + \cdots + \frac{b_1}{(z - z_0)} + p(z)$$

where p is analytic. Also suppose that z_0 is a zero of order m for $g(z)$ and a zero of order $m + k$ for $h(z)$. Then

$$g(z) = \sum_{n=m}^{\infty} \frac{g^{(n)}(z_0)(z - z_0)^n}{n!}$$

and

$$h(z) = \sum_{n=m+k}^{\infty} \frac{h^{(n)}(z_0)(z - z_0)^n}{n!}$$

Thus we can write

$$\sum_{n=m}^{\infty} \frac{g^{(n)}(z_0)(z - z_0)^n}{n!} = \left[\sum_{n=m+k}^{\infty} \frac{h^{(n)}(z_0)(z - z_0)^n}{n!} \right]$$
$$\cdot \left[\frac{b_k}{(z - z_0)^k} + \cdots + \frac{b_1}{(z - z_0)} + p(z) \right]$$

We can then multiply out the right side of the equation as if the factors were polynomials (because of Worked Example 3.2.13) and compare the coefficients of $(z - z_0)^m$, $(z - z_0)^{m+1}$, . . . , $(z - z_0)^{m+k-1}$ to obtain k equations in $b_1, b_2, \ldots , b_k$. Finally, we can solve these equations for b_1. This method may sometimes be more practical than that of Proposition 4.1.6. When $m = 0$ (that is, when $g(z_0) \neq 0$), the explicit formula contained in the following proposition can be used. (The student should prove this result by using the procedure just described.)

4.1.7 PROPOSITION *Let g and h be analytic at z_0, with $g(z_0) \neq 0$, and assume $h(z_0) = 0 = \cdots = h^{(k-1)}(z_0)$ and $h^{(k)}(z_0) \neq 0$. Then g/h has a pole of order k and the* *residue at z_0, Res(g/h, z_0) is given by*

$$
\text{Res } (g/h, z_0) = \left[\frac{k!}{h^{(k)}(z_0)}\right]^k \cdot \begin{vmatrix} \dfrac{h^{(k)}(z_0)}{k!} & 0 & 0 & \cdots & 0 & g(z_0) \\[2mm] \dfrac{h^{(k+1)}(z_0)}{(k+1)!} & \dfrac{h^{(k)}(z_0)}{k!} & 0 & \cdots & 0 & g^{(1)}(z_0) \\[2mm] \dfrac{h^{(k+2)}(z_0)}{(k+2)!} & \dfrac{h^{(k+1)}(z_0)}{(k+1)!} & \dfrac{h^{(k)}(z_0)}{k!} & \cdots & 0 & \dfrac{g^{(2)}(z_0)}{2!} \\[2mm] \cdot & \cdot & \cdot & & & \cdot \\ \cdot & \cdot & \cdot & & & \cdot \\ \cdot & \cdot & \cdot & & & \cdot \\[2mm] \dfrac{h^{(2k-1)}(z_0)}{(2k-1)!} & \dfrac{h^{(2k-2)}(z_0)}{(2k-2)!} & \dfrac{h^{(2k-3)}(z_0)}{(2k-3)!} & \cdots & \dfrac{h^{(k+1)}(z_0)}{(k+1)!} & \dfrac{g^{(k-1)}(z_0)}{(k-1)!} \end{vmatrix} \qquad (6)
$$

where the vertical bars denote the determinant of the enclosed $k \times k$ matrix.

Here are some examples:

(i) $z^2/[(z-1)^3(z+1)]$ at $z_0 = 1$. The pole is of order 3. We use Eq. (5). In this case,

$$
\phi(z) = \frac{z^2}{z+1}
$$

and so

$$
\phi'(z) = \frac{(z+1) \cdot 2z - z^2}{(z+1)^2} = \frac{z^2 + 2z}{(z+1)^2} = 1 - \frac{1}{(z+1)^2}
$$

and

$$
\phi''(z) = \frac{2}{(z+1)^3} \qquad \text{so that} \qquad \phi''(1) = \frac{1}{4}
$$

Thus since $k = 3$, the residue is $(1/2)(1/2^2) = 1/8$.

(ii) $e^z/\sin^3 z$ at $z = 0$. Here $k = 3$ and we shall use Eq. (6), with $g(z) = e^z$ and $h(z) = \sin^3 z$. We need to compute $h'''(0)$, $h^{(iv)}(0)$, and $h^{(v)}(0)$. These are, by straightforward (but laborious) computation, $h'''(0) = 6$, $h^{(iv)}(0) = 0$, and $h^{(v)}(0) = -60$; thus $h'''/3! = 1$, $h^{(iv)}/4! = 0$, and $h^{(v)}/5! = -\frac{1}{2}$. Also $g^{(l)}(0)/l! = 1/l!$, and so the residue is, by Eq. (6),

$$
\left(\frac{3!}{6}\right)^3 \times \begin{vmatrix} 1 & 0 & 1 \\ 0 & 1 & 1 \\ -\frac{1}{2} & 0 & \frac{1}{2} \end{vmatrix} = \begin{vmatrix} 0 & 0 & 1 \\ -1 & 1 & 1 \\ -1 & 0 & \frac{1}{2} \end{vmatrix} = 1
$$

(The last column is subtracted from the first.)

Essential Singularities

In the case of an essential singularity there are no simple formulas like the preceding ones, so we must rely on our ability to find the Laurent expansion. For example, consider

$$f(z) = e^{(z+1/z)} = e^z \cdot e^{1/z} = \left(1 + z + \frac{z^2}{2!} + \cdots\right)\left(1 + \frac{1}{z} + \frac{1}{2!z^2} + \cdots\right)$$

Table 4.1.1   Techniques for Finding Residues

In this table g and h are analytic at z_0 and f has an isolated singularity. The most useful and common tests are indicated by an asterisk.

Function	Test	Type of Singularity	Residue at z_0
1. $f(z)$	$\lim\limits_{z \to z_0} (z - z_0)f(z) = 0$	removable	0
*2. $\dfrac{g(z)}{h(z)}$	g and h have zeros of same order	removable	0
*3. $f(z)$	$\lim\limits_{z \to z_0} (z - z_0)f(z)$ exists and is $\neq 0$	simple pole	$\lim\limits_{z \to z_0} (z - z_0)f(z)$
*4. $\dfrac{g(z)}{h(z)}$	$g(z_0) \neq 0,\ h(z_0) = 0,$ $h'(z_0) \neq 0$	simple pole	$\dfrac{g(z_0)}{h'(z_0)}$
5. $\dfrac{g(z)}{h(z)}$	g has zero of order k, h has zero of order $k + 1$	simple pole	$(k + 1)\dfrac{g^{(k)}(z_0)}{h^{(k+1)}(z_0)}$
*6. $\dfrac{g(z)}{h(z)}$	$g(z_0) \neq 0$ $h(z_0) = 0 = h'(z_0)$ $h''(z_0) \neq 0$	second-order pole	$2\dfrac{g'(z_0)}{h''(z_0)} - \dfrac{2}{3}\dfrac{g(z_0)h'''(z_0)}{[h''(z_0)]^2}$
*7. $\dfrac{g(z)}{(z - z_0)^2}$	$g(z_0) \neq 0$	second-order pole	$g'(z_0)$
8. $\dfrac{g(z)}{h(z)}$	$g(z_0) = 0,\ g'(z_0) \neq 0,$ $h(z_0) = 0 = h'(z_0)$ $= h''(z_0),\ h'''(z_0) \neq 0$	second-order pole	$3\dfrac{g''(z_0)}{h'''(z_0)} - \dfrac{3}{2}\dfrac{g'(z_0)h^{(iv)}(z_0)}{[h'''(z_0)]^2}$
9. $f(z)$	k is the smallest integer such that $\lim\limits_{z \to z_0} \phi(z_0)$ exists where $\phi(z) = (z - z_0)^k f(z)$	pole of order k	$\lim\limits_{z \to z_0} \dfrac{\phi^{(k-1)}(z)}{(k - 1)!}$
*10. $\dfrac{g(z)}{h(z)}$	g has zero of order l, h has zero of order $k + l$	pole of order k	$\lim\limits_{z \to z_0} \dfrac{\phi^{(k-1)}(z)}{(k - 1)!}$ where $\phi(z) = (z - z_0)^k \dfrac{g}{h}$
11. $\dfrac{g(z)}{h(z)}$	$g(z_0) \neq 0,\ h(z_0) =$ $\cdots = h^{k-1}(z_0)$ $= 0,\ h^k(z_0) \neq 0$	pole of order k	see Eq. (6) in Proposition 4.1.7

Gathering terms involving $1/z$, we get

$$\frac{1}{z}\left(1 + \frac{1}{2!} + \frac{1}{2!3!} + \frac{1}{3!4!} + \cdots\right)$$

(We multiply out as in the procedure of Worked Example 3.2.13, a method that is justified by a more general result that is outlined in Exercise 12). The residue is thus

$$\text{Res}\,(f,\,0) = 1 + \frac{1}{2!} + \frac{1}{2!3!} + \frac{1}{3!4!} + \cdots$$

We cannot sum the series explicitly.

Worked Examples

4.1.8   *Compute the residue of $z^2/\sin^2 z$ at $z = 0$.*

Solution. Since both numerator and denominator have a zero of order 2, the singularity is removable, and so the residue is zero.

4.1.9   *Find the residues at all singularities of*

$$\tan z = \frac{\sin z}{\cos z}$$

Solution. The singularities of $\tan z$ occur when $\cos z = 0$. The zeros of $\cos z$ occur at

$$z = \pm\frac{\pi}{2},\, \pm\frac{3\pi}{2},\, \pm\frac{5\pi}{2},\, \cdots$$

The student should recall that these points are the only zeros of $\cos z$. We conclude that the singularities of $\tan z$ occur at the points $z_n = (2n+1)\pi/2$, where n is an integer. We choose $g(z) = \sin z$ and $h(z) = \cos z$. At any z_n, $h'(z_n) = \pm 1 \neq 0$, and so each z_n is a simple pole of $\tan z$. Thus we may use formula 4 of Table 4.1.1 to obtain

$$\text{Res}\,(\tan z,\, z_n) = \frac{g(z_n)}{h'(z_n)} = -1$$

Second Solution. We know that

$$\sin z = (-1)^n \sin\,(z - \pi n) = (-1)^{n+1} \cos\left(z - \pi n - \frac{\pi}{2}\right) = (-1)^{n+1} \cos\,(z - z_n)$$

$$= (-1)^{n+1} \sum_{k=0}^{\infty} \frac{(-1)^k}{(2k)!}\,(z - z_n)^{2k}$$

and

$$\cos z = (-1)^n \cos (z - \pi n) = (-1)^n \sin \left(z - \pi n - \frac{\pi}{2} \right) = (-1)^n \sin (z - z_n)$$

$$= (-1)^n \sum_{k=0}^{\infty} \frac{(-1)^k}{(2k+1)!} (z - z_n)^{2k+1}$$

As before, the poles are simple, so the series for $\tan z$ is of the form $\tan z = b_1/(z - z_n) + \sum_{k=0}^{\infty} u_k(z - z_n)^k$, so that $\sin z = \tan z \cos z$ becomes

$$(-1)^{n+1} \sum_{k=0}^{\infty} \frac{(-1)^k}{(2k)!} (z - z_n)^{2k} = \left[\frac{b_1}{z - z_n} + \sum_{k=0}^{\infty} a_k(z - z_n)^k \right] \left[(-1)^n \sum_{k=0}^{\infty} \frac{(-1)^k}{(2k+1)!} (z - z_n)^{2k+1} \right]$$

that is,

$$-\left[1 - \frac{(z - z_n)^2}{2} + \cdots \right] = \left[\frac{b_1}{z - z_n} + a_0 + a_1(z - z_n) + \cdots \right]$$

$$\times \left[(z - z_n) - \frac{(z - z_n)^3}{6} + \cdots \right]$$

$$= b_1 + a_0(z - z_n) + (a_1 - \tfrac{1}{6})(z - z_n)^2 + \cdots$$

We need only compare the first terms to obtain $b_1 = -1$.

4.1.10   *Evaluate the residue of $(z^2 - 1)/(z^2 + 1)^2$ at $z = i$.*

Solution. $(z^2 + 1)^2$ has a zero of order 2 at i and $i^2 - 1 \neq 0$, so $(z^2 - 1)/(z^2 + 1)^2$ has a pole of order 2; thus to find the residue we use formula 6 of Table 4.1.1. We choose $g(z) = z^2 - 1$, which satisfies $g(i) = -2$ and $g'(i) = 2i$. We also take $h(z) = (z^2 + 1)^2$ and note that $h'(z) = 4z(z^2 + 1)$, so that $h(i) = h'(i) = 0$. Also, $h''(z) = 4(z^2 + 1) + 8z^2 = 12z^2 + 4$, so that $h''(i) = -8$ and $h'''(z) = 24z$; therefore, $h'''(i) = 24i$. Thus the residue is

$$\frac{2 \cdot 2i}{-8} - \frac{2}{3} \cdot \frac{(-2) \cdot 24i}{64} = 0$$

Second Solution. We know from algebra (or integration in calculus) that $(z^2 - 1)/(z^2 + 1)^2$ has a partial-fraction expansion of the form

$$\frac{z^2 - 1}{(z^2 + 1)^2} = \frac{Az + B}{(z - i)^2} + \frac{a}{(z - i)} + \frac{Cz + D}{(z + i)^2} + \frac{b}{(z + i)}$$

Solving for the coefficients gives the identity

$$\frac{z^2 - 1}{(z^2 + 1)^2} = \frac{1}{2} \frac{1}{(z - i)^2} + \frac{1}{2} \frac{1}{(z + i)^2}$$

The second term is analytic at $z = i$, and so the Laurent series is of the form

$$\frac{z^2 - 1}{(z^2 + 1)^2} = \frac{1}{2} \frac{1}{(z - i)^2} + \sum_{n=0}^{\infty} a_n(z - i)^n$$

The residue is the coefficient of $(z - i)^{-1}$, which is 0.

Exercises

1. Find the residues of the following functions at the indicated points:

(a) $\dfrac{e^z - 1}{\sin z}$, $z_0 = 0$

(b) $\dfrac{1}{e^z - 1}$, $z_0 = 0$

(c) $\dfrac{z + 2}{z^2 - 2z}$, $z_0 = 0$

(d) $\dfrac{1 + e^z}{z^4}$, $z_0 = 0$

(e) $\dfrac{e^z}{(z^2 - 1)^2}$, $z_0 = 1$

2. Find the residues of the following functions at the indicated points:

(a) $\dfrac{e^{z^2}}{z - 1}$, $z_0 = 1$

(b) $\dfrac{e^{z^2}}{(z - 1)^2}$, $z_0 = 0$

(c) $\left(\dfrac{\cos z - 1}{z}\right)^2$, $z_0 = 0$

(d) $\dfrac{z^2}{z^4 - 1}$, $z_0 = e^{i\pi/2}$

3. If $f(z)$ has residue b_1 at $z = z_0$, show by example that $[f(z)]^2$ need not have residue b_1^2 at $z = z_0$.

4. Deduce Proposition 4.1.5 from Proposition 4.1.4.

5. Explain what is wrong with the following reasoning. Let

$$f(z) = \frac{1 + e^z}{z^2} + \frac{1}{z}$$

$f(z)$ has a pole at $z = 0$ and the residue at that point is the coefficient of $1/z$, namely, 1. Compute the residue correctly.

6. Complete the proof of Proposition 4.1.7.

7. Find all singular points of the following functions and compute the residues at those points:

(a) $\dfrac{1}{z^3(z + 4)}$

(b) $\dfrac{1}{z^2 + 2z + 1}$

(c) $\dfrac{1}{z^3 - 3}$

8. Find all singular points of the following functions and compute the residues at those points:

(a) $\dfrac{1}{e^z - 1}$

(b) $\sin \dfrac{1}{z}$

9. Find the residue of $1/(z^2 \sin z)$ at $z = 0$.

10. If f_1 and f_2 have residues r_1 and r_2 at z_0, show that the residue of $f_1 + f_2$ at z_0 is $r_1 + r_2$.

11. If f_1 and f_2 have simple poles at z_0, show that $f_1 f_2$ has a second-order pole at z_0. Derive a formula for the residue.

12. Let

$$f(z) = \cdots + \frac{b_k}{(z - z_0)^k} + \cdots + \frac{b_1}{z - z_0} + a_0 + a_1(z - z_0) + \cdots$$

and

$$g(z) = \cdots + \frac{d_k}{(z - z_0)^k} + \cdots + \frac{d_1}{z - z_0} + c_0 + c_1(z - z_0) + \cdots$$

be Laurent expansions for f and g valid for $0 < |z - z_0| < r$. Show that the Laurent expansion for fg is obtained by formally multiplying these series. Do this by proving the following result: If $\Sigma_{n=0}^{\infty} a_n$ and $\Sigma_{n=0}^{\infty} b_n$ are absolutely convergent, then

$$\left(\sum_{n=0}^{\infty} a_n \right) \left(\sum_{n=0}^{\infty} b_n \right) = \sum_{n=0}^{\infty} c_n$$

where $c_n = \Sigma_{j=0}^{n} a_j b_{n-j}$; moreover, the series $\Sigma_{n=0}^{\infty} c_n$ is absolutely convergent. (*Hint.* Show that

$$\sum_{j=0}^{n} |c_j| \leq \sum_{j=0}^{n} \sum_{k=0}^{n} |a_j||b_{k-j}| \leq \left(\sum_{j=0}^{n} |a_j| \right) \left(\sum_{k=0}^{n} |b_k| \right)$$

and use this to deduce that $\Sigma\, c_n$ converges absolutely. Estimate the error between $\Sigma_{k=0}^{n} c_k$, $(\Sigma_{j=0}^{n} a_j)(\Sigma_{k=0}^{n} b_k)$, and $(\Sigma_{j=0}^{n} a_j)(\Sigma_{k=0}^{\infty} b_k))$.

13. Compute the residues of the following functions at their singularities:

(a) $\dfrac{1}{(1 - z)^3}$

(b) $\dfrac{e^z}{(1 - z)^3}$

(c) $\dfrac{1}{z(1 - z)^3}$

(d) $\dfrac{e^z}{z(1 - z)^3}$

14. Find the residues of $(z^2 - 1)/[\cos (\pi z) + 1]$ at each of its singularities. (See Review Exercise 29 of Chap. 3.)

4.2 THE RESIDUE THEOREM

The residue theorem, which is proved in this section, includes Cauchy's theorem and Cauchy's integral formula as special cases. It is one of the main results of complex analysis and leads quickly to interesting applications, some of which are considered in the next section. The main tools needed to prove the theorem are Cauchy's theorem (2.2.1 and 2.3.14) and the Laurent expansion theorem (3.3.1).

The precise proof of the residue theorem is preceded by two intuitive proofs that are based only on the material in Sec. 2.2 and the following property of the residue at z_0:

$$2\pi i \operatorname{Res}(f, z_0) = \int_\gamma f$$

where γ is a small circle around z_0 (see Proposition 3.3.3). For most practical examples the intuitive proofs are in fact perfectly adequate, but, as was evident in Sec. 2.2, it is difficult to formulate a general theorem to which the argument rigorously applies.

Residue Theorem

4.2.1 RESIDUE THEOREM   *Let A be a region and let* $z_1, \ldots, z_n \in A$ *be n distinct points in A. Let f be analytic on* $A \backslash \{z_1, z_2, \ldots, z_n\}$; *that is, let f be analytic on A except for isolated singularities at* $z_1, \ldots, z_n$. *Let* γ *be a closed curve in A homotopic to a point in A. Assume that no* z_i *lies on* γ. *Then*

$$\int_\gamma f = 2\pi i \sum_{i=1}^{n} [\operatorname{Res}(f, z_i)] I(\gamma, z_i) \tag{1}$$

where $\operatorname{Res}(f, z_i)$ *is the residue of f at* z_i (see the preceding section) *and* $I(\gamma, z_i)$ *is the index (or winding number) of* γ *with respect to* z_i (see Sec. 2.4).

In most practical examples, γ will be a simple closed curve traversed counterclockwise, and thus $I(\gamma, z_i)$ will be 1 or 0 according to whether z_i lies inside γ or outside γ. (This is illustrated in Fig. 4.2.1.) The same policy regarding computation of the index $I(\gamma, z)$ as was used in Sec. 2.4 will be followed in this section. An intuitive proof is acceptable so long as such statements can be substantiated with homotopy arguments when asked for. For example, $I(\gamma, z_0) = +1$ if γ can be shown to be homotopic in $\mathbf{C} \backslash \{z_0\}$ to $\gamma(t) = z_0 + re^{i\theta}$, $0 \leq \theta \leq 2\pi$.

Assuming that we accept the Jordan curve theorem, a general formulation of the residue theorem for simple closed curves may be stated as follows: *If* γ *is a simple closed curve in the region A whose inside lies in A and if f is analytic on* $A \backslash \{z_1, \ldots, z_n\}$, *then* $\int_\gamma f$ *is* $2\pi i$ *times the sum of the residues of f inside* γ *when* γ *is traversed in the counterclockwise direction.* This is the classical way of stating the residue theorem, but our

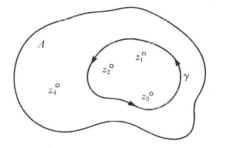

FIGURE 4.2.1 Residue theorem: $\int_\gamma f = 2\pi i[\text{Res}\,(f,\,z_1) + \text{Res}\,(f,\,z_2) + \text{Res}\,(f,\,z_3)]$

original statement (4.2.1) is preferable because it does not restrict us to simple closed curves and does not rely on the difficult Jordan curve theorem.

Two short intuitive proofs of the residue theorem are now given for simple closed curves. They will be illustrated by an example showing that, in practical cases, such proofs can be made quite precise.

FIRST INTUITIVE PROOF OF THE RESIDUE THEOREM FOR SIMPLE CLOSED CURVES Since γ is contractible in A to a point in A, the inside of γ lies in A. Suppose that each z_i lies in the inside of γ. Around each z_i draw a circle γ_i that also lies inside γ. Apply Worked Example 2.2.10 (the generalized deformation theorem) to obtain $\int_\gamma f = \sum_{i=1}^{n} \int_{\gamma_i} f$, since f is analytic on $\gamma, \gamma_1, \ldots, \gamma_n$ and the region between them (Fig. 4.2.2). Suppose that $\gamma, \gamma_1, \ldots, \gamma_n$ are all traversed in the counterclockwise direction. As was proved in Proposition 3.3.3, $\int_{\gamma_i} f = 2\pi i\,\text{Res}\,(f,\,z_i)$, and so $\int_\gamma f = 2\pi i \sum_{i=1}^{n} \text{Res}\,(f,\,z_i)$, which is the statement of the residue theorem since $I(\gamma,\,z) = 1$ for z inside γ and $I(\gamma,\,z) = 0$ for z outside γ. ■

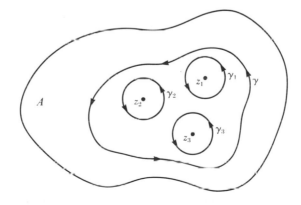

FIGURE 4.2.2 First intuitive proof of the residue theorem.

SECOND INTUITIVE PROOF OF THE RESIDUE THEOREM FOR SIMPLE CLOSED CURVES   This proof proceeds in the same manner as the preceding one except that a different justification is given that $\int_\gamma f = \sum_{i=1}^n \int_{\gamma_i} f$. The circles are connected as shown in Fig. 4.2.3, to obtain a new curve $\tilde{\gamma}$. Thus γ and $\tilde{\gamma}$ are homotopic in $A \backslash \{z_1, \ldots, z_n\}$, and so, by the deformation theorem, $\int_\gamma f = \int_{\tilde{\gamma}} f$. But $\int_{\tilde{\gamma}} f = \sum_{i=1}^n \int_{\gamma_i} f$, since the portions along the connecting curves cancel out. ∎

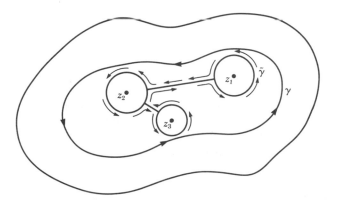

FIGURE 4.2.3   Second intuitive proof of the residue theorem.

It is important that the student clearly understand why these proofs fail to be precise. First, we assumed that γ is simple. Second, we use the Jordan curve theorem (which we did not prove) to be able to discuss the inside and outside of γ and the fact that $I(\gamma, z) = 1$ for z inside γ and $I(\gamma, z) = 0$ for z outside γ. Finally, in the first intuitive proof, we used Worked Example 2.2.10, which was established only informally, to justify that $\int_\gamma f = \sum_{i=1}^n \int_{\gamma_i} f$.

EXAMPLE   Consider $\int_\gamma dz/(z^2 - 1)$, where γ is a circle with center 0 and radius 2. The function $1/(z^2 - 1)$ has simple poles at $-1, 1$ (see Fig. 4.2.4). Note that in this example we can discuss the inside and outside of γ, and we know that -1 and 1 have an index $+1$ with respect to γ. (The Jordan curve theorem is not needed in such a concrete example.)

We draw two circles γ_1 and γ_2 of radius $\frac{1}{4}$ around -1 and 1, respectively. The only statement in the preceding proofs of the residue theorem that was not precise was

$$\int_\gamma f = \int_{\gamma_1} f + \int_{\gamma_2} f$$

The exact justification of $\int_\gamma f = \int_{\gamma_1} f + \int_{\gamma_2} f$ in terms of homotopies was explained in Worked Example 2.3.23.

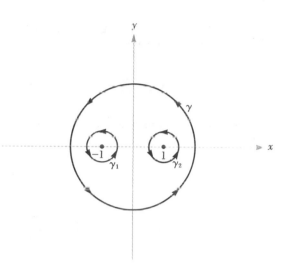

FIGURE 4.2.4   Justifying $\int_\gamma f = \int_{\gamma_1} f + \int_{\gamma_2} f$.

Note that in this case

$$\int_\gamma \frac{dz}{z^2 - 1} = 2\pi i \left[\mathrm{Res}\left(\frac{1}{z^2 - 1}, -1 \right) + \mathrm{Res}\left(\frac{1}{z^2 - 1}, 1 \right) \right]$$

$$= 2\pi i \left[\frac{1}{2(-1)} + \frac{1}{2 \cdot 1} \right] = 0$$

We could also justify $\int_\gamma f = \int_{\gamma_1} f + \int_{\gamma_2} f$ by considering the curve in Fig. 4.2.5(i) and showing that it is homotopic in $\mathbf{C} \backslash \{1, -1\}$ to a point. This is geometrically clear; a homotopy is indicated in Fig. 4.2.5(ii) and (iii). ■

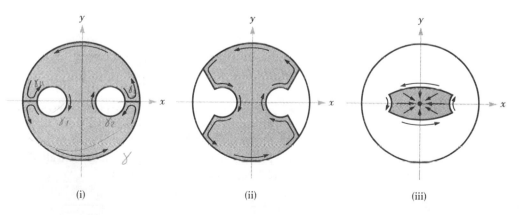

FIGURE 4.2.5   A curve that is homotopic to a point.

PRECISE PROOF OF THE RESIDUE THEOREM   Since z_i is an isolated singularity of f, we can write a Laurent series expansion

$$f(z) = \sum_{n=0}^{\infty} a_n (z - z_i)^n + \sum_{m=1}^{\infty} \frac{b_m}{(z - z_i)^m}$$

in some deleted neighborhood of z_i of the form $\{z \mid r > |z - z_i| > 0\}$ for some $r > 0$. Recall from Proposition 3.3.2 that

$$\sum_{m=1}^{\infty} \frac{b_m}{(z - z_i)^m}$$

is called the singular part of the Laurent series expansion and that it converges on $\mathbf{C} \backslash \{z_i\}$, uniformly outside any circle $|z - z_i| = \epsilon > 0$. Hence

$$\sum_{m=1}^{\infty} \frac{b_m}{(z - z_i)^m}$$

is analytic on $\mathbf{C} \backslash \{z_i\}$ (see Proposition 3.1.8). Denote the singular part of the Laurent expansion of f around z_i by $S_i(z)$.

Consider the function

$$g(z) = f(z) - \sum_{i=1}^{n} S_i(z)$$

Since f is analytic on $A \backslash \{z_1, \ldots, z_n\}$ and since each $S_i(z)$ is analytic on $\mathbf{C} \backslash \{z_i\}$, g is analytic on $A \backslash \{z_1, \ldots, z_n\}$.

All the z_i's are removable singularities of g because on a deleted neighborhood $\{z \mid r > |z - z_i| > 0\}$, which does not contain any of the singularities, we have

$$f(z) = \sum_{n=0}^{\infty} a_n (z - z_i)^n + S_i(z)$$

and so

$$g(z) = \sum_{n=0}^{\infty} a_n (z - z_i)^n - \sum_{j=1}^{i-1} S_j(z) - \sum_{j=i+1}^{n} S_j(z)$$

Since the functions S_j, $j \neq i$, are analytic on $\mathbf{C} \backslash \{z_j\}$, we know that $\lim\limits_{z \to z_i} g(z)$ exists and equals $a_0 - \sum_{\substack{j=1 \\ j \neq i}}^{n} S_j(z_i)$. Consequently, z_i is a removable singularity of g.

Because g can be defined at the points z_i in such a way that g is analytic on all of A, we may apply the Cauchy theorem (2.3.14) to obtain $\int_{\gamma} g = 0$. Hence

$$\int_{\gamma} f = \sum_{i=1}^{n} \int_{\gamma} S_i$$

Next consider the integral $\int_\gamma S_i$. The function $S_i(z)$ is of the form

$$\sum_{m=1}^{\infty} \frac{b_m}{(z - z_i)^m}$$

which, as we have noted, converges uniformly outside a small disk centered at z_i. Thus the convergence is uniform on γ. (Since $\mathbb{C}\backslash\{\gamma([a, b])\}$ is an open set, each z_i has a small disk around it not meeting γ.) By Proposition 3.1.9,

$$\int_\gamma S_i = \sum_{m=1}^{\infty} \int_\gamma \frac{b_m}{(z - z_i)^m} \, dz$$

But for $m > 1$ and $z \neq z_i$,

$$\frac{1}{(z - z_i)^m} = \frac{d}{dz}\left[\frac{(z - z_i)^{1-m}}{1 - m}\right]$$

and so by Proposition 2.1.7 and the fact that γ is a closed curve, all terms are zero except the term in which $m = 1$. Thus

$$\int_\gamma S_i = \int_\gamma \frac{b_1}{z - z_i} \, dz = b_1 \int_\gamma \frac{1}{z - z_i} \, dz$$

By Definition 2.4.1 for the index, this is equal to $b_1 \cdot 2\pi i \cdot I(\gamma, z_i) = 2\pi i[\text{Res}\,(f, z_i)]I(\gamma, z_i)$. Thus

$$\int_\gamma f = \sum_{i=1}^{n} \int_\gamma S_i = \sum_{i=1}^{n} 2\pi i[\text{Res}\,(f, z_i)]I(\gamma, z_i)$$

and the theorem is proved. ■

SUPPLEMENT TO SECTION 4.2: RESIDUES AND BEHAVIOR AT INFINITY

If a function f is analytic for all large enough z (that is, outside some large circle), then it is analytic in a deleted neighborhood of ∞ in the sense of the Riemann sphere and the point at ∞ as defined in Sec. 1.4. We can think of ∞ as an isolated singularity of f, perhaps removable. Let $F(z) = f(1/z)$. If $z = 0$, we set $1/z = \infty$. (Equivalently, $1/z \to \infty$ as $z \to 0$.) Thus it makes sense to discuss the behavior of f at ∞ in terms of the behavior of F at 0.

4.2.2 DEFINITION

(i) *We say f has a pole of order k at ∞ if F has a pole of order k at 0.*
(ii) *We say f has a zero of order k at ∞ if F has a zero of order k at 0.*
(iii) *We define* $\text{Res}\,(f, \infty) = -\text{Res}\,((1/z^2)F(z), 0)$.

Notice in particular that a polynomial of degree k has a pole of order k at ∞. This agrees with what we saw in the proof of the fundamental theorem of algebra in Sec. 2.4. As $z \to \infty$, a polynomial of degree k behaves much like z^k. See also Worked Example 4.2.8. The definition of residue at ∞ may seem a bit strange, but is designed to make the next two propositions work out correctly.

4.2.3 PROPOSITION   *Suppose there is an $R_0 > 0$ such that f is analytic on $\{z \in \mathbf{C}$ such that $|z| > R_0\}$. If $R > R_0$, Γ is the circle of radius R centered at 0, and Γ is traversed once counterclockwise, then $\int_{\Gamma} f = -2\pi i \operatorname{Res}(f, \infty)$.*

PROOF   Let $r = 1/R$, and let γ be the circle of radius r centered at 0, and traversed counterclockwise. If z is inside γ, then $1/z$ is outside Γ, so the function $g(z) = f(1/z)/z^2$ is analytic everywhere inside γ except at 0. Thus

$$2\pi i \operatorname{Res}(g, 0) = \int_{\gamma} [f(1/z)/z^2]\, dz = \int_0^{2\pi} f(r^{-1} e^{-it}) r^{-2} e^{-2it} r e^{it}\, dt$$
$$= \int_0^{2\pi} f(Re^{-it}) Re^{-it}\, dt = \int_{-2\pi}^0 f(Re^{is}) Re^{is}\, ds = \int_0^{2\pi} f(Re^{is}) Re^{is}\, ds$$
$$= \int_{\Gamma} f$$

The next-to-last equality comes from the 2π periodicity of e^{is}. ∎

The choice of the minus sign comes from the fact that as we proceed along a simple closed curve in $\mathbf{C}$ in the counterclockwise direction, the region we normally think of as the inside lies to the left. (Look at any of the figures in this section.) The point at ∞ lies to the left if we proceed in the opposite direction along the curve. Hence the minus sign. For the curves in the last proof, if z proceeds in the counterclockwise direction along Γ, then $1/z$ proceeds in the clockwise direction along γ. Since f is analytic outside Γ except possibly at ∞, Proposition 4.2.3 may be interpreted as saying that $(1/2\pi i) \int_{\Gamma} f$ is the negative of the residue of f outside Γ. This is correct more generally.

4.2.4 PROPOSITION   *Let γ be a simple closed curve in $\mathbf{C}$ traversed once counterclockwise. Let f be analytic along γ and have only finitely many singularities outside γ. Then*

$$\int_{\gamma} f = -2\pi i \sum \{\text{residues of } f \text{ outside } \gamma \text{ including at } \infty\}$$

PROOF   Apply the residue theorem to a composite curve such as that in Fig. 4.2.6. Choose Γ to be a circle large enough to contain γ and all the finite singularities of f in its interior. The reader is asked to supply the remaining details of an informal proof in Exercise 14. ∎

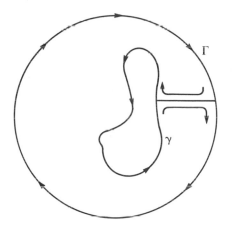

FIGURE 4.2.6 Curve used in the proof of the residue theorem for the exterior of a curve.

Worked Examples

4.2.5 *Let γ be a circle of radius $\frac{1}{2}$ parametrized by $\gamma(t) = e^{it}/2$, $0 \leq t \leq 2\pi$. Evaluate*

$$\int_{\gamma} \frac{dz}{z^2 + z + 1}$$

Solution. The singularities occur at the points for which the denominator vanishes. These points are

$$z_1 = \frac{-1 + \sqrt{1 - 4}}{2} = \frac{-1 + \sqrt{3}i}{2}$$

and

$$z_2 = \frac{-1 - \sqrt{1 - 4}}{2} = \frac{-1 - \sqrt{3}i}{2}$$

It is easy to check that $|z_1| = |z_2| = 1$, so that both z_1 and z_2 lie outside the circle of radius $\frac{1}{2}$. Now γ is homotopic to 0 in $\mathbb{C}\backslash\{z_1\}$, and $1/(z - z_1)$ is analytic on $\mathbb{C}\backslash\{z_1\}$, and so by Cauchy's theorem, $I(\gamma, z_1) = 0$. Similarly, $I(\gamma, z_2) = 0$. Therefore, by the residue theorem,

$$\int_{\gamma} \frac{dz}{z^2 + z + 1} = 0$$

Alternatively, we could note that γ is homotopic to 0 in $\mathbb{C}\backslash\{z_1, z_2\}$ (Fig. 4.2.7) and $1/(z^2 + z + 1)$ is analytic on $\mathbb{C}\backslash\{z_1, z_2\}$; thus, by Cauchy's theorem,

$$\int_{\gamma} \frac{dz}{z^2 + z + 1} = 0$$

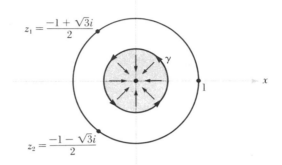

$z_1 = \dfrac{-1 + \sqrt{3}i}{2}$

$z_2 = \dfrac{-1 - \sqrt{3}i}{2}$

FIGURE 4.2.7 $\displaystyle\int_\gamma \frac{dz}{z^2 + z + 1} = 0$

4.2.6   *Evaluate*

$$\int_\gamma \frac{dz}{z^4 + 1}$$

where γ consists of the portion of the x axis from -2 to $+2$ and the semicircle in the upper half plane from 2 to -2 centered at 0.

Solution. The singular points of the integrand occur at the fourth roots of -1: $e^{\pi i/4}$, $e^{(\pi + 2\pi)i/4} = e^{3\pi i/4}$, $e^{(\pi + 4\pi)i/4} = e^{5\pi i/4}$, and $e^{(\pi + 6\pi)i/4} = e^{7\pi i/4}$ (see Fig. 4.2.8). Thus, by the residue theorem,

$$\int_\gamma \frac{dz}{z^4 + 1} = 2\pi i \left[\operatorname{Res}\left(\frac{1}{z^4 + 1}, e^{\pi i/4}\right) I(\gamma, e^{\pi i/4}) + \operatorname{Res}\left(\frac{1}{z^4 + 1}, e^{3\pi i/4}\right) I(\gamma, e^{3\pi i/4}) \right.$$

$$\left. + \operatorname{Res}\left(\frac{1}{z^4 + 1}, e^{5\pi i/4}\right) I(\gamma, e^{5\pi i/4}) + \operatorname{Res}\left(\frac{1}{z^4 + 1}, e^{7\pi i/4}\right) I(\gamma, e^{7\pi i/4}) \right]$$

It is intuitively clear that $I(\gamma, e^{\pi i/4}) = 1$, $I(\gamma, e^{3\pi i/4}) = 1$, whereas the other two indexes are zero. This can be more carefully justified as follows: γ is homotopic to a circle $\tilde{\gamma}$ around $e^{\pi i/4}$ traversed counterclockwise. To see this, reparametrize γ so that it is defined on the interval $[0, 2\pi]$. A suitable homotopy is then $H(s, t) = (1 - t)\gamma(s) + t\tilde{\gamma}(s)$. This homotopy is illustrated in Fig. 4.2.9. We know that $I(\tilde{\gamma}, e^{\pi i/4}) = 1$ by Worked Example 2.1.12, and that $I(\tilde{\gamma}, e^{\pi i/4}) = I(\gamma, e^{\pi i/4})$ by the deformation theorem. Thus $I(\gamma, e^{\pi i/4}) = 1$ and, similarly, $I(\gamma, e^{3\pi i/4}) = 1$. Furthermore, γ can be contracted to the origin along the radii of the semicircle, and so by Cauchy's theorem, $I(\gamma, e^{5\pi i/4}) = 0$ and $I(\gamma, e^{7\pi i/4}) = 0$.

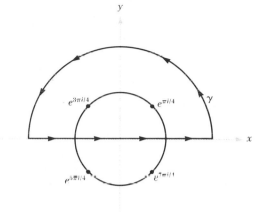

FIGURE 4.2.8   The curve γ in Example 4.2.6.

To calculate

$$\text{Res}\left(\frac{1}{z^4 + 1}, e^{\pi i/4}\right)$$

observe that $e^{\pi i/4}$ is a simple pole of the function $1/(z^4 + 1)$, so we may use formula 4 of Table 4.1.1 to obtain

$$\text{Res}\left(\frac{1}{z^4 + 1}, e^{\pi i/4}\right) = \frac{1}{4(e^{\pi i/4})^3} = \frac{e^{\pi i/4}}{4e^{\pi i}} = -\frac{e^{\pi i/4}}{4}$$

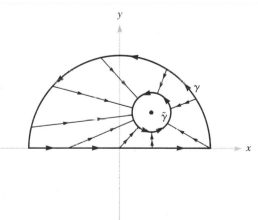

FIGURE 4.2.9   Homotopy between γ and $\tilde{\gamma}$.

Similarly,

$$\text{Res}\left(\frac{1}{z^4+1},\, e^{3\pi i/4}\right) = \frac{1}{4(e^{3\pi i/4})^3} = \frac{e^{-\pi i/4}}{4}$$

Therefore,

$$\int_\gamma \frac{dz}{z^4+1} = \frac{2\pi i}{4}\left(e^{-\pi i/4} - e^{\pi i/4}\right) = \pi \sin\frac{\pi}{4} = \frac{\pi\sqrt{2}}{2}$$

Remark. We do not actually have to use such a detailed method to calculate the indexes (winding numbers). We simply use our intuition to estimate the number of times the curve in question winds around the given point in the counterclockwise direction. But we must bear in mind that the justification for this intuition consists of an argument like the preceding one.

4.2.7 *Evaluate*

$$\int_\gamma \frac{1+z}{1-\cos z}\, dz$$

where γ is the circle of radius 7 around zero.

Solution. The singularities of $(1+z)/(1-\cos z)$ occur where $1-\cos z = 0$. But $(e^{iz} + e^{-iz})/2 = 1$ implies that $(e^{iz})^2 - 2(e^{iz}) + 1 = 0$, that is, that $(e^{iz} - 1)^2 = 0$, and hence $e^{iz} = 1$. Therefore, the singularities occur at $z = 2\pi n$ for $n = \ldots, -2, -1, 0, 1, 2, 3, \ldots$. The only singularities of $(1+z)/(1-\cos z)$ that lie inside the circle of radius 7 are $z_1 = 0$, $z_2 = 2\pi$, and $z_3 = -2\pi$ (see Fig. 4.2.10). Also, $d(1-\cos z)/dz = \sin z$, which is zero at $0, -2\pi, +2\pi$; and $d^2(1-\cos z)/dz^2 = \cos z$, which is nonzero at $0, -2\pi$, and 2π, so these singularities are poles of order 2.

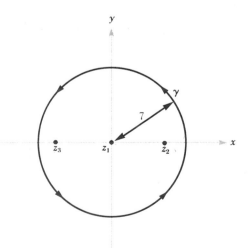

FIGURE 4.2.10 The curve γ contains three singularities.

The residue at one of these poles z_0 is, by formula 6 of Table 4.1.1,

$$2\frac{g'(z_0)}{h''(z_0)} - \frac{2}{3}\frac{g(z_0)h'''(z_0)}{[h''(z_0)]^2}$$

In this case $g(z) = 1 + z$, so that $g'(z) = 1$; and $h(z) = 1 - \cos z$, so that $h'(z) = \sin z$, $h''(z) = \cos z$, and $h'''(z) = -\sin z$. Thus $h'''(z) = 0$ for $z = z_1, z_2, z_3$, and so the formula for the residue becomes $2g'(z_0)/h''(z_0)$. Hence

$$\text{Res } (f, z_1) = \frac{2}{\cos 0} = 2 \qquad \text{Res } (f, z_2) = \frac{2}{\cos (2\pi)} = 2 \qquad \text{Res } (f, z_3) = \frac{2}{\cos (-2\pi)} = 2$$

Thus by the residue theorem,

$$\int_\gamma \frac{1+z}{1 - \cos z}\, dz = 2\pi i[\text{Res } (f, z_1) + \text{Res } (f, z_2) + \text{Res } (f, z_3)] = 12\pi i$$

Notice that we have implicitly used the fact that $I(\gamma, z) = 0$ for z outside γ and $I(\gamma, z) = 1$ for z inside γ. Any student who does not believe his or her intuition should write out a careful proof of this fact.

4.2.8   *Show that if $p(z)$ is a polynomial of degree at least 2, then the sum of the residues of $1/p(z)$ at all the zeros of p must be 0.*

First Solution. Suppose the degree of p is n and $p(z) = \Sigma_{k=0}^n a_k z^k$ with $a_n \neq 0, n \geq 2$. We know p can have at most n different zeros, so that if γ is a circle of large enough radius R centered at 0, it surrounds all the finite singularities of $1/p(z)$. Thus

$$\int_\gamma \frac{1}{p(z)}\, dz = 2\pi i \sum \left(\text{residues of } \frac{1}{p} \right)$$

and this holds for all large enough R. But for large R,

$$\frac{|a_{n-1}|}{R} + \frac{|a_{n-2}|}{R^2} + \cdots + \frac{|a_0|}{R^n} < \frac{|a_n|}{2}$$

so that

$$|p(z)| = \left| z^n \sum_{k=0}^n a_k z^{k-n} \right|$$

$$\geq R^n \left(|a_n| - \left| \frac{a_{n-1}}{R} + \frac{a_{n-2}}{R^2} + \cdots + \frac{a_0}{R^n} \right| \right)$$

$$\geq R^n \left[|a_n| - \left(\frac{|a_{n-1}|}{R} + \frac{|a_{n-2}|}{R^2} + \cdots + \frac{|a_0|}{R^n} \right) \right]$$

$$\geq R^n \frac{|a_n|}{2}$$

and so $|\int_\gamma [1/p(z)] \, dz| \leq 2\pi R/(R^n|a_n|/2) = 4\pi/R^{n-1}|a_n|$. Thus

$$\left| \Sigma \left(\text{residues of } \frac{1}{p} \right) \right| \leq \frac{2}{R^{n-1}|a_n|}$$

Letting $R \to \infty$, we obtain $|\Sigma \text{ (residues of } 1/p)| \leq 0$. Therefore the sum must be 0.

Second Solution (for those who have studied the supplement on residues at infinity). With γ as before, there are no finite singularities of $1/p$ outside γ, and so $\int_\gamma (1/p) = -2\pi i \text{ Res } (1/p, \infty) = 2\pi i \text{ Res } ((1/p(1/z))(1/z^2), 0)$. But

$$\frac{1}{p(1/z)} \cdot \frac{1}{z^2} = \frac{1}{a_0 + \dfrac{a_1}{z} + \cdots + \dfrac{a_n}{z^n}} \cdot \frac{1}{z^2} = \frac{z^n}{a_0 z^n + \cdots + a_n} \cdot \frac{1}{z^2}$$

Since $n \geq 2$, the singularity at $z = 0$ is removable so the residue is 0 and hence the integral is 0. But the integral is equal to the sum of the residues of $1/p$ at the zeros of p.

Exercises

1. Evaluate

$$\int_\gamma \frac{dz}{(z+1)^3}$$

where (a) γ is a circle of radius 2, center 0, and (b) γ is a square with vertices 0, 1, $1 + i$, i.

2. Deduce Cauchy's integral formula from the residue theorem.

3. Evaluate

$$\int_\gamma \frac{z}{z^2 + 2z + 5} \, dz$$

where γ is the unit circle.

4. Evaluate

$$\int_\gamma \frac{1}{e^z - 1} \, dz$$

where γ is the circle of radius 9 and center 0.

5. Evaluate

$$\int_\gamma \tan z \, dz$$

where γ is the circle of radius 8 centered at 0.

6. Show that

$$\int_\gamma \frac{5z - 2}{z(z - 1)} \, dz = 10\pi i$$

where γ is any circle of radius greater than 1 and center 0.

7. Evaluate

$$\int_\gamma \frac{e^{-z^2}}{z^2} \, dz$$

where (a) γ is the square with vertices $-1 - i$, $1 - i$, $-1 + i$, and $1 + i$, and (b) γ is the ellipse $\gamma(t) = a \cos t + ib \sin t$, where $a, b > 0$, $0 \le t \le 2\pi$.

8. Let f be analytic on $\mathbb{C}$ except for poles at 1 and -1. Assume that $\operatorname{Res}(f, 1) = -\operatorname{Res}(f, -1)$. Let $A = \{z \mid z \notin [-1, 1]\}$. Show that there is an analytic function h on A such that $h'(z) = f(z)$.

9. Evaluate the following integrals:

(a) $\displaystyle\int_{|z|=\frac{1}{2}} \frac{dz}{z(1 - z)^3}$ (b) $\displaystyle\int_{|z|=\frac{1}{2}} \frac{e^z \, dz}{z(1 - z)^3}$

10. Evaluate the following integrals:

(a) $\displaystyle\int_{|z|=\frac{1}{2}} \frac{dz}{(1 - z)^3}$ (b) $\displaystyle\int_{|z+1|=\frac{1}{2}} \frac{dz}{(1 - z)^3}$

(c) $\displaystyle\int_{|z-1|=\frac{1}{2}} \frac{dz}{(1 - z)^3}$ (d) $\displaystyle\int_{|z-1|=\frac{1}{2}} \frac{e^z}{(1 - z)^3} \, dz$

11. Let $f : A \to B$ be analytic, one-to-one, and onto and let $f'(z) \ne 0$ for $z \in A$. Let γ be a curve in A and let $\tilde{\gamma} = f \circ \gamma$. Also let g be continuous on $\tilde{\gamma}$. Show that

$$\int_\gamma (g \circ f) \cdot f' = \int_{\tilde{\gamma}} g$$

What does this become in the case where $f(z) = 1/z$?

12. Show that if $\lim_{z \to \infty} [-zf(z)]$ exists, it equals the residue of f at ∞.

13. (a) Find the residue of $(z-1)^3/z(z+2)^3$ at $z=\infty$.

(b) Show two methods of evaluating

$$\int_\gamma \frac{(z-1)^3}{z(z+2)^3}\, dz$$

where γ is the circle with center 0 and radius 3.

14. Show informally that if γ is a simple closed curve travelled counterclockwise, then

$$\int_\gamma f = -2\pi i \sum \{\text{residues of } f \text{ outside } \gamma \text{ including } \infty\}$$

15. Choose a branch of $\sqrt{z^2-1}$ that is analytic on $\mathbf{C}$ except for the segment $[-1, 1]$ on the real axis. Evaluate

$$\int_\gamma \sqrt{z^2-1}\, dz$$

where γ is the circle of radius 2 centered at 0.

4.3 EVALUATION OF DEFINITE INTEGRALS

This section develops systematic methods for using the residue theorem to evaluate certain types of integrals. These techniques are summarized in Table 4.3.1. Some examples and special devices for evaluating integrals involving "multiple-valued" functions such as the square root or logarithm are also given. It is valuable for the reader to understand the techniques and estimates used to establish the formulas since these same ideas can often be used when the formulas obtained here do not directly apply. Worked examples relating directly to the formulas obtained are grouped in the text with the theorems. Miscellaneous examples working out important special cases and illustrating how the methods may be modified to handle nonstandard problems appear at the end of the section.

Integrals of the Type $\int_{-\infty}^{\infty} f(x)\, dx$

This type of improper integral is often most easily evaluated by complex analysis even when the function involved is real-valued for all x. It is important to remember that there are always two questions involved when studying such integrals: Does it converge? If so, to what? Sometimes these must be handled separately. Our first two propositions give conditions guaranteeing convergence and give formulas for the values.

4.3.1 PROPOSITION

(i) *Suppose f is analytic on an open set containing the upper half plane $H = \{z \mid \operatorname{Im} z \geq 0\}$, except for a finite number of isolated singularities none of which are on the real axis, and that there are constants M and $p > 1$ and a number R such that $|f(z)| \leq M/|z|^p$ whenever $z \in H$ and $|z| \geq R$. Then*

$$\int_{-\infty}^{\infty} f(x)\, dx = 2\pi i \sum \{\text{residues of } f \text{ in } H\}$$

(ii) *If the conditions of (i) hold with H replaced by the lower half plane $L = \{z \mid \operatorname{Im} z \leq 0\}$, then*

$$\int_{-\infty}^{\infty} f(x)\, dx = -2\pi i \sum \{\text{residues of } f \text{ in } L\}$$

(iii) *Both of these formulas hold if $f = P/Q$ where P and Q are polynomials, the degree of Q is at least 2 greater than that of P, and Q has no real zeros.*

PROOF (i) Let $r > R$ and consider the curve γ_r shown in Fig. 4.3.1(i). Choose r large enough that all the poles of f in the upper half plane lie inside γ_r. By the residue theorem,

$$\int_{\gamma_r} f = 2\pi i \sum \{\text{residues of } f \text{ in upper half plane}\}$$

The integral over γ_r breaks up into straight and curved portions as follows:

$$\int_{\gamma_r} f = \int_{-r}^{r} f(x)\, dx + \int_{0}^{\pi} f(re^{i\theta}) i r e^{i\theta}\, d\theta$$

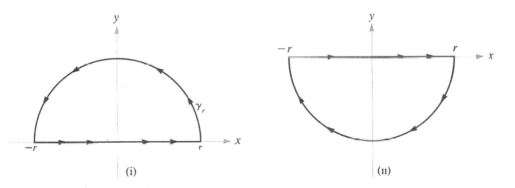

(i)

FIGURE 4.3.1 The curve γ_r.

Suppose that we could show that as $r \to \infty$, the last term approaches zero. Then we would have

$$\lim_{r \to \infty} \int_{-r}^{r} f(x)\, dx = 2\pi i \sum \{\text{residues of } f \text{ in upper half plane}\}$$

From calculus, we know that $f(x)$ is integrable as a function of $x \in \mathbf{R}$ (because f is continuous on $\mathbf{R}$ and because of the condition $|f(z)| \leq M/|x|^p$ for $|x| \geq R$); therefore,

$$\lim_{r \to \infty} \int_{-r}^{r} f(x)\, dx = \int_{-\infty}^{\infty} f(x)\, dx$$

Thus it remains to be proved that

$$\lim_{r \to \infty} \int_{0}^{\pi} f(re^{i\theta}) r i e^{i\theta}\, d\theta = 0$$

But

$$\left| \int_{0}^{\pi} f(re^{i\theta}) r i e^{i\theta}\, d\theta \right| \leq \pi \cdot \frac{M}{r^p} \cdot r = \frac{\pi M}{r^{p-1}}$$

which approaches zero as $r \to \infty$ since $p > 1$. This establishes part (i). Part (ii) follows in a similar way using the curve shown in Fig. 4.3.1(ii). The minus sign comes about since this curve is traveled in the clockwise direction.

Finally, if $f = P/Q$, we shall establish that $|f(z)| \leq M/|z|^2$ for $|z|$ large and thus complete the proof. If P is of degree n and Q is of degree $n + p$, $p \geq 2$, we know there is an $M_1 > 0$ such that $|P(z)| \leq M_1|z|^n$ for $|z| \geq 1$ and an $M_2 > 0$ such that $|Q(z)| \geq M_2|z|^{n+p}$ if $|z| \geq R$ for some $R > 1$ (see the proof of the fundamental theorem of algebra (3.4.9)). Thus

$$\left| \frac{P(z)}{Q(z)} \right| \leq \frac{M_1}{M_2} \cdot \frac{1}{|z|^p} \leq \frac{M_1}{M_2} \cdot \frac{1}{|z|^2}$$

for $|z| \geq R$, since $p \geq 2$. Therefore, we can let $M = M_1/M_2$ in this case. ∎

4.3.2 EXAMPLE   *Evaluate*

$$\int_{-\infty}^{\infty} \frac{dx}{x^4 + 1}$$

SOLUTION   Here $P(x) = 1$ and $Q(x) = x^4 + 1$, so the conditions of Proposition 4.3.1 are met. The poles of P/Q are located at the fourth roots of -1, namely, $e^{\pi i/4}, e^{3\pi i/4}, e^{5\pi i/4},$

$e^{7\pi i/4}$. These poles are simple and only the first two lie in the upper half plane. The residue at such a point z_0 is $1/4z_0^3 = -z_0/4$ (see Table 4.1.1, formula 4), and so

$$\text{Res}\,(f,\,e^{\pi i/4}) + \text{Res}\,(f,\,e^{3\pi i/4}) = -\tfrac{1}{4}(e^{\pi i/4} + e^{3\pi i/4}) = -\tfrac{1}{4}e^{\pi i/4}(1 + e^{\pi i/2})$$

$$= -\frac{1}{4}\left(\frac{1+i}{\sqrt{2}}\right)(1+i) = -\frac{1}{4}\cdot\frac{2i}{\sqrt{2}}$$

$$= -\frac{i}{2\sqrt{2}}$$

Thus the answer is $(2\pi i)(-i)/2\sqrt{2} = \pi/\sqrt{2}$.

Some simple checks such as determining that the answer must be real and positive (because $1/(x^4 + 1) \geq 0$ on $\mathbf{R}$) can often detect basic computational errors. Our integral

$$\int_{-\infty}^{\infty} \frac{dx}{x^4 + 1}$$

could have been evaluated by using the method of partial fractions. However, later in this section we shall encounter integrals that can be evaluated by using residues but for which the method of partial fractions (and, for that matter, all the elementary techniques of integration) fails. ▲

Proposition 4.3.1 fails to apply in some important cases. An obvious source of difficulty is the possibility of singularities on the real axis. We shall shortly see, in the subsection on the Cauchy principal value, how that situation can often be overcome. A more subtle source of trouble is that a function which is very well behaved on the real axis might not have the behavior required by the proposition in either half plane. An important example of this is the *normal probability function*

$$f(x) = \frac{1}{\sqrt{2\pi}}\,e^{-x^2/2}$$

The trouble with this function is that e^{-z^2} does not have the correct limiting behavior as z goes to infinity. In both directions along the real axis it goes to 0 faster than the reciprocal of any polynomial. However, along the 45° lines, where $\arg z = \pm\pi/4$ or $\pm 3\pi/4$, its absolute value is constantly equal to 1, and in both directions along the imaginary axis it grows faster than any polynomial. Nevertheless, we can evaluate the integral.

4.3.3 PROPOSITION (GAUSSIAN INTEGRAL).

$$\sqrt{\pi} = \int_{-\infty}^{\infty} e^{-x^2}\,dx$$

This formula is important in probability and statistics and in other areas of mathematics and applications. We will meet it again in Chap. 7, where we will see one method of evaluating it using the gamma function. Perhaps the most direct method uses a double integral; see Exercise 21, or Chap. 9 of J. Marsden, *Elementary Classical Analysis* (New York: W. H. Freeman and Co., 1974). We will see a method of evaluating it indirectly by residues in the worked examples at the end of this section after relating it to two other interesting integrals. (This method and several others, together with historical comments, are collected in D. Mitrinovic and J. Kečkić, *The Cauchy Method of Residues* (Dordrecht, The Netherlands: D. Reidel Publ. Co., 1984), pp. 158–164.)

Fourier Transforms

Next we consider a technique for evaluating integrals of the form $\int_{-\infty}^{\infty} f(x) \cos(\omega x)\, dx$ and $\int_{-\infty}^{\infty} f(x) \sin(\omega x)\, dx$, which are called the *Fourier sine* and *cosine transforms of f*. If f is a function defined on the real axis for which the integrals make sense, the two preceding integrals are related to the expression

$$F(\omega) = \int_{-\infty}^{\infty} f(x)e^{-i\omega x}\, dx$$

which defines a new function called the *Fourier transform* of f. It is of great importance in differential equations (because of the result of Exercise 24, among other reasons), theoretical physics, quantum mechanics, and many other areas of mathematics and science, and there is an enormous body of literature concerning it. (There are possible variations in its definition. The integral may be multiplied by a constant, or $-2\pi i\omega x$ may appear in the exponent instead of just $-i\omega x$.) If ω and $f(x)$ are real, the Fourier cosine and sine transforms are the real and imaginary parts of the Fourier transform:

$$\int_{-\infty}^{\infty} f(x) \cos(\omega x)\, dx = \operatorname{Re} F(\omega)$$

$$\int_{-\infty}^{\infty} f(x) \sin(\omega x)\, dx = -\operatorname{Im} F(\omega)$$

If ω is real and f satisfies fairly mild conditions, these can be evaluated using the following proposition.

4.3.4 PROPOSITION   *In either situation* (i) *or situation* (ii) *below, the integral $\int_{-\infty}^{\infty} e^{i\omega x} f(x)\, dx$ exists in the sense that $\lim\limits_{c \to \infty} \int_{0}^{c} e^{i\omega x} f(x)\, dx$ and $\lim\limits_{c \to \infty} \int_{-c}^{0} e^{i\omega x} f(x)\, dx$ both exist, and it is given by the corresponding formula. If $f(x)$ is real for real x, then $\int_{-\infty}^{\infty} \cos(\omega x) f(x)\, dx$ and $\int_{-\infty}^{\infty} \sin(\omega x) f(x)\, dx$ are equal respectively to its real and imaginary parts.*

(i) $\omega > 0$. *Suppose that f is analytic on an open set containing the upper half plane $H = \{z \mid \operatorname{Im} z \geq 0\}$, except for a finite number of isolated singularities none of which are on the real axis. Suppose also that $|f(z)| \to 0$ as $z \to \infty$ in H. (That is, suppose for each $\epsilon > 0$ there is an R such that $|f(z)| < \epsilon$ whenever $|z| \geq R$ and $z \in H$.) Then*

$$\int_{-\infty}^{\infty} e^{i\omega x} f(x) \, dx = 2\pi i \sum \{\text{residues of } e^{i\omega z} f(z) \text{ in } H\}$$

(ii) $\omega < 0$: *If the conditions of* (i) *hold with H replaced by the lower half plane $L = \{z \mid \operatorname{Im} z \leq 0\}$, then*

$$\int_{-\infty}^{\infty} e^{i\omega x} f(x) \, dx = -2\pi i \sum \{\text{residues of } e^{i\omega z} f(z) \text{ in } L\}$$

(iii) *Both* (i) *and* (ii) *are valid if $f = P/Q$ where P and Q are polynomials, the degree of Q is greater than that of P, and Q has no zeros on the real axis.*

As with Proposition 4.3.1, we will see in the section on the Cauchy principal value how the integral can sometimes be handled when there are poles on the real axis. Notice that the conditions on f are much weaker than in Proposition 4.3.1. The extra factor of $e^{i\omega x}$ is essential for this, and in fact Proposition 4.3.1 is not true with $p = 1$.

PROOF For $\omega > 0$: Let γ be the curve shown in Fig. 4.3.2, where $y_1 > x_1, x_2 > 0$ and x_1, x_2, y_1 are chosen large enough that γ contains all the poles of f in the upper half plane. By the residue theorem, $\int_\gamma e^{i\omega z} f(z) \, dz = 2\pi i \sum [\text{residues of } f(z)e^{i\omega z} \text{ in upper half plane}]$.

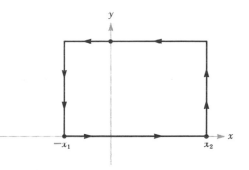

FIGURE 4.3.2 The curve γ used for the proof of Proposition 4.3.4.

Now estimate the absolute values of the three integrals

$$I_1 = \int_0^{y_1} e^{i\omega(x_2+yi)} f(x_2 + yi)i \, dy$$

$$I_2 = \int_{x_2}^{-x_1} e^{i\omega(x+iy_1)} f(x + iy_1) \, dx$$

$$I_3 = \int_{y_1}^0 e^{i\omega(-x_1+yi)} f(-x_1 + iy)i \, dy$$

as follows. Let $\epsilon > 0$, and choose R as in (i). Let

$$M_1 = \max \{|f(x_2 + iy)| \text{ such that } 0 \le y \le y_1\}$$
$$M_2 = \max \{|f(x + iy_1)| \text{ such that } -x_1 \le x \le x_2\}$$
$$M_3 = \max \{|f(-x_1 + iy)| \text{ such that } 0 \le y \le y_1\}$$

If $y_1 > x_1 > R$ and $y_1 > x_2 > R$, then $M_1 < \epsilon$, $M_2 < \epsilon$, and $M_3 < \epsilon$. Then

$$|I_1| \le \int_0^{y_1} e^{-\omega y}|f(x_2 + iy)| \, dy \le M_1 \int_0^{y_1} e^{-\omega y} \, dy = \frac{M_1}{\omega}(1 - e^{-\omega y_1}) \le \frac{M_1}{\omega}$$

Similarly $|I_3| \le M_3/\omega$. Finally,

$$|I_2| \le \int_{-x_1}^{x_2} e^{-\omega y_1}|f(x + iy_1)| \, dx \le M_2 e^{-\omega y_1}(x_2 + x_1)$$

Since $\omega > 0$, we can let y_1 be large enough that $e^{-\omega y_1}(x_1 + x_2) < \epsilon$. Since

$$\int_\gamma e^{i\omega z} f(z) \, dz = I_1 + I_2 + I_3 + \int_{-x_1}^{x_2} e^{i\omega x} f(x) \, dx$$

we have

$$\int_{-x_1}^{x_2} e^{i\omega x} f(x) \, dx - 2\pi i \sum \{\text{residues of } f(z)e^{i\omega z} \text{ in upper half plane}\} = -(I_1 + I_2 + I_3)$$

and so

$$\left| \int_{-x_1}^{x_2} e^{i\omega x} f(x) \, dx - 2\pi i \sum \{\text{residues of } f(z)e^{i\omega z} \text{ in upper half plane}\} \right|$$

$$\le \frac{M_1}{\omega} + \frac{M_3}{\omega} + M_2 e^{-\omega y_1}(x_1 + x_2) < \frac{2\epsilon}{\omega} + \epsilon^2$$

Since ϵ is arbitrary,

$$\underset{\substack{x_1 \to \infty \\ x_2 \to \infty}}{\text{limit}} \int_{-x_1}^{x_2} e^{i\omega x} f(x) \, dx$$

exists and has the required value. The student is left the exercise of proving the fact from calculus that the existence of this double limit is the same as the existence of

$$\underset{x_2 \to \infty}{\text{limit}} \int_0^{x_2} e^{i\omega x} f(x) \, dx$$

and

$$\underset{x_1 \to \infty}{\text{limit}} \int_{-x_1}^0 e^{i\omega x} f(x) \, dx$$

Thus we have proved the existence (in the conditional sense) of

$$\int_{-\infty}^{\infty} e^{i\omega x} f(x) \, dx$$

and that $\int_{-\infty}^{\infty} e^{i\omega x} f(x) \, dx = 2\pi i \, \Sigma$ {residues of $f(z)e^{i\omega z}$ in upper half plane}.

Part (iii), for $f(x) = P(x)/Q(x)$, where P and Q are polynomials with deg $Q(x) \geq 1 + $ deg $P(x)$, follows as in the proof of Proposition 4.3.1: For $|z| \geq 1$ there is an $M_1 > 0$ such that $|P(z)| \leq M_1|z|^n$ where $n = $ deg $P(x)$, and there are an $R > 1$ and an $M_2 > 0$ such that for $|z| \geq R$ we have $|Q(z)| \geq M_2|z|^{n+1}$. Hence

$$\left| \frac{P(z)}{Q(z)} \right| < \frac{M_1}{M_2|z|}$$

for $|z| \geq R$ and parts (i) and (ii) are satisfied.

The proof of part (ii) (the case where $\omega < 0$) is similar to the preceding proof, except that the appropriate curve is a rectangle in the lower half plane. ∎

Note that $\int_{-\infty}^{\infty} \cos \omega x \, f(x) \, dx$ is *not* $2\pi i \, \Sigma$ {residues of $(\cos \omega z) f(z)$ in upper half plane}. (This formula is actually false.) The hypotheses of Propositions 4.3.1 and 4.3.4 simply do not apply, even if $|f(z)| \leq M/|z|^2$.

There is another method of proving this proposition that the interested student can work out. It is based on *Jordan's lemma*: If $f(z) \to 0$ as $|z| \to \infty$, uniformly in arg z, $0 \leq $ arg $z \leq \pi$, and if $f(z)$ is analytic when $|z| > c$, c a constant, $0 < $ arg $z < \pi$, then $\int_{\gamma_\rho} e^{i\omega z} f(z) \, dz \to 0$ as $\rho \to \infty$, where $\gamma_\rho(\theta) = \rho e^{i\theta}$, $0 < \theta < \pi$. (Consult E. T. Whittaker and G. N. Watson, *A Course of Modern Analysis* (New York: Cambridge University Press, 1927), p. 115.)

4.3.5 EXAMPLE   *Show that for $b > 0$*

$$\int_0^\infty \frac{\cos x}{x^2 + b^2}\, dx = \frac{\pi e^{-b}}{2b}$$

SOLUTION   Since $(\cos x)/(x^2 + b^2)$ is an even function, we have

$$\int_0^\infty \frac{\cos x}{x^2 + b^2}\, dx = \frac{1}{2}\int_{-\infty}^\infty \frac{\cos x}{x^2 + b^2}\, dx$$

We must find the residues of $e^{iz}/(z^2 + b^2)$ in the upper half plane. The only pole in the upper half plane is at bi and the pole is simple, so

$$\text{Res}\left(\frac{e^{iz}}{z^2 + b^2}, ib\right) = \frac{e^{-b}}{2ib}$$

(Table 4.1.1 formula 4), and hence

$$\int_{-\infty}^\infty \frac{\cos x}{x^2 + b^2}\, dx = \text{Re}\left[2\pi i\left(\frac{e^{-b}}{2ib}\right)\right] = \frac{\pi e^{-b}}{b}$$

Trigonometric Integrals

4.3.6 PROPOSITION   *Let $R(x, y)$ be a rational function of x, y whose denominator does not vanish on the unit circle. Then*

$$\int_0^{2\pi} R(\cos\theta, \sin\theta)\, d\theta = 2\pi i \sum [\text{residues of } f(z) \text{ inside the unit circle}]$$

where

$$f(z) = \frac{R\left(\frac{1}{2}\left(z + \frac{1}{z}\right), \frac{1}{2i}\left(z - \frac{1}{z}\right)\right)}{iz}$$

PROOF   If $z = x + iy$ is on the unit circle, then

$$x = \tfrac{1}{2}(z + \bar{z}) = \frac{1}{2}\left(z + \frac{1}{z}\right) \qquad \text{and} \qquad y = \frac{1}{2i}\left(z - \frac{1}{z}\right)$$

Since R has no poles on the unit circle, neither does f, so if γ is the unit circle, we have, by the residue theorem,

$$\int_\gamma f = 2\pi i \sum \text{(residues of } f \text{ inside } \gamma)$$

Therefore,

$$\int_0^{2\pi} R(\cos \theta, \sin \theta)\, d\theta = \int_0^{2\pi} R\left(\frac{e^{i\theta} + e^{-i\theta}}{2}, \frac{e^{i\theta} - e^{-i\theta}}{2i}\right) \frac{ie^{i\theta}}{ie^{i\theta}}\, d\theta$$

$$= \int_0^{2\pi} f(e^{i\theta}) ie^{i\theta}\, d\theta = \int_\gamma f$$

which proves the theorem. ∎

If you forget the formula for f, you can always begin with $\int_0^{2\pi} R(\cos \theta, \sin \theta)\, d\theta$ and follow the method of the preceding proof (that is, write $\cos \theta = (e^{i\theta} + e^{-i\theta})/2$, and so on). The formula for f will then be apparent.

4.3.7 EXAMPLE *Evaluate*

$$I = \int_0^{2\pi} \frac{d\theta}{1 + a^2 - 2a \cos \theta} \qquad a > 0, a \neq 1$$

SOLUTION By Proposition 4.3.6,

$$I = \int_\gamma \frac{dz}{iz\left[1 + a^2 - \dfrac{2a}{2}\left(z + \dfrac{1}{z}\right)\right]}$$

$$= \int_\gamma \frac{dz}{i[-az^2 + (1 + a^2)z - a]} = \int_\gamma \frac{i\, dz}{(z - a)(az - 1)}$$

The poles of the integrand are at $z = a$ and $z = 1/a$. First, suppose that $a < 1$; the pole inside the circle thus is at $z = a$. The residue is

$$\frac{i}{a^2 - 1}$$

If we suppose that $a > 1$, the integrand has a pole at $z = 1/a$ and the residue is

$$\frac{i}{a(1/a - a)} = \frac{i}{1 - a^2}$$

Thus

$$I = \begin{cases} \dfrac{2\pi}{1 - a^2} & \text{if } a < 1 \\[2ex] \dfrac{2\pi}{a^2 - 1} & \text{if } a > 1 \end{cases}$$

The last three propositions take care of some of the more common types of integrals. Now we come to more specialized integrals. The first involves the multiple-valued function $z \mapsto z^a$.

Integrals of the Type $\int_0^\infty x^{a-1} f(x)\, dx$: Mellin Transforms

4.3.8 PROPOSITION *Let f be analytic on* **C** *except for a finite number of isolated singularities, none of which lie on the strictly positive real axis (that is, all lie in the complement of the set* $\{x + iy \mid y = 0 \text{ and } x > 0\}$*). Let* $a > 0$ *with the restriction that a is not an integer and suppose that* (i) *there exist constants* M_1, $R_1 > 0$ *and* $b > a$ *such that for* $|z| \geq R_1$, $|f(z)| \leq M_1/|z|^b$; *and* (ii) *there exist constants* M_2, R_2, *and d such that* M_2, $R_2 > 0$ *and* $0 < d < a$ *such that for* $0 < |z| \leq R_2$, $|f(z)| \leq M_2/|z|^d$.
Then the integral $\int_0^\infty x^{a-1} f(x)\, dx$ *exists (in the sense of being absolutely convergent) and*

$$\int_0^\infty x^{a-1} f(x)\, dx = -\frac{\pi e^{-\pi a i}}{\sin (a\pi)} \sum \{\text{residues of } z^{a-1} f(z) \text{ at the singularities of } f, \text{ excluding the residue at } 0\}$$

Here $z^{a-1} = e^{(a-1)\log z}$ *using the branch* $0 < \arg z < 2\pi$.

The proof of this result is typical of the approach taken when dealing with branch points.

PROOF The existence, in the sense of absolute integrability, of $\int_0^\infty x^{a-1} f(x)\, dx$ follows if we use the assumed conditions $|f(x)| \leq M_1/x^b$ for x large and $|f(x)| \leq M_2/x^d$ for x small, together with the comparison test for integrals. The curve that we shall use,

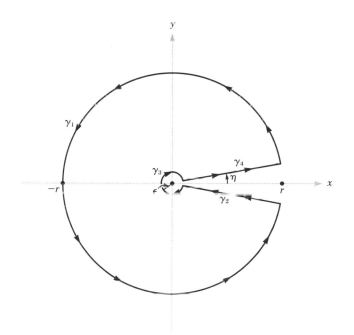

FIGURE 4.3.3 $\gamma = \gamma_1 + \gamma_2 + \gamma_3 + \gamma_4$.

$\gamma = \gamma_1 + \gamma_2 + \gamma_3 + \gamma_4$, is illustrated in Fig. 4.3.3. The radius of the incomplete circle γ_1 is r; the radius of the incomplete circle γ_3 is $\epsilon > 0$; and γ_4, γ_2 each make an angle of $\eta > 0$ with the positive x axis. We choose ϵ small enough, r large enough, and η small enough that (i) $\epsilon \leq R_2$, (ii) $r \geq R_1$; and (iii) $\gamma - \gamma_1 + \gamma_2 + \gamma_3 + \gamma_4$ encloses all the poles of $f(z)$ excluding the pole at 0.

By z^{a-1} we mean $e^{(a-1)\log z}$ where $\log z$ denotes the branch of $\log$ with $0 < \arg z < 2\pi$. Let $I = \int_{\gamma} z^{a-1}f(z)\,dz$, $I_1 = \int_{\gamma_1} z^{a-1}f(z)\,dz$, $I_2 = \int_{\gamma_2} z^{a-1}f(z)\,dz$, $I_3 = \int_{\gamma_3} z^{a-1}f(z)\,dz$, and $I_4 = \int_{\gamma_4} z^{a-1}f(z)\,dz$, so that $I = I_1 + I_2 + I_3 + I_4$. By the residue theorem,

$$I = 2\pi i \sum \{\text{residues of } z^{a-1}f(z), \text{ excluding the residue at } 0\}$$

Clearly, the singularities of $z^{a-1}f(z)$ are the same as those of $f(z)$, except possibly 0.

First we shall show that $I_3 \to 0$ as $\epsilon \to 0$ and $I_1 \to 0$ as $r \to \infty$ independently of η. Recall that when a is real, $|z^{a-1}| = |z|^{a-1}$, so we get

$$|I_3| \leq \int_{\gamma_3} |z^{a-1}||f(z)||dz| \leq M_2 \int_{\gamma_3} \frac{|z|^{a-1}}{|z|^d} |dz|$$
$$= M_2 \epsilon^{a-d-1} l(\gamma_3) < 2\pi M_2 \epsilon^{a-d}$$

Also,

$$|I_1| \le \int_{\gamma_1} |z^{a-1}| |f(z)| |dz| \le M_1 \int \frac{|z|^{a-1}}{|z|^b} |dz|$$
$$< 2\pi M_1 r^{a-b}$$

These estimates show that $I_3 \to 0$ as $\epsilon \to 0$ and $I_1 \to 0$ as $r \to \infty$, both independently of η.

Next we shall study the limiting behavior of I_2 and I_4 for fixed ϵ and r but with $\eta \to 0$. Notice that we have no reason to expect I_2 and $-I_4$ to converge to the same value as $\eta \to 0$ (and in fact they do not); this is because of the discontinuity in the function z^{a-1} as we cross the positive x axis. (The positive x axis is a branch line.) By definition, $I_2 = \int_r^\epsilon (te^{i(2\pi - \eta)})^{a-1} f(te^{i(2\pi - \eta)}) e^{i(2\pi - \eta)} \, dt$.

The student is left the exercise of proving that as $\eta \to 0$, we have

$$(te^{i(2\pi - \eta)})^{a-1} f(te^{i(2\pi - \eta)}) e^{i(2\pi - \eta)} \to t^{a-1} e^{2\pi i a} f(t)$$

uniformly on $[\epsilon, r]$; hence as $\eta \to 0$,

$$I_2 \to \int_r^\epsilon t^{a-1} e^{2\pi i a} f(t) \, dt$$

Similarly,

$$I_4 \to \int_\epsilon^r t^{a-1} f(t) \, dt$$

hence

$$I_2 + I_4 \to (1 - e^{2\pi i a}) \int_\epsilon^r t^{a-1} f(t) \, dt$$
$$= -2i e^{\pi i a} (\sin \pi a) \int_\epsilon^r t^{a-1} f(t) \, dt$$

Suppose that we are given $\delta > 0$. We can choose r sufficiently large and $\epsilon > 0$ sufficiently small that

$$\left| \int_\epsilon^r t^{a-1} f(t) \, dt - \int_0^\infty t^{a-1} f(t) \, dt \right| < \frac{\delta}{4}$$

Furthermore, we can choose r larger and ϵ smaller, if necessary, so that

$$\frac{|I_1|}{|2i e^{\pi i a} \sin \pi a|} < \frac{\delta}{4} \quad \text{and} \quad \frac{|I_3|}{|2i e^{\pi i a} \sin \pi a|} < \frac{\delta}{4}$$

Also, we can choose η small enough that

$$\left| \frac{I_2 + I_4}{-2ie^{\pi ia} \sin \pi a} - \int_{\epsilon}^{r} t^{a-1} f(t) \, dt \right| < \frac{\delta}{4}$$

This implies that

$$\left| \int_{0}^{\infty} t^{a-1} f(t) \, dt - \frac{I}{-2ie^{\pi ia} \sin \pi a} \right| < \delta$$

(Why?). Since δ was arbitrary, we have

$$\int_{0}^{\infty} t^{a-1} f(t) \, dt = \frac{I}{-2ie^{\pi ia} \sin \pi a}$$

$$= \frac{-\pi e^{-\pi ia}}{\sin \pi a} \sum \{\text{residues of } z^{a-1} f(z) \text{ excluding the residue at } 0\} \quad \blacksquare$$

The student should verify the following corollary.

4.3.9 COROLLARY *The hypotheses of Proposition 4.3.8 hold if $f(z) = P(z)/Q(z)$ for polynomials P of degree p and Q of degree q satisfying the following two conditions:*

(i) $0 < a < q - p$.

(ii) *If n_Q is of the order of the zero of Q at $z = 0$ (with the convention that $n_Q = 0$ if $Q(0) \neq 0$) and if n_p is the order of the zero of P at $z = 0$, then $n_Q - n_P < a$. (This condition holds, for instance, if $n_Q = 0$.)*

4.3.10 EXAMPLE *Prove that for $0 < a < 2$*

$$\int_{0}^{\infty} \frac{x^{a-1}}{1 + x^2} \, dx = \frac{\pi}{2 \sin (a\pi/2)}$$

SOLUTION With this restriction on a, the corollary holds (here $q = 2$, $p = 0$), and so

$$\int_{0}^{\infty} \frac{x^{a-1}}{1 + x^2} \, dx = \frac{-\pi e^{-\pi ai}}{\sin (a\pi)} \sum \left(\text{residues of } \frac{z^{a-1}}{1 + z^2} \right)$$

The poles of $1/(1 + z^2)$ are at $\pm i$ and are simple. Thus

$$\text{Res}\left(\frac{z^{a-1}}{1 + z^2}, i \right) = \frac{i^{a-1}}{2i} \quad \text{and} \quad \text{Res}\left(\frac{z^{a-1}}{1 + z^2}, -i \right) = -\frac{(-i)^{a-1}}{2i}$$

and the sum is

$$\frac{i^{a-1} - (-i)^{a-1}}{2i}$$

We compute $i^{a-1} = e^{(a-1)\log i} = e^{(a-1)\pi i/2}$ and $(-i)^{a-1} = e^{(a-1)(3\pi i/2)}$. (Remember that we must choose $\arg(-i)$ such that $0 < \arg(-i) < 2\pi$.) Thus

$$\frac{i^{a-1} - (-i)^{a-1}}{2i} = \frac{1}{2i}(e^{(a-1)\pi i/2} - e^{(a-1)3\pi i/2}) = -\frac{1}{2}(e^{a\pi i/2} + e^{3a\pi i/2})$$

$$= -\tfrac{1}{2}e^{a\pi i}(e^{a\pi i/2} + e^{-a\pi i/2}) = -e^{a\pi i}\cos\frac{a\pi}{2}$$

Hence

$$\int_0^\infty \frac{x^{a-1}}{1+x^2}\,dx = \frac{\pi\cos(a\pi/2)}{\sin(a\pi)}$$

But $\sin(2\phi) = 2\sin\phi\cos\phi$, and so this becomes $\pi/[2\sin(a\pi/2)]$, as required. ∎

Cauchy Principal Value

Suppose that $f(x)$ is continuous on the real line $\mathbf{R}$ except at the point x_0. Then $\int_{-\infty}^\infty f(x)\,dx$ need not be defined. We recall from calculus a way to make a meaningful definition. Consider

$$\int_{-\infty}^{x_0-\epsilon} f(x)\,dx + \int_{x_0+\eta}^\infty f(x)\,dx \qquad \text{for } \epsilon > 0 \text{ and } \eta > 0$$

(assuming that both integrals are convergent for every such ϵ and η) and let $\epsilon \to 0$ and $\eta \to 0$. If each limit exists, we say the integral is *convergent*. We must exercise care when using this definition to solve examples. For instance, suppose that we consider

$$\int_{-\infty}^{-\epsilon} \frac{1}{x^3}\,dx + \int_\epsilon^\infty \frac{1}{x^3}\,dx = -\frac{1}{2\epsilon^2} + \frac{1}{2\epsilon^2} = 0$$

On the other hand,

$$\int_{-\infty}^{-\epsilon} \frac{1}{x^3}\,dx + \int_{2\epsilon}^\infty \frac{1}{x^3}\,dx = -\frac{1}{2\epsilon^2} + \frac{1}{2(2\epsilon)^2} = -\frac{3}{8\epsilon^2} \to -\infty$$

as $\epsilon \to 0$. Thus we see that we can get different values for $\int_{-\infty}^\infty (1/x^3)\,dx$ depending on how we let ϵ and η approach zero; in this example the integrals $\int_0^\infty (1/x^3)\,dx$ and $\int_{-\infty}^0 (1/x^3)\,dx$ are not convergent.

We shall choose a particular, more restrictive way of letting ϵ and η approach zero: the symmetric way, taking $\epsilon = \eta$. We will find that by doing so, we can apply the residue theorem to the evaluation of such integrals. The definition that follows is slightly more general than the preceding one in that any finite number of discontinuities are allowed on the real axis. Let f be continuous on $\mathbf{R}$ except for a finite number of points $x_1 < x_2 < \cdots < x_n$. If $\int_{-\infty}^{x_1-\epsilon} f(x)\, dx$ is convergent for every $\epsilon > 0$, if $\int_{x_n+\epsilon}^{\infty} f(x)\, dx$ is convergent for every $\epsilon > 0$, and if

$$\lim_{\epsilon \to 0} \left[\int_{-\infty}^{x_1-\epsilon} f(x)\, dx + \int_{x_1+\epsilon}^{x_2-\epsilon} f(x)\, dx + \cdots + \int_{x_{n-1}+\epsilon}^{x_n-\epsilon} f(x)\, dx + \int_{x_n+\epsilon}^{\infty} f(x)\, dx \right]$$

exists and is finite, then we call this limit the *Cauchy principal value*, abbreviated as P.V. $\int_{-\infty}^{\infty} f(x)\, dx$. Observe that if the integral is convergent at all these points, we recover the usual value for the integral $\int_{-\infty}^{\infty} f(x)\, dx$. However, as the previous example shows, the Cauchy principal value can exist even though the integrals are not convergent in the usual sense.

The general technique to be used for evaluating a large class of such Cauchy principal value integrals is the following. We require that $f(z)$ be defined and analytic on $\mathbf{C}$ with a finite number of isolated singularities, some of which may lie on the real axis, say, at points $x_1, \ldots, x_n$ where $x_1 < x_2 < \cdots < x_n$. Those remaining lie off the real axis. Let us consider the curve γ shown in Fig. 4.3.4. In this figure, the radius r of the large semicircle is chosen sufficiently large and the radius $\epsilon > 0$ of the small semicircles is chosen sufficiently small that γ encloses all the poles of f in the upper half plane, not including those on the real axis. We also require that the integral around the large

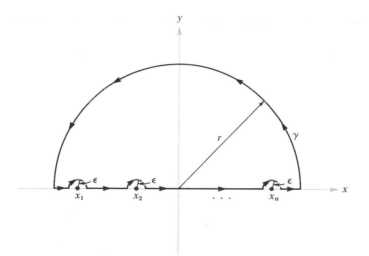

FIGURE 4.3.4 The curve γ for evaluating the Cauchy principal value.

semicircle approach zero as $r \to \infty$ and that the limits of the integrals around the small semicircles exist and are finite as $\epsilon \to 0$. These requirements, together with the fact that $\int_\gamma f = 2\pi i \sum$ (residues in upper half plane, off the real axis), ensure the existence of P.V. $\int_{-\infty}^{\infty} f(x)\, dx$ and enable us to calculate its value.

These requirements are met in Proposition 4.3.11, in which an explicit formula is derived for P.V. $\int_{-\infty}^{\infty} f(x)\, dx$. This theorem reduces to Proposition 4.3.1 in the event that no poles lie on the real axis.

4.3.11 PROPOSITION   *Let f be analytic in an open set containing the upper half plane $H = \{z \,|\, \mathrm{Im}\ z \geq 0\}$ except for a finite number of isolated singularities. Let $x_1, \ldots, x_m$ be singularities on the real axis and suppose they are all simple poles. If either*

(i) *f satisifes the condition of part* (i) *of Proposition 4.3.1 (except for the poles on the axis)*

or

(ii) *$f(z) = e^{iaz} g(z)$ with $a > 0$ and g satisfying the condition of part* (i) *of Proposition 4.3.4,*

then P.V. $\int_{-\infty}^{\infty} f(x)\, dx$ exists and

$$\text{P.V.} \int_{-\infty}^{\infty} f(x)\, dx = 2\pi i \sum (\text{residues of } f \text{ in upper half plane } \mathrm{Im}\ z > 0)$$

$$+ \pi i \sum \{\text{residues of } f \text{ on } x \text{ axis}\}$$

Naturally there are corresponding results for the lower plane.

4.3.12 PROPOSITION   *Let f be analytic on an open set containing the lower half plane $L = \{z \,|\, \mathrm{Im}\ z \leq 0\}$ except for a finite number of isolated singularities and suppose that those on the real axis are all simple poles. If either*

(i) *f satisfies the condition of part* (ii) *of Proposition 4.3.1*

or

(ii) *$f(z) = e^{iaz} g(z)$ with $a < 0$ and g satisfying the condition of part* (iii) *of Proposition 4.3.4,*

then

$$\text{P.V.} \int_{-\infty}^{\infty} f(x)\, dx = -2\pi i \sum \{\text{residues of } f \text{ in the open half plane } \mathrm{Im}\ z < 0\}$$

$$- \pi i \sum \{\text{residues of } f \text{ on the real axis}\}$$

REMARK The following proof assumes that condition (i) holds. The proof that assumes condition (ii) differs in one way: the three sides of a large rectangle in the upper half plane would be more suitable than the large semicircle, just as in Proposition 4.3.4. Note that $\int_{-\infty}^{x_1-\epsilon} f(x)\,dx$ and $\int_{x_n+\epsilon}^{\infty} f(x)\,dx$ are convergent integrals for every $\epsilon > 0$ in the event that condition (ii) is fulfilled. This conclusion follows from the proof of Proposition 4.3.4.

If we consider the curve γ in Fig. 4.3.4, we see that to prove Proposition 4.3.11 we need to know how to handle integrals over small circles. This result is provided by the following lemma.

4.3.13 LEMMA *Let $f(z)$ have a simple pole at z_0 and let γ_ϵ be a portion of a circular arc of radius ϵ and angle α (see Fig. 4.3.5). Then*

$$\lim_{\epsilon \to 0} \int_{\gamma_\epsilon} f = \alpha i \operatorname{Res}(f, z_0)$$

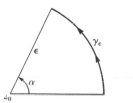

FIGURE 4.3.5 The curve γ_ϵ.

PROOF OF LEMMA Near z_0 we can write $f(z) = b_1/(z - z_0) + h(z)$ where h is analytic and $b_1 = \operatorname{Res}(f, z_0)$ (Why?). Thus

$$\int_{\gamma_\epsilon} f = \int_{\gamma_\epsilon} \frac{b_1}{z - z_0}\,dz + \int_{\gamma_\epsilon} h(z)\,dz$$

Therefore,

$$\int_{\gamma_\epsilon} \frac{b_1}{z - z_0}\,dz = b_1 \int_{\alpha_0}^{\alpha_0+\alpha} \frac{\epsilon i e^{i\theta}}{\epsilon e^{i\theta}}\,d\theta = b_1 \alpha i$$

Here, $\gamma_\epsilon(\theta) = z_0 + \epsilon e^{i\theta}, \alpha_0 \leq \theta \leq \alpha_0 + \alpha$. Also, since h is analytic, it is bounded near z_0, say, by M, and so

$$\left| \int_{\gamma_\epsilon} h(z) \, dz \right| \leq Ml(\gamma_\epsilon) = M\alpha\epsilon \to 0$$

as $\epsilon \to 0$. The lemma follows. ▼

PROOF OF PROPOSITION 4.3.11   Let $\gamma = \gamma_r + \gamma_1 + \cdots + \gamma_m + \tilde{\gamma}$ where γ_r is the semicircular portion of radius r; $\gamma_1, \ldots, \gamma_m$ are the semicircular portions of radius ϵ; and $\tilde{\gamma}$ consists of the straight-line portions of γ along the real axis.

By the residue theorem, $\int_\gamma f = 2\pi i \sum$ {residues in upper half plane}. Next observe that $\int_{-\infty}^{x_1-\epsilon} f(x) \, dx$ and $\int_{x_m+\epsilon}^{\infty} f(x) \, dx$ are convergent integrals for every $\epsilon > 0$ by the comparison test for integrals, together with the condition that $|f(z)| \leq M/|z|^p$ for large $|z|$. Thus $\lim_{r \to \infty} \int_{\tilde{\gamma}} f$ exists. Therefore, $2\pi i \sum$ {residues in upper half plane} $= \int_\gamma f = \int_{\tilde{\gamma}} f + \int_{\gamma_r} f + \sum_{i=1}^m \int_{\gamma_i} f$. As in Proposition 4.3.1, $\int_{\gamma_r} f \to 0$ as $r \to \infty$ because of the condition $|f(z)| \leq M/|z|^p$ for large $|z|$ where Im $z \geq 0$. Thus $\sum_{i=1}^m \int_{\gamma_i} f + \lim_{r \to \infty} \int_{\tilde{\gamma}} f = 2\pi i \sum$ {residues in upper half plane}. By the lemma, $\lim_{\epsilon \to 0} \int_{\gamma_j} f = -\pi i$ Res (f, x_j), $j = 1, 2, \ldots, m$. Hence $\lim_{\epsilon \to 0} (\lim_{r \to \infty} \int_{\tilde{\gamma}} f)$ exists and is equal to $2\pi i \sum$ {residues in upper half plane} $+$ $\pi i \sum$ {residues on real axis}. But by definition the Cauchy principal value is precisely $\lim_{\epsilon \to 0} (\lim_{r \to \infty} \int_{\tilde{\gamma}} f)$, and so the proposition is proved. ∎

The proof of Proposition 4.3.12 is analogous except that it uses a curve through the lower half plane, as in Fig. 4.3.6. Notice that although the contour as a whole is nega-

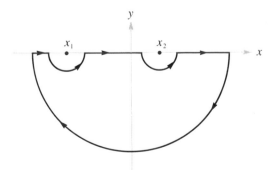

FIGURE 4.3.6   Curve used for the proof of Proposition 4.3.12.

tively oriented (that is, oriented clockwise) with respect to its interior, the small semicircles around the poles on the real axis proceed *counterclockwise* with respect to those points.

4.3.14 EXAMPLE *Consider* $(\sin x)/x$, *which is defined to have value* 1 *when* $x = 0$. *Then* $(\sin x)/x$ *is defined and continuous on* **R**. *Show that* $\int_0^\infty [(\sin x)/x]\, dx$ *exists and compute its value.*

SOLUTION The integral

$$I = \text{P.V.} \int_{-\infty}^{\infty} \frac{e^{ix}}{x}\, dx$$

exists by Proposition 4.3.11. Therefore,

$$\text{Im}\, I = \text{P.V.} \int_{-\infty}^{\infty} \frac{\sin x}{x}\, dx$$

exists. But

$$\text{P.V.} \int_{-\infty}^{\infty} \frac{\sin x}{x}\, dx = \int_{-\infty}^{\infty} \frac{\sin x}{x}\, dx$$

by the continuity of $(\sin x)/x$ at $x = 0$. The existence of $\int_{-\infty}^{\infty} [(\sin x)/x]\, dx$ implies the existence of $\int_0^\infty [(\sin x)/x]\, dx$ since $(\sin x)/x$ is an even function, and thus

$$\text{Im}\, I = 2 \int_0^\infty \frac{\sin x}{x}\, dx$$

From the result in Proposition 4.3.11 we have $I = \pi i$ (residue of e^{iz}/z at $z = 0$) $= \pi i$. Hence

$$\int_0^\infty \frac{\sin x}{x}\, dx = \frac{\pi}{2} \quad \blacktriangle$$

Notice that if we used the integrand e^{-iz}/z, then this example would satisfy the conditions of Proposition 4.3.12, but not of Proposition 4.3.11. Although it still has residue 1 at $z = 0$, the minus sign comes into play and gives

$$\text{P.V.} \int_{-\infty}^{\infty} \frac{e^{-ix}}{x}\, dx = -\pi i$$

But this is reasonable since Im $(e^{-ix}/x) = [\sin(-x)]/x = -(\sin x)/x$, and so we would expect the integral to be negative.

Further Analysis of Integrals Involving Multiple-Valued Functions

By the result of Proposition 4.3.8, we have seen that when dealing with multiple-valued functions, we must choose a curve suitable to a branch of the function. This requirement can be illustrated further with an example.

4.3.15 EXAMPLE   *Use residues to prove that*

$$\int_1^\infty \frac{dx}{x\sqrt{x^2-1}} = \frac{\pi}{2}$$

This integral cannot be directly evaluated by any of the formulas that we have developed. The basic techniques used in the solution of this problem, however, are similar to those we have already applied. Following is an outline, the details of which are left to the student.

SOLUTION   Recall that a suitable domain of $\sqrt{z^2-1}$ consists of **C** minus the half lines $x \geq 1$ and $x \leq -1$. Consider the curve γ in Fig. 4.3.7, consisting of the incomplete circles of radius r around 0 and radius ϵ around 1 and -1 and horizontal lines a distance δ from the real axis. The function $1/z\sqrt{z^2-1}$ is defined and analytic in the region **C** minus the half lines $x \geq 1$ and $x \leq -1$ except for a simple pole at 0.

This is accomplished as follows: Consider $\sqrt{z^2-1}$ as a product $\sqrt{z-1}\,\sqrt{z+1}$ in which the first factor uses a branch of the square root defined with a branch cut from $+1$ to $-\infty$ by

$$f(z) = \sqrt{z-1}$$
$$= \sqrt{|z-1|}e^{i[\arg(z-1)]/2} \qquad \text{for} \qquad -\pi < \arg(z-1) \leq \pi$$

and the second factor uses a branch of square root defined with a branch cut from -1 to $+\infty$ by

$$g(z) = \sqrt{z+1}$$
$$= \sqrt{|z+1|}e^{i[\arg(z+1)]/2} \qquad \text{for} \qquad 0 \leq \arg(z+1) < 2\pi$$

(See Fig. 4.3.8.) The product $f(z)g(z)$ gives a square root for z^2-1 which appears to be analytic only on the plane with the whole real axis deleted. Crossing the branch cut for

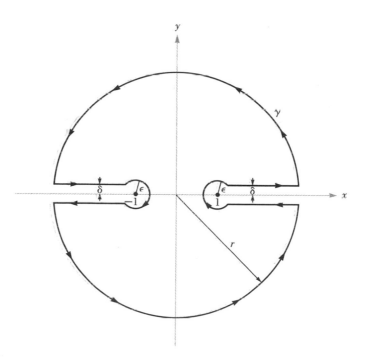

FIGURE 4.3.7 The curve γ.

either factor changes the sign of that factor. Thus, the product changes sign if we cross the axis at a point x with $|x| > 1$. However, crossing in the region $-1 < x < 1$ changes both factors, so that the product does not change but is continuous across this segment. Thus, it is analytic across this segment by the corollary to Morera's theorem established in Worked Example 2.4.17. We may use this function to define our integrand in a way

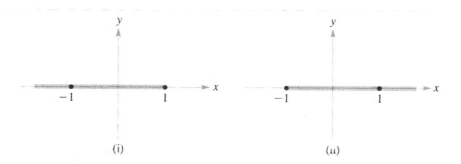

FIGURE 4.3.8 Branch cuts needed for $\sqrt{z^2 - 1}$: (i) for $\sqrt{z - 1}$; (ii) for $\sqrt{z + 1}$.

which is analytic on $\mathbf{C}\backslash\{z \mid \text{Im } z = 0 \text{ and } |\text{Re } z| \geq 1\}$. By the residue theorem,

$$\int_\gamma \frac{dz}{z\sqrt{z^2 - 1}} = 2\pi i \text{ Res}\left(\frac{1}{z\sqrt{z^2 - 1}}, 0\right) = 2\pi$$

The student should verify that (a) the integral over the incomplete circle of radius r approaches zero as $r \to \infty$ (the integrand is less than or equal to $M/|z|^2$ for $|z|$ large), (b) the integral over the incomplete circles of radius ϵ approaches zero as $\epsilon \to 0$ (the integral is bounded by a constant times $\epsilon/\sqrt{\epsilon} = \sqrt{\epsilon}$ on those circles), and (c) for fixed ϵ and r the (continued on page 318)

Table 4.3.1 Evaluation of Definite Integrals

Type of Integral	Conditions	Formula
1. $\displaystyle\int_{-\infty}^{\infty} f(x)\, dx$	No poles of $f(z)$ on real axis; finite number of poles in $\mathbf{C}$; $\|f(z)\| \leq M/\|z\|^2$ for large $\|z\|$.	$\displaystyle\int_{-\infty}^{\infty} f(x)\, dx = 2\pi i \sum \begin{Bmatrix} \text{residues of } f \text{ in} \\ \text{upper half plane} \end{Bmatrix}$
2. $\displaystyle\int_{-\infty}^{\infty} \frac{P(x)}{Q(x)}\, dx$	P, Q polynomials; $\deg Q \geq 2 + \deg P$; no real zeros of Q.	$\displaystyle\int_{-\infty}^{\infty} \frac{P(x)}{Q(x)}\, dx = 2\pi i \sum \begin{Bmatrix} \text{residues of } P/Q \text{ in} \\ \text{upper half plane} \end{Bmatrix}$
3a. $\displaystyle\int_{-\infty}^{\infty} e^{i\omega x} f(x)\, dx$	$\omega > 0$; $\|f(z)\| \leq M/\|z\|$ for $\|z\|$ large and no poles of f on real axis; or $f(z) = P(z)/Q(z)$ where $\deg Q(z) \geq 1 + \deg P(z)$ and Q has no real zeros.	$\displaystyle\int_{-\infty}^{\infty} e^{i\omega x} f(x)\, dx = I$ $ = 2\pi i \sum \begin{Bmatrix} \text{residues of } e^{i\omega z} f(z) \\ \text{in upper half plane} \end{Bmatrix}$ If $\omega < 0$, use $-\sum$ {residues in lower half plane}
b. $\displaystyle\int_{-\infty}^{\infty} \cos(\omega x) f(x)\, dx$ $\displaystyle\int_{-\infty}^{\infty} \sin(\omega x) f(x)\, dx$	f real on real axis.	$\displaystyle\int_{-\infty}^{\infty} \cos(\omega x) f(x)\, dx = \text{Re } I$ $\displaystyle\int_{-\infty}^{\infty} \sin(\omega x) f(x)\, dx = \text{Im } I$ If $\omega < 0$, use lower half plane as above.
4. $\displaystyle\int_0^{2\pi} R(\cos\theta, \sin\theta)\, d\theta$	R rational and $R(\cos\theta, \sin\theta)$ continuous in θ. (No poles on unit circle.)	$\displaystyle\int_0^{2\pi} R(\cos\theta, \sin\theta)\, d\theta$ $ = 2\pi i \sum \begin{Bmatrix} \text{residues of } f \\ \text{inside unit circle} \end{Bmatrix}$ $f(z) = \dfrac{1}{iz} R\left(\dfrac{1}{2}\left(z + \dfrac{1}{z}\right), \dfrac{1}{2i}\left(z - \dfrac{1}{z}\right)\right)$

Table 4.3.1 *continued*

Type of Integral	Conditions	Formula
5. $\displaystyle\int_0^\infty x^{a-1}f(x)\,dx$	$a > 0$ and f has finite number of poles, none on positive real axis; $\|f(z)\| \le M/\|z\|^b$, $b > a$, for $\|z\|$ large; and $\|f(z)\| \le M/\|z\|^d$, $d < a$, for $\|z\| \to 0$. or $f = P/Q$, and Q has no zeros on positive real axis. $0 < a < \deg Q - \deg P$ and $n_Q - n_P < a$, where $n_Q = $ order of the zero of Q at 0 and $n_P = $ order of the zero of P at 0.	$\displaystyle\int_0^\infty x^{a-1}f(x)\,dx$ $= \dfrac{-\pi e^{\pi a i}}{\sin(\pi a)}\sum \left\{\begin{array}{l}\text{residues of } z^{a-1}f(z) \text{ at} \\ \text{poles of } f \text{ excluding } 0\end{array}\right\}$ using the branch $0 < \arg z < 2\pi$
6. $\displaystyle\int_{-\infty}^\infty f(x)\,dx$	Same as entry 1 except that simple poles are allowed on x axis.	$\displaystyle\int_{-\infty}^\infty f(x)\,dx = 2\pi i \sum \left\{\begin{array}{l}\text{residues in upper} \\ \text{half plane}\end{array}\right\}$ $+\, \pi i \sum \{\text{residues on } x \text{ axis}\}$
7. $\displaystyle\int_{-\infty}^\infty \dfrac{P(x)}{Q(x)}\,dx$	Same as entry 2 except that simple poles are allowed on x axis	$\displaystyle\int_{-\infty}^\infty \dfrac{P(x)}{Q(x)}\,dx = 2\pi i \sum \left\{\begin{array}{l}\text{residues in upper} \\ \text{half plane}\end{array}\right\}$ $+\, \pi i \sum \{\text{residues on } x \text{ axis}\}$
8a. $\displaystyle\int_{-\infty}^\infty e^{i\omega x}f(x)\,dx$	Same as entry 3 except that simple poles are allowed on x axis.	(i) $\omega > 0$: $\displaystyle\int_{-\infty}^\infty e^{i\omega x}f(x)\,dx = I$ $= 2\pi i \sum \left\{\begin{array}{l}\text{residues of } e^{i\omega z}f(z) \\ \text{in upper half plane}\end{array}\right\}$ $+\, \pi i \sum \left\{\begin{array}{l}\text{residues of } e^{i\omega z}f(z) \\ \text{on } x \text{ axis}\end{array}\right\}$ (ii) $\omega < 0$: $I = -2\pi i \sum \left\{\begin{array}{l}\text{residues of } e^{i\omega z}f(z) \\ \text{in lower half plane}\end{array}\right\}$ $-\, \pi i \sum \left\{\begin{array}{l}\text{residues of } e^{i\omega z}f(z) \\ \text{on } x \text{ axis}\end{array}\right\}$
b. $\displaystyle\int_{-\infty}^\infty \cos(\omega x)f(x)\,dx$ $\displaystyle\int_{-\infty}^\infty \sin(\omega x)f(x)\,dx$	f real on real axis; simple poles allowed on real axis as in entry 8a.	$\displaystyle\int_{-\infty}^\infty \cos(\omega x)f(x)\,dx = \operatorname{Re} I$ $\displaystyle\int_{-\infty}^\infty \sin(\omega x)f(x)\,dx = \operatorname{Im} I$ If $\omega < 0$, use lower half plane as in entry 8a.

integral over the horizontal lines approaches

$$4 \int_{1+\epsilon}^{r} \frac{dx}{x\sqrt{x^2 - 1}}$$

These three facts, together with the fact that

$$\int_{\gamma} \frac{dz}{z\sqrt{z^2 - 1}} = 2\pi$$

show that

$$\int_{1}^{\infty} \frac{dx}{x\sqrt{x^2 - 1}}$$

exists and equals $\pi/2$. ▲

Worked Examples

4.3.16   *Evaluate*

$$\int_{-\infty}^{\infty} \frac{1}{1 + x^{2n}} dx \qquad \text{for} \qquad n \geq 1$$

Solution. This integral could be evaluated using Proposition 4.3.1, but we would have to consider all the poles in the upper half plane. If we use instead the contour indicated in Fig. 4.3.9, we need

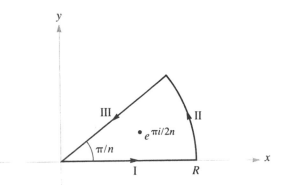

FIGURE 4.3.9   Contour for Worked Example 4.3.16.

consider only one pole. The only singularity of $f(z) = 1/(1 + z^{2n})$ inside this contour is a simple pole at $e^{\pi i/2n}$, where the residue is $-e^{\pi i/2n}/2n$. Thus,

$$-\frac{\pi i}{n} e^{\pi i/2n} = \int_I f + \int_{II} f + \int_{III} f$$

$$= \int_0^R \frac{1}{1 + x^{2n}} \, dx + \int_0^{\pi/n} \frac{1}{1 + R^{2n}e^{2ni\theta}} iRe^{i\theta} \, d\theta + \int_R^0 \frac{1}{1 + r^{2n}e^{2\pi i}} e^{\pi i/n} \, dr$$

$$= (1 - e^{\pi i/n}) \int_0^R \frac{1}{1 + x^{2n}} \, dx + iR \int_0^{\pi/n} \frac{1}{1 + R^{2n}e^{2ni\theta}} e^{i\theta} \, d\theta$$

The second integral is no larger in absolute value than $(\pi/n)R/(R^{2n} - 1)$, which goes to 0 as $R \to \infty$. Letting $R \to \infty$, we obtain

$$\int_0^\infty \frac{1}{1 + x^{2n}} \, dx = -\frac{\pi i}{n} \frac{e^{\pi i/2n}}{1 - e^{\pi i/n}} = \frac{\pi}{2n} \csc \frac{\pi}{2n}$$

4.3.17 *Fresnel Integrals. Show that $\int_{-\infty}^\infty \cos(x^2) \, dx$ and $\int_{-\infty}^\infty \sin(x^2) \, dx$ both exist and equal $\sqrt{\pi/2}$.*

Solution. First we show the integrals exist. Observe that $\sin(x^2)$ has zeros at $x_n = \sqrt{\pi n}$ for integers n. Since $\sqrt{n+1} - \sqrt{n} = 1/(\sqrt{n+1} + \sqrt{n})$, the distance between these zeros shrinks to zero as n increases, and so the quantities $a_n = |\int_{x_{n-1}}^{x_n} \sin(x^2) \, dx|$ decrease monotonically to 0. Thus $\Sigma_0^\infty (-1)^n a_n$ converges by the alternating series test to some number A. If R is any real number, then $x_{N-1} \leq R < x_N$ for a unique N, and $\int_0^R \sin(x^2) \, dx$ is between the partial sums $\Sigma_0^{N-1} (-1)^n a_n$ and $\Sigma_0^N (-1)^n a_n$. Thus $\lim_{R \to \infty} \int_0^R \sin(x^2) \, dx$ exists and is equal to A. Similarly, $\lim_{R \to \infty} \int_0^R \cos(x^2) \, dx$ exists.

Consider the integral of $f(z) = e^{iz^2}/\sin(\sqrt{\pi}z)$ around the contour $\gamma = I + II + III + IV$ shown in Fig. 4.3.10. The function f has a simple pole at 0 inside γ with residue $1/\sqrt{\pi}$, and so

$$\int_\gamma f = 2\sqrt{\pi}i$$

Along I, $z = x - Ri$, so that

$$|e^{iz^2}| = |e^{i(x^2 - 2Rix - R^2)}| = e^{2Rx}$$

and

$$|\sin \sqrt{\pi}z| = \tfrac{1}{2}|e^{i\sqrt{\pi}x - R\sqrt{\pi}} - e^{-i\sqrt{\pi}x + R\sqrt{\pi}}| \geq \tfrac{1}{2}(e^{R\sqrt{\pi}} - 1)$$

Thus, along I we have

$$\left| \int_I f \right| < \frac{2}{e^{R\sqrt{\pi}} - 1} \int_{-\sqrt{\pi}/2}^{\sqrt{\pi}/2} e^{2Rx} \, dx = \frac{1}{R} \frac{e^{R\sqrt{\pi}} - e^{-R\sqrt{\pi}}}{e^{R\sqrt{\pi}} - 1}$$

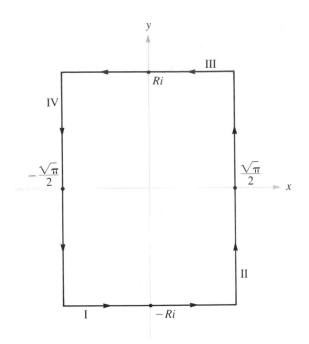

FIGURE 4.3.10   The contour used for evaluating the Fresnel integrals.

which goes to 0 as $R \to \infty$. Similarly $\int_{III} f \to 0$ as $R \to \infty$. The contribution from the vertical sides is

$$\int_{II} f + \int_{IV} f = \int_{-R}^{R} \frac{e^{i(\sqrt{\pi}/2 + iy)^2}}{\sin (\pi/2 + \sqrt{\pi} yi)} i \, dy + \int_{R}^{-R} \frac{e^{i(\sqrt{\pi}/2 + iy)^2}}{\sin (-\pi/2 + \sqrt{\pi} yi)} i \, dy$$

$$= \int_{-R}^{R} \frac{e^{i(\pi/4 - y^2)}(e^{-\sqrt{\pi} y} + e^{\sqrt{\pi} y})}{\cos (i\sqrt{\pi} y)} i \, dy = 2i \int_{-R}^{R} e^{i(\pi/4 - y^2)} \, dy$$

$$= 2e^{3\pi i/4} \int_{-R}^{R} e^{-iy^2} \, dy = \sqrt{2}(-1 + i) \left[\int_{-R}^{R} \cos (x^2) \, dx - i \int_{-R}^{R} \sin (x^2) \, dx \right]$$

Letting $R \to \infty$ we obtain

$$2\sqrt{\pi} i = \sqrt{2}(-1 + i) \left[\int_{-\infty}^{\infty} \cos (x^2) \, dx - i \int_{-\infty}^{\infty} \sin (x^2) \, dx \right]$$

and

$$\sqrt{2}\pi i = \left[-\int_{-\infty}^{\infty} \cos (x^2) \, dx + \int_{-\infty}^{\infty} \sin (x^2) \, dx \right] + i \left[\int_{-\infty}^{\infty} \cos (x^2) \, dx + \int_{-\infty}^{\infty} \sin (x^2) \, dx \right]$$

The real part of this equation shows that our integrals are equal, while the imaginary part shows that their common value is $\sqrt{2\pi}/2 = \sqrt{\pi}/2$.

4.3.18 *Show that* $\int_{-\infty}^{\infty} e^{-x^2}\,dx = \sqrt{\pi}$

*Solution.** Let $f(z) = e^{-z^2}$ and consider the integral of f along the contour $\gamma = I + II + III$ shown in Fig. 4.3.11. Notice that

$$\int_{I} f = \int_{0}^{R} e^{-x^2}\,dx$$

and

$$\int_{III} f = \int_{R}^{0} e^{-ir^2} e^{\pi i/4}\,dr = e^{5\pi i/4} \int_{0}^{R} (\cos r^2 - i\sin r^2)\,dr$$

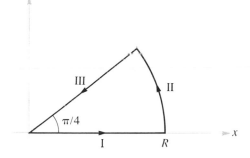

y

III

II

$\pi/4$

I

R

x

FIGURE 4.3.11 Contour for $\int_{-\infty}^{\infty} e^{-x^2}\,dx$.

Along II, $z = Re^{i\theta}$, and so

$$|f(z)| = |e^{-R^2(\cos 2\theta + 2\sin 2\theta)}| = e^{-R^2\cos 2\theta}$$

But for $0 \le \theta \le \pi/4$, we have $\cos 2\theta \ge 1 - 4\theta/\pi$ (see Fig. 4.3.12) and so $|f(z)| \le e^{-R^2} e^{4R^2\theta/\pi}$, and thus

$$\left| \int_{II} f \right| \le \int_{0}^{\pi/4} e^{-R^2} e^{4R^2\theta/\pi} |iRe^{i\theta}|\,d\theta = Re^{-R^2} \frac{\pi}{4R^2} (e^{R^2} - 1) = \frac{\pi}{4R} (1 - e^{-R^2})$$

---

* The method that follows is usually attributed to R. Courant. See Exercise 25 as well.

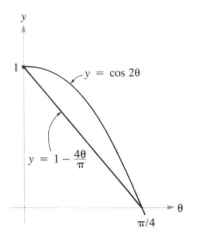

FIGURE 4.3.12   Proof that $\cos 2\theta \geq 1 - 4\theta/\pi$.

This certainly goes to 0 as $R \to \infty$. Since f is entire,

$$0 = \int_{\mathrm{I}} f + \int_{\mathrm{II}} f + \int_{\mathrm{III}} f$$

Letting $R \to \infty$ we obtain

$$0 = \int_0^\infty e^{-x^2}\, dx - \frac{1+i}{\sqrt{2}}\left[\int_0^\infty \cos(x^2)\, dx - i\int_0^\infty \sin(x^2)\, dx\right]$$

since we already know from the last example that both of these integrals exist. Both integrands are even, and by the last example, both integrals are $\sqrt{\pi}/2\sqrt{2}$. We are left with

$$\int_0^\infty e^{-x^2}\, dx = \frac{1+i}{\sqrt{2}}(1-i)\frac{\sqrt{\pi}}{2\sqrt{2}} = \frac{\sqrt{\pi}}{2}$$

Again, the integrand is even, and so

$$\int_{-\infty}^\infty e^{-x^2}\, dx = 2\int_0^\infty e^{-x^2}\, dx = \sqrt{\pi}$$

4.3.19   **Fourier Transforms** *Suppose that the Fourier transform of a function f is defined with a multiplicative constant $1/\sqrt{2\pi}$ and sign conventions as*

$$\hat{f}(\omega) = \frac{1}{\sqrt{2\pi}}\int_{-\infty}^\infty f(x)e^{-i\omega x}\, dx$$

Show that the normal probability function $f(x) = (1/\sqrt{2\pi})e^{-x^2/2}$ satisfies $\hat{f}(\omega) = f(\omega)$.

Solution. Let $g(z) = e^{-z^2/2}e^{-i\omega z}$. Then g is an entire function, and so its integral around the contour $\gamma = I + II + III + IV$ shown in Fig. 4.3.13 is 0 for any real R and τ. (If $\tau < 0$, draw the contour in the lower half plane.) Thus,

$$\int_I g = \int_{-R}^{R} e^{-x^2/2}e^{-i\omega x}\, dx$$

and

$$\int_{III} g = \int_{R}^{-R} e^{-(x+\tau i)^2/2}e^{-i\omega(\tau+xi)}\, dx$$

$$= -e^{-(\tau^2/2)+\omega\tau} \int_{-R}^{R} e^{-x^2/2}e^{-i(\tau+\omega)x}\, dx$$

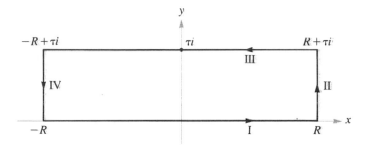

FIGURE 4.3.13 Contour for evaluating the Fourier transform of $e^{-x^2/2}/\sqrt{2\pi}$.

Along II and IV,

$$|g(z)| = |e^{-(\pm R+iy)^2/2}e^{-i\omega(\pm R+iy)}|$$
$$= e^{-R^2/2+y^2/2}e^{\omega y} \le e^{-R^2/2}e^{\tau^2/2+|\omega\tau|}$$

Thus $|\int_{II} g|$ and $|\int_{IV} g|$ are each bounded above by $e^{-R^2/2}|\tau|e^{\tau^2/2+|\omega\tau|}$. With τ and ω fixed, this goes to 0 as $R \to \infty$. Letting $R \to \infty$ in $\int_\gamma g = 0$, we obtain

$$\int_{-\infty}^{\infty} e^{-x^2/2}e^{-i\omega x}\, dx - e^{\tau^2/2+\omega\tau} \int_{-\infty}^{\infty} e^{-x^2/2}e^{-i(\tau+\omega)}\, dx = 0$$

This holds for any real τ and ω. Setting $\tau = -\omega$ gives

$$\int_{-\infty}^{\infty} e^{-x^2/2}e^{-i\omega x}\, dx = e^{-\omega^2/2} \int_{-\infty}^{\infty} e^{-x^2/2}\, dx$$

Making the change of variables $(x \mapsto x/\sqrt{2})$ in the last example (or in Proposition 4.3.3), this becomes

$$\sqrt{2\pi} \int_{-\infty}^{\infty} f(x)e^{-i\omega x}\, dx = e^{-\omega^2/2}\sqrt{2\pi}$$

that is,

$$\frac{1}{\sqrt{2\pi}} \int_{-\infty}^{\infty} f(x)e^{-i\omega x}\, dx = \frac{1}{\sqrt{2\pi}} e^{-\omega^2/2}$$

or

$$\hat{f}(\omega) = f(\omega)$$

as desired.

4.3.20   *If $p > 0$ and $q > 0$, show that*

$$\int_0^{\infty} \frac{\log (px)}{q^2 + x^2}\, dx = \frac{\pi}{2q} \log (pq)$$

Solution. $f(z) = [\log (pz)]/(q^2 + z^2)$ is analytic on the plane with the negative imaginary axis deleted if we define a branch of the logarithm by

$$\log (\rho e^{i\phi}) = \log \rho + i\phi \qquad -\frac{\pi}{2} \le \phi < \frac{3\pi}{2}$$

We consider the integral of f along the contour $\gamma = \text{I} + \text{II} + \text{III} + \text{IV}$ shown in Fig. 4.3.14. Note that

$$\int_{\text{I}} f = \int_{\epsilon}^{R} \frac{\log (px)}{q^2 + x^2}\, dx \quad \text{and} \quad \int_{\text{III}} f = \int_{R}^{\epsilon} \frac{\log (-px)}{q^2 + (-x)^2}(-dx) = \int_{\epsilon}^{R} \frac{\log (px) + \pi i}{q^2 + x^2}\, dx$$

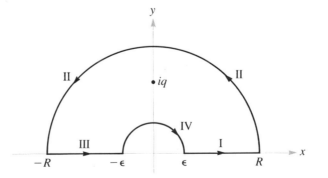

FIGURE 4.3.14   The contour used in Example 4.3.20.

Along II,

$$|f(x)| \le \frac{|\log(pR)| + \pi}{R^2 - q^2} \quad \text{and so} \quad \left|\int_{\text{II}} f\right| \le \frac{|\log(pR)| + \pi}{R^2 - q^2} \pi R$$

This goes to 0 as $R \to \infty$ (use L'Hôpital's rule on $(\log R)/R$). Along IV,

$$|f(z)| \le \frac{|\log(p\epsilon)| + \pi}{q^2 - \epsilon^2} \pi\epsilon$$

This also goes to 0 as $\epsilon \to 0$ since $\lim_{\epsilon \to 0} \epsilon \log \epsilon = 0$. Since F has a simple pole inside γ at $z = iq$ with residue $[\log(pqi)]/2qi = [\log(pq)]/2qi + \pi/4q$, we can let $R \to \infty$ and $\epsilon \to 0$ to obtain

$$2\int_0^\infty \frac{\log(px)}{q^2 + x^2} dx + \pi i \int_0^\infty \frac{1}{q^2 + x^2} dx = 2\pi i \left[\frac{\log(pq)}{2qi} + \frac{\pi}{4q}\right]$$

Comparing real and imaginary parts gives

$$\int_0^\infty \frac{\log(px)}{q^2 + x^2} dx = \frac{\pi}{2q} \log(pq) \quad \text{and} \quad \int_0^\infty \frac{1}{q^2 + x^2} dx = \frac{\pi}{2q}$$

The latter integral, of course, may be evaluated by elementary calculus using the inverse tangent function.

Exercises

1. Evaluate

$$\int_{-\infty}^\infty \frac{dx}{x^2 - 2x + 4}$$

2. Prove that

$$\int_0^\infty \frac{\sin^2 x}{x^2} dx = \frac{\pi}{2}$$

(*Hint.* Consider $\displaystyle\int_{-\infty}^\infty \frac{1 - e^{2ix}}{x^2} dx$ and apply Proposition 4.3.11.)

3. Evaluate

$$\int_0^\pi \frac{d\theta}{(a + b \cos \theta)^2} \quad \text{for} \quad 0 < b < a$$

4. Evaluate

$$\int_0^\infty \frac{dx}{1+x^6}$$

5. Evaluate

$$\int_0^\infty \frac{\cos mx}{1+x^4}\,dx$$

6. Evaluate

$$\int_0^\infty \frac{x^{a-1}}{1+x^3}\,dx \qquad \text{for} \qquad 0 < a < 3$$

7. Evaluate

$$\int_0^\infty \frac{x \sin x}{1+x^2}\,dx$$

8. (a) Prove that

$$\int_{-\infty}^\infty \frac{\cos x}{e^x + e^{-x}}\,dx = \frac{\pi}{e^{\pi/2} + e^{-\pi/2}}$$

by integrating $e^{iz}/(e^z + e^{-z})$ around the rectangle with vertices $-r, r, r + \pi i, -r + \pi i$; let $r \to \infty$.
(b) Use the same technique to show that

$$\int_{-\infty}^\infty \frac{e^{-x}}{1 + e^{-2\pi x}}\,dx = \frac{1}{2 \sin \frac{1}{2}}$$

9. Evaluate

$$\text{P.V.} \int_{-\infty}^\infty \frac{dx}{(x-a)^2(x-1)}$$

where Im $a > 0$.

10. Show that

$$\int_0^\pi \sin^{2n}\theta\,d\theta = \frac{\pi(2n)!}{(2^n n!)^2}$$

11. Show that for $a > 0, b > 0$

$$\int_0^\infty \frac{\cos ax}{(x^2 + b^2)^2}\,dx = \frac{\pi}{4b^3}(1 + ab)e^{-ab}$$

12. Show that for $0 < b < 1$

$$\int_0^\infty \frac{1}{x^b(x+1)}\,dx = \frac{\pi}{\sin(b\pi)}$$

13. Find

$$\text{P.V.} \int_{-\infty}^\infty \frac{dx}{x(x^2-1)}$$

14. Prove that

$$\int_0^\infty \frac{\log x}{(x^2+1)^2}\,dx = -\frac{\pi}{4}$$

15. Find

$$\int_0^1 \frac{dx}{\sqrt{x^2-1}}$$

by (a) changing the variables to $y = 1/(x + \sqrt{x^2-1})$, and (b) considering the curve in Fig. 4.3.15 and finding the residue of a branch of $1/\sqrt{z^2-1}$ at ∞.

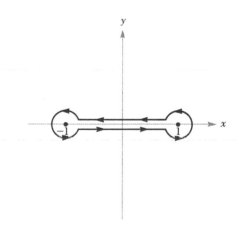

FIGURE 4.3.15 Contour for evaluating $\displaystyle\int_0^1 \frac{dx}{\sqrt{x^2-1}}$

16. Let $P(z)$ and $Q(z)$ be polynomials with $\deg Q(z) \ge 2 + \deg P(z)$. Show that the sum of the residues of $P(z)/Q(z)$ is zero.

17. Evaluate

$$\int_{-\infty}^{\infty} \frac{\cos bx}{x^2 + a^2} \, dx \qquad \text{for} \qquad a > 0, b > 0$$

18. Let $f(z)$ be as in formula 5 of Table 4.3.1 except allow f to have a finite number of simple poles on the (strictly) positive real axis. Show that

$$\text{P.V.} \int_0^\infty x^{a-1} f(x) \, dx = \frac{-\pi e^{-\pi a i}}{\sin(\pi a)} \sum \{\text{residues of } (-z)^{a-1} f(z) \text{ at poles of } f \\ \text{off the nonnegative real axis}\}$$

$$+ \frac{\pi e^{-\pi a i} \cos \pi a}{\sin \pi a} \sum \{\text{residues of } (-z)^{a-1} f(z) \text{ at} \\ \text{poles of } f \text{ on the positive real} \\ \text{axis}\}$$

19. Use Exercise 18 to show that

$$\text{P.V.} \int_0^\infty \frac{x^{a-1}}{1 - x} \, dx = -\pi \cot(\pi a) \qquad \text{for} \qquad 0 < a < 1$$

20. Establish the following formulas:

(a) $\displaystyle\int_0^\pi \frac{d\theta}{1 + \sin^2 \theta} = \frac{\pi}{\sqrt{2}}$

(b) $\displaystyle\int_0^\infty \frac{x^2 \, dx}{(x^2 + a^2)^2} = \frac{\pi}{4a}$ for $a > 0$

(c) $\displaystyle\int_{-\infty}^\infty \frac{x^3 \sin x}{(x^2 + 1)^2} \, dx = \frac{\pi}{2} e^{-1}$

(d) $\displaystyle\int_0^\infty \frac{x \sin x}{x^4 + 1} \, dx = \frac{\pi}{2} e^{-1/\sqrt{2}} \sin \frac{1}{\sqrt{2}}$

21. Prove Proposition 4.3.3 by evaluating a double integral over the whole plane in polar coordinates.

22. In Worked Example 4.3.16, may the exponent $2n$ be replaced by any other power $p \geq 2$?

23. Evaluate $\displaystyle\int_0^{2\pi} \frac{1}{2 + \cos \theta} \cos (4\theta) \, d\theta$ by considering the real part of $\displaystyle\int_0^{2\pi} \frac{1}{2 + \cos \theta} e^{4i\theta} \, d\theta$ and then converting to an integral around the unit circle.

24. Suppose that the Fourier transform of a function $g(x)$ is defined as in Worked Example 4.3.19 by

$$\hat{g}(\omega) = \frac{1}{\sqrt{2\pi}} \int_{-\infty}^\infty g(x) e^{-i\omega x} \, dx$$

Show that if f is differentiable and the integrals for $\hat{f}$ and $(f')\hat{\ }$ converge, then

$$(f')\hat{\ }(\omega) = \frac{1}{i\omega} \hat{f}(\omega)$$

(*Hint.* $f(x)$ must go to 0 in both directions along the x axis. Try integrating by parts.)

25. (a) Evaluate

$$\text{limit}_{R \to \infty} \int_{\gamma_R} \frac{e^{-z^2 + \sqrt{\pi i} z}}{e^{2\sqrt{\pi i} z} - 1} \, dz$$

where $\sqrt{\pi i} = \sqrt{\pi} e^{i\pi/4}$ and γ_R is as shown in Fig. 4.3.16.

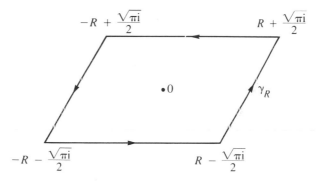

FIGURE 4.3.16 The contour used for $\int_{-\infty}^{\infty} e^{-x^2} \, dx$.

(b) Show that the integrals along the horizontal parts partially cancel to give a multiple of $\int_{-\infty}^{\infty} e^{-x^2} \, dx$. Use this to show $\int_{-\infty}^{\infty} e^{-x^2} \, dx = \sqrt{\pi}$.

4.4 EVALUATION OF INFINITE SERIES AND PARTIAL-FRACTION EXPANSIONS

In the last section we saw how to use sums of residues to evaluate integrals. In this section we give a brief discussion of some applications in the other direction: using integrals to evaluate sums. For instance, we shall see that by applying these theorems we can prove that

$$\sum_{n=1}^{\infty} \frac{1}{n^2} = \frac{\pi^2}{6}$$

This is a famous formula of Leonhard Euler, who discovered it in the eighteenth century by using other techniques.

Infinite Series

We shall develop a general method for evaluating series of the form $\sum_{n=-\infty}^{\infty} f(n)$, where f is a given function. Suppose that we restrict f to being a meromorphic function with a finite number of poles, none of which are integers. Suppose that $G(z)$ is a meromorphic function whose only poles are simple poles at the integers, where the residues are all 1. Thus at the integers, the residues of $f(z)G(z)$ are $f(n)$. Then if γ is a closed curve enclosing $-N, -N+1, \ldots, 0, 1, \ldots, N$, the residue theorem gives

$$\int_{\gamma} G(z)f(z)\, dz = 2\pi i \left\{ \left[\sum_{n=-N}^{N} f(n) \right] + \sum \{\text{residues of } G(z)f(z) \text{ at poles of } f\} \right\}$$

If $\int_{\gamma} G(z)f(z)\, dz$ exhibits a controllable limiting behavior as γ becomes large, we will have information about the limiting behavior of $\sum_{n=-N}^{N} f(n)$ as $N \to \infty$ in terms of the residues of $G(z)f(z)$ at the poles of f. A suitable $G(z)$ is $\pi \cot \pi z$.

Of course, we always have

$$\int_{\gamma} G(z)f(z)\, dz = 2\pi i \sum \{\text{all residues of } G(z)f(z) \text{ inside } \gamma\}$$

so that if some of the poles of f happen to be at integers, we need only move terms around:

$$\int_{\gamma} G(z)f(z)\, dz = 2\pi i \left\{ \sum_{n=-N}^{N} \{f(n) \mid n \text{ is not a singularity of } f\} \right.$$
$$\left. + \sum \{\text{residues of } G(z)f(z) \text{ at singularities of } f\} \right\}$$

4.4.1 SUMMATION THEOREM   *Let f be analytic in $\mathbf{C}$ except for finitely many isolated singularities. Let C_N be a square with vertices at $(N + \frac{1}{2}) \times (\pm 1 \pm i)$, $N = 1, 2, 3, \ldots$ (Fig. 4.4.1). Suppose that $\int_{C_N} (\pi \cot \pi z) f(z)\, dz \to 0$ as $N \to \infty$. Then we have the summation formula*

$$\lim_{N \to \infty} \sum_{-N}^{N} \{f(n) \mid n \text{ is not a singularity of } f\}$$
$$= -\sum \{\text{residues of } \pi \cot \pi z\, f(z) \text{ at singularities of } f\}$$

If none of the singularities of f are at integers, then $\lim\limits_{N \to \infty} \sum\limits_{n=-N}^{N} f(n)$ *exists, is finite, and*

$$\lim_{N \to \infty} \sum_{n=-N}^{N} f(n) = -\sum \{\text{residues of } \pi \cot \pi z\, f(z) \text{ at singularities of } f\}$$

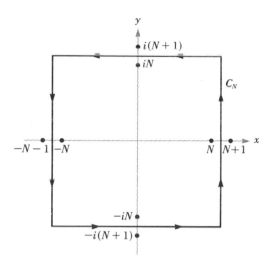

FIGURE 4.4.1 Contour for evaluating $\sum_{-\infty}^{\infty} f(n)$.

PROOF In this argument we assume none of the singularities of f are at integers. (For the more general case simply insert the qualifying phrase in the first sum and move appropriate terms. There can be only finitely many of them.) By the residue theorem,

$$\int_{C_N} \pi \cot \pi z\, f(z)\, dz = 2\pi i \sum \left\{ \begin{array}{l} \text{residues of } \pi \cot \pi z\, f(z) \text{ at the} \\ \text{integers } -N, -N+1, \ldots, 0, 1, \ldots, N \end{array} \right\}$$

$$+\, 2\pi i \sum \left\{ \begin{array}{l} \text{residues of } \pi \cot \pi z\, f(z) \\ \text{at the singularities of } f \end{array} \right\}$$

for N sufficiently large that C_N encloses all of the singularities of f. Since $\cot \pi z = (\cos \pi z)/(\sin \pi z)$ and $(\sin \pi z)' \neq 0$ at $z = n$, we see that n is a simple pole of $\cot \pi z$ and that $\operatorname{Res}(\cot \pi z, n) = (\cos \pi n)/(\pi \cos \pi n) = 1/\pi$ (use formula 4 of Table 4.1.1). Therefore, $\operatorname{Res}(\pi \cot \pi z\, f(z), n) = \pi f(n) \operatorname{Res}(\cot \pi z, n) = f(n)$. Thus $\sum$ {residues of $\pi \cot \pi z\, f(z)$ at the integers $-N, -N+1, \ldots, 0, 1, \ldots, N$} $= \sum_{n=-N}^{N} f(n)$. Taking limits on both sides of the preceding equation for $\int_{C_N} \pi \cot \pi z\, f(z)\, dz$ and using the fact that $\int_{C_N} \pi \cot \pi z\, f(z)\, dz \to 0$ as $N \to \infty$, we obtain

$$\lim_{N\to\infty} \sum_{n=-N}^{N} f(n) = -\sum \{\text{residues of } \pi \cot \pi z\, f(z) \text{ at the singularities of } f\} \qquad \blacksquare$$

It is important to notice that what we have obtained is a formula for the limit of the symmetric partial sums of $\sum_{-\infty}^{\infty} f(n)$. This is not the same as the doubly infinite series itself, which demands that the upper and lower limits grow independently: $\sum_{-\infty}^{\infty} f(n) =$

limit $\Sigma^N_{-M} f(n) = \lim_{\substack{M \to \infty}} \Sigma^{-1}_{n=-M} f(n) + \lim_{N \to \infty} \Sigma^N_{n=0} f(n)$. If the doubly infinite series is
$\lim_{\substack{N \to \infty \\ M \to \infty}}$

known to converge, then our limit must give the same answer, but $\lim_{N \to \infty} \Sigma^N_{-N} f(n)$ may

exist when the more general limit does not. The situation is somewhat analogous to the computation of an improper integral by a Cauchy principal value. Nevertheless, our formula is often sufficient. We may check independently that the double limit exists, or as in our first example, we may be interested in a singly infinite series. If f is an even function, then $\Sigma^N_{-N} f(n) = f(0) + 2 \Sigma^N_{n=1} f(n)$.

The cotangent is not the only candidate for a useful function for G. Others are $2\pi i/(e^{2\pi i z} - 1)$ and $-2\pi i/(e^{-2\pi i z} - 1)$. We indicate in the exercises a way of using $\pi \csc \pi z$ which is particularly useful for alternating series. (A more complete exposition and extensive references may be found in D. S. Mitrinovic and J. D. Kečkić, *The Cauchy Method of Residues* (Dordrecht, The Netherlands: D. Reidel Publ. Co., 1984).)

Next we establish a criterion by which f can be judged to satisfy the hypotheses of the summation theorem (4.4.1).

4.4.2 PROPOSITION   *Suppose f is analytic on $\mathbf{C}$ except for isolated singularities. If there are constants R and $M > 0$ such that $|zf(z)| \le M$ whenever $|z| \ge R$, then the hypotheses of the summation theorem (4.4.1) are satisfied.*

PROOF   Since $|zf(z)|$ is bounded outside R, all singularities of f are in the region $|z| \le R$. Since they are isolated, there must be a finite number of them (Why?). Furthermore, $|f(1/z)/z|$ is bounded by M in the region $|z| < 1/R$, and so 0 is a removable singularity of $f(1/z) \cdot 1/z$ and we can therefore write $f(1/z) \cdot 1/z = a_0 + a_1 z + a_2 z^2 + \cdots$ for $|z| < 1/R$; hence

$$f(z) = \frac{a_0}{z} + \frac{a_1}{z^2} + \frac{a_2}{z^3} + \cdots$$

for $|z| > R$. Now consider the integral

$$\int_{C_N} \frac{\pi \cot \pi z}{z} \, dz$$

By the residue theorem,

$$\int_{C_N} \frac{\pi \cot \pi z}{z} \, dz = 2\pi i \left\{ \text{residues of } \frac{\pi \cot \pi z}{z} \text{ at } z = 0 \right\}$$

$$+ 2\pi i \sum \left\{ \begin{array}{l} \text{residue of } \dfrac{\pi \cot \pi z}{z} \text{ at} \\ n = \pm 1, \pm 2, \ldots, \pm N \end{array} \right\}$$

Since the pole at 0 is of order 2, we can write

$$\frac{\pi \cot \pi z}{z} = \frac{b_{-2}}{z^2} + \frac{b_{-1}}{z} + b_0 + b_1 z + b_2 z^2 + \cdots$$

Because $(\pi \cot \pi z)/z$ is an even function of z (that is, $[\pi \cot \pi(-z)]/(-z) = (\pi \cot \pi z)/z)$, uniqueness of the Laurent expansion shows that coefficients of odd powers of z are zero; in particular, $b_{-1} = 0$. But b_{-1} is exactly Res $[(\pi \cot \pi z)/z, 0]$. (Instead of this trick we could have used formula 9 of Table 4.1.1.) Also, Res $[(\pi \cot \pi z)/z, n] = 1/n$ for $n = \pm 1$, $\pm 2, \ldots, \pm N$ (Why?), so Σ {residues of $(\pi \cot \pi z)/z$ at $n = \pm 1, \pm 2, \ldots, \pm N$} $= 0$. Consequently,

$$\int_{C_N} \frac{\pi \cot \pi z}{z} \, dz = 0$$

Thus we can write

$$\int_{C_N} \pi \cot \pi z \, f(z) \, dz = \int_{C_N} \pi(\cot \pi z) \left[f(z) - \frac{a_0}{z} \right] dz$$

To estimate this latter integral, we observe that

$$f(z) - \frac{a_0}{z} = \frac{a_1}{z^2} + \frac{a_2}{z^3} + \cdots$$

for $|z| > R$. Since $a_1 + a_2 z + a_3 z^2 + \cdots$ represents an analytic function for $|z| < 1/|R|$, it is bounded, say, by M' on the closed disk $|z| \leq 1/R'$, where $R' > R$. This implies that

$$\left| f(z) - \frac{a_0}{z} \right| \leq \frac{M'}{|z|^2}$$

for $|z| \geq R'$. Suppose that N is sufficiently large that all points on C_N satisfy $|z| \geq R'$. Then

$$\left| \int_{C_N} \pi(\cot \pi z) \left[f(z) - \frac{a_0}{z} \right] dz \right| \leq \frac{\pi M' \cdot 8(N + \frac{1}{2})}{(N + \frac{1}{2})^2} \left(\sup_{z \, \text{on} \, C_N} |\cot \pi z| \right)$$

The student should verify that

$$\sup \{|\cot \pi z| \text{ such that } z \text{ lies on } C_N\} = \frac{e^{2\pi(N+1/2)} + 1}{e^{2\pi(N+1/2)} - 1}$$

(noting that on the vertical sides, $|\cot \pi z| \leq 1$; on the horizontal sides, the maximum occurs at $x = 0$). Hence for all N sufficiently large we have $\sup_{z \, \text{on} \, C_N} |\cot \pi z| \leq 2$. The previous inequality then shows that

$$\int_{C_N} \pi(\cot \pi z) \left[f(z) - \frac{a_0}{z} \right] dz$$

approaches zero as $N \to \infty$, which, in turn, shows that

$$\int_{C_N} \pi \cot \pi z \, f(z) \, dz \to 0 \qquad \text{as} \qquad N \to \infty \quad \blacksquare$$

We give here one famous example of this technique. One usually learns in calculus that the p series $\Sigma_{n=1} (1/n^p)$ converges if $p > 1$ and diverges if $p \le 1$, but usually with no indication of just what that sum might be. We encountered this sum in Chap. 3 as $\zeta(p)$ where ζ is the Riemann zeta function, an important ingredient in number theory. The case $p = 2$ is interesting and there are many ways to evaluate $\zeta(2)$. Here is one.

4.4.3 PROPOSITION

$$\sum_{n=1}^{\infty} \frac{1}{n^2} = \frac{\pi^2}{6}$$

PROOF   We apply the summation theorem (or its corollary) with $f(z) = 1/z^2$. Since $\tan z$ has a simple zero at $z = 0$, $\cot z$ has a simple pole there. If the Laurent expansion is $\cot z = b_1/z + a_0 + a_1 z + \cdots$, then

$$\left(1 - \frac{z^2}{2!} + \frac{z^4}{4!} - \cdots \right) = \left(z - \frac{z^3}{3!} + \frac{z^5}{5!} - \cdots \right) \left(\frac{b_1}{z} + a_0 + a_1 z + \cdots \right)$$

Multiplying, collecting terms, and comparing coefficients, we find $b_1 = 1$, $a_0 = 0$, and $a_1 = -\frac{1}{3}$. Thus

$$\frac{\pi \cot \pi z}{z^2} = \frac{\pi(1/\pi z - \pi z/3 + \cdots)}{z^2} = \frac{1}{z^3} - \frac{\pi^2}{z} \cdot \frac{1}{3} + \cdots$$

Hence

$$\text{Res} \left(\frac{\pi \cot \pi z}{z^2}, 0 \right) = \frac{-\pi^2}{3}$$

Since the only singularity of f is at $z = 0$, the summation formula becomes

$$\lim_{N \to \infty} \left(\sum_{n=-N}^{-1} \frac{1}{n^2} + \sum_{n=1}^{\infty} \frac{1}{n^2} \right) = \frac{\pi^2}{3}$$

and, since $1/(-n)^2 = 1/n^2$, we obtain

$$\lim_{N \to \infty} \sum_{n=1}^{N} \frac{1}{n^2} = \frac{\pi^2}{6}$$

We conclude that

$$\sum_{n=1}^{\infty} \frac{1}{n^2} = \frac{\pi^2}{6} \quad \blacksquare$$

Partial-Fraction Expansions

If $f(z) = p(z)/q(z)$ is a rational function, we know a trick from algebra which is often useful in calculus: The function f can be expanded in "partial fractions" in terms of the zeros of the denominator. A meromorphic function can sometimes be thought of as somewhat like a rational function with possibly infinitely many zeros in the denominator, and one might wonder if a similar expansion is possible. Although one should not take this analogy too seriously, something along these lines can be done. First we give a specific example which shows how the summation theorem can be used and which will be used in Chap. 7. Then we will give a somewhat more general result.

4.4.4 PROPOSITION   *Let z be any complex number not equal to an integer; then both*

$$\sum_{n=1}^{\infty} \left(\frac{1}{z-n} + \frac{1}{n} \right) \quad and \quad \sum_{n=1}^{\infty} \left(\frac{1}{z+n} - \frac{1}{n} \right)$$

are absolutely convergent series and

$$\pi \cot \pi z = \frac{1}{z} + \sum_{n=1}^{\infty} \left(\frac{1}{z-n} + \frac{1}{n} \right) + \sum_{n=1}^{\infty} \left(\frac{1}{z+n} - \frac{1}{n} \right)$$

This equation can also be written

$$\pi \cot \pi z = \frac{1}{z} + \sum_{n=-\infty}^{\infty}{}' \left(\frac{1}{z-n} + \frac{1}{n} \right)$$

where the prime indicates that the term corresponding to $n = 0$ is omitted.

PROOF   For n sufficiently large, $|z - n| > n/2$. Therefore,

$$\left| \frac{1}{z-n} + \frac{1}{n} \right| = \left| \frac{z}{(z-n)n} \right| \le \frac{2|z|}{n^2}$$

By comparison with the convergent series

$$2|z| \cdot \left(\frac{1}{n^2} + \frac{1}{(n+1)^2} + \cdots \right)$$

we see that

$$\sum_{n=1}^{\infty} \left(\frac{1}{z-n} + \frac{1}{n} \right)$$

is absolutely convergent. Similarly,

$$\sum_{n=1}^{\infty} \left(\frac{1}{z+n} - \frac{1}{n} \right)$$

is absolutely convergent. Fix z and consider the function $f(w) = 1/(w - z)$. This function is meromorphic; its only pole is at z, which is not an integer, and it is easy to see that $|wf(w)|$ is bounded for w sufficiently large (as in Proposition 4.3.1). By Proposition 4.4.2, we see that the hypotheses of the summation theorem are satisfied, and so

$$\lim_{N \to \infty} \sum_{n=-N}^{N} \frac{1}{n-z} = - \left\{ \text{residue of } \frac{\pi \cot \pi w}{w-z} \text{ at } w = z \right\} = -\pi \cot \pi z$$

We note that

$$\sum_{n=-N}^{N} \frac{1}{z-n} = \frac{1}{z} + \sum_{n=1}^{N} \left(\frac{1}{z-n} + \frac{1}{n} \right) + \sum_{n=1}^{N} \left(\frac{1}{z+n} - \frac{1}{n} \right)$$

and so

$$\frac{1}{z} + \sum_{n=1}^{\infty} \left(\frac{1}{z-n} + \frac{1}{n} \right) + \sum_{n=1}^{\infty} \left(\frac{1}{z+n} - \frac{1}{n} \right) = \pi \cot \pi z \qquad \blacksquare$$

We could also have obtained the expansion for cotangent from the following theorem.

4.4.5 PARTIAL-FRACTION THEOREM   *Suppose that f is meromorphic with simple poles at $a_1, a_2, a_3, \ldots$ with $0 < |a_1| \le |a_2| \le \cdots$ and residues b_k at a_k. (We are assuming f is analytic at 0.) Suppose there is a sequence $R_1, R_2, R_3, \ldots$ with the property $\lim_{n \to \infty} R_n = \infty$ and simple closed curves C_N satisfying*

(i) *$|z| \ge R_N$ for all z on C_N.*
(ii) *There is a constant S with length $(C_N) \le SR_N$ for all N.*
(iii) *There is a constant M with $|f(z)| \le M$ for all z on C_N and for all N. (The same M should work for all N.)*

Then

$$f(z) = f(0) + \sum_{n=1}^{\infty} \left(\frac{b_n}{z - a_n} + \frac{b_n}{a_n} \right)$$

PROOF   If $z_0 \ne 0$ is not a pole of f, let $F(z) = f(z)/(z - z_0)$. Then F has simple poles at z_0 and at $a_1, a_2, a_3, \ldots$. Clearly

$$\text{Res } (F; z_0) = \lim_{z \to z_0} (z - z_0)F(z) = f(z_0)$$

and

$$\text{Res } (F; a_n) = \lim_{z \to a_n} (z - a_n) \frac{f(z)}{z - z_0} = \frac{b_n}{a_n - z_0}$$

By the residue theorem,

$$\frac{1}{2\pi i} \int_{C_N} \frac{f(z)}{z - z_0} \, dz = f(z_0) + \sum \left\{ \frac{b_n}{a_n - z_0} \, \middle| \, a_n \text{ is inside } C_N \right\}$$

$$\frac{1}{2\pi i} \int_{C_N} \frac{f(z)}{z} \, dz = f(0) + \sum \left\{ \frac{b_n}{a_n} \, \middle| \, a_n \text{ is inside } C_N \right\}$$

Subtracting,

$$\frac{z_0}{2\pi i} \int_{C_N} \frac{f(z)}{z(z - z_0)} \, dz = f(z_0) - f(0) + \sum \left\{ \frac{b_n}{a_n - z_0} - \frac{b_n}{a_n} \, \middle| \, a_n \text{ is inside } C_N \right\}$$

Along C_N, $|z| \ge R_N$ and $|z - z_0| \ge |R_N - |z_0||$, and so the integral in the last equality is bounded above by

$$\frac{|z_0|}{2\pi} \frac{M}{R_N |R_N - |z_0||} \text{ [length } (C_N)] \le \frac{|z_0| MS}{2\pi |R_N - |z_0||}$$

This goes to 0 as $N \to \infty$, and each of the a_n is eventually inside C_N. Therefore

$$f(z_0) = f(0) - \lim_{N \to \infty} \left(\sum \left\{ \frac{b_n}{a_n - z_0} - \frac{b_n}{a_n} \middle| a_n \text{ is inside } C_N \right\} \right)$$

$$= f(0) - \sum_{n=1}^{\infty} \left(\frac{b_n}{a_n - z_0} - \frac{b_n}{a_n} \right) = f(0) + \sum_{n=1}^{\infty} \left(\frac{b_n}{z_0 - a_n} + \frac{b_n}{a_n} \right)$$

Since this formula holds at all z_0 for which f is analytic, we have established the theorem. ∎

Contours commonly used for the C_N are circles of radius R_N or large squares such as those in Fig. 4.4.1. The expansion given in the partial-fraction theorem is a special case of a more general result known as the Mittag-Leffler theorem,* named after the famous Swedish mathematician Gösta Mittag-Leffler (1846–1927).

Exercises

1. Show that

$$\sum_{n=1}^{\infty} \frac{1}{n^4} = \frac{\pi^4}{90}$$

2. Show that

$$\sum_{n=1}^{\infty} \left[\frac{(-1)^{n-1}}{(2n-1)^3} \right] = \frac{\pi^3}{32}$$

3. Show that

$$\sum_{n=0}^{\infty} \frac{1}{n^2 + a^2} = \frac{\pi}{2a} \coth \pi a + \frac{1}{2a^2} \qquad \text{for} \qquad a > 0$$

4. Show that

$$\frac{\pi^2}{\sin^2 \pi z} = \sum_{n=-\infty}^{\infty} \frac{1}{(z - n)^2}$$

(*Hint.* Start with the expansion for $\pi \cot \pi z$.)

---

* It may be found, for example, in Peter Henrici, *Applied and Computational Complex Analysis,* Vol. 1 (New York: Wiley-Interscience, 1974), pp. 655–660 and (New York: Springer-Verlag, 1986).

5. Develop a method for evaluating series of the form $\Sigma_{n=-\infty}^{\infty} (-1)^n f(n)$ where f is a meromorphic function in $\mathbb{C}$ with a finite number of poles none of which lie at the integers. In other words, develop theorems analogous to the summation theorem (4.4.1) and Proposition 4.4.2. (*Hint.* $\pi/\sin \pi z$ has poles at the integers with Res $(\pi/\sin \pi z, n) = (-1)^n$. Discuss how you would handle the summation if some of the poles of f did lie at the integers; see Proposition 4.4.4.)

6. Show that if $2z - 1$ is not an integer, then

$$\frac{1}{\cos \pi z} = 1 + \frac{4}{\pi} \sum_{n=1}^{\infty} \left[\frac{2z-1}{(2z-1)^2 - 4n^2} + \frac{4}{1 - 4n^2} \right]$$

(*Hint.* $\cos (x + iy) = \cos x \cosh y + i \sin x \sinh y$. Use the square with corners $\pm N \pm Ni$ for C_N given in the partial-fraction theorem (4.4.5). Finally, combine the n and $-n$ terms.)

7. Use the partial-fraction theorem to show that

$$\cot z = \frac{1}{z} + \Sigma' \left(\frac{1}{z - n\pi} + \frac{1}{n\pi} \right) = \frac{1}{z} + \sum_{n=1}^{\infty} \frac{1}{z^2 - n^2 \pi^2}$$

where Σ' means the sum is over all $n \neq 0$.

8. Prove that $1 - 1/2^2 + 1/3^2 - 1/4^2 + \cdots = \pi^2/12$.

9. Try to evaluate the sum $\Sigma_{n=1}^{\infty} (1/n^3)$. (Don't be too discouraged if you are not successful. See the remarks in the answers to odd-numbered exercises at the back of the book.)

REVIEW EXERCISES FOR CHAPTER 4

1. Evaluate

$$\int_0^{2\pi} \frac{d\theta}{2 - \sin \theta}$$

2. Evaluate $\displaystyle\int_\gamma \frac{1}{(z - 1)(z - 2)} \, dz$ where

(a) γ is the circle with center 0 and radius $\frac{1}{2}$ traveled once counterclockwise.
(b) Same as (a) but radius $\frac{3}{2}$.
(c) Same as (a) but radius $\frac{5}{2}$.

3. Evaluate

$$\int_{-\infty}^{\infty} \frac{x^2}{x^4 + 1} \, dx$$

4. Evaluate $\int_C z^n e^{1/z} \, dz$ if C is the unit circle centered at 0 and n is a positive integer.

5. Compute

$$\int_{-\infty}^{\infty} \frac{\sin x}{x(x+1)(x^2+1)} \, dx$$

6. Evaluate

$$\int_{-\infty}^{\infty} \frac{1}{x^6 + 1} \, dx$$

7. Evaluate

$$\int_0^{\pi} \frac{d\theta}{2 \cos \theta + 3}$$

8. Let f be analytic on a region containing the upper half plane $\{z \mid \text{Im } z \geq 0\}$. Suppose that for some $a > 0$, $|f(z)| \leq M/|z|^a$ for $|z|$ large. Show that for Im $z > 0$,

$$f(z) = \frac{1}{2\pi i} \int_{-\infty}^{\infty} \frac{f(x)}{x - z} \, dx$$

9. Evaluate

$$\int_0^{2\pi} \exp\left(e^{i\theta}\right) d\theta$$

10. Show that

$$\frac{2}{\pi} \int_0^{\infty} \frac{\sin kt}{t} \, dt = \begin{cases} 1 & k > 0 \\ 0 & k = 0 \\ -1 & k < 0 \end{cases}$$

11. Evaluate

$$\int_{|z|=1} \frac{\cos (e^{-z})}{z^2} \, dz$$

12. Show that

$$\int_0^{\infty} \frac{x^{m-1}}{1 + x^n} \, dx = \frac{\pi}{n \sin (m\pi/n)}$$

where $0 < m < n$.

13. Find the Laurent expansions of

$$f(z) = \frac{1}{(z-1)(z-2)}$$

that are valid for (a) $0 < |z| < 1$ and (b) $|z| > 2$. Choose $z_0 = 0$.

14. Show that

$$\int_0^\infty \text{sech } x \, dx = \frac{\pi}{2}$$

(*Hint.* Consider the rectangle with corners at $(\pm R, \pm R + \pi i)$.)

15. What is the radius of convergence of the Taylor series of $1/\cos z$ around $z = 0$?

16. Explain what is wrong with the following reasoning. We know that $a^z = e^{z \log a}$, so $da^z/dz = (\log a)a^z$. On the other hand, $da^z/dz = za^{z-1}$. Thus $za^{z-1} = a^z(\log a)$, so $z = a \log a$.

17. Find the residues of the following at each singularity:

(a) $\dfrac{z}{1 - e^{z^2}}$ (b) $\dfrac{\sin (z^2)}{(\sin z)^7}$

(c) $\sin (e^{1/z})$

18. Where is $\sum_{n=0}^{\infty} z^n e^{-izn}$ analytic?

19. Let $f(z)$ have a zero of order k at z_0. Show that Res $(f'/f, z_0) = k$. Find $(f''/f', z_0)$ and Res $(f''/f, z_0)$.

20. Let f be entire and suppose that Re f is a polynomial in x, y. Prove that f is a polynomial.

21. Explain what is wrong with the following argument, then compute the residue correctly. The expansion

$$\frac{1}{z(z-1)^2} = \frac{1}{(z-1)^2} \cdot \frac{1}{1 + (z-1)} = \cdots + \frac{1}{(z-1)^5} - \frac{1}{(z-1)^4} + \frac{1}{(z-1)^3}$$

is the Laurent expansion; since there is no term in $1/(z-1)$, the residue at $z = 1$ is zero.

22. Verify the maximum principle for harmonic functions and the minimum principle for harmonic functions for the harmonic function $u(x, y) = x^2 - y^2$ on $[0, 1] \times [0, 1]$.

23. Evaluate

$$\int_{\gamma} \frac{1}{z(z-1)(z-2)} \, dz$$

where γ is the circle centered at 0 with radius $\frac{3}{2}$.

24. Same as Exercise 23 but with radius $\frac{1}{2}$.

25. Determine the radius of convergence of the following series:

(a) $\displaystyle\sum_{1}^{\infty} \frac{\log{(n^n)}}{n!} z^n$

(b) $\displaystyle\sum_{1}^{\infty} \left(1 - \frac{1}{n}\right)^n z^n$

26. Establish the following:

$$\int_{0}^{\infty} \frac{\sinh ax}{\sinh \pi x} \, dx = \frac{1}{2} \tan \frac{a}{2} \qquad \text{for} \qquad -\pi < a < \pi$$

(*Hint.* Integrate $e^{az}/\sinh{(\pi z)}$ over a "square" with sides $y = 0$, $y = 1$, $x = -R$, $x = +R$, and circumvent the singularities at 0, i.)

27. Expand the following in Laurent series as indicated: $f(z) = \left(\dfrac{1}{1-z}\right)^3$

(a) for $|z| < 1$; $z_0 = 0$

(b) for $|z| > 1$; $z_0 = 0$

(c) for $|z + 1| < 2$; $z_0 = -1$

(d) for $0 < |z - 1| < \infty$; $z_0 = 1$

28. Show that

$$\int_{0}^{\pi/2} \frac{d\theta}{(a + \sin^2 \theta)^2} = \frac{\pi(2a + 1)}{4(a^2 + a)^{3/2}} \qquad \text{for} \qquad a > 0$$

29. Establish the following formulas:

(a) $\displaystyle\int_{0}^{\infty} \frac{\sin^3 x}{x^3} \, dx = \frac{3\pi}{8}$

(b) $\displaystyle\int_{0}^{\infty} \frac{x^a}{x^2 + b^2} \, dx = \frac{\pi b^{a-1}}{2 \cos{(\pi a/2)}} \qquad \text{for} \qquad -1 < a < 1$

30. Prove that

$$\tan z = 2z \sum_{0}^{\infty} \frac{1}{(n + \frac{1}{2})^2 \pi^2 - z^2} \qquad \text{for} \qquad z \neq (n + \frac{1}{2})\pi$$

(*Hint.* Start with the identity for $\cot z$ in Proposition 4.4.4 and use $\tan z = \cot z - 2 \cot 2z$.)

31. Prove that

$$\sum_{-\infty}^{\infty} \frac{(-1)^n}{(a + n)^2} = \pi^2 \csc{(\pi a)} \cot{(\pi a)}$$

32. Evaluate

$$\sum_{n=1}^{\infty} \frac{1}{n^6}$$

33. Explain what is wrong with the following reasoning:

$$\int_0^\infty \frac{\sin x}{x}\,dx = \frac{1}{2} \int_{-\infty}^\infty \frac{\sin x}{x}\,dx = \frac{1}{2} \lim_{R\to\infty} \int_{\gamma_R} \frac{\sin z}{z}\,dz$$

where γ_R is the x axis from $-R$ to R plus the circumference $\{z = Re^{i\theta} \mid 0 \le \theta \le \pi\}$. But $(\sin z)/z$ is analytic everywhere, including zero, and so by Cauchy's theorem,

$$\int_{\gamma_R} \frac{\sin z}{z}\,dz = 0$$

Hence

$$\int_0^\infty \frac{\sin x}{x}\,dx = 0$$

34. Let $f(z)$ be analytic inside and on a simple closed contour γ. For z_0 on γ, and γ differentiable near z_0, show that

$$f(z_0) = \frac{1}{\pi i}\,\mathrm{P.V.} \int_\gamma \frac{f(\zeta)}{\zeta - z_0}\,d\zeta$$

35. Use Exercise 34 to find sufficient conditions under which

$$f(x, 0) = \frac{1}{\pi i}\,\mathrm{P.V.} \int_{-\infty}^\infty \frac{f(\zeta, 0)}{\zeta - x}\,d\zeta$$

for $f(x, y) = f(z)$ analytic. Deduce that

$$u(x, 0) = \frac{1}{\pi}\,\mathrm{P.V.} \int_{-\infty}^\infty \frac{v(\zeta, 0)}{\zeta - x}\,d\zeta \quad \text{and} \quad v(x, 0) = -\frac{1}{\pi}\,\mathrm{P.V.} \int_{-\infty}^\infty \frac{u(\zeta, 0)}{\zeta - x}\,d\zeta$$

(*Note. u* and *v* are called *Hilbert transforms* of one another.)

36. Show that

$$\frac{1}{\sin z} = \frac{1}{z} + \sum'(-1)^n \left(\frac{1}{z - n\pi} + \frac{1}{n\pi}\right) = \frac{1}{z} + 2z \sum_{n=1}^{\infty} \frac{(-1)^n}{z^2 - n^2\pi^2}$$

where Σ' means the sum is taken over all $n \neq 0$.

37. Evaluate $\displaystyle\int_0^\infty \frac{\sin \omega x}{\sinh bx}\,dx.$

38. When a nonlinear oscillator is forced with a frequency ω, a measure of the oscillator's "chaotic response" is given by*

$$M = \int_0^\infty \operatorname{sech} bt \cos \omega t\, dt$$

Show that $M = (\pi/2b) \operatorname{sech}(\omega\pi/2b)$.

---

* See J. Guckenheimer and P. Holmes, *Nonlinear Oscillations, Dynamical Systems and Bifurcations of Vector Fields* (New York: Springer-Verlag, 1983), sec. 4.5.

CHAPTER 5

CONFORMAL MAPPINGS

Chapter 1 included a brief investigation of some geometric aspects of analytic functions. Now we return to this topic to develop some further techniques. We want to be able to map one given region to another given region by a one-to-one, onto, analytic function. Section 5.2 discusses several concrete cases for which such mappings can be written explicitly. That such mappings always exist in theory is the statement of the famous Riemann mapping theorem, which is discussed but not proved in Sec. 5.1. A proof is given in a supplement to Chap. 6, but that material is not required for understanding the main theory and applications in this chapter or the rest of the text, so may be omitted if desired.

The theory of conformal mappings has several important applications to the Dirichlet problem and to harmonic functions. These applications are used in problems of heat conduction, electrostatics, and hydrodynamics, which will be discussed in Sec. 5.3. The basic idea of such applications is that a conformal mapping can be used to map a given region to a simpler region on which the problem can be solved by inspection. By transforming back to the original region, the desired answer is obtained.

5.1 BASIC THEORY OF CONFORMAL MAPPINGS

Conformal Transformations

The following definition was presented in Sec. 1.5: A mapping $f: A \to B$ is called *conformal* if, for each $z_0 \in A$, f rotates tangent vectors to curves through z_0 by a definite angle θ and stretches them by a definite factor r. Let us also recall the following theorem proved in Sec. 1.5.

5.1.1 CONFORMAL MAPPING THEOREM *Let* $f: A \to B$ *be analytic and let* $f'(z_0) \neq 0$ *for each* $z_0 \in A$. *Then* f *is conformal.*

Actually, if f merely preserves angles and if certain conditions of regularity hold, then f must be analytic and $f'(z_0) \neq 0$ (see Exercise 8). Therefore, we can say that "conformal" means *analytic with a nonzero derivative.* We shall find it convenient to assume this meaning in the remainder of this text.

Let $A = \{z | \text{Re } z > 0 \text{ and Im } z > 0\}$ and $B = \{z | \text{Im } z > 0\}$. Then the map $f : A \to B$ defined by $z \mapsto z^2$ is conformal. (The student can easily check this.) Figure 5.1.1 illustrates the theorem by showing preservation of angles in this case. If $f'(z_0) = 0$, angles need not be preserved. For instance, for the map $z \mapsto z^2$, the x and y axes intersect at an angle $\pi/2$ but the images intersect at an angle π. Such a point where $f'(z_0) = 0$ for an analytic function f is called a *singular point.* Singular points are studied in greater detail in Chap. 6.

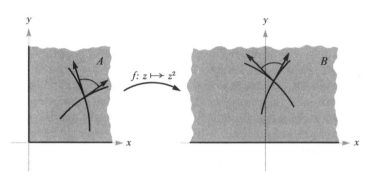

FIGURE 5.1.1   A conformal map.

5.1.2 PROPOSITION

(i) *If* $f : A \to B$ *is conformal and bijective (that is, one-to-one and onto), then* $f^{-1} : B \to A$ *is also conformal.*

(ii) *If* $f : A \to B$ *and* $g : B \to C$ *are conformal and bijective, then* $g \circ f : A \to C$ *is conformal and bijective.*

PROOF

(i) Since f is bijective, the mapping f^{-1} exists. By the inverse function theorem (1.5.10), f^{-1} is analytic with $df^{-1}(w)/dw = 1/[df(z)/dz]$ where $w = f(z)$. Thus $df^{-1}(w)/dw \neq 0$, so f^{-1} is conformal.

(ii) Certainly $g \circ f$ is bijective and analytic, since g and f are. (The inverse of $g \circ f$ is $f^{-1} \circ g^{-1}$.) The derivative of $g \circ f$ at z is $g'(f(z)) \cdot f'(z) \neq 0$. Therefore, $g \circ f$ is conformal by definition. ■

Because of the two properties in Proposition 5.1.2 (and the obvious fact that the identity map $z \mapsto z$ is conformal), we can justifiably refer to the set of bijective conformal maps of a fixed region to itself as a *group.*

Property (i) is important because we will be able to use it to solve various problems (such as the Dirichlet problem) associated with a given region A. The method will be to find a bijective conformal map $f: A \to B$ where B is a simpler region on which the problem can be solved. To obtain the answer on A we then transform our answer from B to A by f^{-1}. The Dirichlet problem involves harmonic functions, so we should check that harmonic functions remain harmonic when we compose them with a conformal map. To do so we prove the following result.

5.1.3 PROPOSITION *Let u be harmonic on a region D and let $f: A \to D$ be analytic. Then $u \circ f$ is harmonic on A.*

PROOF Let $z \in A$ and $w = f(z)$. Let U be an open disk in B around w and let $V = f^{-1}(U)$. It suffices to show that $u \circ f$ is harmonic on V (Why?). By Proposition 2.5.8, there is an analytic function g on U such that $u = \text{Re } g$. Then $u \circ f = \text{Re } (g \circ f)$ (Why?), and we know that $g \circ f$ is analytic by the chain rule. Thus $\text{Re } (g \circ f)$ is harmonic. ∎

Riemann Mapping Theorem

There is a basic but more sophisticated theorem concerning conformal mappings that guarantees the existence of such mappings between given regions A and B. The validity of this theorem in several special cases is verified in the next section. The general theorem is not always of immediate practical value, because it does not tell us explicitly how to find conformal maps. Nevertheless, it is an important theorem that we should be aware of. We shall prove uniqueness here, but leave existence to the supplement to Chap. 6; see also E. Hille, *Analytic Function Theory*, Vol. II (Boston: Ginn, 1959), p. 322, or L. Ahlfors, *Complex Analysis* (New York: McGraw-Hill, 1966), p. 222.

5.1.4 RIEMANN MAPPING THEOREM *Let A be a simply connected region such that $A \neq \mathbb{C}$. Then there exists a bijective conformal map $f: A \to D$ where $D = \{z$ such that $|z| < 1\}$. Furthermore, for any fixed $z_0 \in A$, we can find an f such that $f(z_0) = 0$ and $f'(z_0) > 0$. With such a specification, f is unique.*

From this result we see that if A and B are any two simply connected regions with $A \neq \mathbb{C}$, $B \neq \mathbb{C}$, then there is a bijective conformal map $g: A \to B$. Indeed, if $f: A \to D$ and $h: B \to D$ are conformal, we can set $g = h^{-1} \circ f$ (see Fig. 5.1.2). Two regions A and B are called *conformal* if there is a bijective conformal map from A to B. Thus the Riemann mapping theorem implies that two simply connected regions (unequal to $\mathbb{C}$) are conformal.

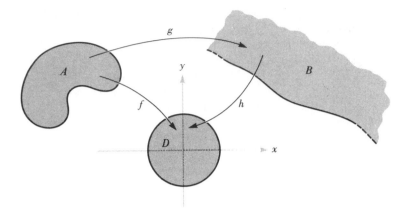

FIGURE 5.1.2   To transform A to B we compose h^{-1} with f.

PROOF OF UNIQUENESS IN THEOREM 5.1.4   Suppose f and g are bijective conformal maps of A onto D with $f(z_0) = g(z_0) = 0$, $f'(z_0) > 0$, and $g'(z_0) > 0$. We want to show that $f(z) = g(z)$ for all z in A. To do this, define h on D by $h(w) = g(f^{-1}(w))$ for $w \in D$. Then $h: D \to D$ and $h(0) = g(f^{-1}(0)) = g(z_0) = 0$. By the Schwarz lemma (2.5.7), $|h(w)| \le |w|$ for all $w \in D$. Exactly the same argument applies to $h^{-1} = f \circ g^{-1}$, so that $|h^{-1}(\zeta)| \le |\zeta|$ for all $\zeta \in D$. With $\zeta = h(w)$, this gives $|w| \le |h(w)|$. Combining these inequalities, we get $|h(w)| = |w|$ for all $w \in D$. The Schwarz lemma now tells us that $h(w) = cw$ for a constant c with $|c| = 1$. Thus $cw = g(f^{-1}(w))$. With $z = f^{-1}(w)$ we obtain $cf(z) = g(z)$ for all $z \in A$. In particular, $cf'(z_0) = g'(z_0)$. Since both $f'(z_0)$ and $g'(z_0)$ are positive real numbers, so is c. Thus $c = 1$ and so $f(z) = g(z)$, as desired. ∎

The condition $f'(z_0) > 0$ is equivalent to saying that $\arg f'(z_0) = 0$. Using the preceding argument, one can modify the uniqueness assertion so that $f(z_0)$ and $\arg f'(z_0)$ are specified. The student is asked to prove this in Exercise 7.

Here is another useful fact we should know about conformal maps. Let A and B be two (connected) regions with boundaries bd (A) and bd (B). Suppose that $f: A \to f(A)$ is conformal. If $f(A)$ *has boundary* bd (B) *and if, for some $z_0 \in A$, we have $f(z_0) \in B$, then $f(A) = B$. In other words, to determine the image of a conformal map, we merely need to check the boundaries and a single point inside.* To prove this we argue as follows. Since B is open, $B \cap$ bd $(B) = \varnothing$. The closure of B is $B \cup$ bd (B), so we can decompose the plane as a disjoint union $\mathbb{C} = B \cup$ bd $(B) \cup$ ext B, where ext B is open. Since f' never vanishes on A, the inverse function theorem shows that $f(A)$ is open. Thus $f(A) \cap$ bd $(f(A)) = \varnothing$. But bd $(f(A)) =$ bd (B), so $f(A)$ is contained in the union of the disjoint open sets B and ext B. Since f is continuous on the connected set A, $f(A)$ is connected. Therefore either

$f(A) \subset B$ or $f(A) \subset$ ext B. As $f(z_0) \in B$, we must have $f(A) \subset B$. Since $f(A)$ is open, it is open relative to B. Finally,

$$f(A) = f(A) \cap B = [f(A) \cap B] \cup [\text{bd } (B) \cap B]$$
$$= [f(A) \cap B] \cup [\text{bd } (f(A)) \cap B]$$
$$= [f(A) \cup \text{bd } (f(A))] \cap B = \text{cl } (f(A)) \cap B$$

so that $f(A)$ is closed relative to B. Since B is connected, $f(A) = B$ (see Proposition 1.4.13).

Simple connectivity is essential in the Riemann mapping theorem. It is easy to show (see Worked Example 5.1.7) that only a simply connected region can be mapped bijectively by an analytic map onto D. A related result that can be shown is that the annuli $0 < |z| < 1$ and $1 < |z| < 2$ are not conformal; see Worked Example 6.1.14.

Behavior on the Boundary

The Riemann mapping theorem and most of our other remarks about conformal maps have discussed behavior on *regions* which are *open, connected* sets. In particular the Riemann mapping theorem does not say what happens on the boundary of A or of D. Many of the applications, however, involve finding something inside a region from information on the boundary. Thus the behavior of conformal maps at the boundary may be important. In the next section we will look at many concrete examples involving such regions as disks, halfplanes, quarter planes, and so on, and the maps will usually be well behaved on the boundary. This is no accident, as the next theorem shows, but it is not automatic. The connected, simply connected open sets to which the Riemann mapping theorem applies can be rather complicated. For example, consider the set A obtained by deleting from the square $S = \{z | 0 < \text{Re } z < 2 \text{ and } 0 < \text{Im } z < 2\}$ the vertical segments $J_n = \{z = 1/n + yi | 0 \le y \le 1\}$, $n = 1, 2, 3, \ldots$. (See Fig. 5.1.3.) The

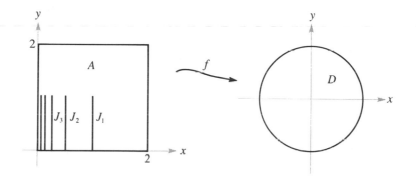

FIGURE 5.1.3 Even though A has a wild boundary, it can be mapped conformally to D.

Riemann mapping theorem guarantees that there is a conformal map of A onto D, but attempting to extend it continuously to the boundary of A, particularly to 0, creates problems. (A detailed description of the boundary behavior of conformal maps may be found in the book *Theory of Functions of a Complex Variable* by A. I. Markushevich (New York: Chelsea Publ. Co., 1977), Vol. 3, Chap. 2.) For well-behaved regions there is a nice result, which we state without proof:

5.1.5 OSGOOD-CARATHEODORY THEOREM   *If A_1 and A_2 are bounded simply connected regions whose boundaries γ_1 and γ_2 are simple continuous closed curves, then any conformal map of A_1 one-to-one onto A_2 can be extended to a continuous map of $A_1 \cup \gamma_1$ one-to-one onto $A_2 \cup \gamma_2$.*

Once the boundaries are known to be mapped continuously we can get information about the regions themselves. The next theorem outlines such a procedure. The conditions are restrictive enough that we do not need to check a point $z_0 \in A$.

5.1.6 THEOREM   *Let A be a bounded region with $f : A \to \mathbf{C}$ a bijective conformal map onto its image $f(A)$. Suppose that f extends to be continuous on cl (A) and that f maps the boundary of A onto a circle of radius R. Then $f(A)$ equals the inside of that circle. More generally, if B is a bounded region that, together with its boundary, can be mapped conformally onto the unit disk and its boundary and if f maps* bd (A) *onto* bd (B), *then $f(A) = B$.*

PROOF   By composing f with the conformal map h that takes B to the unit disk it is sufficient to consider the special case in which B equals $D = \{z \text{ such that } |z| < 1\}$. On bd (A), $|f(z)| = 1$, and so by the maximum modulus theorem, $|f(z)| \leq 1$ on A. Since f cannot be constant, $f(A) \subset D$. In other words, at no $z \in A$ is the maximum $|f(z)| = 1$ reached. We have assumed that $f(\text{bd}(A)) = \text{bd}(D)$, but this is also equal to bd $(f(A))$. To see this, use compactness of cl (A), continuity of f, and $D \cap \text{bd}(D) = f(A) \cap \text{bd}(f(A)) = \varnothing$. Thus our earlier argument applies to show that $f(A) = D$. ∎

Worked Examples

5.1.7   Find a bijective conformal map that takes a bounded region to an unbounded region. Can you find one that takes a simply connected region to a region that is not simply connected?

Solution. Consider $f(z) = 1/z$ on $A = \{z \mid 0 < |z| < 1\}$. Clearly, A is bounded. Also, $B = f(A) = \{z \text{ such that } |z| > 1\}$; f is conformal from A to B and has an inverse $g^{-1}(w) = 1/w$. But B is unbounded.

The answer to the second part of the question is no. If A is simply connected and $f : A \to B$ is a bijective conformal map, then B must be simply connected. To show this, let γ be a closed curve in B and let $\tilde{\gamma} = f^{-1} \circ \gamma$. Then if $H(t, s)$ is a homotopy shrinking $\tilde{\gamma}$ to a point, $f \circ H(t, s)$ is a homotopy shrinking γ to a point.

5.1.8 *Consider the harmonic function $u(x, y) = x + y$ on the region $A = \{z \mid 0 < \operatorname{Im} z < 2\pi\}$. What is the corresponding harmonic function in $B = \mathbb{C} \backslash (\text{positive real axis})$ when A is transformed by $z \mapsto e^z$?*

Solution. Let $f(z) = e^z$. We know from Chap. 1 that f is one-to-one, onto B, and that $f'(z) = e^z \neq 0$. Thus f is conformal from A to B, and therefore, by Proposition 5.1.3, the corresponding function on B is harmonic. This function is:

$$v(x, y) = u(f^{-1}(x, y)) = u(\log(x + iy))$$

$$= u\left(\log \sqrt{x^2 + y^2} + i \tan^{-1} \frac{y}{x}\right)$$

$$= \log \sqrt{x^2 + y^2} + \tan^{-1} \frac{y}{x}$$

where $\tan^{-1}(y/x) = \arg(x + iy)$ lies in $]0, 2\pi[$. Note that to check directly that v is harmonic would be slightly tedious, but we know it must be so by Proposition 5.1.3.

5.1.9 *What is the image of the region $A = \{z \mid (\operatorname{Re} z)(\operatorname{Im} z) > 1 \text{ and } \operatorname{Re} z > 0, \operatorname{Im} z > 0\}$ under the transformation $z \mapsto z^2$?*

Solution. On the right half plane $\{z \mid \operatorname{Re} z > 0\}$, we know that $f(z) = z^2$ is conformal (Why?). To find the image of A we first find the image of the curve $xy = 1$. Let $w = z^2 = u + iv$. Then $u = x^2 - y^2$, $v = 2xy$. Thus the image of $xy = 1$ is the curve $v = 2$. We must check the location of the image of a point in A, say, $z = 2 + 2i$. Here $z^2 = 8i$, and therefore the image is the shaded region B in Fig. 5.1.4.

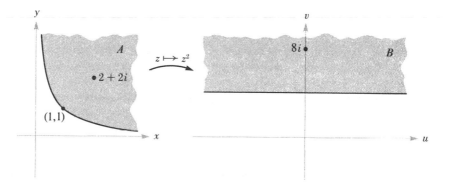

FIGURE 5.1.4 Image of the set A under the conformal map $z \mapsto z^2$.

Exercises

1. What is the image of the first quadrant under the mapping $z \mapsto z^3$?

2. Consider $f = u + iv$ where $u(x, y) = 2x^2 + y^2$ and $v = y^2/x$. Show that the curves $u = $ constant and $v = $ constant intersect orthogonally but that f is not analytic.

3. Near what points are the following maps conformal?
    (a) $f(z) = z^3 + z^2$ (b) $f(z) = z/(1 + 5z)$

4. Near what points are the following maps conformal?
    (a) $f(z) = \bar{z}$ (b) $f(z) = (\sin z)/(\cos z)$

5. Consider the harmonic function $u(x, y) = 1 - y + x/(x^2 + y^2)$ on the upper half plane $y > 0$. What is the corresponding harmonic function on the first quadrant $x > 0, y > 0$, under the transformation $z \mapsto z^2$?

6. Let A and B be regions whose boundaries are smooth arcs. Let f be conformal on a region including $A \cup \mathrm{bd}\,(A)$ and map A onto B and $\mathrm{bd}\,(A)$ onto $\mathrm{bd}\,(B)$. Let u be harmonic on B and $u = h(z)$ for z on the boundary of B. Let $v = u \circ f$ so that v equals $h \circ f$ on the boundary of A. Prove that $\partial v/\partial n = 0$ at z_0 iff $\partial u/\partial n = 0$ at $f(z_0)$ where $z_0 \in \mathrm{bd}\,(A)$ and $\partial/\partial n$ denotes the derivative in the normal direction to the boundary.

7. Let A and B be regions as in the Riemann mapping theorem. Given $z_0 \in A$, $w_0 \in B$, and an angle θ_0, and by assuming this theorem, show that there exists a conformal map $f : A \to B$ with $f(z_0) = w_0$ and $\arg f'(z_0) = \theta_0$; also show that such an f is unique.

8. Let $f : A \to B$ be a function such that $\partial f/\partial x$ and $\partial f/\partial y$ exist and are continuous. Suppose that f is one-to-one and onto and preserves angles; prove that f is analytic and conformal. Can the map in Exercise 2 preserve all angles? (*Hint.* Let $c(t)$ be a curve with $c(0) = z_0$ and let $d(t) = f(c(t))$. Prove

$$d'(t) = \frac{1}{2}\left(\frac{\partial f}{\partial x} - i\frac{\partial f}{\partial y}\right)c'(t) + \frac{1}{2}\left(\frac{\partial f}{\partial x} + i\frac{\partial f}{\partial y}\right)\overline{c'(t)}$$

and examine the statement that $d'(0)/c'(0)$ has constant argument in order to establish the Cauchy-Riemann equations for f.)

9. If $f : A \to B$ is bijective and analytic with an analytic inverse, prove that f is conformal.

10. Let $f : \mathbf{C} \to \mathbf{C}$, $z \mapsto az + b$. Show that f can be written as a rotation followed by a magnification followed by a translation.

11. The Riemann mapping theorem explicitly excludes the case $A = \mathbf{C}$ from consideration.
    (a) Is there a conformal map of $\mathbf{C}$ one-to-one onto the unit disk D?
    (b) Is there a conformal map of D one-to-one onto $\mathbf{C}$?

12. Show that every bijective conformal transformation of $\mathbb{C}$ onto $\mathbb{C}$ is of the type described in Exercise 10.

13. Suppose that f is a conformal map from a bounded region A onto an unbounded region B. Show that f cannot be extended in such a way as to be continuous on $A \cup \text{bd}(A)$. (Note: The full force of conformality is not needed in this problem.)

5.2 FRACTIONAL LINEAR AND SCHWARZ-CHRISTOFFEL TRANSFORMATIONS

This section investigates some ways of obtaining specific conformal maps between two given regions. No general prescription can be given for obtaining these maps; however, after a little practice the student will be able to combine fractional linear transformations (studied in this section) with other familiar transformations (like z^2, e^z, or $\sin z$) and thus be able to handle many useful situations. To aid in this effort, some common transformations are illustrated in Fig. 5.2.11 at the end of the section. In addition, the Schwarz-Christoffel formula will be studied briefly, even though it yields answers that usually can only be given in terms of integrals.

Fractional Linear Transformations

The simplest and one of the most useful kinds of conformal mappings will be discussed first. A *fractional linear transformation* (also called a bilinear transformation or Möbius transformation) is a mapping of the form

$$T(z) = \frac{az + b}{cz + d} \tag{1}$$

where a, b, c, d are fixed complex numbers. We shall assume that $ad - bc \neq 0$, because otherwise T would be a constant (Why?) and we want to omit that case. The properties of these transformations will be developed in the next four propositions.

5.2.1 PROPOSITION *The map T defined by Eq. (1) is bijective and conformal from*

$$A = \left\{ z \,\middle|\, cz + d \neq 0, \text{ that is, } z \neq \frac{-d}{c} \right\} \text{ onto } B = \left\{ w \,\middle|\, w \neq \frac{a}{c} \right\}$$

In fact, the inverse of T is also a fractional linear transformation given by

$$T^{-1}(w) = \frac{-dw + b}{cw - a} \tag{2}$$

PROOF   Certainly T is analytic on A and $S(w) = (-dw + b)/(cw - a)$ is analytic on B. The map T will be bijective if we can show that $T \circ S$ and $S \circ T$ are the identities since this means that T has S as its inverse. Indeed, this is seen in this computation:

$$T(S(w)) = \frac{a\left(\dfrac{-dw + b}{cw - a}\right) + b}{c\left(\dfrac{-dw + b}{cw - a}\right) + d}$$

$$= \frac{-adw + ab + bcw - ab}{-cdw + bc + dcw - da}$$

$$= \frac{(bc - ad)w}{bc - ad} = w$$

We can cancel because $cw - a \neq 0$ and $bc - ad \neq 0$. Similarly, $ST(z) = z$. Finally, $T'(z) \neq 0$ because

$$\frac{d}{dz} S(T(z)) = \frac{d}{dz} z = 1$$

and so

$$S'(T(z)) \cdot T'(z) = 1$$

Therefore, $T'(z) \neq 0$. ∎

It is sometimes convenient to write $T(-d/c) = \infty$ (although we must, as always, be careful to avoid the erroneous answers that we would obtain if we canceled ∞/∞ or $0/0$). In fact, we can show that all fractional linear transformations are conformal maps of the extended plane $\overline{\mathbf{C}}$ to itself. Some special cases should be noted. For example, if $a = 1$, $c = 0$, and $d = 1$, we get $T(z) = z + b$, which is a translation or "shift" that merely translates by the vector b (see Fig. 5.2.1). In case $b = c = 0, d = 1$, T becomes $T(z) = az$.

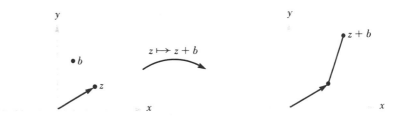

FIGURE 5.2.1   Translation.

This map, multiplication by a, is a rotation by arg a and magnification by $|a|$. The student should review the geometric meaning in this case. Finally, $T(z) = 1/z$ is an inversion. It is pictured in Fig. 5.2.2.

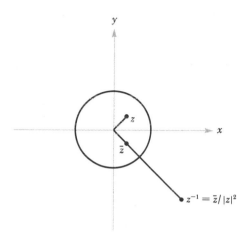

FIGURE 5.2.2   Inversion.

5.2.2 PROPOSITION   *Any conformal map of $D = \{z$ such that $|z| < 1\}$ onto itself is a fractional linear transformation of the form.*

$$T(z) = e^{i\theta} \frac{z - z_0}{1 - \bar{z}_0 z} \tag{3}$$

for some fixed $z_0 \in D$ and $\theta \in [0, 2\pi[$; moreover, any T of this form is a conformal map of D onto D.

PROOF   First we check that for T of this form, $|z| = 1$ implies that $|T(z)| = 1$. Indeed,

$$|T(z)| = \left| \frac{z - z_0}{1 - \bar{z}_0 z} \right| = \frac{|z - z_0|}{|z| \, |z^{-1} - \bar{z}_0|}$$

But $|z| = 1$ and so $z^{-1} = \bar{z}$. Hence we get

$$|T(z)| = \frac{|z - z_0|}{|\bar{z} - \bar{z}_0|} = 1$$

since $|w| = |\bar{w}|$. The only singularity of T is at $z = \bar{z}_0^{-1}$, which lies outside the unit circle.

Thus by the maximum modulus theorem, T maps D to D. But by Proposition 5.2.1,

$$T^{-1}(w) = e^{-i\theta} \left[\frac{w - (-e^{i\theta}z_0)}{1 - (-e^{-i\theta}\bar{z}_0)w} \right]$$

which, since it has the same form as T, is also a map from D to D. Thus T is conformal from D onto D.

Let $R : D \to D$ be any conformal map. Let $z_0 = R^{-1}(0)$ and let $\theta = \arg R'(z_0)$. The map T defined by Eq. (3) also has $T(z_0) = 0$ and $\theta = \arg T'(z_0)$; indeed,

$$T'(z) = e^{i\theta} \left[\frac{1 - |z_0|^2}{(1 - \bar{z}_0 z)^2} \right]$$

which, at $z = z_0$, equals

$$e^{i\theta} \left(\frac{1}{1 - |z_0|^2} \right)$$

a real constant times $e^{i\theta}$. Thus, by uniqueness of conformal maps (see the Riemann mapping theorem (5.1.4) and Exercise 7, Sec. 5.1), $R = T$. ∎

The result of this is that the only way to map a disk onto itself conformally is by means of a fractional linear transformation. These transformations have two additional properties, as will be shown in the two results that follow.

5.2.3 PROPOSITION *Let T be a fractional linear transformation. If $L \subset \mathbf{C}$ is a straight line and $S \subset \mathbf{C}$ is a circle, then $T(L)$ is either a straight line or a circle and $T(S)$ is either a straight line or a circle.*

A line can map either to a circle or to a line. If we regard lines as circles of infinite radius, then this result can be summarized by saying that circles transform into circles.

PROOF We can write $T = T_4 \circ T_3 \circ T_2 \circ T_1$, where $T_1(z) = z + d/c$, $T_2(z) = 1/z$, $T_3(z) = (bc - ad)z/c^2$, and $T_4(z) = z + a/c$. (If $c = 0$, we merely write $T(z) = (a/d)z + b/d$.) This is easily verified (see Exercise 11). It is obvious that T_1, T_3, and T_4 map lines to lines and circles to circles. Thus if we can verify the conclusion for $T(z) = 1/z$, the proof will be complete. We know from analytic geometry that a line or circle is determined by the equation

$$Ax + By + C(x^2 + y^2) = D$$

for constants A, B, C, D, with not all A, B, C zero. Let $z = x + iy$, suppose that $z \neq 0$, and let $1/z = u + iv$, so that $u = x/(x^2 + y^2)$ and $v = -y/(x^2 + y^2)$. Thus the preceding equation is equivalent to

$$Au - Bv - D(u^2 + v^2) = -C$$

which is also a line or a circle. ∎

Another property of fractional linear transformations is described in the next result.

5.2.4 CROSS-RATIOS *Given two sets of distinct points z_1, z_2, z_3 and w_1, w_2, w_3 (that is, $z_1 \neq z_2$, $z_1 \neq z_3$, $z_2 \neq z_3$ and $w_1 \neq w_2$, $w_2 \neq w_3$, $w_1 \neq w_3$, but we could have $z_1 = w_2$, and so on), there is a unique fractional linear transformation T taking $z_i \mapsto w_i$, $i = 1, 2, 3$. In fact, if $T(z) = w$, then*

$$\frac{w - w_1}{w - w_2} \cdot \frac{w_3 - w_2}{w_3 - w_1} = \frac{z - z_1}{z - z_2} \cdot \frac{z_3 - z_2}{z_3 - z_1} \tag{4}$$

The student will find that instead of trying to remember Eq. (4) it is often easier to proceed directly (see Worked Example 5.2.13).

PROOF Equation (4) defines a fractional linear transformation $w = T(z)$ (Why?). By direct substitution we see that it has the desired properties $T(z_i) = w_i$, $i = 1, 2, 3$. (See Exercise 20.) Let us show that it is unique. Define

$$S(z) = \frac{z - z_1}{z - z_2} \cdot \frac{z_3 - z_2}{z_3 - z_1}$$

Then S is a fractional linear transformation taking z_1 to 0, z_3 to 1, and z_2 to ∞. (z_2 is the singularity of S.) Let R be any other fractional linear transformation $R(z) = (az + b)/(cz + d)$ with $R(z_1) = 0$, $R(z_3) = 1$, and $R(z_2) = \infty$ (that is, $cz_2 + d = 0$). Then $az_1 + b = 0$, $cz_2 + d = 0$, and $(az_3 + b)/(cz_3 + d) = 1$. Thus we get $a = -b/z_1$ and $c = -d/z_2$, so that the last condition gives $b(z_1 - z_3)/z_1 = d(z_2 - z_3)/z_2$. Substituting in R we see, after simplification (which the student should do), that $R = S$.

We use this result to prove that T is unique as follows. Let T be any fractional linear transformation taking z_i to w_i, $i = 1, 2, 3$. The fractional linear transformation ST^{-1} takes $w_1 = Tz_1$ to 0, $w_3 = Tz_3$ to 1, and $w_2 = Tz_2$ to ∞. Therefore, ST^{-1} is uniquely determined by the preceding computation. Hence T is uniquely determined since $T = (ST^{-1})^{-1}S$. ∎

It follows that we can use a fractional linear transformation to map any three points to any other three. Three points lie on a unique circle or line, and so by Proposition 5.2.3, the transformation takes the circle (or line) through z_1, z_2, z_3 to the circle (or line) through w_1, w_2, w_3. For example, we could have the situation depicted in Fig. 5.2.3. The

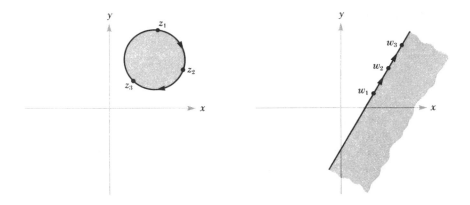

FIGURE 5.2.3   Effect of a fractional linear transformation.

inside of the disk maps to one of the two half planes. To determine which, one can check to see where the center of the circle goes (or any other point, especially if the center happens to go to ∞). Another way to do this is by checking orientation. As we proceed from z_1 through z_2 to z_3, located as in the figure, we go clockwise around the circle with the disk to the right. The image must proceed from w_1 through w_2 to w_3 along the line with the image of the disk to the right as shown. The half plane which is the image can be switched by interchanging z_1 and z_3. Suppose z_1, z_2, z_3 and w_1, w_2, w_3 determine circles C_1 and C_2 bounding disks D_1 and D_2. If the fractional linear transformation taking z_1, z_2, and z_3 to w_1, w_2, and w_3 is analytic on D_1 then it must map D_1 onto D_2 and the exterior of C_1 onto the exterior of C_2. If the zero of the denominator is in D_1 then it maps D_1 to the exterior of C_2 together with the point at infinity. Again the situation may be determined by examining the orientation of the points along the circles and may be reversed by changing the orientation of one of the triples of points. Many of these ideas and techniques are illustrated in Worked Example 5.2.15.

As mentioned earlier, fractional linear transformations can be combined with other transformations to obtain a fairly large class of conformal maps. This is also illustrated in the worked examples.

Reflection in a Circle

The idea of reflection in a circle which was used in the proof of the Poisson formula in Sec. 2.5 can readily be generalized to circles with centers other than 0. It can be discussed purely geometrically and works well with fractional linear transformations. In the spirit of this section, straight lines can be thought of as circles of infinite radius. In this case the new notion of reflection becomes the usual reflection. In particular, reflection in the real axis is complex conjugation. The key proposition is a nice illustration of the use of complex analysis in an apparently completely geometric setting.

5.2.5 PROPOSITION *Let C be a circle (or straight line) and z a point not on C. Then all the circles (or lines) through z which cross C at right angles intersect each other at a single point $\tilde{z}$. (If z happens to be the center of C, then $\tilde{z}$ is the point at infinity.)*

PROOF Let f be any fractional linear transformation which takes C to the real line and the interior of C to the upper half plane. The family of circles which pass through z and cross C at right angles must map to the family of circles which pass through $w = f(z)$ and cross the real axis at right angles, since f maps circles to circles and preserves angles. But the latter family clearly all intersect at $\bar{w}$. Thus the first family must all cross each other at the single point $\tilde{z} = f^{-1}(\bar{w})$. See Fig. 5.2.4. ∎

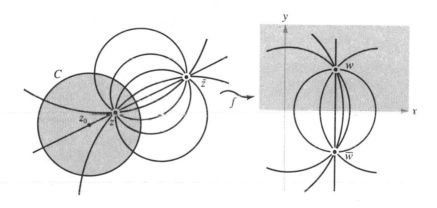

FIGURE 5.2.4 The circles emanating from z which cross C at right angles; all pass through $\tilde{z}$.

5.2.6 DEFINITION *Let C be a circle or straight line and z a point not on C. The unique point $\tilde{z}$ obtained in Proposition 5.2.5 is called the **reflection** of z in C. If z is on C, put $\tilde{z} = z$.*

Since fractional linear transformations take circles to circles and preserve angles, the next assertion should not be surprising.

5.2.7 PROPOSITION   *If g is a fractional linear transformation and C is a circle (or line), then g takes the reflection of z in C to the reflection of g(z) in g(C).*

This assertion may be paraphrased in the following somewhat imprecise but easily remembered form: *A fractional linear transformation preserves reflection in circles; that is,*

$$g(\tilde{z}) = [g(z)]^{\sim}$$

PROOF   The family of circles through z orthogonal to C is carried over to the family of circles through $g(z)$ orthogonal to $g(C)$ since g takes circles to circles and preserves angles. Thus the intersection of the first family, which is $\tilde{z}$, must map to the intersection of the second family, which is $g(z)$. ∎

In fact, reflection is almost a fractional linear transformation itself.

5.2.8 PROPOSITION   *If C is a circle (or line), then the map $z \mapsto \tilde{z}$ of reflection in C is a composition of linear fractional transformations and complex conjugation. If C is the circle with center z_0 and radius R, then*

$$\tilde{z} = \overline{\left(\frac{z_0 z + R^2 - |z_0|^2}{z - z_0} \right)}$$

PROOF   Let f be as in the proof of Proposition 5.2.5. The end of that proof was the equation $\tilde{z} = f^{-1}(\overline{w}) = f^{-1}(\overline{f(z)})$, which is already the general assertion of the proposition. To obtain the concrete formula we construct a function f taking C to the unit circle by $z \mapsto (z - z_0)/R$ and then composing with the map from the unit disk to the upper half plane given in Fig. 5.2.11 (vi). (See also Worked Example 5.2.13.) The result is

$$f(z) = i \frac{R + z - z_0}{R - z + z_0}$$

Set $f(\tilde{z}) = \overline{f(z)}$ and solve for $\tilde{z}$ to obtain the desired formula. ∎

From the formula of Proposition 5.2.8 we can readily calculate another geometric description of $\tilde{z}$.

5.2.9 PROPOSITION *If C is a circle with center z_0 and radius R, and if $z \neq z_0$, then $\tilde{z}$ is the point on the same ray from z_0 as z, and is selected such that the product of the distances from z_0 is R^2: i.e.,*

$$|z - z_0| \cdot |\tilde{z} - z_0| = R^2$$

PROOF Use the fact that $|\tilde{z} - z_0| = |\bar{\tilde{z}} - \bar{z}_0|$ and compute $|z - z_0| \cdot |\bar{\tilde{z}} - z_0|$ using the formula of Proposition 5.2.8. (See Exercise 23 and Fig. 5.2.5.) ∎

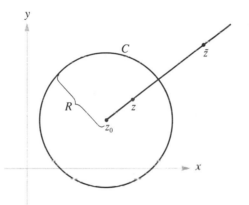

FIGURE 5.2.5 The reflection of z in C.

From the characterizations above most of the following should now be clear.

5.2.10 PROPOSITION

(i) $\tilde{\tilde{z}} = z$.

(ii) *The map $z \mapsto \tilde{z}$ is not conformal, but angles are preserved in magnitude and reversed in direction (just as in complex conjugation).*

(iii) *If C is a straight line, $\tilde{z}$ is the point on the line perpendicular to C through z and at an equal distance on the opposite side of C.*

(iv) *The map $z \mapsto \tilde{z}$ takes circles to circles (straight lines count as circles of infinite radius).*

The Schwarz-Christoffel Formula

The Schwarz-Christoffel formula gives an integral expression for mapping the upper half plane or unit circle to the interior of a given polygon. The case of the upper half plane will be discussed here; the case of a circle is left as an exercise.

5.2.11 SCHWARZ-CHRISTOFFEL FORMULA *Suppose that P is a polygon in the w plane with vertices at $w_1, w_2, \ldots, w_n$ and with exterior angles $\pi\alpha_i$, where $-1 < \alpha_i < 1$ (see Fig. 5.2.6). Then conformal maps from $A = \{z \mid \text{Im } z > 0\}$ onto B, the interior of P, have the form*

$$f(z) = a\left(\int_{z_0}^{z} (\zeta - x_1)^{-\alpha_1} \cdots (\zeta - x_{n-1})^{-\alpha_{n-1}} \, d\zeta\right) + b \tag{5}$$

where a and b are constants and the integration is along any path in A joining $z_0 \in A$ to z; the principal branch is used for the powers in the integrand. Furthermore,

(i) *Two of the points $x_1, \ldots, x_n$ may be chosen arbitrarily;*
(ii) *a and b determine the size and position of P;*
(iii) *$f(x_i) = w_{i+1}, i = 1, \ldots, n - 1$;*
   and
(iv) *f takes the point at infinity to w_1.*

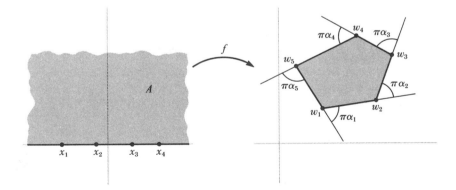

FIGURE 5.2.6 Schwarz-Christoffel formula.

The geometric meaning of the constants a and b here is explained in more detail in the following proof. It can be shown that the function f can be extended to be continuous on the x axis and that it maps the x axis to the polygon P. However, the function f is not analytic on the x axis, because it does not preserve angles at x_i. But f will be analytic on A itself. Only the main ideas of the proof of Eq. (5) will be given here, because to make the proof absolutely precise would be rather tedious.

SKETCH OF PROOF OF THE SCHWARZ-CHRISTOFFEL FORMULA   The first step is to show that if $x_1, \ldots, x_{n-1}$ have already been chosen, then f maps the real axis to a polygon having the correct angles. Let

$$g(z) = a(z - x_1)^{-\alpha_1} \cdots (z - x_{n-1})^{-\alpha_{n-1}}$$

so that on A, $f'(z) = g(z)$. Then

$$\begin{aligned}
\arg f'(z) - \arg g(z) \\
= \arg a - \alpha_1 \arg (z - x_1) - \cdots - \alpha_{n-1} \arg (z - x_{n-1})
\end{aligned}$$

At a point where $f'(z)$ exists, $\arg f'(z)$ represents the amount f rotates tangent vectors. Thus, as z moves along the real axis, $f(z)$ moves along a straight line for z on each of the segments $]-\infty, x_1[, \ldots,]x_i, x_{i+1}[, \ldots,]x_{n-1}, \infty[$. As z crosses x_i, $\arg f(z)$ jumps by an amount $\alpha_i \pi$. (If $z - x_i < 0$, $\arg (z - x_i) = \pi$; if $z - x_i > 0$, $\arg (z - x_i) = 0$.) Thus the real axis is mapped to a polygon with the correct angles. The last angle of the polygon is determined since we must have $\sum_{i=1}^{n} \alpha_i \pi = 2\pi$.

Next, we adjust this polygon to obtain P. Equality of angles forces similarity of polygons only for triangles. (For example, not all rectangles are squares.) This is the basic reason why two of the points x_i may be chosen arbitrarily (three if we count the point at infinity). The positions of the other points relative to these points control the ratios of the lengths of the sides of the image polygon. By choosing the x_i correctly we thus obtain a polygon similar to P. Another way to understand this problem is to consider mapping the upper half plane to a disk. We know that this can be accomplished by a linear fractional transformation and that this transformation is completely determined by its value at three of the boundary points. (Two of the finite points and the value at infinity are specified.) Choosing a and b properly means performing a scaling, a rotation, and a translation to bring this polygon to P. ∎

Worked Examples

5.2.12   *Find a conformal map taking* $A = \{z \mid 0 < \arg z < \pi/2, 0 < |z| < 1\}$ *to* $D = \{z$ *such that* $|z| < 1\}$.

Solution. The answer is not given by $z \mapsto z^4$, because this map does not map A onto D; its image omits the positive real axis.

First consider $z \mapsto z^2$. This maps A to B where B is the intersection of D and the upper half plane (Figure 5.2.7). Next (consult Fig. 5.2.11(iv)) map B to the first quadrant by $z \mapsto (1 + z)/(1 - z)$ and square to get the upper half plane; then map $z \mapsto (z - i)/(z + i)$ to give the unit circle.

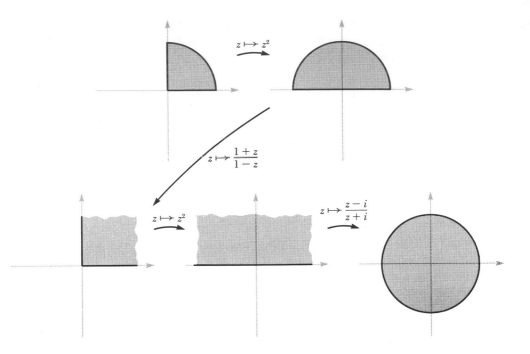

FIGURE 5.2.7   Successive transformations taking the quarter circle to a full circle.

Thus we obtain our transformation by successive substitution:

$$w_1 = z^2; \quad w_2 = \frac{1 + w_1}{1 - w_1} = \frac{1 + z^2}{1 - z^2}; \quad w_3 = w_2^2 = \left(\frac{1 + z^2}{1 - z^2}\right)^2;$$

$$w_4 = \frac{w_3 - i}{w_3 + i} = \frac{\left(\dfrac{1 + z^2}{1 - z^2}\right)^2 - i}{\left(\dfrac{1 + z^2}{1 - z^2}\right)^2 + i}$$

Therefore,

$$f(z) = \frac{(1 + z^2)^2 - i(1 - z^2)^2}{(1 + z^2)^2 + i(1 - z^2)^2}$$

is the required transformation.

5.2.13   *Verify Fig. 5.2.11(vi).*

Solution. We seek a fractional linear transformation $T(z) = (az + b)/(cz + d)$ such that $T(-1) = i$, $T(0) = -1$, $T(1) = -i$. Thus $(-a + b)/(-c + d) = i$, $b = -d$, and $(a + b)/(c + d) = -i$. Solving gives $-a - d = i(-c + d)$, $b = -d$, $a - d = -i(c + d)$. Adding the first and last equations, we get $-2d = i(-2c)$ or $d = ic$, and subtracting gives us $a = -id$. We can set, say, $b = 1$ (because numerator and denominator can be multiplied by a constant), so that $d = -1$, $a = i$, $c = i$, and thus

$$T(z) = \frac{iz + 1}{iz - 1} = \frac{z - i}{z + i}$$

We must check that $T(i)$ lies inside the unit circle. This is true because $T(i) = 0$. (If it lay outside, we would interchange $A = -1$ and $B = 0$.)

5.2.14   *Find a conformal map that takes the half plane shown in Fig. 5.2.8 onto the unit disk.*

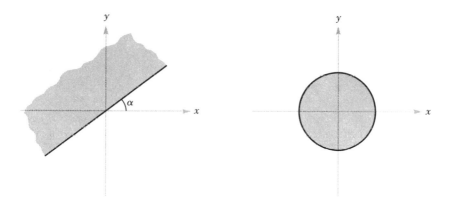

FIGURE 5.2.8   Mapping a rotated half plane to the disk.

Solution. Consider $S(z) = e^{-i\alpha}z$. This maps the region A to the upper half plane (Why?). Then, using Fig. 5.2.11(vi), we get

$$T(z) = \frac{e^{-i\alpha}z - i}{e^{-i\alpha}z + i}$$

as the required transformation.

5.2.15   *Study the action of the functions $f(z) = (z - 1)/(z - 3)$ and $g(z) = (z + 1)/(3z + 1)$ on the unit circle, the unit disk, and the real axis.*

Solution. First we compute the images of a few likely points:

(a) $f(1) = 0$ $g(1) = \frac{1}{2}$
(b) $f(i) = \frac{2}{5} - \frac{1}{5}i$ $g(i) = \frac{2}{5} - \frac{1}{5}i$
(c) $f(-1) = \frac{1}{2}$ $g(-1) = 0$
(d) $f(0) = \frac{1}{3}$ $g(0) = 1$
(e) $f(3) = \infty$ $g(-\frac{1}{3}) = \infty$

Thus, f takes the unit circle through $1, i, -1$ to the circle through $0, \frac{2}{3} - \frac{1}{3}i, \frac{1}{2}$. The map g takes the unit circle to the same circle but with the orientation reversed. f takes the unit disk to the interior of the image circle while g takes it to the exterior. This can be determined by examining the images of 0 or by noticing that $g(-\frac{1}{3}) = \infty$.

It may not be obvious what the image circle is, but it is easier after noticing that both f and g take the real axis onto the real axis. (The line through $-1, 0, 1$ goes to the line through $\frac{1}{2}, \frac{1}{3}, 0$ in the case of f and through $0, 1, \frac{1}{2}$ in the case of g. Think about where the various pieces of the line go.) The unit circle crosses the real axis at right angles at ± 1 and so the image circle must cross the axis at right angles at 0 and $\frac{1}{2}$. Thus, it is the circle of radius $\frac{1}{4}$ centered at $\frac{1}{4}$; check that this goes through $\frac{2}{3} - \frac{1}{3}i$. The effects of these maps are indicated in Fig. 5.2.9.

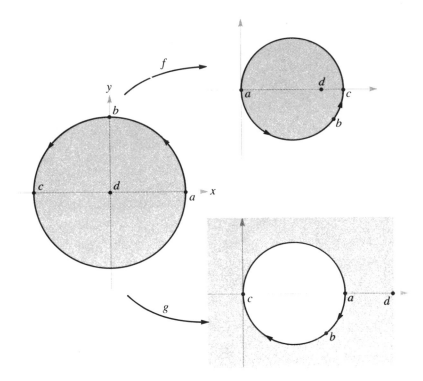

FIGURE 5.2.9   The maps for Worked Example 5.2.15.

5.2.16   *Let* $A = \{z \mid \operatorname{Im} z > 0\} \backslash \{z \mid \operatorname{Re} z = 0 \text{ and } 0 \le \operatorname{Im} z \le 1\}$.
(a) *Find a conformal map f of A one-to-one onto the upper half plane.*
(b) *Find a conformal map g of A one-to-one onto the unit disk.*

First Solution Consider the following functions:

$$f_1(z) = -iz$$
$$f_2(z) = z^2$$
$$f_3(z) = z^2 - 1$$
$$f_4(z) = \sqrt{z} \quad \text{(branch cut on negative real axis)}$$
$$f_5(z) = iz$$
$$f_6(z) = \frac{1-z}{1+z}$$

The map f_1 rotates A by $90°$ to the right half plane cut from 0 to 1. f_2 opens this out to the whole plane cut from $-\infty$ to 1, and f_3 translates this to the plane cut from $-\infty$ to 0. f_4 takes this back to the right half plane. Finally, f_5 rotates the right half plane to the upper half plane while f_6 takes the right half plane to the unit disk. Thus $f(z) = f_5(f_4(f_3(f_2(f_1(z)))))$ should do for part (a), and $g(z) =$

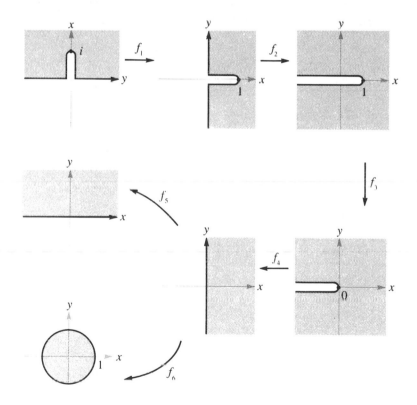

$f_6(f_4(f_3(f_2(f_1(z)))))$ should do for part (b). We calculate

$$f_4(f_3(f_2(f_1(z)))) = \sqrt{(-iz)^2 - 1} = \sqrt{-1} \sqrt{z^2 + 1}$$

Choose $\sqrt{-1}$ with care: -1 is in the branch cut. With $\sqrt{-1} = -i$ we have $f(z) = i(-i) \sqrt{z^2 + 1} = \sqrt{z^2 + 1}$, and the image is the upper half plane. To get to the unit disk, take

$$g(z) = f_6(-i\sqrt{z^2 + 1}) = \frac{1 + i\sqrt{z^2 + 1}}{1 - i\sqrt{z^2 + 1}}$$

See Figure 5.2.10 on p. 367.

Second Solution (for part (a)). The region A could be considered as a (strange) polygon with exterior angles of 90° at 0, $-180°$ at i, 90° a second time at 0, and 0° at ∞. Try the Schwarz-Christoffel formula using points $-\sqrt{2}, 0$, and $\sqrt{2}$ on the x axis. The Schwarz-Christoffel formula should give a map from z in the upper half plane to w in A. Thus, we try

$$w = f(z) = a \int (\zeta + \sqrt{2})^{-1/2} \zeta (\zeta - \sqrt{2})^{-1/2} \, d\zeta + b$$

$$= a \int \frac{\zeta}{\sqrt{\zeta^2 - 2}} \, d\zeta + b = a\sqrt{z^2 - 2} + b$$

If $a = 1/\sqrt{2}$ and $b = 0$, this takes the upper half plane to A with 0 going to i. Thus we get $w = \sqrt{z^2 - 2}/\sqrt{2}$. Solving for z gives $z = \sqrt{2} \sqrt{w^2 + 1}$ as a map from A to the upper half plane. The function f obtained in the first solution has been multiplied by $\sqrt{2}$. But that is all right since it maps the upper half plane to itself. ▲

Figure 5.2.11 summarizes some of the common transformations.

Exercises

1. Let $f(z) = (z - 1)/(z + 1)$. What is the image under f of
(a) The real line?
(b) The circle with center 0 and radius 2?
(c) The circle with center 0 and radius 1?
(d) The imaginary axis?

2. Let $f(z) = (z - i)/(z + i)$. What is the image under f of
(a) The real line?
(b) The circle with center 0 and radius 2?
(c) The circle with center 0 and radius 1?
(d) The imaginary axis?

3. Find fractional linear transformations f satisfying $f(z_i) = w_i$ for $i = 1, 2, 3$ if

(a) $z_1 = -1$, $z_2 = 1$, $z_3 = 2$; $w_1 = 0$, $w_2 = -1$, $w_3 = -3$

(b) $z_1 = -1$, $z_2 = 1$, $z_3 = 2$; $w_1 = -3$, $w_2 = -i$, $w_3 = 0$

4. Find fractional linear transformations f satisfying $f(z_i) = w_i$ for $i = 1, 2, 3$ if

(a) $z_1 = i$, $z_2 = 0$, $z_3 = -1$; $w_1 = 0$, $w_2 = -i$, $w_3 = \infty$

(b) $z_1 = i$, $z_2 = 0$, $z_3 = -1$; $w_1 = -i$, $w_2 = 0$, $w_3 = \infty$

5. Find a fractional linear transformation which takes the unit disk to the upper half plane with $f(0) = 2 + 2i$.

6. Find a fractional linear transformation which takes the unit disk to the right half plane with $f(0) = 3$.

7. Find a conformal map of the unit disk onto itself which takes $\frac{1}{2}$ to $\frac{1}{3}$.

8. Find a conformal map of the unit disk onto itself which takes $\frac{1}{4}$ to $-\frac{1}{3}$.

9. Find a conformal map of $A = \{z$ such that $|z - 1| < \sqrt{2}$ and $|z + 1| < \sqrt{2}$ one-to-one onto the open first quadrant.

10. Map the region in Exercise 9 to the upper half plane.

11. Prove: Any fractional linear transformation with $c \neq 0$ can be written $T = T_4 \circ T_3 \circ T_2 \circ T_1$, where $T_1(z) = z + d/c$, $T_2(z) = 1/z$, $T_3(z) = [(bc - ad)/c^2]z$, and $T_4(z) = z + a/c$. Interpret T geometrically.

12. Prove that if both T and R are fractional linear transformations, then so is $T \circ R$.

13. Find a conformal map of the unit disk onto itself that maps $\frac{1}{2}$ to 0.

14. Show that $K(z) = z/(1 - z)^2$ takes the open unit disk one to one onto $\mathbb{C}\backslash\{z \mid \text{Re } z = 0$ and $\text{Im } z < -\frac{1}{4}\}$.

15. Find all conformal maps that take the disk of radius R and center 0 onto the unit disk.

16. Establish parts (iii), (iv), and (v) of Fig. 5.2.11.

17. Prove: The most general conformal transformation that takes the upper half plane onto the unit disk is

$$T(z) = e^{i\theta}\left(\frac{z - \lambda}{z - \bar{\lambda}}\right)$$

where $\text{Im } \lambda > 0$.

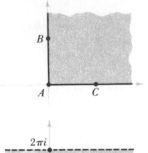

$z \mapsto z^2$

$\sqrt{z} \longleftarrow\!\shortmid z$

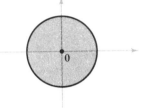

(i)

$z \mapsto e^z$

$\log z \longleftarrow\!\shortmid z$

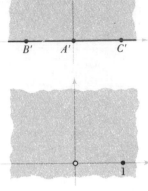

(ii)

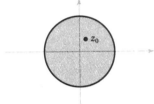

$z \mapsto e^{i\theta} \dfrac{z - z_0}{1 - \bar{z}_0 z}$

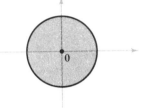

(iii)

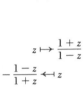

$z \mapsto \dfrac{1 + z}{1 - z}$

$-\dfrac{1 - z}{1 + z} \longleftarrow\!\shortmid z$

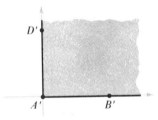

(iv)

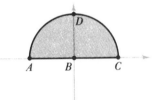

$z \mapsto \dfrac{z - 1}{z + 1}$

$-\left(\dfrac{z + 1}{z - 1}\right) \longleftarrow\!\shortmid z$

(v)

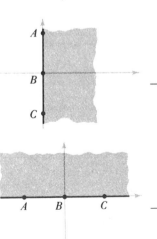

$z \mapsto \dfrac{z - i}{z + i}$

$-i\left(\dfrac{z + 1}{z - 1}\right) \longleftarrow\!\shortmid z$

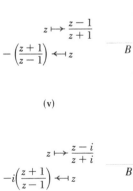

(vi)

FIGURE 5.2.11 Some common transformations.

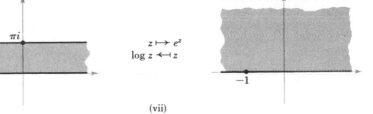
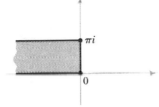

$$z \mapsto e^z$$
$$\log z \mapsfrom z$$

(vii)

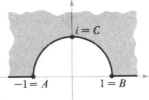
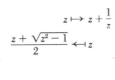
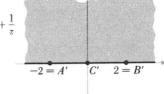

$$z \mapsto e^z$$
$$\log z \mapsfrom z$$

(viii)

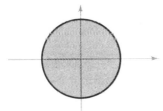

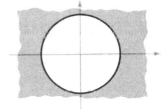

$$z \mapsto z + \frac{1}{z}$$
$$\frac{z + \sqrt{z^2 - 1}}{2} \mapsfrom z$$

(ix)

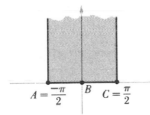

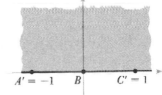

$$z \mapsto \frac{1}{z}$$
$$\frac{1}{z} \mapsfrom z$$

(x)

$$z \mapsto \sin z$$
$$\sin^{-1} z \mapsfrom z$$

(xi)

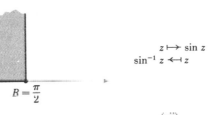

$$z \mapsto \sin z$$
$$\sin^{-1} z \mapsfrom z$$

(xii)

18.   Suppose that a, b, c, d are real and that $ad > bc$; show that $T(z) = (az + b)/(cz + d)$ leaves the upper half plane invariant. Show that any conformal map of the upper half plane onto itself is of this form.

19.   Find a conformal map that takes $\{z \mid 0 < \arg z < \pi/8\}$ onto the unit disk.

20.   The *cross ratio* of four distinct points z_1, z_2, z_3, z_4 is defined by

$$[z_1, z_2, z_3, z_4] = \frac{z_4 - z_1}{z_4 - z_2} \cdot \frac{z_3 - z_2}{z_3 - z_1}$$

Show that any fractional linear transformation has the property that $[T(z_1), T(z_2), T(z_3), T(z_4)] = [z_1, z_2, z_3, z_4]$. (*Hint.* Use Exercise 11.)

21.   Let γ_1 and γ_2 be two circles that intersect orthogonally. Let T be a fractional linear transformation. What can be said about $T(\gamma_1)$ and $T(\gamma_2)$?

22.   (See Exercise 20.) Show that $[z_1, z_2, z_3, z_4]$ is real iff z_1, z_2, z_3, z_4 lie on a line or circle. Use Exercise 20 to give another proof of Proposition 5.2.3.

23.   Complete the calculation in the proof of Proposition 5.2.9.

24.   Show that a fractional linear transformation T that is not the identity map has at most two fixed points (that is, points z for which $T(z) = z$). Give an example to show that T need not have any fixed points. Find the fixed points of $T(z) = z/(z + 1)$.

25.   Conformally map $A = \{z \mid \operatorname{Re} z < 0, 0 < \operatorname{Im} z < \pi\}$ onto the first quadrant.

26.   Conformally map $A = \{a \text{ such that } |z - 1| < 1\}$ onto $B = \{z \mid \operatorname{Re} z > 1\}$.

27.   Conformally map $\mathbb{C}\backslash\{\text{nonpositive real axis}\}$ onto the region $A = \{z \mid -\pi < \operatorname{Im} z < \pi\}$.

28.   Argue that the conformal maps that take $|z| < 1$ to the interior of a polygon with vertices $w_1, \ldots, w_n$ and points $z_1, \ldots, z_n$ on the unit circle $|z| = 1$ to the points $w_1, \ldots, w_n$ are given by

$$f(z) = a \left[\int_0^z (\zeta - z_1)^{-\alpha_1} \cdots (\zeta - z_n)^{-\alpha_n} \, d\zeta \right] + b$$

where the α_i's are as in the Schwarz Christoffel formula (5.2.11).

29.   Show that

$$f(z) = \int_0^z \frac{d\zeta}{\sqrt{\zeta(\zeta - 1)(\zeta - c)}}$$

where $c > 0$, maps the upper half plane to a rectangle. (The integrand is called an *elliptic integral* and generally cannot be computed explicitly.)

30. Verify part (ix) of Fig. 5.2.11.

31. Is it possible to map the upper half plane conformally to a triangle using fractional linear transformations? Devise a formula that is based on the Schwarz-Christoffel formula.

32. Verify from the Schwarz-Christoffel formula that a conformal map from the upper half plane to $\{z \mid \text{Im } z > 0 \text{ and } -\pi/2 < \text{Re } z < \pi/2\}$ is $z \mapsto \sin^{-1} z$.

33. Show that $f(z) = 4/z$ maps the region $A = \{z \text{ such that } |z - 1| > 1 \text{ and} |z - 2| < 2\}$ one to one onto the strip $B = \{z \mid 1 < \text{Re } z < 2\}$.

34. Suppose C_1 and C_2 are two tangent circles with C_2 in the interior of C_1. Show that an infinite number of circles can be placed in the region between C_1 and C_2, each tangent to C_1 and C_2 and each tangent to the next as shown in Fig. 5.2.12. Show that the points of tangency of these circles each with the next lie on a circle. (*Hint:* Consider Exercise 33.)

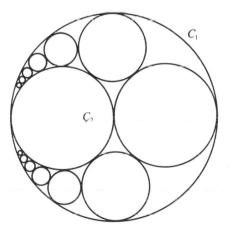

FIGURE 5.2.12 A fractional linear transformation can be used to pack the disk with circles.

35. Consider a fractional linear transformation of the form

$$f(z) = a\left(\frac{z - b}{z - d}\right)$$

Show that
(a) Circles through the points b and d are mapped to lines through the origin.

(b) The *circles of Apollonius* with equation $|(z - b)/(z - d)| = r/|a|$ are mapped to circles with center 0, radius r.

(c) The circles in (a) and (b), when located in the z plane, are called *Steiner circles*. Sketch them and verify that both these circles and their images meet in right angles.

5.3 APPLICATIONS OF CONFORMAL MAPPING TO LAPLACE'S EQUATION, HEAT CONDUCTION, ELECTROSTATICS, AND HYDRODYNAMICS

We are now in a position to apply the theory of conformal maps developed in Secs. 5.1 and 5.2 to some physical problems. In doing so we will solve the Dirichlet problem* and related problems for several types of two-dimensional regions. We will then apply these results to the three classes of physical problems mentioned in the title of this section. Only a very meager knowledge of elementary physics is needed to understand these examples. The student is cautioned that the variety of problems that can explicitly be solved in this way is somewhat limited and that the methods discussed apply only to two-dimensional problems.

The Dirichlet and Neumann Problems

Let us recall that $u(x, y)$ is said to satisfy *Laplace's equation* (or be *harmonic*) on a region A when

$$\nabla^2 u = \frac{\partial^2 u}{\partial x^2} + \frac{\partial^2 u}{\partial y^2} = 0$$

In addition to this condition, some boundary behavior determined by the physical problem to be solved is generally specified. This boundary behavior (or boundary conditions) usually determines u uniquely. For example, the uniqueness theorem for the Dirichlet problem (2.5.12) indicated that a harmonic function $u(x, y)$ whose value on the boundary of A is specified is uniquely determined. We shall also have occasion to consider the boundary condition in which $\partial u/\partial n = \text{grad } u \cdot n$ is specified on the boundary. ($\partial u/\partial n$ equals the derivative in the direction normal to bd (A)). (For $\partial u/\partial n$ to be well defined, the boundary of A should be at least piecewise smooth, so that it has a well-defined normal direction.) The *outward* normal direction of n can be defined precisely, but since the general emphasis of this section is on computational methods, a mathematically precise treatment of such a question will not be given here. Thus we accept as clear what is meant by the outward unit normal, n (see Fig. 5.3.1).

The problem of finding a harmonic function u with $\partial u/\partial n$ specified on the boundary is called the *Neumann problem*. We cannot specify $\phi = \partial u/\partial n$ arbitrarily because if such a

---

* The problem of finding a harmonic function on a region A whose values are specified on the boundary of A is called the *Dirichlet problem*. This problem was discussed in Sec. 2.5.

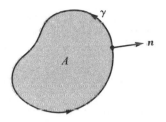

FIGURE 5.3.1 Neumann problem.

u exists, then we claim that

$$\int_\gamma \frac{\partial u}{\partial n} = 0$$

where γ is the boundary of A. To prove this, we apply Green's theorem (see Sec. 2.2), which can be written in the divergence form (often called Gauss' theorem)

$$\int_\gamma X \cdot n \, ds = \int_A \operatorname{div} X \, dx \, dy$$

where X is a given vector function with components (X^1, X^2) and where the divergence of X is defined by

$$\operatorname{div} X = \frac{\partial X^1}{\partial x} + \frac{\partial X^2}{\partial y} \tag{1}$$

Applying Eq. (1) to $X = \operatorname{grad} u$ gives

$$\int_\gamma \frac{\partial u}{\partial n} \, ds = \int_\gamma (\operatorname{grad} u) \cdot n \, ds = \int_A \operatorname{div} \operatorname{grad} u \, dx \, dy = \int_A \nabla^2 u \, dx \, dy = 0$$

because $\operatorname{div} \operatorname{grad} u = \nabla^2 u = 0$. This proves our claim.

If we are given a boundary condition ϕ on γ with $\int_\gamma \phi = 0$, then it can be shown that the Neumann problem indeed has a solution. However, we can prove the following fact: *The solution of the Neumann problem on a bounded simply connected region is unique up to the addition of a constant.* Let u_1 and u_2 be two solutions with $\partial u_1/\partial n = \partial u_2/\partial n$ on $\gamma = \operatorname{bd}(A)$. Let v_1 and v_2 be harmonic conjugates of u_1 and u_2 and set $u = u_1 - u_2$, $v = v_1 - v_2$. Now $\partial u/\partial n = 0$, and so, by Proposition 1.5.12, v is constant along γ. Thus by uniqueness of the solution to the Dirichlet problem, v is constant on A. Therefore, since $-u$ is the harmonic conjugate of v, u is constant on A as well. This proves our claim.

If the boundary values specified in the Dirichlet and Neumann problems are not continuous, the uniqueness results are still valid, in a sense, but are more difficult to obtain. However, the student is cautioned that on an unbounded region we do not have uniqueness. For example, let A be the upper half plane. Then $u_1(x, y) = x$ and $u_2 = x + y$ have the same boundary values (at $y = 0$) and are harmonic but are not equal. To recover uniqueness for unbounded regions, a "condition at ∞" must also be specified; "u bounded on all of A" is such a condition. Some of these conditions will be illustrated in the examples that are integrated into this section.

The Dirichlet and Neumann problems can also be combined; for example, u can be specified on one part of the boundary and $\partial u/\partial n$ can be specified on another.

Method of Solution

The basic method for solving the Dirichlet and Neumann problems in a given region A is as follows. Take the given region A and transform it by a conformal map to a "simpler" region B on which the problem can be solved. This procedure is justified by the fact that under a conformal map f, harmonic functions are transformed again into harmonic functions (see Proposition 5.1.3). When we have solved the problem on B, we can transform the answer back to A.

For the Dirichlet problem we are given the boundary values on bd (A). These values obviously get mapped to the corresponding boundary values on B. (We assume that the conformal map f is defined on the boundary.) The specification of $\partial u/\partial n$ is a bit more complicated. However, the special case $\partial u/\partial n = 0$ is easy to understand. Let $u \circ f = u_0$ be the solution we seek; that is, $u_0(x, y) = u(f(x, y))$ (see Fig. 5.3.2). Then we claim that

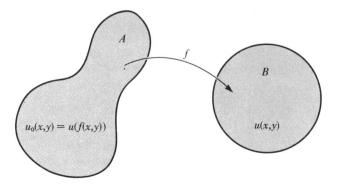

FIGURE 5.3.2   Transformation of harmonic functions.

$\partial u_0/\partial n = 0$ iff $\partial u/\partial n = 0$ *on corresponding portions.* This follows because $\partial u_0/\partial n = 0$ and $\partial u/\partial n = 0$ mean that the conjugates are constant on those portions, and if v is the conjugate of u, then $v \circ f = v_0$ is the conjugate of u_0. This proves the claim. These are the only types of boundary conditions for $\partial u/\partial n$ that will be dealt with in this text.

To use this method we need to be able to solve the problem in some simple region B. We already saw in Sec. 2.5 that the unit disk is suitable for this purpose because in that case we have the Poisson formula for the solution of the Dirichlet problem. However, we can sometimes get more explicit solutions than those yielded by that formula.

The following situation is used to illustrate the method, and will be used in subsequent examples. We consider the upper half plane H and the problem of finding a harmonic function that takes the constant boundary values c_0 on $]-\infty, x_1[$, c_1 on $]x_1, x_2[$, $\ldots$, and c_n on $]x_n, \infty[$ where $x_1 < x_2 < \cdots < x_n$ are points on the real axis. We claim that a solution is given by

$$u(x, y) = c_n + \frac{1}{\pi}[(c_{n-1} - c_n)\,\theta_n + \cdots + (c_0 - c_1)\theta_1] \tag{2}$$

where $\theta_1, \ldots, \theta_n$ are as indicated in Fig. 5.3.3; $0 \le \theta_i \le \pi$.

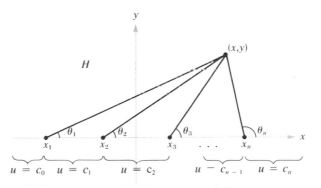

FIGURE 5.3.3   Equation (2) gives the solution of this Dirichlet problem in the upper half plane.

This is easy to see. First, since u is the real part of

$$c_n + \frac{1}{\pi i}[(c_{n-1} - c_n)\log(z - x_n) + \cdots + (c_0 - c_1)\log(z - x_1)]$$

it is harmonic. Also, on $]x_i, x_{i+1}[$, u reduces to c_i. (The student should check this.) As mentioned previously, the Dirichlet problem does not have a unique solution, so the question arises: Why was this solution chosen? Another solution could have been obtained by adding $u(x, y) = y$ to the solution given by Eq. (2). The answer is that the u that is given by Eq. (2) is bounded (Why?). The student will find this answer physically reasonable after studying the examples that follow.

Thus if a problem can be transformed to the one described by Fig. 5.3.3, we can use Eq. (2) to find a solution. This will be done in the examples in this section.

Heat Conduction

Physical laws tell us that if a two-dimensional region is maintained at a steady temperature T (that is, a temperature not changing in time, accomplished by fixing the temperature at the walls, or by insulating them), then T should be harmonic.*

The negative of the gradient of T represents the direction in which heat flows. Thus we can, by using Proposition 1.5.12, *interpret the level surfaces of the harmonic conjugate ϕ of T as the lines along which heat flows and the temperature is decreasing.* Lines of constant T are called *isotherms;* lines on which the conjugate ϕ are constant are called *flux lines* (Fig. 5.3.4).

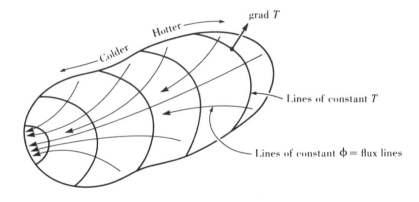

FIGURE 5.3.4   Heat conduction.

Thus to say that T is prescribed on a portion of the boundary means that the portion is maintained at a preassigned temperature (for example, by a heating device). The condi-

---

* This is a consequence of conservation of energy and Gauss' theorem (see Eq. (1)). The heat flows in the direction of the vector field $= \kappa\nabla T$ ($\kappa =$ conductivity) and the energy density is $c\rho T$ ($c =$ specific heat, $\rho =$ density). Then the law of conservation of energy states that the rate of change of energy in any region V equals the rate at which energy enters V; that is,

$$\frac{d}{dt}\int_V c\rho T\, dx = -\int_{bd(V)} -\kappa\nabla T \cdot \boldsymbol{n}\, ds$$

By Gauss' theorem, this condition is equivalent to

$$\frac{\partial}{\partial t}(c\rho T) = \kappa\nabla^2 T$$

If c, ρ, T are independent of t, we conclude that T is harmonic: $\nabla^2 T = 0$.

tion $\partial T/\partial n = 0$ means that the flux line (or $-$grad T) is parallel to the boundary; in other words, the boundary is *insulated*. (No heat flows across the boundary.)

5.3.1 EXAMPLE *Let A be the first quadrant; the x axis is maintained at $T = 0$ while the y axis is maintained at $T = 100$. Find the temperature distribution in A. (Physically, the region may be approximated by a thin metal sheet.)*

SOLUTION We map the first quadrant to the upper half plane by $z \mapsto z^2$ (Fig. 5.3.5).

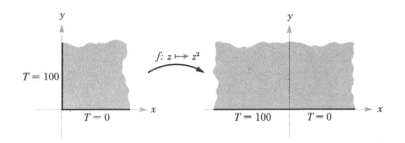

FIGURE 5.3.5 Map the region A to the upper half plane.

It is physically reasonable that the temperature should be a bounded function; otherwise we would obtain arbitrarily high (or low) temperatures. Therefore, the solution in the upper half plane is given by Eq. (2):

$$u(x, y) = \frac{1}{\pi}(100 \arg z) = \frac{100}{\pi} \tan^{-1}\left(\frac{y}{x}\right)$$

Thus the solution we seek is

$$u_0(x, y) = u(f(x, y))$$

where $f(x, y) = z^2 = x^2 - y^2 + 2ixy = (x^2 - y^2, 2xy)$. Hence

$$u_0(x, y) = \frac{100}{\pi} \tan^{-1}\left(\frac{2xy}{x^2 - y^2}\right)$$

is the desired answer. It is understood that $\tan^{-1}$ is taken in the interval $[0, \pi]$. Another form of the answer may be obtained as follows:

$$u_0(x, y) = u(z^2) = \frac{100}{\pi} \arg (z^2) = \frac{200}{\pi} \arg z = \frac{200}{\pi} \tan^{-1}\left(\frac{y}{x}\right)$$

The isotherms and flux lines are indicated in Fig. 5.3.6. ▲

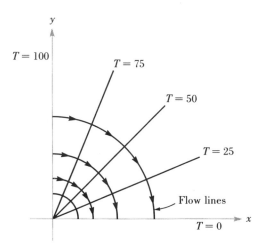

FIGURE 5.3.6   Isotherms and flow lines for Example 5.3.1.

5.3.2 EXAMPLE   *Let A be the upper half of the unit disk $|z| \leq 1$. Find the temperature inside if the circular portion is insulated; $T = 0$ for $x > 0$ and $T = 10$ for $x < 0$ on the real axis.*

SOLUTION   For this type of problem where there is a portion of the boundary where $\partial T/\partial n = 0$ (insulated), it is convenient to map the region to a half strip. This can be done for A by means of log z (using the principal branch); see Fig. 5.3.7 (and 5.2.11(viii)). For strip B we obtain, by inspection, the solution

$$T_0(x, y) = \frac{10y}{\pi}$$

(Note that along the y axis $\partial T_0/\partial n = \partial T_0/\partial x = 0$.)
    Thus our answer is

$$T(x, y) = T_0(\log (x + iy)) = \frac{10}{\pi} \tan^{-1}\left(\frac{y}{x}\right) ▲$$

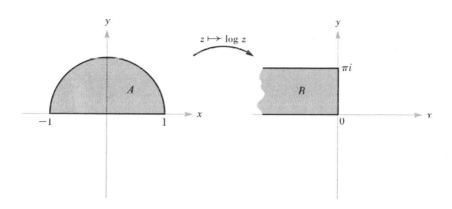

FIGURE 5.3.7 Mapping the semicircular region A to the half strip B in Example 5.3.2.

Electric Potential

From physics we learn that if an electric potential ϕ is determined by static electric charges, ϕ must satisfy Laplace's equation (that is, be harmonic). The conjugate function Φ of ϕ is interpreted as follows: Lines along which Φ is constant are lines along which a small test charge would travel. They are called *flux lines*. Tangent vectors to such lines are $-\operatorname{grad}\phi = E$, called the *electric field* (see Fig. 5.3.8). Thus *the flux lines and the equipotential lines (lines of constant ϕ) intersect orthogonally*.

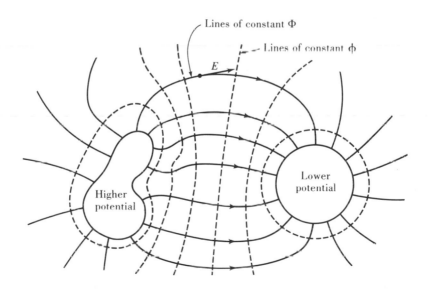

FIGURE 5.3.8 Electrical potential.

The Dirichlet problem arises in electrostatics when the boundary is maintained at a given potential (for example, by means of a battery or by grounding).

5.3.3 EXAMPLE *Consider the unit circle. The electric potential is maintained at $\phi = 0$ on the lower semicircle and at $\phi = 1$ on the upper semicircle. Find ϕ inside.*

SOLUTION We use the general procedure for solving the Dirichlet problem by mapping our given region to the upper half plane. In the present case we may use a fractional linear transformation (see Figs. 5.2.11(vi) and 5.3.9).

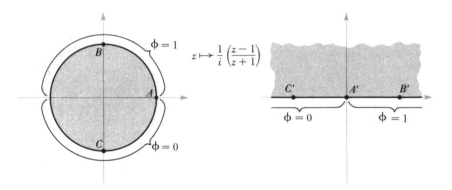

FIGURE 5.3.9 The conformal mapping used to solve the Dirichlet problem on the disk.

As with temperature, it is physically reasonable that the potential be bounded. Thus, by Eq. (2), the solution in the upper half plane is

$$\phi_0(x, y) = 1 - \frac{1}{\pi} \tan^{-1}\left(\frac{y}{x}\right)$$

and so the solution on the unit circle is

$$\phi(x, y) = \phi_0(f(x, y))$$

where $f(z) = (z - 1)/i(z + 1)$. If we let $(z - 1)/i(z + 1) = u + iv$, then

$$u = \frac{2y}{x^2 + y^2 + 2x + 1} \qquad \text{and} \qquad v = -\frac{x^2 + y^2 - 1}{x^2 + y^2 + 2x + 1}$$

Thus the solution is

$$\phi(x, y) = 1 - \frac{1}{\pi} \tan^{-1} \frac{1 - x^2 - y^2}{2y}$$

The equipotential and flux lines are shown in Fig. 5.3.10.

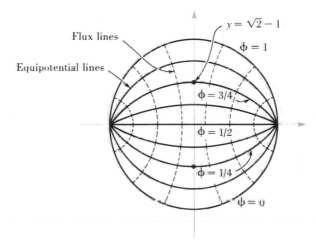

FIGURE 5.3.10 Equipotential and flux lines for the potential.

This example could also be solved by using Poisson's formula. The two answers would be equal, although this would not be obvious from their form. ▲

5.3.4 EXAMPLE *The harmonic function $\phi(z) = (Q/2\pi) \log|z - z_0| + K$ for a constant K, which is the real part of $(Q/2\pi) \log(z - z_0) + K$, represents the potential of a charge Q located at z_0. (This is because ϕ is a radial field such that if $E = -\text{grad } \phi$ is the electric force field, then the integral of $E \cdot n$ around a curve surrounding z_0 is Q by Gauss' theorem).* [*] *The constant K may be adjusted to make any convenient place such as infinity or some grounded object have potential 0. (This is reasonable since only the force E is actually observed, and it is not changed by changing K.) Sketch the equipotential lines for two charges of like or opposite signs.*

---

[*] This is the potential for a charge in the *plane*. In space it corresponds to the potential produced by a charged line.

SOLUTION   The potential of two charges is obtained by adding the respective potentials. Thus two charges $Q > 0$ located at z_1 and z_2 have the electrostatic potential $(Q/2\pi) \log (|z - z_1||z - z_2|)$; a charge $Q > 0$ at z_1 and a charge $-Q$ at z_2 have potential $(Q/2\pi) \log (|z - z_1|/|z - z_2|)$. The equipotential lines are sketched in Fig. 5.3.11. The

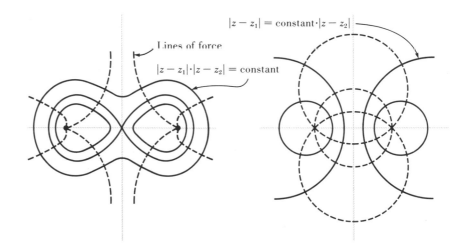

FIGURE 5.3.11   The field of like charges (*left*) and the field of opposite charges (*right*).

curves $\phi =$ constant in the drawing on the left are called lemniscates; in the drawing on the right they are called circles of Apollonius. The lines of force are the family of circles orthogonal to them which pass through the two points. Together they form the Steiner circles discussed in Exercise 35 of Sec. 5.2. ▲

5.3.5 EXAMPLE   *Suppose a point charge of $+1$ is located at $z_0 = \frac{1}{2}$ and the unit circle is a grounded conductor maintained at potential 0. Find the potential at every point $z \neq z_0$ inside the unit circle.*

FIRST SOLUTION   The function $f(z) = (2z - 1)/(2 - z)$ maps the unit disk D to itself taking $z_0 = \frac{1}{2}$ to 0. The function $u(z) = (1/2\pi) \log |z|$ is a solution on the image disk (point charge of $+1$ at 0 and 0 potential around the unit circle). Thus $\phi(z) = u(f(z)) = (1/2\pi) \log |(2z - 1)/(2 - z)|$ solves the original problem. (See Fig. 5.3.12.)

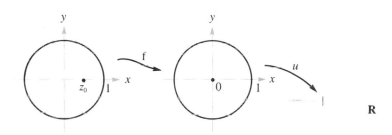

FIGURE 5.3.12 The conformal map f shifts the point charge from $z_0 = \frac{1}{2}$ to 0.

SECOND SOLUTION We give a second solution which illustrates the method of reflection in a circle from Sec. 5.2 and an interesting idea from electrostatics called *image charges*. We need a field ϕ inside the unit disk D which has the unit circle C as an equipotential curve. The electric force lines must be a family of curves ending at z_0 which cross C at right angles. This can be accomplished by placing an artificial "image charge" of -1 at the reflection $\tilde{z}_0$ of z_0 in C. As in the last example, the electric force lines for such a pair of charges are the family of circles through z_0 and $\tilde{z}_0$. We know from the last section that these cross C at right angles as desired. See Fig. 5.3.13. With charges of $+1$ at

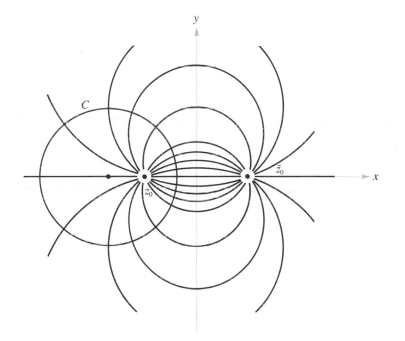

FIGURE 5.3.13 Reflection and image charge.

$z_0 = \frac{1}{2}$ and -1 at $\tilde{z}_0 = 2$, we have

$$\phi(z) = \frac{1}{2\pi} \log |z - \tfrac{1}{2}| - \frac{1}{2\pi} \log |z - 2| + K$$

$$= \frac{1}{2\pi} \log \left| \frac{z - \tfrac{1}{2}}{z - 2} \right| + K$$

$$= \frac{1}{2\pi} \log \left| \frac{2z - 1}{2 - z} \right| - \frac{1}{2\pi} \log 2 + K$$

Setting the constant K to $(1/2\pi) \log 2$ makes the potential 0 around C and gives the same answer as the first solution. ▲

Hydrodynamics

If we have a (steady-state) incompressible, nonviscous fluid, we are interested in finding its velocity field, $V(x, y)$. From elementary vector analysis we know that "incompressible" means that the divergence div $V = 0$. (We say V is *divergence free.*) We shall assume that V is also a *potential flow* and hence is circulation-free; that is, $V = \text{grad } \phi$ for some ϕ called the *velocity potential*. Thus ϕ is harmonic because $\nabla^2 \phi = \text{div grad } \phi = \text{div } V = 0$. Thus when we solve for ϕ we can obtain V by taking $V = \text{grad } \phi$.

The conjugate ψ of the harmonic function ϕ (ψ will exist on any simply connected region) is called the *stream function,* and the analytic function $F = \phi + i\psi$ is called the *complex potential.* Lines of constant ψ have V as their tangents (Why?), and so *lines of constant ψ may be interpreted as the lines along which particles of fluid move;* hence the name stream function (see Fig. 5.3.14).

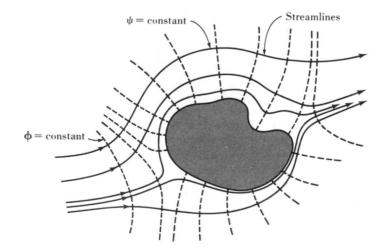

FIGURE 5.3.14   Fluid flow.

The natural boundary condition is that V should be parallel to the boundary. (The fluid flows parallel to the walls.) This means that $\partial\phi/\partial n = 0$, so we are led to the Neumann problem for ϕ.

Let us again consider the upper half plane. A physically acceptable motion is obtained by setting $V(x, y) = \alpha = (\alpha, 0)$ or $\phi(x, y) = \alpha x = \text{Re }(\alpha z)$, where α is real. The flow corresponding to V is parallel to the x axis, with velocity α. Notice that now ϕ is not bounded; thus the behavior at ∞ for fluids, for temperature, and for electric potential is different because of the different physical circumstances (see Fig. 5.3.15).

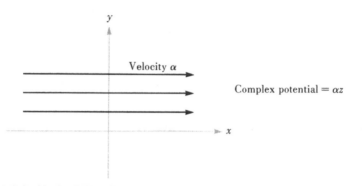

FIGURE 5.3.15   Flow in the upper half plane.

Thus to find the flow in a region we should map the region to the upper half plane and use the solution $\phi(x, y) = \alpha x$. α may be specified as the velocity at infinity. It should be clear that if f is the conformal map from the given region to the upper half plane, the required complex potential is given by $F(z) = \alpha f(z)$.

5.3.6 EXAMPLE   *Find the flow around the upper half of the unit circle if the velocity is parallel to the x axis and is α at infinity.*

SOLUTION   We shall map the exterior of the given region to the upper half plane. Such a conformal map is $z \mapsto z + 1/z$ (Fig. 5.3.16). Thus $F_0(z) = \alpha z$ is the complex potential in the upper half plane, and so the required complex potential is

$$F(z) = \alpha\left(z + \frac{1}{z}\right)$$

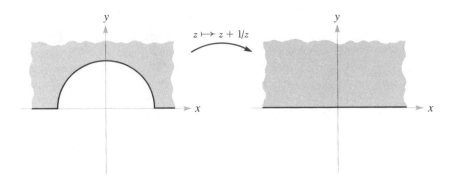

FIGURE 5.3.16   Effect of $z \mapsto z + 1/z$.

It is convenient to use polar coordinates r and θ to express ϕ and ψ. Then we get

$$\phi(r, \theta) = \alpha \left(r + \frac{1}{r} \right) \cos \theta \quad \text{and} \quad \psi(r, \theta) = \alpha \left(r - \frac{1}{r} \right) \sin \theta$$

A few streamlines are shown in Fig. 5.3.17.

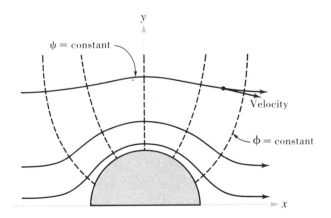

FIGURE 5.3.17   Streamlines for flow around a cylinder.

NOTE   By slightly modifying the transformation $z \mapsto z + 1/z$ by the addition of appropriately chosen higher-order terms, the half circle can be replaced by something more closely resembling an airplane wing; these are called *Joukowski transformations*. ▲

Exercises

1. Find a formula for determining the temperature in the region with the indicated boundary values shown in Fig. 5.3.18.

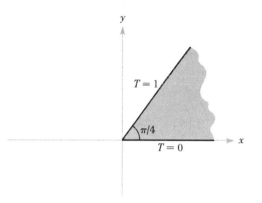

FIGURE 5.3.18 Find T for this region.

2. Find a formula for determining the temperature in the region illustrated in Fig. 5.3.19. (*Hint.* Consider the map $z \mapsto \sin z$.)

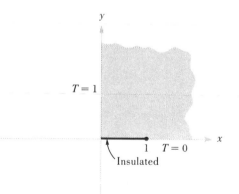

FIGURE 5.3.19 Find the temperature in this region.

3. Find the electric potential in the region illustrated in Fig. 5.3.20 on p. 390. Sketch a few equipotential curves.

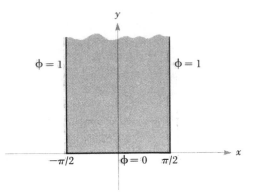

FIGURE 5.3.20 Find the electric potential.

4. Find the electric potential in the region illustrated in Fig. 5.3.21.

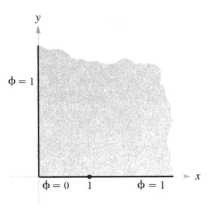

FIGURE 5.3.21 Find ϕ for this region.

5. Find the flow around a circular disk if the flow is at an angle θ to the x axis with velocity α at infinity (Fig. 5.3.22).

6. Suppose a point charge of $+1$ is located at $z_0 = (1 + i)/\sqrt{2}$ and the positive real and imaginary axes (boundaries of the first quadrant $A = \{z \mid \text{Re } z > 0 \text{ and Im } z > 0\}$) are a grounded conductor maintained at potential 0. Find the potential at every point $z \neq z_0$ inside the region A.

7. Obtain a formula for determining the flow of a fluid in the region illustrated in Fig. 5.3.23. (The velocity at ∞ is α.)

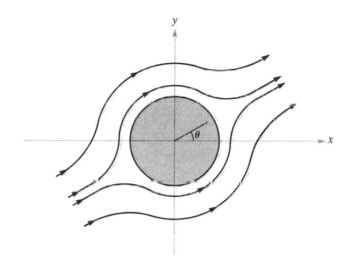

FIGURE 5.3.22   Flow around a disk (Exercise 5).

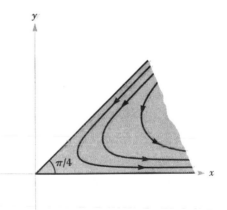

FIGURE 5.3.23   Fluid flow in a wedge (Exercise 7).

REVIEW EXERCISES FOR CHAPTER 5

1.   Consider the map $z \mapsto z^3$. On what sets $A \subset \mathbb{C}$ is this map conformal (onto its image)?

2.   Verify directly that the map $z \mapsto z^n$ preserves the orthogonality between rays from 0 and circles around 0.

3.   Find a conformal map that takes the unit disk onto itself and maps $i/2$ to 0.

4. Find a conformal map of the unit disk onto itself with $f(\frac{1}{4}) = -\frac{1}{3}$ and $f'(\frac{1}{4}) > 0$.

5. Find a conformal map that takes the region $A = \{z \text{ such that } |z - 1| > 1 \text{ and } |z - 2| < 2\}$ onto $B = \{z \,|\, 0 < \text{Re } z < 1\}$.

6. Let $z_1, z_2 \in \mathbf{C}$ and $a \in \mathbf{R}, a > 0$. Show that

$$\left| \frac{z - z_1}{z - z_2} \right| = a$$

defines a circle and z_1, z_2 are inverse points in that circle (that is, they are collinear with the center z_0 and $|z_1 - z_0| \cdot |z_2 - z_0| = \rho^2$ where ρ is the radius of the circle).

7. Examine the image of the set $\{z \in \mathbf{C} \,|\, \text{Im } z \geq 0, 0 \leq \text{Re } z \leq \pi/2\}$ under the map $z \mapsto \sin z$ by considering it to be the composition of the maps $z \mapsto e^{iz}$, $z \mapsto z - 1/z$, $z \mapsto z/2i$.

8. Let $f: A \to B$ be a conformal map, let γ be a curve in A, and let $\tilde{\gamma} = f \circ \gamma$. Show that

$$l(\tilde{\gamma}) = \int_a^b |f'(\gamma(t))| \cdot |\gamma'(t)| \, dt$$

If f preserves the lengths of all curves, argue that $f(z) = e^{i\theta}z + a$ for some $a \in \mathbf{C}$ and for $\theta \in [0, 2\pi[$.

9. Find a conformal map that takes $A = \{z \text{ such that } |z - i| < 1\}$ onto $B = \{z \text{ such that } |z - 1| < 1\}$.

10. Show that the function $f(z) = (z - 1)/(z + 1)$ maps the region $A = \{z \text{ such that } |z| > 1 \text{ and } |z - 1| < 2\}$ one to one onto $B = \{z \,|\, 0 < \text{Re } z < \frac{1}{2}\}$.

11. The region A in Exercise 10 is bounded by two circles, as is the region $\{z \,|\, 1 < |z| < 2\}$. Can a conformal map from this region to B be accomplished by a fractional linear transformation? If so, display the function. If not, why not?

12. Let $T(z) = (az + b)/(cz + d)$. Show that $T(T(z)) = z$ (that is, $T \circ T = \text{identity}$) if and only if $a = -d$.

13. Is it possible to find a conformal map of the interior of the unit circle onto its exterior? Is $f(z) = 1/z$ such a map?

14. Find a conformal map of the quarter plane $A = \{z \,|\, \text{Re } z > 0 \text{ and } \text{Im } z > 0\}$ one-to-one onto the unit disk which takes $1 + i$ to 0 with positive derivative at $1 + i$.

15. Let F_1 and F_2 be conformal maps of the unit disk onto itself and let $F_1(z_0) = F_2(z_0) = 0$ for some fixed $z_0, |z_0| < 1$. Show that there is a $\theta \in [0, 2\pi[$ such that $F_1(z) = e^{i\theta}F_2(z)$.

16. Suppose f is a conformal map of the upper half plane one-to-one onto itself with $f(-1) = 0$, $f(0) = 2$, and $f(1) = 8$. Find $f(i)$.

17. Give a complete list of all conformal maps of the first quadrant $A = \{z \mid \mathrm{Re}\ z > 0$ and $\mathrm{Im}\ z > 0\}$ onto itself.

18. Describe the region $A = \left\{ z \text{ such that } \left| \dfrac{z+3}{z-1} \right| < 3 \right\}$. (*Hint.* $f(z) = (z+3)/(z-1)$ takes what points to the circle $|w| = 3$?)

19. Find a conformal map that takes the region in Fig. 5.R.1 to the upper half plane. Use this map to find the electric potential ϕ with the stated boundary conditions. (*Hint.* Consider a branch of $z \mapsto \sqrt{z^2 - 1}$ after rotating the figure by $90°$. See also Worked Example 5.2.16.)

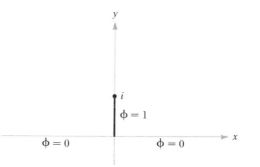

FIGURE 5.R.1   Boundary data for Exercise 19.

20. Find the flow of a fluid in the region shown in Fig. 5.R.2.

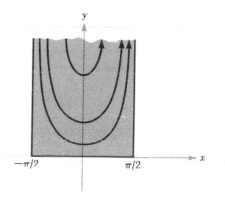

FIGURE 5.R.2   Region for Exercise 20.

21. Use Exercise 19 or Worked Example 5.2.16 to find the fluid flow over the obstacle in Fig. 5.R.3 and plot a few streamlines.

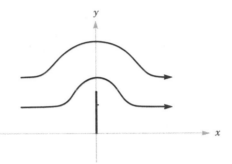

FIGURE 5.R.3 Flow over an obstacle.

22. Let B be the open first quadrant, that is, $B = \{z \mid \operatorname{Re} z > 0 \text{ and } \operatorname{Im} z > 0\}$, and let $S = \{z \mid 0 < \operatorname{Im} z < \pi\}$.
(a) Find an analytic function which maps B one to one onto S.
(b) Find a function u harmonic on B and continuous on the closure of B except at $(0, 0)$ which satisfies $u(x) = 0$ and $u(iy) = \pi$ for $y > 0$.

23. Find the electric potential in the region shown in Fig. 5.R.4.

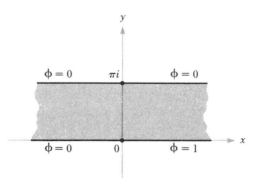

FIGURE 5.R.4 Boundary data for Exercise 23.

24. Suppose a point charge of $+1$ is placed at $z_0 = i$ in the upper half plane and the real axis is a grounded conductor maintained at constant potential 0. Find the potential at every point $z \neq i$ in the upper half plane.

25. Use the Schwarz-Christoffel formula to find a conformal map between the two regions shown in Fig. 5.R.5. ($A = -1$, $B = 1$, $B' = 0$.)

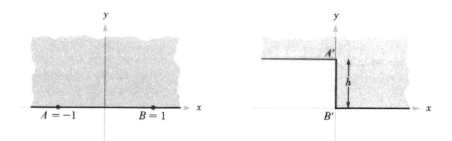

FIGURE 5.R.5 Regions for Exercises 25 and 26.

26. Use Exercise 25 to find the flow lines over the step in the bed of the deep channel shown in Fig. 5.R.5.

27. Find the temperature on the region illustrated in Fig. 5.R.6. (*Hint.* Use $z \mapsto \sin^{-1} z$.)

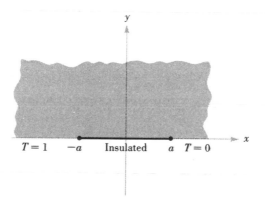

FIGURE 5.R.6 Boundary data for Exercise 27.

28. Let g_n be a sequence of analytic functions defined on a region A. Suppose that $\sum_{n=1}^{\infty} |g_n(z)|$ converges uniformly on A. Prove that $\sum_{n=1}^{\infty} |g_n'(z)|$ converges uniformly on closed disks in A.

29. Evaluate by residues:

$$\int_0^{\infty} \frac{\cos x}{x^2 + 3} \, dx$$

30. Let f be analytic on $\mathbb{C}\backslash\{0\}$. Suppose that $f(z) \to \infty$ as $z \to 0$ and $f(z) \to \infty$ as $z \to \infty$. Prove that f can be written in the form

$$f(z) = \frac{c_k}{z^k} + \cdots + \frac{c_1}{z} + c_0 + d_1 z + \cdots + d_l z^l$$

for constants c_i and d_j.

31. If $\sum_0^\infty a_n z^n$ has radius of convergence ρ, what is the radius of convergence of $\sum_{n=0}^\infty a_n z^{2n}$? Of $\sum_{n=0}^\infty a_n^2 z^n$?

32. Find the Laurent expansion of $f(z) = z^4/(1 - z^2)$ that is valid on the annulus $1 < |z| < \infty$.

CHAPTER 6

FURTHER DEVELOPMENT OF THE THEORY

This chapter continues the development of the theory of analytic functions that was begun in Chaps. 3 and 4. The main tools used in this development are Taylor series and the residue theorem.

The first topic in this chapter is analytic continuation; that is, the attempt to make the domain of an analytic function as large as possible. Further investigation of analytic continuation leads naturally to the concept of a Riemann surface, which is briefly discussed. Additional properties of analytic functions are developed in subsequent sections. Some of these properties deal with such topics as counting zeros of an analytic function; others are generalizations of the inverse function theorem.

6.1 ANALYTIC CONTINUATION AND ELEMENTARY RIEMANN SURFACES

The first theorem to be proved in this section is called the principle of analytic continuation, which is also referred to as the identity theorem. This theorem and its proof lead to a discussion of Riemann surfaces, which facilitates a more satisfactory treatment of what were previously referred to as "multiple-valued functions," such as $\log z$ and $\sqrt{z}$. The discussion is heuristic, being primarily intended to motivate more advanced work.

Analytic Continuation

If two analytic functions agree on a small portion of a (connected) region, then they agree on the whole region on which they are both analytic. This is stated precisely in the following theorem.

6.1.1 PRINCIPLE OF ANALYTIC CONTINUATION, OR IDENTITY THEOREM
Let f and g be analytic in a region A. Suppose that there is a sequence z_1, $z_2, \ldots$ of distinct points of A converging to $z_0 \in A$, such that $f(z_n) = g(z_n)$ for all $n = 1, 2, 3, \ldots$. Then $f = g$ on all of A (see Fig. 6.1.1). The conclusion is valid, in particular, if $f = g$ on some neighborhood of some point in A.

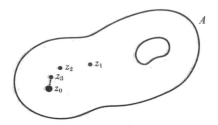

FIGURE 6.1.1   Identity theorem: $\{f = g \text{ on } z_1, z_2, \ldots\} \Rightarrow \{f(z) = g(z) \text{ for all } z \in A\}$.

PROOF   We must show that $f - g = 0$ on A. We shall do this by proving that for a given analytic function h on A, the following four assertions are equivalent:

(i) For some z_0, the nth derivative at z_0 vanishes: $h^{(n)}(z_0) = 0$, $n = 0, 1, 2, \ldots$.
(ii) $h = 0$ on some neighborhood of z_0.
(iii) $h(z_k) = 0$ where z_k is some sequence of distinct points converging to z_0.
(iv) $h = 0$ on all of A.

After these equivalences are proved, the theorem follows by taking $f - g = h$ and applying (iv).

First, (i) $\Longleftrightarrow$ (ii) by Taylor's theorem. (The student should write out the details if this assertion is not clear.) Next we show that (iii) $\Rightarrow$ (ii) by showing that the zeros of an analytic function h are isolated, unless $h = 0$ on a neighborhood of z_0. Indeed, if h is not identically zero on a disk around z_0, we can write $h(z) = (z - z_0)^k \phi(z)$ where $\phi(z_0) \neq 0$ and k is an integer ≥ 1 (Why?). Now $\phi(z) \neq 0$ in a whole neighborhood of z_0 by continuity, so in that neighborhood $h(z) \neq 0$ except at $z = z_0$. This statement contradicts $h(z_n) = 0$ since, for n large enough, z_n lies in the neighborhood, and we can suppose that $z_n \neq z_0$.

Clearly, (iv) $\Rightarrow$ (iii). The proof will be complete when we show that (ii) $\Rightarrow$ (iv). Let $B = \{z \in A \mid h \text{ is zero in a neighborhood of } z\}$. By its definition, B is open and is nonempty by hypothesis. We shall show that B is closed in A as well by showing that if $z_k \to z$, $z_k \in B$, then $z \in B$. It suffices (by the previous result (i) $\Rightarrow$ (ii) applied to the point z) to show that $h^{(n)}(z) = 0$ for all n. But $h^{(n)}(z) = \lim_{k \to \infty} h^{(n)}(z_k) = 0$. Thus B is closed, open, and nonempty, and so $B = A$, because A is assumed to be connected (recall that a region is open and connected by definition). ∎

An interesting application of this theorem is the following. There is exactly one analytic function on $\mathbb{C}$ that agrees with e^x on the x axis, namely, e^z. This is an immediate consequence of the identity theorem because the x axis contains a convergent sequence of distinct points (for example, $1/n$).

The following consequences of the identity theorem are sufficiently important to be worth stating explicitly.

6.1.2 COROLLARY *The zeros (or, more generally, points where a specified value w_0 is assumed) of a nonconstant analytic function are isolated in the following sense. If f is analytic and not constant in a region A and $f(z_0) = w_0$ for a point z_0 in A, then there is a number $\epsilon > 0$ such that $f(z)$ is not equal to w_0 for any z in the deleted neighborhood $\{z \mid 0 < |z - z_0| < \epsilon\}$.*

PROOF If there were no such ϵ then f would agree with the constant function $h(z) = w_0$ at least on a sequence of points converging to z_0. But then it would agree with h everywhere on A by the identity theorem and so be constant. ∎

Notice that there can be a limit point of zeros on the boundary of the region of analyticity. (This is illustrated in Worked Example 6.1.11 with the function $\sin(1/z)$.) The identity theorem says that a nonconstant function cannot have a limit point of zeros in the *interior* of the region of analyticity.

6.1.3 COROLLARY *Let $f: A \to C$ and $g: B \to C$ be analytic on regions A and B. Suppose that $A \cap B \neq \varnothing$ and $f = g$ on $A \cap B$. Define*

$$h(z) = \begin{cases} f(z) & \text{if } z \in A \\ g(z) & \text{if } z \in B \end{cases}$$

*Then h is analytic on $A \cup B$ and is the only analytic function on $A \cup B$ equaling f on A (or g on B). We say that h is an **analytic continuation** of f (or g) (see Fig. 6.1.2).*

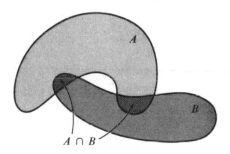

FIGURE 6.1.2 Analytic continuation.

PROOF That h is analytic is obvious because f and g are. Uniqueness of h results at once from the identity theorem and from the facts that $A \cup B$ is a region and that $A \cap B$ is open. ∎

Analytic continuation is important because it provides a method for making the domain of an analytic function as large as possible. However, the following phenomenon can occur. Let f on A be continued to a region A_1 and let A_2 be as pictured in Fig. 6.1.3. If we continue f to be analytic on A_1, then continue this new function from A_1 to

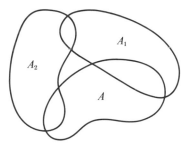

FIGURE 6.1.3   Continuation of a function from A to A_1 and from A_1 to A_2.

A_2, the result need not agree with the original function f on A. A specific example should clarify this point. Consider $\log z$, the principal branch $(-\pi < \arg z < \pi)$ on the region A consisting of the right half plane union the lower half plane. The log function may be continued uniquely so as to include $A_1 =$ the upper half plane in its domain. Similarly, we can continue the log again from the upper half plane so as to include $A_2 =$ the left half plane in its domain by choosing the branch $0 < \arg z < 2\pi$. But these branches do not agree on the third quadrant; they differ by $2\pi i$ (see Fig. 6.1.4). Therefore, in continuing a

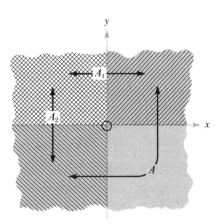

FIGURE 6.1.4   Continuing the log.

function we must be sure that the function on the extended region B agrees with the original on the whole intersection $A \cap B$ and not merely on part of it.

It is not always possible to extend an analytic function to a larger domain. The reader is asked in Exercise 5 to confirm that the power series $\sum_{n=0}^{\infty} z^{n!}$ converges to an analytic function $f(z)$ on the open unit disk but that this function cannot be analytically continued to any larger open set. The unit circle is called a *natural boundary* for this function. In the next two subsections we will examine techniques by which analytic continuation may sometimes be accomplished.

Schwarz Reflection Principle

There is a special case of analytic continuation that can be dealt with directly as follows.

6.1.4 SCHWARZ REFLECTION PRINCIPLE *Let A be a region in the upper half plane whose boundary bd (A) intersects the real axis in an interval $[a, b]$ (or finite union of disjoint intervals). Let f be analytic on A and continuous on $A \cup\,]a, b[$. Let $\bar{A} = \{z \mid \bar{z} \in A\}$, the reflection of A (see Fig. 6.1.5), and define g on $\bar{A}$ by $g(z) = \overline{f(\bar{z})}$. Assume that f is real on $]a, b[$. Then g is analytic and is the unique analytic continuation of f to $A \cup\,]a, b[\, \cup \bar{A}$.*

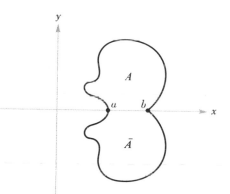

FIGURE 6.1.5 $\bar{A}$ is the reflection of A.

PROOF Uniqueness is implied by the identity theorem because f equals g on a set (namely, $]a, b[$) containing a convergent sequence of distinct points; note that $f = g$ on the real axis because there, $z = \bar{z}$ and $\bar{f} = f$. (f is assumed real on the real axis.) The analyticity of g on $\bar{A}$ follows directly from the Cauchy-Riemann equations and was established in Worked Example 1.5.18. If h is defined on $A \cup\,]a, b[\, \cup \bar{A}$ by setting it equal to $f(z)$ on $A \cup\,]a, b[$ and to $g(z)$ on $\bar{A}$, then h is continuous since $f = g$ on the real axis. ($\bar{z} = z$ and $\bar{f} = f$ there, since f is real on the real axis.) Thus h is analytic on A and $\bar{A}$ and is continuous on $A \cup]a, b[\cup \bar{A}$. Analyticity on the whole set follows from Morera's theorem and was established in Worked Example 2.4.17. ∎

This result is remarkable in that we required only that f be continuous and real on $]a, b[$. It followed automatically that f is analytic on $]a, b[$ when continued across the real axis. To help see that g (and thus h) is analytic on $\bar{A}$, consider the map in three steps:

$$z \mapsto \bar{z}; \qquad \bar{z} \mapsto f(\bar{z}); \qquad f(\bar{z}) \mapsto \overline{f(\bar{z})}$$

The middle map is conformal; the first and last are anticonformal in the sense that they reverse angles. Since angles are reversed twice, the net result is to preserve angles. The whole map is thus conformal.

A related reflection principle can be formulated using circles in place of the real axis and replacing complex conjugation by reflection in the circle. The Schwarz reflection theorem (6.1.4) is a special case if lines are treated as circles of infinite radius, as in Chap. 5.

6.1.5 SCHWARZ REFLECTION PRINCIPLE FOR A CIRCLE *Let A be a region in the interior or exterior of a circle C_1 (or on one side of a line) with part of its boundary an arc γ of C_1. Suppose f is analytic on A and continuous on $A \cup \gamma$ and $f(\gamma)$ is an arc Γ of another circle (or line) C_2. Let $\tilde{A} = \{z \mid \tilde{z} \in A\}$ be the reflection of A in C_1 and define g on $\tilde{A}$ by $g(z) = [f(\tilde{z})]^{\sim}$ (the second $\sim$ denotes reflection in C_2.) Then g is analytic and is the unique analytic continuation of f to $A \cup \gamma \cup \tilde{A}$.*

PROOF We assume A is interior to C_1 and $f(A)$ is interior to C_2. The other cases are similar. Let T_i, $i = 1, 2$, be fractional linear transformations taking C_i to the real axis and their interiors to the upper half plane. For w in $T_1(A)$, $h(w) = T_2(f(T_1^{-1}(w)))$ is analytic and by the Schwarz reflection principle (6.1.4), $h(\bar{w})$ gives an analytic continuation to $\overline{T_1(A)}$. Using the fact that fractional linear transformations preserve reflection in circles (Proposition 5.2.7) (and that complex conjugation is reflection in the real axis) we find that $[f(\tilde{z})]^{\sim} = T_2^{-1}(h(\overline{T_1(z)}))$ and so is an analytic continuation of f. (See Fig. 6.1.6.) ∎

An argument similar to that used above to establish Worked Example 2.4.17 and the Schwarz reflection principle (6.1.4) from Morera's theorem can be used to establish the following.

6.1.6 ANALYTIC CONTINUATION BY CONTINUITY *Let A and B be disjoint simply connected regions whose boundaries intersect in a simple smooth curve γ. Let*

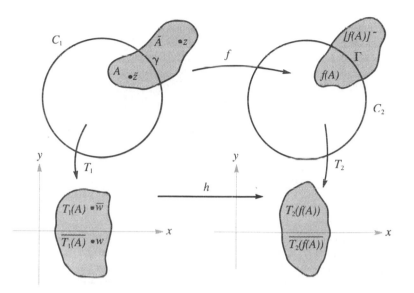

FIGURE 6.1.6 Analytic continuation by reflection.

$C = A \cup$ *(interior γ) $\cup$ B (where interior γ means the image of γ without its endpoints) and suppose that*

(i) *Each point in interior γ has a neighborhood in C.*
(ii) *f is analytic in A and continuous on $A \cup \gamma$.*
(iii) *g is analytic in B and continuous on $B \cup \gamma$.*
(iv) *For $t \in \gamma$, $\displaystyle\lim_{\substack{z \to t \\ z \in A}} f(z) = \lim_{\substack{z \to t \\ z \in B}} g(z)$.*

Then there is a function h analytic on C that agrees with f on A and g on B.

Analytic Continuation by Power Series along Curves

Suppose that f is analytic in a neighborhood U of z_0 and that γ is a curve joining z_0 to another point z' (as in Fig. 6.1.7). If we want to continue f to z' we can proceed as follows. For z_1 on γ in U consider the Taylor series of f expanded around z_1:

$$\sum_{n=0}^{\infty} \frac{f^{(n)}(z_1)}{n!} (z - z_1)^n$$

This power series may have a radius of convergence such that the power series is analytic farther along γ than the portion of γ in U. The power series so obtained then

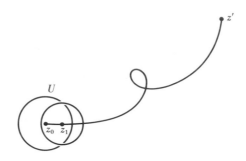

FIGURE 6.1.7   Continuation by power series.

defines an analytic continuation of f. We can continue this way along γ in hopes of reaching z', which will be possible if the successive radii f convergence do not shrink to 0 before we reach z'. If we succeed, we say f can be *analytically continued* along γ. However, we must be careful because the analytic continuation of f so defined might not be single-valued if γ intersects itself (as in Fig. 6.1.8).

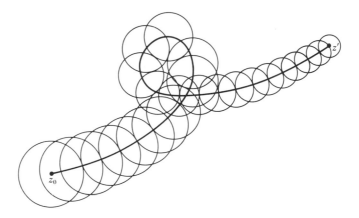

FIGURE 6.1.8   Continuation can lead to self-intersections.

The coefficients of the power series around the new center z_1 can be computed in terms of those for the Taylor series of f around the original center z_0. (See Worked Example 6.1.13.) If z' can be reached at all by this process, then it can be reached in a finite number of steps. This is essentially because of the path covering lemma. (See Exercise 7.) Thus the continuation at z' can be computed in terms of the original function. (A discussion of this including numerical aspects of the computation may be found in Chap. 3 of *Applied and Computational Complex Analysis* by Peter Henrici New York: Wiley-Interscience, 1974.)

The example $\Sigma\, z^{n!}$ mentioned earlier shows that it can happen that there is no direction in which a power series can be continued. Fortunately this is not usually the case. However, there must always be at least one direction in which continuation is not possible.

6.1.7 PROPOSITION *Suppose that $f(z) = \Sigma_{n=0}^{\infty} a_n(z - z_0)^n$ has radius of convergence $R < \infty$. Then there must be at least one point z_1 with $|z_0 - z_1| = R$ such that f cannot be analytically continued to any open set containing z_1.*

PROOF Let $B = \{z$ such that $|z - z_0| < R\}$ and let C be its boundary circle $\{z$ such that $|z - z_0| = R\}$. We will show that if the assertion were false then f could be analytically continued to an open set A containing the closed disk $B \cup C$. If this were done, Worked Example 1.4.28 would show that A contains a larger disk $B_\epsilon = \{z$ such that $|z - z_0| < R + \epsilon\}$. (See Fig. 6.1.9.) We would have continued f to a larger disk with the same center.

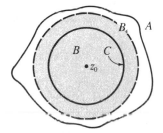

FIGURE 6.1.9 If A is an open set containing B and its boundary, then A contains a slightly larger disk.

This is not possible, since it implies a radius of convergence larger than R. (See Worked Example 6.1.12.)

To obtain A we proceed as follows. For each w on C there is a neighborhood B_w of w and an analytic continuation f_w of f to $A_w = B \cup B_w$. (See Fig. 6.1.10.) Then $A =$ (union of all the A_w) is an open set containing $B \cup C$. We try to define a continuation of f to A by setting $g(z) = f_w(z)$ for z in A_w. If this makes unambiguous sense it will certainly be analytic on A since f_w is analytic on A_w. For g to make sense we need to know that if z is in $A_{w_1} \cap A_{w_2}$, then $f_{w_1}(z) = f_{w_2}(z)$. But this is true. The two functions are both analytic on the region $A_{w_1} \cap A_{w_2}$ and they are both equal to f on the open set $B \subset A_{w_1} \cap A_{w_2}$. Therefore they must agree on the whole region, by the identity theorem. Thus the definition of g makes sense. It does not depend on which A_w we happen to select containing z. ∎

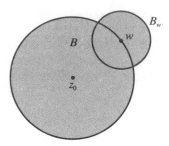

FIGURE 6.1.10   The set $A_w = B \cup B_w$.

Consequently, the radius of convergence of an analytic continuation is largely independent of the method used to obtain it. This agrees with what we were seeing in Chap. 3: the radius of convergence should be the distance to the nearest unavoidable singularity.

6.1.8 PROPOSITION   *Suppose f is analytic on a neighborhood of a point z_0 of a region A and that f can be analytically continued along every curve joining z_0 to every other point z_1 of A. Then the radius of convergence of the Taylor series at z_1 for each such continuation to z_1 is the same, and is at least as great as the distance from z_1 to the complement of A.*

PROOF   Suppose not. Then by extending the curve radially from z_1 to any point on the circle of convergence, the continuation could be analytically continued still further in every direction from z_1, contrary to Proposition 6.1.7. ∎

This proposition does *not* claim that the continuations are all the same. They might not be, as the example of the logarithm shows. We may merely obtain local functions defined on disks but they need not agree on overlaps. This construction is one basic way in which multiple-valued functions arise. A point is called a *branch point* if analytic continuation around a closed curve surrounding it can produce a different value upon return to the starting point. The following basic result says that multiple-valued functions do not arise from continuation along curves in simply connected regions.

6.1.9 MONODROMY PRINCIPLE   *Let A be simply connected and let $z_0 \in A$. Let f be analytic in a neighborhood of z_0. Suppose that f can be analytically continued along any arc joining z_0 to another point $z \in A$. Then this continuation defines a (single-valued) analytic continuation of f on A.*

PROOF We need to show that if z_1 is another point of A, then the process of continuation along a curve γ from z_0 to z_1 through A will always produce the same value at z_1 regardless of what curve is used. To this end, let γ_0 and γ_1 be two curves from z_0 to z_1 in A. Since A is simply connected, they are homotopic with fixed endpoints in A. That is, there is a continuous function H: $[0, 1] \times [0, 1] \to A$ from the unit square into A such that $H(0, t) = \gamma_0(t)$, $H(1, t) = \gamma_1(t)$, $H(s, 0) = z_0$, and $H(s, 1) = z_1$ for all s and t between 0 and 1, inclusive. The functions $\gamma_s(t) = H(s, t)$ are a family of curves from z_0 to z_1 in A deforming continuously from γ_0 to γ_1. See Fig. 6.1.11. There is an analytic continuation

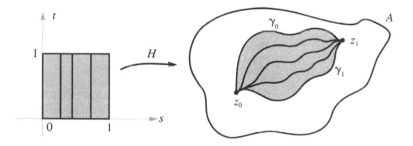

FIGURE 6.1.11 Homotopy between γ_0 and γ_1.

f_s of f from z_0 to z_1 along each curve γ_s. We will show that $f_s(z_1)$ cannot change as s is shifted continuously from 0 to 1, and therefore that $f_0(z_1) = f_1(z_1)$. This is exactly what we need to establish the theorem.

The image of the square is a closed bounded subset of A. Thus by the distance lemma (1.4.21) it lies at a positive distance ρ from the complement of A. By Proposition (6.1.8) the radius of convergence always remains at least ρ as we analytically continue f along any of the curves γ_s. By the path covering lemma (1.4.24) the continuation along any γ_s may be completed to z_1 in a finite number of steps using disks of radius ρ. For each s this procedure produces an analytic continuation of f to a function f_s analytic on a "tube" A_s around γ_s, as in Fig. 6.1.12. With a bit of care we can select a finite number of points $0 = s_0 < s_1 < s_2 < \cdots < s_N = 1$ and the values of t defining the centers of the disks making up the tubes A_s close enough together that γ_{s_k} is contained in the preceding tube $A_{s_{k-1}}$ and in the succeeding tube $A_{s_{k+1}}$. This is done using the uniform continuity of H exactly as in the proof of the deformation theorems (2.3.12); see, in particular, Fig. 2.3.14. The functions f_{s_k} are each analytic on the region $A_{s_k} \cap A_{s_{k+1}}$ and agree on the open set $D(z_0; \rho) \subset A_{s_k} \cap A_{s_{k+1}}$, and so they agree on the whole region by the identity theorem. In particular, $f_{s_k}(z_1) = f_{s_{k+1}}(z_1)$, so that $f_0(z_1) = f_{s_1}(z_1) = f_{s_2}(z_1) = \cdots = f_{s_N}(z_1) = f_1(z_1)$. The continuation of f along γ_0 to z_1 agrees with that along γ_1 to z_1 at the point z_1. This is what we needed to show. ∎

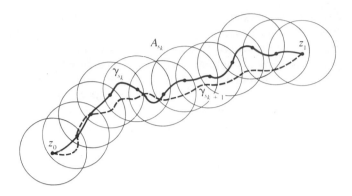

FIGURE 6.1.12   Each $\gamma_{s_{k+1}}$ is contained in A_{s_k}.

For regions that are not simply connected, we can get different values for the continuation of f when we traverse two different paths. This fact was already mentioned at the beginning of this section in connection with log z. For example, in Fig. 6.1.13, starting with log defined near 1 and continuing along γ_1, we get log $(-1) = \pi i$, whereas along γ_2, we get log $(-1) = -\pi i$. This is because the region $\mathbf{C}\backslash\{0\}$ is not simply connected.

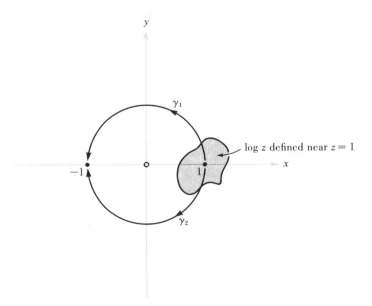

FIGURE 6.1.13   Continuation of log z along two different arcs from 1 to -1.

Riemann Surfaces of Some Elementary Functions

The phenomenon just described may lead the student to ask if there is a definition of log that does not introduce any artificial branch lines (which, after all, can be chosen arbitrarily). The answer is given by a brilliant idea of Georg Riemann in his doctoral thesis in 1851 that is briefly described here.

For the logarithm, if log z is to be single-valued, we should merely regard γ_1 and γ_2 in Fig. 6.1.13 as ending up in different places. This can be pictured as in Fig. 6.1.14. Only a

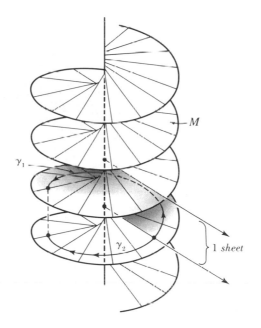

FIGURE 6.1.14   Riemann surface for log z.

core of the spiral staircase M, with axis over the origin, is shown—it should have infinite extent laterally. If we cut from 0 outward at any level and the one directly below it, we get a part of the surface called a *sheet* (the shaded portion in Fig. 6.1.14). This can be identified with the domain for a branch of log. Thus we have stacked up infinitely many copies of the complex plane **C** joined through 0 and glued together as shown in Fig. 6.1.14. The arcs γ_1 and γ_2 now go to different points so we can assign different values of log z to each without ambiguity.

The main property of this surface that enables us to define log $z - \log |z| + i$ arg z as a single-valued function is that on this surface arg z is well defined, and the different sheets correspond to different intervals of length 2π in which arg z takes its values.

Thus we can take care of multiple-valued functions by introducing an enlarged domain on which the function becomes single-valued.

Let us briefly consider another example, the square root function: $z \mapsto \sqrt{z} = \sqrt{r}e^{i\theta/2}$. Here the situation is slightly different from that for the log function. If we go around the origin once, $\sqrt{z}$ takes on a different value, but if we go around twice (increase θ by 4π), we arrive back at the same value, so we want to be at the same point on the Riemann surface. The surface is illustrated in Fig. 6.1.15.

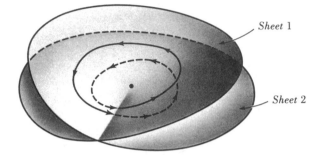

Sheet 1

Sheet 2

FIGURE 6.1.15   Riemann surface for $\sqrt{z}$.

Though the sheets in this figure appear to intersect, they are not supposed to. At fault is our attempt at visualization in $\mathbf{R}^3$. One can consider the Riemann surface to be in $\mathbf{R}^4$ or $\mathbf{C}^2$. Figure 6.1.15 is a picture of its "shadow" in $\mathbf{R}^3$. Here is another way to think about how the surface is related to analytic continuation. Let γ be the unit circle traveled twice counterclockwise by letting t change smoothly from 0 to 4π in $\gamma(t) = e^{it}$. Then $f(t) = e^{it/2}$ gives a smoothly changing square root for $\gamma(t)$. At the start, $\gamma(0) = 1$ and $f(0) = 1 = $ "$\sqrt{1}$." As we make the first transit around the circle, $\gamma(t)$ successively hits points B, C, and D (i, -1, and $-i$), and $f(t)$ hits the corresponding points on the image circle. At $t = 2\pi$, $\gamma(t)$ has returned to 1, but $f(t)$ has reached the other "square root," -1. In the second transit around the circle, $\gamma(t)$ revisits the points it hit on the first circuit, while $f(t)$ goes through the other possible square roots in the lower half plane. At the end of the second circuit $\gamma(4\pi) = 1$, and $f(4\pi) = 1$ has returned to the original value. (See Fig. 6.1.16.)

For more complicated functions like $\cos^{-1}(z)$ the Riemann surface can be constructed as follows. On certain regions of $\mathbf{C}$, $\cos z$ is one-to-one and we define $\cos^{-1}(z)$ to be the inverse function. The period strips defined in Sec. 1.3 are examples of such regions for e^z and $\log z$. Such a region for $\cos z$ is shown in Fig. 6.1.17.

The interior of each such strip is mapped conformally onto $\mathbf{C}$ minus the portions $]-\infty, -1]$ and $[1, \infty[$ of the real axis with half planes corresponding to half strips as shown in Fig. 6.1.18. Each of the deleted portions is the image of two different portions

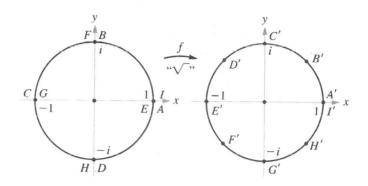

FIGURE 6.1.16 Tracking $\sqrt{z}$ as one traverses the unit circle.

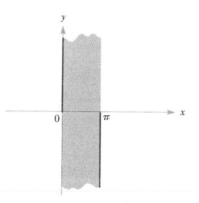

FIGURE 6.1.17 A region on which $\cos z$ is one-to-one.

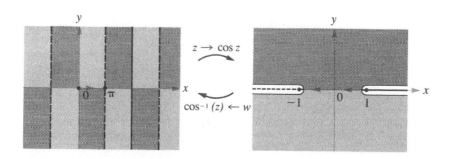

FIGURE 6.1.18 Construction of the Riemann surface for $\cos^{-1} z$.

of the boundary of each strip. Each sheet of the Riemann surface is a copy of $\mathbf{C}$ slit along these portions of the real axis. The surface is then constructed by "gluing" the sheets together along these slits in such a way that half planes are joined in the same way as the corresponding preimage half strips.

A cross section of the surface over the circle $C = \{z$ such that $|z| = 2\}$ might be diagramed somewhat as in Fig. 6.1.19. The black dots in the diagram on the right indicate

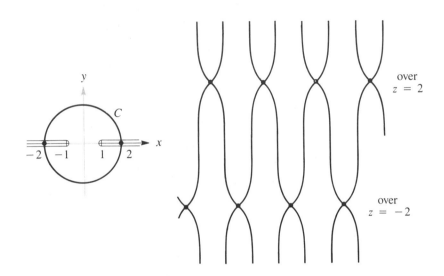

FIGURE 6.1.19   Cross section of Riemann surface for $\cos^{-1} z$ over the circle $|z| = 2$.

the places on the surface over 2 and -2 where the circle C crosses the slits along which the sheets are glued. To construct the model one would roll the diagram on the right into a cylinder joining the top and bottom edges so that the labels on the sheets match. Then one would stand the cylinder over the circle C so that the rows of black dots are over 2 and -2. If we follow a suitably chosen curve winding around 1 and -1 passing sometimes between them and sometimes over the branch cuts, we may pass from any sheet to any other and obtain all possible values of $\cos^{-1} z$.

Worked Examples

6.1.10   *Let f be an entire function equaling a polynomial on* [0, 1] *on the real axis. Show that f is a polynomial.*

Solution. Let $f(x) = a_0 + a_1 x + \cdots + a_n x^n$ on [0, 1]. Then $f(z)$ and $a_0 + a_1 z + \cdots + a_n z^n$ agree for $z \in [0, 1]$, and both are analytic on $\mathbf{C}$ (that is, both are entire). Then by the identity theorem they are equal on all of C, since $[0, 1]$ contains a convergent sequence of distinct points (for example, $z_n = 1/n$). This proves the assertion.

6.1.11 *Prove that if $z_n \to 0$ and $z_n \neq 0$, and if f is defined in a deleted neighborhood of 0, with $f(z_n) = 0$, then f has a nonremovable singularity at $z = 0$ unless f is identically zero. Illustrate with $\sin(1/z)$.*

Solution. If the singularity of f at 0 were removable, then (by definition) we could define $f(0)$ so that f was analytic at 0. Thus if $f(z_n) = 0$, the identity theorem would imply that f is identically zero. (The z_n have infinitely many distinct values—Why?) This is true with $\sin(1/z) = f(z)$ because for $z_n = 1/\pi n$, $z_n \to 0$ but $f(z_n) = 0$. Thus the singularity is not removable.

We can go one step further. For such an f the singularity must be essential, because if f had a pole at 0, then $f(z)$ would go to infinity as $z \to 0$ (see Exercise 7, Sec. 3.3).

6.1.12 *Let $f(z) = \sum_{n=0}^{\infty} a_n(z - z_0)^n$ have a radius of convergence $R > 0$.*
(a) *Is there always a sequence z_n with $|z_n - z_0| < R$ for $n = 1, 2, 3, \ldots$ and $|z_n - z_0| \to R$ such that $f(z_n) \to \infty$?*
(b) *Can f be continued analytically to a disk $|z - z_0| < R + \epsilon$ for some $\epsilon > 0$?*

Solution.
(a) Such a sequence does not necessarily exist. Consider the series $\sum_0^{\infty} z_n/n^2$. By the ratio test, the radius of convergence is

$$\lim_{n \to \infty} \left| \frac{a_n}{a_{n+1}} \right| = \lim_{n \to \infty} \frac{(n+1)^2}{n^2} = 1$$

But for $|z| \leq 1$ we have

$$\left| \sum_0^{\infty} \frac{z^n}{n^2} \right| \leq \sum_0^{\infty} \left| \frac{z^n}{n^2} \right| = \sum_0^{\infty} \frac{|z|^n}{n^2} \leq \sum_0^{\infty} \frac{1}{n^2} < \infty$$

Thus $|f(z)|$ is bounded by $\sum_0^{\infty} 1/n^2$ on $\{z \text{ such that } |z| < 1\}$, and so $f(z_n) \to \infty$ is impossible.
(b) No. Suppose that there is an analytic function g on $|z - z_0| < R + \epsilon$ with $g(z) = f(z)$ for $|z - z_0| < R$. Since f and g are analytic and agree on $|z - z_0| < R$, the Taylor series of g, $\sum_0^{\infty} a_n(z - z_0)^n$, is valid for $|z - z_0| < R + \epsilon$. Hence the radius of convergence of the given series is greater than R, which is impossible (since it equals R).

6.1.13 (a) *Suppose f is given by the power series $f(z) = \sum_{n=0}^{\infty} a_n(z - z_0)^n$ valid for $|z - z_0| < R$. Show that if $|z_1 - z_0| < R$, then the Taylor series for f centered at z_1 is $\sum_{k=0}^{\infty} b_k(z - z_1)^k$, where*
$$b_k = \sum_{m=0}^{\infty} \left[\frac{(k+m)!}{k!m!} a_{k+m}(z_1 - z_0)^m \right].$$
(b) *Work out the first few terms, starting with the principal branch of $\log z$ at $z_0 - 1$ and $z_1 = (1 + i)/2$.*

Solution.

(a) By taking the kth derivative of the series expansion for f about z_0, we find for $|z - z_0| < R$ that

$$f^{(k)}(z) = \sum_{m=0}^{\infty} (k+m)(k+m-1) \cdots (m+1)a_{k+m}(z-z_0)^m$$

$$= \sum_{m=0}^{\infty} \frac{(k+m)!}{m!} a_{k+m}(z-z_0)^m$$

Thus the Taylor series for f around z_1 is

$$\sum_{k=0}^{\infty} \frac{1}{k!} f^{(k)}(z_1)(z-z_1)^k = \sum_{k=0}^{\infty} \left[\sum_{m=0}^{\infty} \frac{(k+m)!}{k!m!} a_{k+m}(z_1-z_0)^m \right] (z-z_1)^k$$

This converges to $f(z)$ when $|z - z_1| < R - |z_1 - z_0|$, but may actually have a larger radius of convergence. If it does, then it gives an analytic continuation of f by power series.

(b) The principal branch of $\log (1 + w)$ for $|w| < 1$ has the expansion $\log (1 + w) = \sum_{n=1}^{\infty} \frac{(-1)^{n-1}}{n} w^n$.
Setting $w = z - 1$ gives $\log z = \sum_{n=1}^{\infty} [(-1)^{n-1}/n](z-1)^n$, valid for $|z - 1| < 1$. Thus $a_0 = 0$ and $a_n = (-1)^{n-1}/n$ for $n > 0$. Since $z_1 - z_0 = (i - 1)/2$, we get

$$b_0 = \sum_{m=1}^{\infty} \frac{(-1)^{m-1}}{m} \left(\frac{i-1}{2}\right)^m = -\frac{1}{2} \log 2 + \frac{\pi i}{4}$$

$$b_1 = \sum_{m=0}^{\infty} \frac{(m+1)!}{m!} \frac{(-1)^{m-1+1}}{m+1} \left(\frac{i-1}{2}\right)^m = \sum_{m=0}^{\infty} \left(\frac{1-i}{2}\right)^m$$

$$= \frac{1}{1 - (1-i)/2} = 1 - i$$

and

$$b_2 = \sum_{m=0}^{\infty} \frac{(m+2)!}{2!m!} \frac{(-1)^{m+1}}{m+2} \left(\frac{i-1}{2}\right)^m = -\frac{1}{2} \sum_{m=0}^{\infty} (m+1) \left(\frac{1-i}{2}\right)^m$$

$$= -\frac{1}{2} \frac{1}{\left[1 - \left(\frac{1-i}{2}\right)\right]^2} = i$$

6.1.14 *Conformal Maps of Annuli. (There is a rich literature about conformal maps of regions which are not simply connected. The subject is complicated since there is no theorem as broad in scope as the Riemann mapping theorem. In some sense the presence of more than one boundary component restricts the possible maps. This example shows how to use the Schwarz reflection principle to study the situation for an annulus.) If $0 < r < 1$, let $A_r = \{z \mid r < |z| < 1\}$, let $C_r = \{z$ such that $|z| = r\}$, and let $C_1 = \{z$ such that $|z| = 1\}$, so that the closure of A_r is $\text{cl}(A_r) = C_r \cup A_r \cup C_1$. Prove the following: Suppose $0 < r < 1$ and $0 < R < 1$ and that f is a one-to-one analytic map of A_r onto A_R which extends to a one-to-one continuous map of $\text{cl}(A_r)$ onto $\text{cl}(A_R)$. Then $r = R$ and f must* *be one of two types: either* (i) *a rotation, where there is a real constant* θ *such that* $f(z) = e^{i\theta}z$ *for all z in* A_r; *or* (ii) *a rotation and inversion, where there is a real constant* θ *such that* $f(z) = re^{i\theta}/z$ *for all z in* A_r.

Solution f must either map C_1 to C_1 and C_r to C_R or interchange the inner and outer circles. If the latter holds, then $f(r/z)$ is another map of A_r onto A_R which does not interchange them. Thus, we may assume that f takes C_1 to C_1 and C_r to C_R continuously. The extended Schwarz reflection principle (6.1.5) shows how to continue f analytically to a map from the larger annulus A_{r^2} onto A_{R^2} so that the continuation is again continuous on the boundary and takes C_{r^2} to C_{R^2}. This process may be repeated indefinitely, to extend f to an increasing sequence of annuli

$$A_r \subset A_{r^2} \subset A_{r^4} \subset \cdots$$

mapping respectively onto

$$A_R \subset A_{R^2} \subset A_{R^4} \subset \cdots$$

(See Fig. 6.1.20.) Each extension maps $C_{r^{2n}}$ to $C_{R^{2n}}$ and the annuli between them correspondingly.

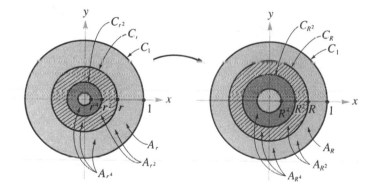

FIGURE 6.1.20 Conformal maps of annuli.

Since $R^{2n} \to 0$ as $n \to \infty$, $\lim_{z \to 0} f(z) = 0$. Thus $z = 0$ is a removable singularity and setting $f(0) = 0$ serves to complete the extension of f to an analytic function of the disk $D = \{z \text{ such that } |z| < 1\}$ to itself with $f(0) = 0$. The extended function satisfies the conditions of the Schwarz lemma, and so $|f(z)| \le |z|$ for all z. Since C_r goes to C_R, this forces $R \le r$. The process could just as well have been applied to f^{-1}, which takes A_R to A_r. This would give $r \le R$, and so $r = R$. Finally, this shows $|f(z)| = |z|$ on each of the circles $C_{r^{2n}}$, so f must be a rotation by the Schwarz lemma. ∎

Exercises

1. (a) Let $f(z) = e^{1/z} - 1$. If $z_n = 1/2\pi ni$, then $z_n \to 0$ and $f(z_n) = 0$, yet f is not identically zero. Does this contradict the identity theorem? Why or why not?
(b) Is the identity theorem true for harmonic functions?

2. Let $h(x)$ be a function of a real variable $x \in \mathbf{R}$. Suppose that $h(x) = \Sigma_{n=0}^{\infty} a_n x^n$, which converges for x in some interval $]-\eta, \eta[$ around 0. Prove that h is the restriction of some analytic function defined in a neighborhood of 0.

3. Let f be analytic in a region A and let $z_1, z_2 \in A$. Let $f''(z_1) \neq 0$. Show that f is not constant on a neighborhood of z_2.

4. Let f be analytic and not identically zero on A. Show that if $f(z_0) = 0$, there is an integer k such that $f(z_0) = 0 = \cdots = f^{(k-1)}(z_0)$ and $f^{(k)}(z_0) \neq 0$.

5. Prove the following result of Karl Weierstrass. Let $f(z) = \Sigma_{n=0}^{\infty} z^{n!}$. Then f cannot be analytically continued to *any* open set properly containing $A = \{z$ such that $|z| < 1\}$. (*Hint.* First consider $z = re^{2\pi i p/q}$ where p and q are integers.)

6. Formulate a Schwarz reflection principle for harmonic functions.

7. Suppose that f can be continued analytically along a curve γ in the manner shown in Fig. 6.1.21. Show that f can be continued by power series (in a finite number of steps).

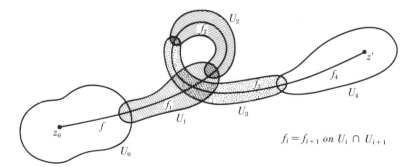

FIGURE 6.1.21   Analytic continuation of f along a curve from z_0 to z'.

8. Discuss the Reimann surface for $\sqrt{z^2 - 1}$.

9. Discuss the Riemann surface for $\sqrt[3]{z}$.

10. Discuss the relationship between Proposition 2.2.6 and the monodromy theorem.

11. Consider the power series $\sum_0^\infty (-1)^n z^n$ defined in $|z| < 1$. To what domain in $\mathbf{C}$ can you analytically continue this function?

12. Show that if f is an analytic map of $\{z \mid r_1 < |z| < R_1\}$ one-to-one onto $\{z \mid r_2 < |z| < R_2\}$ which extends to a continuous map of $\{z \mid r_1 \leq |z| \leq R_1\}$ one-to-one onto $\{z \mid r_2 \leq |z| \leq R_2\}$, then $R_1/r_1 = R_2/r_2$. Give a description of all such functions.

13. Let A be a region, let $f: A \to \mathbf{C}$, and let $\gamma: \,]a, b[\, \to A$ be a smooth non-self-intersecting curve with $\gamma'(t) \neq 0$. Assume f is continuous on A and analytic on $A \backslash \gamma$.
(a) Show that f is analytic.
(b) Use (a) to prove the Schwarz reflection principle.

6.2 ROUCHÉ'S THEOREM AND PRINCIPLE OF THE ARGUMENT

In this section we develop some properties of analytic functions which are used to locate roots of equations within curves. The main tool will be the residue theorem.

A Root and Pole Counting Formula

The main results of this section will be those mentioned in the title. It is convenient to begin with a formula which counts the roots of an equation within a closed curve. A more intuitive version will be given as a corollary of the following precise version.

6.2.1 ROOT-POLE COUNTING THEOREM *Let f be analytic on a region A except for poles at $b_1, \ldots, b_m$ and zeros at $a_1, \ldots, a_n$, counted with their multiplicities (that is, if b_1 is a pole of order k, b_1 is to be repeated k times in the list, and similarly for the zeros a_j). Let γ be a closed curve homotopic to a point in A and passing through none of the points a_i or b_j. Then*

$$\int_\gamma \frac{f'(z)}{f(z)} \, dz = 2\pi i \left[\sum_{j=1}^n I(\gamma, a_j) - \sum_{l=1}^m I(\gamma, b_l) \right] \tag{1}$$

NOTE Equation (1) applies in particular to *meromorphic functions,* that is, functions defined on $\mathbf{C}$ except for poles (see Sec. 3.3). There can be only a finite number of poles in any bounded region, since poles are isolated.

PROOF First, it is clear that $f'(z)/f(z) = g(z)$ is analytic except at $a_1, \ldots, a_n$, $b_1, \ldots, b_m$. If f has a zero of order k at a_j, f' has a zero of order $k - 1$, and so $f'/f = g$

has a simple pole at a_j and the residue there is k. This is because we can write $f(z) = (z - a_j)^k \phi(z)$, as shown in Sec. 3.2, where ϕ is analytic and $\phi(a_j) \neq 0$; therefore,

$$g(z) = \frac{k(z - a_j)^{k-1}\phi(z)}{(z - a_j)^k \phi(z)} + \frac{(z - a_j)^k \phi'(z)}{(z - a_j)^k \phi(z)} = \frac{k}{z - a_j} + \frac{\phi'(z)}{\phi(z)}$$

Thus the residue at a_j is clearly k. Similarly, if b_1 is a pole of order k, we can write, near b_l,

$$f(z) - \frac{\phi(z)}{(z - b_l)^k}$$

where ϕ is analytic and $\phi(b_l) \neq 0$ (see Proposition 3.3.4(iv)). Then, as above, we see that near b_l,

$$g(z) = \frac{-k}{z - b_l} + \frac{\phi'(z)}{\phi(z)}$$

so the residue is $-k$.

By the Residue Theorem,

$$\int_\gamma g(z)\, dz = 2\pi i \left\{ \sum_j{}' [\text{Res}\,(g, a_j)]I(\gamma, a_j) + \sum_l{}' [\text{Res}\,(g, b_l)]I(\gamma, b_l) \right\}$$

where Σ' means the sum over the distinct points. But since the residue equals the number of times a_j occurs and minus that number for the b_l, this expression becomes

$$2\pi i \left[\sum_{j=1}^n I(\gamma, a_j) - \sum_{l=1}^n I(\gamma, b_l) \right]$$

as required. ∎

The formula in the last theorem may be best understood for a simple closed curve, when it may be used for counting zeros and poles.

6.2.2 COROLLARY   *Let γ be a simple closed curve:*

(i) *If f is analytic on an open set containing γ and its interior except for finitely many zeros and poles none of which lie on γ, then*

$$\int_\gamma \frac{f'(z)}{f(z)}\, dz = 2\pi i(Z_f - P_f) \tag{2}$$

where Z_f is the number of zeros inside γ and P_f the number of poles inside γ each counted with their multiplicities (orders).

(ii) **Root Counting Formula:** If f is analytic on an open set containing γ and its interior and $f(z)$ is never equal to w on γ, then

$$\int_\gamma \frac{f'(z)}{f(z) - w}\, dz = 2\pi i N_w$$

where N_w is the number of roots of the equation $f(z) = w$ inside γ counted with their multiplicities as zeros of $f(z) - w$.

PROOF Since γ is simple, the index of γ with respect to a_j is 1 if a_j is inside γ and 0 if it is outside. Theorem 6.2.1 thus gives part (i). Part (ii) follows by applying part (i) to $g(z) = f(z) - w$. ∎

Principle of the Argument

We now consider a useful consequence of the root-pole counting theorem. For a closed curve γ and z_0 not on γ we can legitimately say that the change in argument of $z - z_0$ as z traverses γ is $2\pi \cdot I(\gamma, z_0)$. This is actually the intuitive basis on which the index was developed; it is written $\Delta_\gamma \arg (z - z_0) = 2\pi \cdot I(\gamma, z_0)$ (see Fig. 6.2.1).

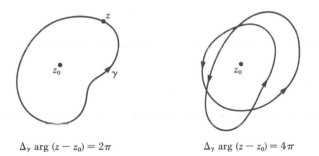

$$\Delta_\gamma \arg (z - z_0) = 2\pi \qquad\qquad \Delta_\gamma \arg (z - z_0) = 4\pi$$

FIGURE 6.2.1 Change in the argument of $z - z_0$ when the two curves are traversed.

Next we want to define $\Delta_\gamma \arg f$, that is, the change in $\arg f(z)$ as z goes once around γ. Intuitively, and for practical computations, the meaning is clear; we merely compute $\arg f(\gamma(t))$ and let t run from a to b if $\gamma: [a, b] \to \mathbb{C}$, then look at the difference $\arg f(\gamma(b)) - \arg f(\gamma(a))$. We choose a branch of the argument such that $\arg f(\gamma(t))$ varies continuously with t. Equivalently, by changing variables, we can let $\tilde{\gamma} = f \circ \gamma$ and compute $\Delta_{\tilde{\gamma}} \arg z$. This leads to formulation of the following definition.

6.2.3 DEFINITION   *Let f be analytic on a region A and let γ be a closed curve in A homotopic to a point and passing through no zero of f. Then set*

$$\Delta_\gamma \arg f = 2\pi \cdot I(f \circ \gamma, 0)$$

(The index makes sense because 0 does not lie on f ∘ γ.)

In examples, we can make use of our previous intuition about the index to compute $\Delta_\gamma \arg f$. The argument principle is as follows.

6.2.4 PRINCIPLE OF THE ARGUMENT   *Let f be analytic on a region A except for poles at $b_1, \ldots, b_m$ and zeros at $a_1, \ldots, a_n$ counted according to their multiplicity. Let γ be a closed curve homotopic to a point and passing through no a_j or b_l. Then*

$$\Delta_\gamma \arg f = 2\pi \left[\sum_{j=1}^{n} I(\gamma, a_j) - \sum_{l=1}^{m} I(\gamma, b_l) \right] \tag{3}$$

PROOF   By the root-pole counting theorem, it suffices to show that

$$i \, \Delta_\gamma \arg f = \int_\gamma \frac{f'(z)}{f(z)} \, dz \tag{4}$$

since f has no zeros or poles on γ. Indeed,

$$i \, \Delta_\gamma \arg f = 2\pi i \cdot I(f \circ \gamma, 0) = \int_{f \circ \gamma} \frac{dz}{z}$$

by the formula for the index (see Sec. 2.4). Letting $\gamma: [a, b] \to \mathbf{C}$, we have

$$\int_{f \circ \gamma} \frac{dz}{z} = \int_a^b \frac{\frac{d}{dt} f(\gamma(t))}{f(\gamma(t))} \, dt = \int_a^b \frac{f'(\gamma(t))}{f(\gamma(t))} \, \gamma'(t) \, dt$$

by the definition of the integral and the chain rule. The latter integral is equal to $\int_\gamma [f'(z)/f(z)] \, dz$ by definition. (If γ is only piecewise C^1, this holds only on each interval where γ' exists, but we get the result by addition.) ∎

Equation (3) is usually applied in the case where γ is a *simple* closed curve. Then we may conclude that *the change in* arg f(z) *as we go once around γ (in a counterclockwise direction) is* $2\pi(Z_f - P_f)$ *where Z_f (or P_f) is the number of zeros (or poles) inside γ counted* *with their multiplicities.* It is somewhat surprising, a priori, that $Z_f - P_f$ and the argument change of f are even related.

This may sound familiar to the alert reader who remembers a trick from calculus called logarithmic differentiation. If γ is a small segment of curve short enough so that $f(\gamma)$ is a curve segment which lies in a half plane as in Fig. 6.2.2, we can define a branch of

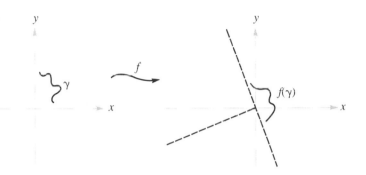

FIGURE 6.2.2 Logarithmic differentiation and the principle of the argument.

logarithm with the branch cut leading away from that half plane by an appropriate choice of the reference angle for defining arg z. Then

$$\frac{d}{dz}\left[\log f(z)\right] = \frac{f'(z)}{f(z)} \qquad \text{along } \gamma$$

and so

$$\int_\gamma \frac{f'(z)}{f(z)}\, dz = \Delta \log f(z) = \Delta \log |f(z)| + i\,\Delta \arg f(z)$$

For a closed curve γ we can do this along successive short parts of the curve using an appropriate choice of logarithm for each. When we return to the starting point, the contributions for $\Delta \log f(z)$ will all have canceled out, but not those for $\Delta \arg f(z)$, since we have kept changing determinations of argument.

Rouché's Theorem

The argument principle can be used to prove a very useful theorem that has many applications, some of which will be given throughout the remainder of this chapter.

6.2.5 ROUCHÉ'S THEOREM *Let f and g be analytic on a region A except for a finite number of zeros and poles in A. Let γ be a closed curve in A homotopic to a point and* *passing through no zero or pole of f or g. Suppose that on γ,*

$$|f(z) - g(z)| < |f(z)|$$

Then (i) $\Delta_\gamma \arg f = \Delta_\gamma \arg g$; and (ii) $Z_f - P_f = Z_g - P_g$ where $Z_f = \sum_{j=1}^n I(\gamma, a_j)$, a_j being the zeros of f, counted with multiplicities, and P_f, Z_g, P_g being defined similarly.

PROOF   Since f and g have no zeros on γ, we can write our assumption as

$$\left|\frac{g(z)}{f(z)} - 1\right| < 1 \qquad \text{on } \gamma$$

Thus $g(z)/f(z) = h(z)$ maps γ into the unit disk centered at 1 (see Fig. 6.2.3). Therefore,

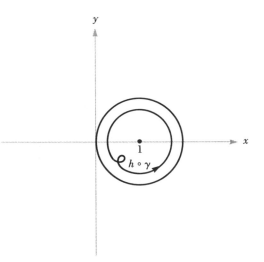

FIGURE 6.2.3   The image of γ under h.

$I(h \circ \gamma, 0) = 0$, since $h \circ \gamma$ is homotopic to 1 in that disk (which does not contain 0). By Eq. (4) applied to h and the definition of $\Delta_\gamma \arg h$, we get

$$\int_\gamma \frac{h'(z)}{h(z)}\, dz = 0$$

But we compute that

$$\frac{h'(z)}{h(z)} = \frac{g'(z)}{g(z)} - \frac{f'(z)}{f(z)}$$

and thus

$$\int_\gamma \frac{g'(z)}{g(z)} \, dz = \int_\gamma \frac{f'(z)}{f(z)} \, dz$$

Hence the result follows from the root-pole counting theorem and the principle of the argument. ∎

An important special case of Rouché's theorem is the following. *Let γ be a simple closed curve and let f and g be analytic inside and on γ with γ passing through no zeros of f or g; suppose that $|f(z) - g(z)| < |f(z)|$ on γ. Then f and g have the same number of zeros inside γ.* Note that if $|f(z) - g(z)| < |f(z)|$ on γ, then γ automatically can pass through no zeros of f or g (Why?).

Rouché's theorem can be used to locate the zeros of a polynomial. An illustration is given in Worked Example 6.2.11. Rouché's theorem can also be used to give a simple proof of the fundamental theorem of algebra, including the fact that an nth-degree polynomial has exactly n roots (see Exercise 9).

Hurwitz' Theorem

One of the theoretical applications of Rouché's theorem is the following result of Hurwitz.

6.2.6 HURWITZ' THEOREM *Let f_n be a sequence of analytic functions on a region A converging uniformly on every closed disk in A to f. Assume that f is not identically zero, and let $z_0 \in A$. Then $f(z_0) = 0$ iff there is a sequence $z_n \to z_0$ and there is an integer N such that $f_n(z_n) = 0$ whenever $n \geq N$ (that is, a zero of f is a limit of zeros of the functions f_n).*

The theorem will follow from the next proposition.

6.2.7 PROPOSITION *Let f_n be a sequence of functions analytic on a region A which converge uniformly on every closed disk in A to f. Assume f is not identically 0 and that γ is a simple closed curve which together with its interior is contained in A and which passes through no zeros of f. Then there is an integer $N(\gamma)$ such that each f_n with $n \geq N(\gamma)$ has the same number of zeros inside γ as does f (counted according to multiplicity).*

PROOF Since $|f|$ is continuous and never 0 on the compact set γ, it has a nonzero minimum m on γ; let us say $|f(z)| \geq m > 0$ for all z on γ. The curve is covered by a finite number of closed disks, and so the convergence of f_n to f is uniform on γ. Accordingly, there is an integer $N(\gamma)$ such that $|f_n(z) - f(z)| < m \leq f(z)$ for all z on γ whenever $n \geq N(\gamma)$. Rouché's theorem applies and we conclude that f_n and f have the same number of zeros inside γ, as desired. (Note that f is analytic on A by the analytic convergence theorem (3.1.8).) ∎

PROOF OF THEOREM 6.2.6   Again, f is analytic on A by the analytic convergence theorem (3.1.8). Suppose $f(z_0) = 0$. Since f is not identically 0, the zeros are isolated by the identity theorem. There is a number $\delta > 0$ such that $f(z)$ is never 0 in the deleted neighborhood $\{z \mid 0 < |z - z_0| < \delta\}$. For each positive integer k, let γ_k be the circle $\{z$ such that $|z - z_0| = \delta/k\}$. Pick N_k as $N(\gamma_k)$ by Proposition 6.2.7. Then $n \geq N_k$ implies that f_n has at least one zero z_n inside γ_k. That is $f_n(z_n) = 0$. For $n \geq N_k$ we have $|z_n - z_0| < \delta/k$. This proves the theorem with $N = N_1$ provided we make sure to begin choosing the z_n inside γ_k as soon as $n \geq N_k$ guarantees their availability. ∎

We must assume that f is not identically zero. Consider, for example, the function $f_n(z) = e^z/n$, which approaches zero uniformly on closed disks (Why?) but for which f_n has no zeros.

6.2.8 COROLLARY   *Let f_n be a sequence of functions analytic on a region A which converge uniformly on closed disks in A to f. If each f_n is one-to-one on A and f is not constant, then f is one-to-one on A.*

PROOF   Suppose a and b are in A and $f(a) = f(b)$. We want to show that $a = b$. Consider $g_n(z) = f_n(z) - f_n(a)$ and $g(z) = f(z) - f(a)$. Then $g_n \to g$ uniformly on closed disks in A and $g(b) = 0$. Since g is not identically 0, Hurwitz' theorem says there is a sequence $z_n \to b$ with $g_n(z_n) = 0$. That is, $f_n(z_n) = f_n(a)$. But f_n is one-to-one, and so $z_n = a$. Since $z_n \to b$, we must have $a = b$, as desired. ∎

It is perfectly possible for one-to-one functions to converge uniformly on closed disks to a constant function. For example, the functions $f_n(z) = z/n$ converge uniformly on the unit disk to the constant function $f(z) = 0$.

One-to-One Functions

Analytic functions that are one-to-one find many useful applications. The term *schlicht* (simple) function is often used. We now relate one-to-one functions with the inverse mapping theorem. Again Rouché's theorem is the appropriate tool.

6.2.9 PROPOSITION *If $f : A \to \mathbb{C}$ is analytic and locally one-to-one, then $f'(z_0) \neq 0$ for all $z_0 \in A$. It follows from the inverse function theorem that $f(A)$ is open and, if f is globally one-to-one, that f^{-1} is analytic from $f(A)$ to A.*

PROOF Suppose that, to the contrary, for some point z_0 we have $f'(z_0) = 0$. Then $f(z) - f(z_0)$ has a zero of order $k \geq 2$ at z_0. Now f is not constant and thus the zeros of f' are isolated. Thus there are a $\delta > 0$ and an $m > 0$ such that on the circle $|z - z_0| = \delta$, $|f(z) - f(z_0)| \geq m > 0$ and $f'(z) \neq 0$ for $0 < |z - z_0| \leq \delta$. For $0 < \eta < m$, we conclude that $f(z) - f(z_0) - \eta$ has k zeros inside $|z - z_0| = \delta$, by Rouché's theorem. A zero cannot be a double zero, since $f'(z) \neq 0$ for $|z - z_0| \leq \delta$, $z \neq z_0$. Thus $f(z) = f(z_0) + \eta$ for two distinct points z and therefore is not one-to-one. This contradiction means that $f'(z_0) \neq 0$, as was to be shown. ∎

Another basic property of one-to-one functions is the following.

6.2.10 ONE-TO-ONE THEOREM *Let f be analytic on a region A and let γ be a closed curve homotopic to a point in A. Suppose that $I(\gamma, z) = 0$ or 1. Define the set $B = \{z \in A \mid I(\gamma, z) \neq 0\}$ (the "inside" of γ). If f is such that each point of $f(B)$ has index 1 with respect to the curve $\tilde{\gamma} = f \circ \gamma$, then f is one-to-one on B.*

PROOF Consider, for $z_0 \in B$ and $w_0 = f(z_0)$,

$$N = \frac{1}{2\pi i} \int_\gamma \frac{f'(z)}{f(z) - w_0}\, dz$$

By Corollary 6.2.2, N equals the number of times that $f(z) = w_0$ on B. We therefore must show that it equals 1. Letting $\tilde{\gamma} = f \circ \gamma$, we conclude, as in the principle of the argument, that

$$N = \frac{1}{2\pi i} \int_{\tilde{\gamma}} \frac{1}{z - w_0}\, dz$$

which is the index of w_0 with respect to $\tilde{\gamma}$. Thus $N = 1$ and therefore $f(z) = w_0$ has exactly one solution, $z = z_0$. This means that f is one-to-one. ∎

The one-to-one theorem becomes more intuitive if we use the Jordan curve theorem. Let γ be a simple closed curve and let B be its interior. Suppose that the set $f(B)$ is bounded by the curve $\tilde{\gamma} = f \circ \gamma$. The hypothesis of the one-to-one theorem will be fulfilled if $\tilde{\gamma}$ is a simple closed curve (since this means that f should be one-to-one on γ). Therefore, the result may be rephrased as follows: *To see if an analytic function is one-to-one on a region, it is sufficient to check that it is one-to-one on the boundary.*

Worked Examples

6.2.11   *Use Rouché's theorem to determine the quadrants in which the zeros of $z^4 + iz^2 + 2$ lie and the number of zeros that lie inside circles of varying radii.*

Solution. Let $g(z) = z^4$, $f(z) = z^4 + iz^2 + 2$, and note that

$$|f(z) - g(z)| = |iz^2 + 2| \leq |z|^2 + 2$$

and that

$$|g(z)| = |z|^4$$

Hence if $r = |z| > \sqrt{2}$, we have

$$|f(z) - g(z)| < |g(z)|$$

Since g does not vanish on any circle of positive radius, the preceding inequality shows that f does not vanish on circles with radius $> \sqrt{2}$. Rouché's theorem then shows that all four roots of f lie inside these circles, that is, inside the closed disk $|z| \leq \sqrt{2}$.

Next, let $h(z) = z^4 + 2iz^2 = z^2(z^2 + 2i)$. Clearly, h has a double root at 0 and two additional roots on the circle $|z| = \sqrt{2}$. Furthermore,

$$|f(z) - h(z)| = |-iz^2 + 2| = |z^2 + 2i| = \frac{|h(z)|}{|z|^2}$$

For any choice of r with $1 < r < \sqrt{2}$, h and hence f do not vanish on the circle $|z| = r$ and $|f(z) - h(z)| < |h(z)|$. Rouché's theorem shows that f has precisely two zeros in $|z| < r$ for any of these values of r. Letting r approach 1 and $\sqrt{2}$, we see that f has two roots in the closed disk $|z| \leq 1$ and two on the circle $|z| = \sqrt{2}$.

Finally, let $k(z) = 2$. Then

$$|f(z) - k(z)| = |z^4 + iz^2| \leq |z|^4 + |z|^2 < 2 = |k(z)|$$

whenever $|z| < 1$. Arguing as before, for any r with $0 < r < 1$, k and hence f do not vanish in $|z| < r$. Combining these three results we find that f has two zeros on $|z| = 1$ and two on $|z| = \sqrt{2}$.

Now we turn to an analysis of the quadrants in which the roots lie. For z either real or purely imaginary, $f(z) = z^4 + iz^2 + 2$ has a nonzero imaginary part unless $z = 0$. Thus f has no roots on the axes. Consider a large quarter circle as shown in Fig. 6.2.4. We shall compute $\Delta_\gamma \arg (z^4 +$

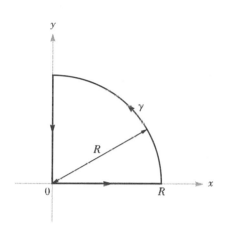

FIGURE 6.2.4 The curve γ used to locate the quadrants in which the zeros of the polynomial $z^4 + iz^2 + 2$ lie.

$iz^2 + 2$) and use the principle of the argument. Along the x axis z is real and $f(z)$ lies in the first quadrant. Also $f(0) = 2$, and as $R \to \infty$, $\arg f(R) \to 0$ since

$$\arg f(R) = \arg R^4 \left(1 + \frac{i}{R^2} + \frac{2}{R^4}\right) = \arg \left(1 + \frac{i}{R^2} + \frac{2}{R^4}\right)$$

tends to 0 as $R \to \infty$. Since f takes its values in the first quadrant, we conclude that the change in the argument is zero as z moves from 0 to ∞. Along the curved portion of γ, z^4 clearly changes argument by $2\pi (= 4 \times \pi/2)$. As $R \to \infty$, 2π is the limiting change in argument for $f(z)$ as well, as we see by writing

$$f(z) = z^4 \left(1 + \frac{i}{z^2} + \frac{2}{z^4}\right)$$

Similarly, coming down the imaginary axis there is, in the limit of large R, no change in the argument of f. (If $f(0)$ were not real, this device would still give the limiting behavior of the argument at infinity and the value at zero, and so the change in the argument, at least up to multiples of 2π, can be inferred.) We conclude that the change in argument as we traverse γ is 2π. From the principle of the argument, there is exactly one zero in the first quadrant. By inspection, $f(z) = f(-z)$, so that $-z$ is a root when z is. Thus there must be a root in each quadrant. Therefore, we must have one of the two possibilities shown in Fig. 6.2.5. The methods used here do not enable us to tell which of these possibilities actually occurs without more detailed analysis. We can check this example by finding the roots directly using the quadratic formula twice; however, in other examples a direct computation may be impossible or impractical whereas the methods described here can nevertheless be used.

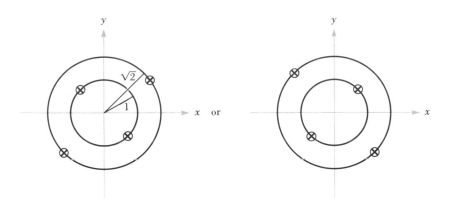

FIGURE 6.2.5   Locating the roots of the polynomial $z^4 + iz^2 + 2$.

6.2.12   *If $a > e$, show that the equation $e^z = az^n$ has n solutions inside the unit circle.*

Solution. Let $f(z) = e^z - az^n$ and let $g(z) = -az^n$. Then g has exactly n roots. We shall show that f and g have the same number of roots inside the unit circle $|z| = 1$. To do so we must show that

$$|f(z) - g(z)| < |g(z)|$$

for $|z| = 1$. But

$$|f(z) - g(z)| = |e^z| = e^x \le e$$

since $|x| \le 1$. Also $|g(z)| = |az^n| = a > e$, and so the result follows by Rouché's theorem.

6.2.13   *Let $f(z) = \sum_{n=0}^{\infty} a_n z^n$. Assume that $a_0 = 0$ and $a_1 = 1$. Prove that f is one-to-one on the unit disk $\{z$ such that $|z| < 1\}$ if $\sum_{l=2}^{\infty} l|a_l| \le 1$.*

Solution. The series for f converges for $|z| < 1$ since, as a consequence of $\sum_{l=2}^{\infty} l|a_l| \le 1$, we get $|a_n| \le 1$, and thus $|a_n z^n| \le |z|^n$; we know that $\sum |z|^n$ converges for $|z| < 1$. Thus f is analytic on $\{z$ such that $|z| < 1\}$.

Let $|z_0| < 1$. We want to show that $f(z) = f(z_0)$ has exactly one solution, z_0. Let $g(z) = z - z_0$, which has exactly one zero. If we set $h(z) = f(z) - f(z_0)$, then

$$h(z) - g(z) = \sum_{n=2}^{\infty} a_n z^n - \sum_{n=2}^{\infty} a_n z_0^n$$

To estimate this we can use the following trick. Let $\phi(z) = \sum_{n=2}^{\infty} a_n z^n$. Then

$$|\phi(z) - \phi(z_0)| \le [\max |\phi'(\zeta)|] \cdot |z - z_0|$$

where the maximum is over those ζ on the line joining z_0 to z (Why?). However, $|\phi'(\zeta)| = |\Sigma_{n=2}^{\infty} n a_n \zeta^{n-1}| < \Sigma_{n=2}^{\infty} n |a_n| \le 1$, since $|\zeta| < 1$. Hence

$$|h(z) - g(z)| = |\phi(z) - \phi(z_0)| < |z - z_0| = |g(z)|$$

Thus, by Rouché's theorem, $h(z) = f(z) - f(z_0)$ has exactly one solution, namely, $z = z_0$; this proves the assertion.

6.2.14 *Use Rouché's theorem to show that the roots of an nth-degree polynomial (the roots being counted with multiplicities) depend continuously on the coefficients of the polynomial.*

Solution: Part of the problem here is to state precisely what the question means. Suppose the polynomial is $p(z) = a_n z^n + a_{n-1} z^{n-1} + \cdots + a_1 z + a_0$ with $a_n \ne 0$ and the zeros are w_1, w_2, . . . , w_n. (Some of these may be the same.) After normalizing by dividing by a_n, the coefficients certainly depend continuously on the roots. In fact they are polynomial expressions in the w_j's. Simply multiply out $p(z)/a_n = (z - w_1)(z - w_2) \ldots (z - w_n)$ to obtain $a_{n-1}/a_n = -(w_1 + w_2 + \cdots + w_n)$, etc. Our present problem is in the other direction. It can be thought of as a question of *stability*. If the equation is changed a small amount by changing the coefficients, can we conclude that the solutions do not change very much? Such questions can be important; for example, the coefficients may be known only approximately. They may be estimated by experiment and be subject to experimental error or known within a statistical confidence interval. Can a small error in measurement cause a very large error in the solutions? This example says that in some sense the answer is no. Here is a possible formulation.

PROPOSITION *Suppose $p(z) = a_n z^n + a_{n-1} z^{n-1} + \cdots + a_1 z + a_0$ with $a_n \ne 0$ has zeros at $w_1, w_2, \ldots, w_k$ with multiplicities $n_1, n_2, \ldots, n_k$, and that c is a positive number less than half the minimum distance between the points w_j. Then there is a number $\delta > 0$ such that the polynomial $q(z) = b_n z^n + b_{n-1} z^{n-1} + \cdots + b_1 z + b_0$ has exactly n_j zeros (counted with multiplicity) in the disk $D(w_j; \varepsilon)$ for each $j = 1, 2, \ldots, k$ provided that $|b_m - a_m| \le \delta$ for each $m = 0, 1, 2, \ldots, n$.*

PROOF Let γ_j be the circle $\{z \text{ such that } |z - w_j| = \varepsilon\}$. Then $|p(z)|$ is never 0 on $\gamma_1 \cup \cdots \cup \gamma_k$, and since it is continuous, it has a nonzero minimum A on this compact set. Let $M = \max (|w_1|, \ldots, |w_k|) + \varepsilon$ and choose $\delta < A/(2 \Sigma_{m=0}^{n} M^m)$. Then if $|b_m - a_m| \le \delta$ for all m and z is on $\gamma_1 \cup \cdots \cup \gamma_k$, we have

$$|p(z) - q(z)| \le \sum_{m=0}^{n} |a_m - b_m| \cdot |z|^m \le \sum_{m=0}^{n} \delta M^m < A \le p(z)$$

Thus Rouché's theorem shows that the number of zeros of q inside γ_j is the same as the number of zeros of p inside γ_j for $j = 1, 2, \ldots, k$. ∎

6.2.15 *Find the largest disk centered at $z_0 = 1$ on which the function $f(z) = z^4$ is one-to-one.*

Solution. This problem is intended to provide a warning against a common error. The derivative $f'(z) = 4z^3$ is 0 only at $z = 0$. In particular, $f'(z)$ is never 0 on the disk $D(1; 1)$. However, we *cannot* conclude that f is one-to-one on this disk. In fact it is not. $f((1 + i)/\sqrt{2}) = f((1 - i)/\sqrt{2}) = -1$. If f is to be one-to-one near a point, the derivative must not be 0 at that point, and $f'(z) \neq 0$ is enough to guarantee that f is one-to-one in some neighborhood of z. But f' being never 0 on a large region is *not* enough to force f to be globally one-to-one on the whole region. In the present example $f(re^{i/4}) = f(re^{-i/4})$ for any r. Therefore the function will cease to be one-to-one as soon as the disk hits these $45°$ lines. This occurs for $D(1; R)$ when $R = 1/\sqrt{2}$. See Fig. 6.2.6.

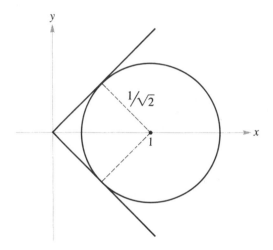

FIGURE 6.2.6   The function $f(z) = z^4$ is one-to-one on this disk.

Methods based on the one-to-one theorem (6.2.10) which involve looking at the boundary are usually more useful than examining the derivative. If $z_1 = r_1 e^{i\theta_1}$ and $z_2 = r_2 e^{i\theta_2}$ are on the circle of radius R around 1 with $0 < R < \sqrt{2}$, then $-\pi/4 < \theta_1$, $\theta_2 < \pi/4$. $z_1^4 = z_2^4$ forces $r_1 = r_2$ and $e^{i4\theta_1} = e^{i4\theta_2}$, and so $4(\theta_1 - \theta_2) = 2\pi n$. This cannot happen with θ_1 and θ_2 both between $-\pi/4$ and $\pi/4$ unless $\theta_1 = \theta_2$. But then $z_1 = z_2$. (We have actually shown that f is one-to-one on the open quarter plane $\{z \mid -\pi/4 < \arg z < \pi/4\}$.)

Exercises

1.   How many zeros does $z^6 - 4z^5 + z^2 - 1$ have in the disk $\{z$ such that $|z| < 1\}$?

2.   How many zeros does $z^4 - 5z + 1$ have in the annulus $\{z \mid 1 < |z| < 2\}$?

3.   Show that there is exactly one point z in the right half plane $\{z \mid \text{Re } z > 0\}$ at which $z + e^{-z} = 2$. (*Hint:* Consider contours such as the one in Fig. 6.2.7.)

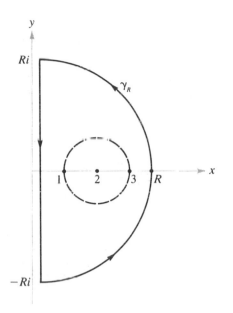

FIGURE 6.2.7 Contour for Exercise 3.

4. Show that if $p(z) = z^n + a_{n-1}z^{n-1} + \cdots + a_1z + a_0$, then there must be at least one point z with $|z| = 1$ and $|p(z)| \geq 1$. (*Hint:* If $|p(z)| < 1$ everywhere on $\{|z| = 1\}$, how many zeros has $a_{n-1}z^{n-1} + \cdots + a_1z + a_0$?)

5. Let f be analytic inside and on the unit circle $|z| = 1$. Suppose that $0 < |f(z)| < 1$ if $|z| = 1$. Show that f has exactly one fixed point (a point z_0 such that $f(z_0) = z_0$) inside the unit circle.

6. Show that $e^z = 5z^3 - 1$ has three solutions in the disk $\{z$ such that $|z| < 1\}$. (*Hint:* Think about Worked Example 6.2.12.)

7. Show that the conclusion of Exercise 5 still holds if the assumption $0 < |f(z)| < 1$ is replaced by $0 < |f(z)| \leq 1$ (Except that the fixed point might be on the unit circle.)

8. Let $g_n = \sum_{k=0}^{n} z^k/k!$. Let $D(0, R)$ be the disk of radius $R > 0$. Show that for n large enough, g_n has no zeros in $D(0, R)$.

9. (*Fundamental theorem of algebra*) Use Rouché's theorem to prove that if $f(z) = a_0 + a_1z + \cdots + a_nz^n$, $n \geq 1$, and $a_n \neq 0$, then f has exactly n roots.

10. Supply the details of the following proof of Rouché's theorem: Under the hypotheses of Theorem 6.2.5, the function $H(s, t) = sg(\gamma(t)) + (1 - s)f(\gamma(t))$ is a closed-curve homotopy between the curves $f \circ \gamma$ and $g \circ \gamma$ in $\mathbf{C} \backslash \{0\}$. It follows that $I(f \circ \gamma; 0) = I(g \circ \gamma; 0)$. The conclusion of Rouché's theorem follows from this and the argument principle.

11. Extend the root-pole counting theorem (6.2.1) to include the following result. If f is analytic on A except for zeros at $a_1, \ldots, a_n$ and poles at $b_1, \ldots, b_m$ (each repeated according to its multiplicity), if h is analytic on A, and if γ is a closed curve homotopic to a point in A, passing through none of $a_1, \ldots, a_n, b_1, \ldots, b_m$, then

$$\int_\gamma \frac{f'(z)}{f(z)} h(z) \, dz = 2\pi i \left[\sum_{i=1}^n h(a_i) I(\gamma, a_i) - \sum_{k=1}^m h(b_k) I(\gamma, b_k) \right]$$

12. Supply the details of the following proof of Rouché's theorem (due to Caratheodory): The function

$$F(\lambda) = \frac{1}{2\pi i} \int_\gamma \frac{\lambda g'(z) + (1 - \lambda) f'(z)}{g(z) + (1 - \lambda) f(z)} \, dz$$

is a continuous function of λ for $0 \le \lambda \le 1$. But its value is always an integer, and so

$$Z_f - P_f = F(0) = F(1) = Z_g - P_g$$

13. If $f(z)$ is a polynomial, use Exercise 11 to prove that

$$\frac{1}{2\pi i} \int_\gamma \frac{f'(z)}{f(z)} z \, dz$$

is the sum of the zeros of f if the circle γ is large enough.

14. (a) Let $f: A \to B$ be analytic, one-to-one, and onto. Let $w \in B$ and let γ be a small circle centered at z_0 in A. Use Exercise 11 to prove that

$$f^{-1}(w) \doteq \frac{1}{2\pi i} \int_\gamma \frac{f'(z) z}{f(z) - w} \, dz$$

for w sufficiently close to $f(z_0)$.
(b) Explain the meaning of

$$\frac{1}{2\pi i} \int_\gamma \frac{f'(z)}{f(z) - w} \, dz$$

15. Let $f(z)$ be a polynomial of degree n, $n \ge 1$. Show that f maps $\mathbf{C}$ *onto* $\mathbf{C}$.

16. Suppose $g_n(z) = \sum_{k=0}^{n} 1/(k!z^k)$, and let $\varepsilon > 0$. For large enough n, are all the zeros of g_n in the disk $D(0; \varepsilon)$?

17. If $f(z)$ is analytic and has n zeros inside the simple closed curve γ, must it follow that $f'(z)$ has $n-1$ zeros inside γ?

18. Locate the zeros (as was done in Worked Example 6.2.11) for the polynomial $z^4 - z + 5 = 0$.

19. Find an $r > 0$ such that the polynomial $z^3 - 4z^2 + z - 4$ has exactly two roots inside the circle $|z| = r$.

20. Let f be analytic inside and on $|z| = R$ and let $f(0) \neq 0$. Let $M = \max |f(z)|$ on $|z| = R$. Show that the number of zeros of f inside $|z| = R/3$ does not exceed

$$\frac{1}{\log 2} \cdot \log \frac{M}{|f(0)|}$$

(*Hint.* Let $h(z) = f(z)/[(z - z_1) \cdots (z - z_n)]$ where z_n are the zeros of f inside $|z| = R/3$ and apply the maximum modulus theorem to h.)

21. Show that $z \mapsto z^2 + 3z$ is one-to-one on $\{z$ such that $|z| < 1\}$.

22. What is the largest disk around $z_0 = 0$ on which the function in Exercise 21 is one-to-one?

23. Prove that the following statement is false: For every function f analytic on the annulus $\frac{1}{2} < |z| < \frac{3}{2}$, there is a polynomial p such that $|f(z) - p(z)| < \frac{1}{2}$ for $|z| = 1$.

24. Let f be analytic on $\mathbb{C}$ and let $|f(z)| \leq 5\sqrt{|z|}$ for all $|z| \geq 1$. Prove that f is constant.

6.3 MAPPING PROPERTIES OF ANALYTIC FUNCTIONS

Further properties of analytic functions that are of a local nature (that is, that depend only on the values of $f(z)$ for z in a neighborhood of a given point z_0) will be proved in this section. Additional proofs will be given here of the inverse function theorem (1.5.10), the maximum modulus theorem (2.5.6), and the open mapping theorem (stated formally for the first time in this section, but previously mentioned in Exercise 8, Sec. 1.5). We can prove these theorems and also obtain information concerning the behavior of a function near a point by using the root counting formula (see Corollary 6.2.2):

$$\frac{1}{2\pi i} \int_\gamma \frac{f'(z)}{f(z) - w} \, dz = \text{number of roots of } f(z) = w \text{ inside } \gamma, \text{ counting multiplicities}$$

Local Behavior of Analytic Functions

If $f(z_0) = w_0$ with multiplicity k in the sense that $f(z) - w_0$ has a zero of order k at z_0, then we shall show that f is locally k-to-one near z_0. The precise statement of the theorem is somewhat cumbersome and confusing. We begin with the motivating and typical example and a somewhat imprecise paraphrase of the theorem before stating and proving the more precise version. (The reader should keep the example in mind and follow the precise argument by referring to Fig. 6.3.1.) Consider the special case in which $f(z) = z^k$.

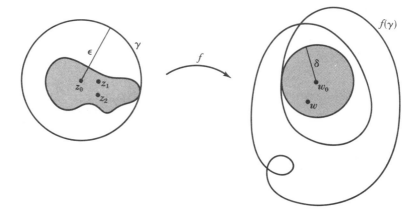

FIGURE 6.3.1   This function is two-to-one near z_0.

This function has a zero of order k at $z_0 = 0$ (here $w_0 = 0$). Then for all w near 0, $z^k = w$ has exactly k solutions near 0. To see that this behavior is inherited by a more general function f for which $f(z_0) = w_0$ with multiplicity k, consider the power series expansion of f around z_0:

$$f(z) - w_0 = \sum_{n=k}^{\infty} a_n(z - z_0)^n$$

For $|z - z_0|$ that are small enough, we might guess (correctly) that the behavior of the lowest-degree nonvanishing term, $a_k(z - z_0)^k$, will dominate.

6.3.1 MAPPING THEOREM: INFORMAL VERSION   *Suppose f takes on the value w_0 at z_0 with multiplicity k. Then for all w sufficiently near w_0, f takes on the value w*

exactly k times near z_0 (counting multiplicities). For all w still nearer w_0, the k roots of $f(z) = w$ near z_0 are distinct.

The more precise statement is the following.

6.3.2 MAPPING THEOREM *Let f be analytic and not constant on a region A and let $z_0 \in A$. Suppose that $f(z) - w_0$ has a zero of order $k \geq 1$ at z_0. Then there is an $\eta > 0$ such that for any $\epsilon \in]0, \eta]$, there is a $\delta > 0$ such that if $|w - w_0| < \delta$, then $f(z) - w$ has exactly k roots (counted with their multiplicities) in the disk $|z - z_0| < \epsilon$ (see Fig. 6.3.1). In fact, there is a $\lambda > 0$ (probably smaller than η) such that for any $\epsilon \in]0, \lambda]$, there is a $\delta > 0$ such that if $0 < |w - w_0| < \delta$, then $f(z) - w$ has exactly k distinct roots in the disk $0 < |z - z_0| < \epsilon$.*

PROOF Since f is not constant, the zeros of $f(z) - w_0$ are isolated. Thus there is an $\eta > 0$ such that for $|z - z_0| \leq \eta$, $f(z) - w_0$ has no zeros other than z_0. On the compact set $\{z$ such that $|z - z_0| = \epsilon\}$ (the circle γ in Fig. 6.3.1), $f(z) - w_0$ is continuous and never zero. Hence there is a $\delta > 0$ such that $|f(z) - w_0| \geq \delta > 0$ for $|z - z_0| = \epsilon$. Thus if w satisfies $|w - w_0| < \delta$, then for $|z - z_0| = \epsilon$, the following hold:

(i) $f(z) - w_0 \neq 0$
(ii) $f(z) - w \neq 0$ (since $f(z) = w$ would mean that $|w - w_0| \geq \delta$)
(iii) $|(f(z) - w) - (f(z) - w_0)| = |w - w_0| < \delta \leq |f(z) - w_0|$

By Rouché's theorem, $f(z) - w$ has the same number of zeros, counting multiplicities, as $f(z) - w_0$ inside the circle $|z - z_0| = \epsilon$. Thus we have proved the first part of the theorem. To prove the second part, notice that f' is not identically zero on A. The zeros of f' are thus isolated. Therefore, for some $\lambda \leq \eta$, neither $f(z) - w_0$ nor $f'(z)$ is zero in $|z - z_0| \leq \lambda$ except at z_0. Observe that $f(z) - w$ still has the same number of roots as $f(z) - w_0$ for any w near enough to w_0, but now the roots must be first-order, hence distinct, since f' is nonzero. ■

The Open Mapping and Inverse Function Theorems

The mapping theorem tells us that on some disk centered at z_0, f is exactly k-to-one. The theorem may not be directly helpful in finding the size of this disk (see the examples and exercises at the end of this section), but often knowledge of its existence can lead to interesting results.

A function $f : A \to \mathbf{C}$ is called *open* iff, for every open set $U \subset A$, $f(U)$ is open. By the definition of an open set, this statement is equivalent to: For every $\epsilon > 0$ sufficiently small, there is a $\delta > 0$ such that $|w - w_0| < \delta$ implies that there is a z, $|z - z_0| < \epsilon$ with $w = f(z)$. In other words, if f hits w_0, it hits every w sufficiently near w_0. Careful reading of the definition of open set and examination of Fig. 6.3.1 show that the mapping theorem implies the next theorem:

6.3.3 OPEN MAPPING THEOREM *Let $A \subset \mathbf{C}$ be open and $f : A \to \mathbf{C}$ be nonconstant and analytic. Then f is an open mapping; that is, the image of any open set under f is open.*

Using the mapping theorem (6.3.2), we can also get an alternative proof of the inverse function theorem (1.5.10).

6.3.4 INVERSE FUNCTION THEOREM *Let $f : A \to \mathbf{C}$ be analytic, let $z_0 \subset A$, and let $f'(z_0) \neq 0$. Then there is a neighborhood U of z_0 and a neighborhood V of $w_0 = f(z_0)$ such that $f : U \to V$ is one-to-one and onto and $f^{-1} : V \to U$ is analytic.*

PROOF $f(z) - w_0$ has a simple zero at z_0 since $f'(z_0) \neq 0$. We can use Theorem 6.3.2 to find $\epsilon > 0$ and $\delta > 0$ such that each w with $|w - w_0| < \delta$ has exactly one preimage x with $|z - z_0| < \epsilon$. Let $V = \{w \text{ such that } |w - w_0| < \delta\}$ and let U be the inverse image of V under the map f restricted to $\{z \text{ such that } |z - z_0| < \epsilon\}$ (the shaded region of Fig. 6.3.1). By the mapping theorem, f maps U one-to-one onto V. Since f is continuous, U is a neighborhood of z_0. By the open mapping theorem, $f = (f^{-1})^{-1}$ is an open map, and so f^{-1} is continuous from V to U. To show that it is analytic, use

$$f^{-1}(w) = \frac{1}{2\pi i} \int_{|z - z_0| = \epsilon} \frac{f'(w)}{f(z) - w} z \, dz$$

(see Exercise 14 at the end of the preceding section). This is analytic in w from Worked Example 2.4.15. ∎

These ideas can be used as the basis for another proof of the maximum modulus theorem (see Sec. 2.5), as follows.

6.3.5 MAXIMUM MODULUS THEOREM *Let f be analytic on a region (open connected set) A. If $|f|$ has a local maximum at $z_0 \in A$, then f is constant.*

PROOF Suppose that f is not constant and that $z_0 \in A$. Since f is an open map, for $|w - f(z_0)|$ sufficiently small there is a z near z_0 with $w = f(z)$. Choose w with $|w| > |f(z_0)|$. Specifically, choose $w = (1 + \delta/2)f(z_0)$ if $f(z_0) \neq 0$ and $w = \delta/2$ if $f(z_0) = 0$ for δ small. Then it is clear that f does not have a relative maximum at z_0. ∎

A similar proof shows that if $f(z_0) \neq 0$, then f has no minimum at z_0 unless f is constant. The maximum modulus principle (2.5.6) follows as in Sec. 2.5.

Worked Examples

6.3.6 *Determine the largest disk around $z_0 = 0$ on which $f(z) = 1 + z + z^2$ is one-to-one.*

Solution. Since $f'(0) = 1$, $f(z) - 1$ has a simple zero at 0, and the mapping theorem (6.3.2) shows that f is one-to-one on some disk around $z_0 = 0$. Because $f(z) - 1 = z + z^2 = z(1 + z)$, which has roots at 0 and -1, we know that $f(z) - 1$ has only one root in the disk {z such that $|z| < 1$}. This disk is the disk in the first part of the mapping theorem, but that does not guarantee that f is one-to-one on the disk; in fact, it is not. The mapping theorem shows only that f is one-to-one on the subregion of the disk shaded in Fig. 6.3.1, the preimage of {w such that $|w - w_0| < \delta$}. We can find out what causes this phenomenon by plotting the image of the unit circle. In this case $f(z) = 1 + z + z^2$, $z_0 = 0$, and $w_0 = 1$. Thus

$$f(0) = 1 \qquad\qquad f(e^{2\pi i/3}) = 0$$
$$f(1) = 3 \qquad\qquad f(e^{4\pi i/3}) = 0$$
$$f(i) = i \qquad\qquad f\left(\frac{1}{\sqrt{2}} + \frac{1}{\sqrt{2}}i\right) = \left(1 + \frac{1}{\sqrt{2}}\right) + \left(1 + \frac{1}{\sqrt{2}}\right)i$$
$$f(-1) = 1 \qquad\qquad f\left(-\frac{1}{\sqrt{2}} - \frac{1}{\sqrt{2}}i\right) = \left(1 - \frac{1}{\sqrt{2}}\right) + \left(1 - \frac{1}{\sqrt{2}}\right)i$$
$$f(-i) = -i$$
$$f\left(\frac{1}{\sqrt{2}} - \frac{1}{\sqrt{2}}i\right) = \left(1 + \frac{1}{\sqrt{2}}\right) - \left(1 + \frac{1}{\sqrt{2}}\right)i \qquad f\left(-\frac{1}{\sqrt{2}} + \frac{1}{\sqrt{2}}i\right) = \left(1 - \frac{1}{\sqrt{2}}\right) - \left(1 - \frac{1}{\sqrt{2}}\right)i$$

By plotting these points, we find that the image of the unit circle is as shown in Fig. 6.3.2. The index

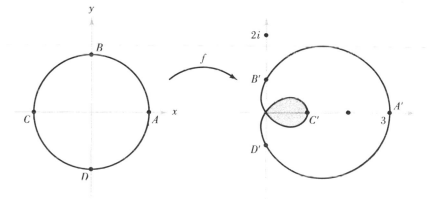

FIGURE 6.3.2 Image of the unit circle under $f(z) = 1 + z + z^2$.

of the image curve with respect to the small shaded region is 2. Therefore, each point here is hit twice by points in the unit disk; for example, $f'(-\frac{1}{2}) = 0$ and $f(-\frac{1}{2}) = \frac{3}{4}$. The mapping theorem shows that f is two-to-one on small neighborhoods of $-\frac{1}{2}$. Thus f will not be one-to-one on any disk containing a neighborhood of $-\frac{1}{2}$.

Consider the disk $D(0, r) = \{z \text{ such that } |z| < r\}$. The boundary curve is the circle $\gamma_r = \{z \text{ such that } |z| = r\}$. As r gets smaller, the troublemaking loop in the image curve shrinks. For some critical r_0 it disappears. For $r > r_0$, f is not one-to-one on γ_r. For $r < r_0$, f is one-to-one on γ_r. By the one-to-one theorem (6.2.10), f is thus one-to-one on $D(0, r)$, and the desired disk is $D(0, r_0)$. To find r_0, suppose that $re^{i\theta}$ and $re^{i\psi}$ lie on γ, and that $f(re^{i\theta}) = f(re^{i\psi})$. Then

$$1 + re^{i\theta} + r^2e^{i2\theta} = 1 + re^{i\psi} + r^2e^{i2\psi}$$

Hence $e^{i\theta} + re^{i2\theta} = e^{i\psi} + re^{i2\psi}$, and so

$$re^{i(\theta+\psi)} (e^{i(\theta-\psi)} - e^{i(\psi-\theta)}) = e^{i(\theta+\psi)/2} (e^{i(\psi-\theta)/2} - e^{i(\theta-\psi)/2})$$

Thus

$$re^{i(\theta+\psi)/2} \sin (\theta - \psi) = -\sin \frac{\theta - \psi}{2}$$

In other words,

$$2re^{i(\theta+\psi)/2} \sin \frac{\theta - \psi}{2} \cos \frac{\theta - \psi}{2} = -\sin \frac{\theta - \psi}{2}$$

Now one of two things must happen: either $\sin [(\theta - \psi)/2] = 0$, in which case $\theta - \psi = 2\pi n$ for some integer n, and thus $re^{i\theta} = re^{i\psi}$, or $\cos [(\theta - \psi)/2] = -(1/2r)e^{-i(\theta+\psi)/2}$. If $r > \frac{1}{2}$, the latter can happen for $\psi = -\theta$; for example, at $r = 1$, it occurs at the points $e^{2\pi i/3}$ and $e^{4\pi i/3}$. If $r < \frac{1}{2}$, this same condition cannot hold, since $|\cos [(\theta - \psi)/2]| \leq 1$. If $r = \frac{1}{2}$, it can happen only for $\theta = \psi = \pi$. The critical radius is therefore $r_0 = \frac{1}{2}$. Hence f is one-to-one on the disk $D(0, \frac{1}{2}) = \{z \text{ such that } |z| < \frac{1}{2}\}$ but not on any larger open disk. ($D(0, \frac{1}{2})$ is the largest disk around $z_0 = 0$ on which $f'(z)$ is never zero. It is not generally true that this will also be the disk on which f is one-to-one (see Exercise 3)).

6.3.7 Prove the following: *If f is analytic near $z_0 \in A$ and if $f(z) - f(z_0)$ has a zero of order k at z_0, $1 \leq k < \infty$, then there is an analytic function $h(z)$ such that $f(z) = f(z_0) + [h(z)]^k$ for z near z_0, and h is locally one-to-one.*

Solution. Since $k < \infty$, f is not constant. We can write $f(z) - f(z_0) = (z - z_0)^k \phi(z)$ where $\phi(z_0) \neq 0$ and ϕ is analytic. For z near z_0, $\phi(z)$ lies in a small disk around $\phi(z_0)$ not containing 0, by continuity. On such a disk we can define $\sqrt[k]{\phi(z)}$ and let $h(z) = (z - z_0)\sqrt[k]{\phi(z)}$. Then $h'(z_0) \neq 0$, and so by the inverse function theorem, h is locally one-to-one. See Fig. 6.3.3.

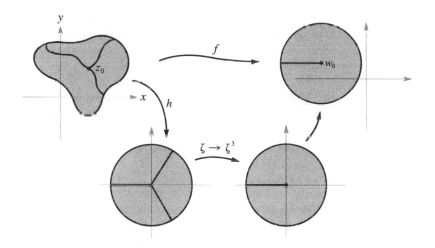

FIGURE 6.3.3 Worked Example 6.3.7 with $k = 3$.

Exercises

1. Let $f(z) = z + z^2$. For each z_0 specified, find the largest disk centered at z_0 on which f is one-to-one:

(a) $z_0 = 0$ (b) $z_0 = 1$

2. What is the largest disk around $z_0 = 1$ on which $f(z) = e^z$ is one-to-one?

3. Let f be analytic on $D = \{z$ such that $|z - z_0| < r\}$. Let $f(z_0) = w_0$ and suppose that $f(z) - w_0$ has no roots in D other than z_0 and that $f'(z)$ is never zero in D. Show that it is not necessarily true that f is one-to-one on D. (*Hint.* Consider z^3.)

4. What is the largest disk centered at $z_0 = 1$ on which $f(z) = z^3$ is one-to-one? (*Hint:* See Exercise 3.)

5. If f is analytic on A, $0 \in A$, and $f'(0) \neq 0$, then prove that near 0 we can write $f(z^n) = f(0) + [h(z)]^n$ for some analytic function h that is one-to-one near 0. (*Hint.* Use Worked Example 6.3.7.)

6. Let $u: A \to \mathbf{R}$ be harmonic and nonconstant on a region A. Prove that u is an open mapping.

7. Use Exercise 6 to prove the maximum and minimum principles for harmonic functions (see Sec. 2.5).

8. Let f be entire and have the property that if $B \subset \mathbf{C}$ is any bounded set, then $f^{-1}(B)$ is bounded (or perhaps empty). Show that for any $w \in \mathbf{C}$, there exists $z \in \mathbf{C}$ such that $f(z) = w$. (*Hint.*

Show that $f(\mathbf{C})$ is both open and closed and deduce that $f(\mathbf{C}) = \mathbf{C}$.) Apply this result to polynomials to deduce yet another proof of the fundamental theorem of algebra.

9. Show that the equation $z = e^{z-a}$, $a > 1$, has exactly one solution inside the unit circle.

10. Consider Worked Example 6.3.7 and take the case where $k = 4$. Visualize the local mapping in three steps as follows:

$$z \mapsto t = (z - z_0) \sqrt[4]{\phi(z)}; \qquad t \mapsto s = t^4; \qquad s \mapsto w = s + f(z_0)$$

Sketch this mapping.

11. Suppose f is analytic in a region A containing the closed unit disk $D = \{z \text{ such that } |z| \le 1\}$ and that $|f(z)| > 2$ whenever $|z| = 1$. If $f(0) = 1$, show that f has a zero in D.

12. Let $f(z) = \Sigma_0^\infty a_n z^n$ have a radius of convergence R. Suppose that $|a_1| \ge \Sigma_{n=2}^\infty n|a_n|r^{n-1}$ for some $0 < r \le R$. Show that f is one-to-one on $\{z \text{ such that } |z| < r\}$ unless f is constant. Compare your method with that used to solve Worked Example 6.2.13.

SUPPLEMENT A TO CHAPTER 6: NORMAL FAMILIES AND THE RIEMANN MAPPING THEOREM

The main objective of this supplement is to outline a proof of the Riemann mapping theorem. The material is set off since it is a bit more advanced than the rest of the chapter and is not needed for understanding or using this or succeeding chapters. However, it does illustrate several powerful tools and techniques of complex analysis.

Throughout the supplement, G will represent a connected, simply connected open set properly contained in the complex plane $\mathbf{C}$, and D the open unit disk $D = D(0; 1) = \{z \text{ such that } |z| < 1\}$. Given $z_0 \in \mathbf{C}$, the Riemann mapping theorem asserts that *there is a function f which is analytic on G and maps G one-to-one onto D with $f(z_0) = 0$. Furthermore, if it is required that $f'(z_0) > 0$, then there is exactly one such function.* The uniqueness has already been established in Chap. 5; that is, there can be no more than one such function. We still need to show there is at least one. The idea of the proof is to look at all the analytic functions which map G one-to-one *into* D taking z_0 to 0 with positive derivative at z_0; find one among them which maximizes $f'(z_0)$, and show that this function must take G *onto* D.

MONTEL'S THEOREM ON NORMAL FAMILIES

The proof of the existence of a function which maximizes $f'(z_0)$ will rest on the material of Sec. 3.1 concerning uniform convergence on closed disks. We learned there that if a sequence of analytic functions on a region converges uniformly on closed disks contained in the region, then the limit function must be analytic. The existence of such sequences is addressed by the theorem of Montel on normal families.

6.S.A.1 DEFINITION OF NORMAL FAMILY *If A is an open subset of* **C**, *a set $\mathscr{S}$ of functions analytic on A is called a **normal family** if every sequence of functions in $\mathscr{S}$ has a subsequence which converges uniformly on closed disks in A.*

Notice that by the analytic convergence theorem (3.1.8), the limit of such a subsequence must be analytic on A.

6.S.A.2 MONTEL'S THEOREM *If A is an open subset of* **C** *and $\mathscr{S}$ is a set of functions analytic on A which is uniformly bounded on closed disks in A, then every sequence of functions in $\mathscr{S}$ has a subsequence which converges uniformly on closed disks in A. That is, $\mathscr{S}$ is a normal family.*

PROOF* The plan of attack is as follows:

(i) Select a countable set of points $C = \{z_1, z_2, z_3, \ldots\}$ which are scattered densely throughout A in the sense that $A \subset \text{cl}(C)$.
(ii) Show that there is a subsequence of the original sequence of functions which converges at all of these points.
(iii) Show that convergence on this dense set of points is enough to force the subsequence to converge at all points of A.
(iv) Check that this convergence is uniform on every closed disk in A.

The first step may be accomplished by taking those points with both real and imaginary parts rational. There are only countably many of these, and so they may be arranged in a sequence, and they are scattered densely in A in the sense that some of them are arbitrarily close to anything in A.

Let $f_1, f_2, f_3, \ldots$ be a sequence of functions in $\mathscr{S}$. The assumption of uniform boundedness on closed disks is that for each closed disk $B \subset A$, there is a number $M(B)$ such that $|f_n(z)| < M(B)$ for all n and for all z in B. In particular, the numbers $f_1(z_1), f_2(z_1), f_3(z_1), \ldots$ are all smaller than $M(\{z_1\})$. Thus there must be a subsequence of them

---

* The student who has seen the Arzela-Ascoli theorem (see, for example, J. Marsden, *Elementary Classical Analysis* (New York: W. H. Freeman and Company, 1974)) can give a quick proof of Montel's theorem by using the assumed uniform boundedness and Worked Example 3.1.19 of this book to prove equicontinuity.

which converges to a point w_1 with $|w_1| \leq M(\{z_1\})$. Relabel this subsequence as

$$f_{1,1}(z_1), f_{1,2}(z_1), f_{1,3}(z_1), \ldots \to w_1$$

Evaluating these functions at z_2 gives another sequence of numbers $f_{1,1}(z_2)$, $f_{1,2}(z_2)$, $f_{1,3}(z_2)$, ... which are bounded by $M(\{z_2\})$. Some subsequence of these must converge to a point w_2. Relabel this sub-subsequence as

$$f_{2,1}(z_2), f_{2,2}(z_2), f_{2,3}(z_2), \ldots \to w_2$$

It is important to notice that the functions $f_{2,1}, f_{2,2}, f_{2,3}, \ldots$ are selected from among $f_{1,1}, f_{1,2}, f_{1,3}, \ldots$. Continuing in this way, selecting subsequences of subsequences, produces an array

$$
\begin{array}{l}
f_{1,1}(z_1), f_{1,2}(z_1), f_{1,3}(z_1), \ldots \to w_1 \\
f_{2,1}(z_2), f_{2,2}(z_2), f_{2,3}(z_2), \ldots \to w_2 \\
f_{3,1}(z_3), f_{3,2}(z_3), f_{3,3}(z_3), \ldots \to w_3 \\
f_{4,1}(z_4), f_{4,2}(z_4), f_{4,3}(z_4), \ldots \to w_4
\end{array}
$$

$$
\begin{array}{ccccc}
\cdot & \cdot & \cdot & \cdot & \cdot \\
\cdot & \cdot & \cdot & \cdot & \cdot \\
\cdot & \cdot & \cdot & \cdot & \cdot
\end{array}
$$

in which the kth horizontal row converges to some complex number w_k and the functions used in each row are selected from among those in the row above. The proof uses a procedure, called the *diagonal construction,* which is sometimes useful in other contexts. Let $g_n = f_{n,n}$. Then $g_1, g_2, g_3, \ldots$ is a subsequence of the original sequence of functions, and $\lim_{l\to\infty} g_l(z_k) = w_k$ for each k. This is because $g_n = f_{n,n}$ is a subsequence of $f_{k,1}$,

$f_{k,2}, f_{k,3}, \ldots$ as soon as $n > k$. Thus the subsequence g_n converges at a set of points which are scattered densely throughout A. Steps (iii) and (iv) of the program are to show that the fact that the g_n's are uniformly bounded on closed disks in A is enough to force them to converge everywhere in G and in fact to do so uniformly on closed disks in A. We accomplish this by showing that the sequence satisfies the Cauchy condition uniformly on closed disks.

Let B be a closed disk contained in A, and let $\epsilon > 0$. By Worked Example 3.1.19, the functions g_n are uniformly equicontinuous on B; that is, there is a number $\delta > 0$ such that $|g_l(\zeta) - g_l(\xi)| < \epsilon/3$ for all l whenever ζ and ξ are in B and $|\zeta - \xi| < \delta$. By using only finitely many of the points z_k we can guarantee that everything in B is within a distance δ of at least one of them. That is, there is an integer $K(B)$ such that for each $z \in B$ there is at least one $k \in \{1, 2, 3, \ldots, K(B)\}$ with $|z - z_k| < \delta$ and hence $|g_l(z) - g_l(z_k)| < \epsilon/3$ for all l. One way to do this would be to take a square grid of points with rational coordinates and separation less that δ. See Fig. 6.S.A.1. Since $\lim_{l\to\infty} g_l(z_k) = w_k$ for each k, each of these sequences satisfies the Cauchy condition and as there are only finitely many of

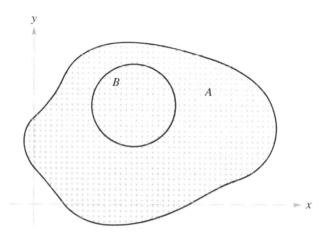

FIGURE 6.S.A.1 Finitely many of the z_k's give one within δ of anything in B.

them there is an integer $N(B)$ such that $|g_n(z_k) - g_m(z_k)| < \epsilon/3$ whenever $n \geq N(B)$, $m \geq N(B)$, and $1 \leq k \leq K(B)$.

Putting all this together, suppose $n \geq N(B)$ and $m \geq N(B)$. If $z \in A$, then z is within δ of z_k for some $k \leq K(B)$, and so

$$|g_n(z) - g_m(z)| \leq |g_n(z) - g_n(z_k)| + |g_n(z_k) - g_m(z_k)| + |g_m(z_k) - g_m(z)| \leq \frac{\epsilon}{3} + \frac{\epsilon}{3} + \frac{\epsilon}{3} = \epsilon$$

The sequence g_n thus uniformly satisfies the Cauchy condition on B and so converges uniformly on B to some limit function, as desired. ∎

PROOF OF THE RIEMANN MAPPING THEOREM

We are now in a position to prove the Riemann mapping theorem. Let G be a connected, simply connected, open set properly contained in the complex plane $\mathbf{C}$. Let $z_0 \in G$, and let $D = D(0; 1)$ be the open unit disk. We must show that there is a function f analytic on G which maps G one-to-one onto D with $f(z_0) = 0$ and $f'(z_0) > 0$. To do this, let

$$\mathscr{S} = \{f : G \to D \mid f \text{ is analytic and one-to-one on } G, f(z_0) = 0, \text{ and } f'(z_0) > 0\}$$

The main steps of the proof are:

(i) Show that $\mathscr{S}$ is not empty.
(ii) Show that the numbers $\{f'(z_0) \mid f \in \mathscr{S}\}$ are bounded above and so have a finite least upper bound M.

(iii) Use Montel's theorem to extract from a sequence of functions in $\mathscr{S}$ whose derivatives at z_0 converge to M a subsequence which converges uniformly on closed disks in G. The limit function f is analytic in G and $f'(z_0) = M$.

(iv) Show that $f \in \mathscr{S}$.

(v) Show that f must map G onto D.

To show that $\mathscr{S}$ is not empty, it is enough to show that we can map G analytically into the unit disk. Once that is done, we need only compose with a linear fractional transformation of the disk onto itself which takes z_0 to 0 and then multiply by a constant $e^{i\theta}$ chosen so the derivative of the resulting map at z_0 is positive. If G is bounded, for example, if $|z - z_0| < R$ for all z in G, the map $z \mapsto (z - z_0)/R$ does the job. If G is not bounded, it at least omits a point a. The translation $z \mapsto z - a$ takes G to a simply connected region G_1 not containing 0. By Theorem 2.2.6, there is a branch of logarithm defined on G_1 which we will call F. Then the map g defined by $z \mapsto e^{(1/2)F(z)}$ is a branch of the square root function; by the open mapping or inverse mapping theorem, one sees that $G_2 = g(G_1)$ contains some disk $D(b; r)$. By properties of the square root function, $D(-b; r)$ fails to meet G_2. The map $f(z) = r/[b + z]$ then maps G_2 into the unit disk. See Fig. 6.S.A.2.

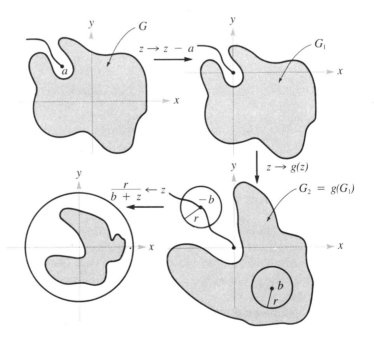

FIGURE 6.S.A.2 Mapping G into the unit disk.

Having shown that $\mathscr{S}$ is not empty, we must establish part (ii). The family $\mathscr{S}$ is uniformly bounded by 1 on G, and so by Worked Example 3.1.18, the derivatives are uniformly bounded on closed disks in G. In particular, there is a finite number $M(\{z_0\})$ such that $f'(z_0) \le M(\{z_0\})$ for all f in $\mathscr{S}$. Let M be the least upper bound of these derivatives. There must be a sequence $f_1, f_2, f_3, \ldots$ of functions in $\mathscr{S}$ with the property that $\lim_{n\to\infty} f'_n (z_0) = M$. Since the family $\mathscr{S}$ is uniformly bounded, it is normal by Montel's theorem and there must be a subsequence which converges uniformly on closed disks in G. We may as well throw away the functions we don't need and assume that we have a sequence which converges uniformly on closed disks in G. By the analytic convergence theorem (3.1.8) they converge to a limit function f which is analytic on G and $f'(z_0) = M$.

We next want to know that f is a member of $\mathscr{S}$. Each of the functions f_n maps G into the open unit disk, and so f certainly maps G into the closed unit disk. Since f is not constant, the maximum modulus principle says that $|f(z)|$ cannot have a maximum anywhere in G, and so the image never touches the boundary of the disk and f maps G into D. Certainly $f(z_0) = \lim_{n\to\infty} f_n(z_0) = 0$. Finally, the corollary of Hurwitz' theorem (6.2.8) shows that f must be one-to-one since it is a nonconstant limit of one-to-one functions which converge uniformly on closed disks. Thus $f \in \mathscr{S}$.

The final step, (iv), is to show that f must actually map G onto D. This follows from the following assertion.

CLAIM *If A is a connected, simply connected open set properly contained in D and $0 \in A$, then there is a function F analytic on A which maps A one-to-one into D with $F(0) = 0$ and $F'(0) > 1$.*

To see how (iv) follows from this assertion, suppose that f does not map G onto D. Then $A = f(G)$ satisfies the conditions of the claim. (That it is open follows from the open mapping theorem (6.3.3). Consider $g(z) = F(f(z))$. Then $g \subset \mathscr{S}$, but $g'(z_0) = F'(f(z_0))f'(z_0) = F'(0)M > M$, contradicting the maximality of M.

Thus it remains to check the claim. The construction is a bit like that used in step (i), and is perhaps best traced through by following the diagrams in Fig. 6.S.A.3. The region A is shaded by diagonal lines in the first diagram. It misses a point a indicated by an open circle in that diagram. The successive images of a and 0 are indicated by open dots and solid dots respectively in each of the following diagrams. Map F_1 is a linear fractional transformation of the disk to itself taking a to 0 and 0 somewhere. The purpose of map F_2 is to guarantee a situation in which the image of A misses a neighborhood of a point on the boundary circle. This is done just as in step (i) by using a branch of logarithm on the simply connected region $F_1(A)$ which misses 0. Map F_3 is another linear fractional transformation which returns the image of 0 to 0. At this stage the image of A misses a small circle γ which intersects the unit circle C at right angles at two points. An appropriate linear fractional transformation F_4 taking these points to 0 and ∞ will take the circles to lines through 0 and ∞ and the region between them to a quarter plane. Squaring F_5 opens this up to a half plane. Finally another linear fractional transformation takes the half plane to the unit disk with the black dot going to 0 and the correct rotation making the derivative of the whole thing at 0 positive. The function F is $F(z) =$

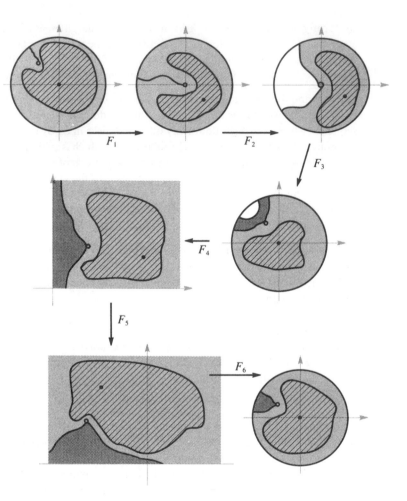

FIGURE 6.S.A.3   Construction for the claim in the proof of the Riemann mapping theorem.

$F_6(F_5(F_4(F_3(F_2(F_1(z)))))) = w$. The inverse function $g(w) = F^{-1}(w) = z$ satisfies the conditions of the Schwarz lemma. Since it is not a rotation, we have strict inequality $|g'(0)| < 1$ by Worked Example 2.5.19, but $F'(0) = 1/g'(0)$. Therefore $F'(0) > 1$, as required. All the pieces have been assembled, and so the proof of the Riemann mapping theorem is now complete. ■

SUPPLEMENT B TO CHAPTER 6: THE DYNAMICS OF COMPLEX ANALYTIC MAPPINGS

The pictures shown in Fig. 6.S.B.1 are representations of the dynamics of complex analytic mappings. The purpose of this supplement is to provide a brief introduction to

(a)

(b)

FIGURE 6.S.B.1 The different shadings represent the rate of approach of points to infinity under iteration of the mapping; the black region consists of "stable" points which remain bounded under iteration. In part (a) the mapping is $(1 + 0.1i) \sin z$, while in (b) it is $(1 + 0.2i) \sin z$. *(Courtesy of R. Devany of Boston University, with the assistance of C. Mayberry, C. Small, and S. Smith)*

this subject — mainly to inspire the reader to find out more by consulting a reference on the subject, such as R. L. Devany, *An Introduction to Chaotic Dynamical Systems,* (Reading, Mass.: Addison-Wesley, 1985); P. Blanchard, Complex Dynamics on the Riemann Sphere, *Bulletin of the American Mathematical Society,* Vol. 11 (1984), pp. 85–141; or B. Mandelbrot, *The Fractal Geometry of Nature* (New York: W. H. Freeman and Co., 1982). The subject we will be looking at has to do with the way points in the complex plane behave under iteration of an analytic function. It has its origins in the classical and beautiful work of G. Julia ("Memoire sur l'iterations des fonctions ration-elles," *J. Math.,* Vol. 8 (1918) pp. 47–245) and P. Fatou ("Sur l'iteration des fonctions transcendantes entières," *Acta Math.,* Vol. 47 (1926), pp. 337–370). In this study normal families (see Supplement A) play an important role. In fact, Montel himself was interested in these questions; see his *Leçons sur les familles normales de fonctions analytiques et leurs applications* (1927; rpt. New York: Chelsea, 1974), Chap. VIII.

Let us fix an entire function $f : \mathbf{C} \to \mathbf{C}$. We need a little terminology to get going. Given a point $z \in \mathbf{C}$, the *orbit* of z is the sequence of points z, $f(z)$, $f(f(z))$, $f(f(f(z)))$, . . . , which we also write as $z, f(z), f^2(z), f^3(z),$ . . . . We think of the point z as moving successively under the mapping f to new locations. A *fixed point* is a point z such that $f(z) = z$, that is, a point z that does not move when we apply f. A *periodic point* is a point z such that $f^n(z) = z$ for some integer n (called the period), where f^n means f composed with itself n times.

A fixed point z is called an *attracting fixed point* if $|f'(z)| < 1$. The reason for this terminology is that the orbits of nearby points converge to z; this is so because near z, f behaves like a mapping that rotates by an amount $\arg f'(z)$ and magnifies by an amount $|f'(z)|$, so that every time that f is applied, points will be pulled toward z by a factor $|f'(z)|$, and so as this is repeated, the point tends to z. Likewise, a point z is called a *repelling fixed point* if $|f'(z)| > 1$; points near repelling points will be pushed away under iteration of the function f. Similarly, a periodic point z with period n is called an *attracting periodic point* if $|(f^n)'(z)| < 1$; such points have the property that the orbits of points close to z tend to the orbit of z. Likewise, a *repelling periodic point* has the property that $|(f^n)'(z)| > 1$; orbits of points near such points will be shoved away from the orbit of z.

The *Julia set* $J(f)$ of f is defined to be the closure of the set of repelling periodic points of f. This set can be of remarkable and beautiful complexity usually called a *fractal;* in fact, in the picture in Fig. 6.S.B.1 the nonblack region is the Julia set. This statement rests on a theorem which we shall not prove which states that the Julia set is the closure of the points which go to infinity under iteration of f. It is this characterization which is useful for computational purposes. Figure 6.S.B.2 shows two more Julia sets for quadratic maps.

As far as complex analysis is concerned, one of the most important results is the following:

The Julia set of f is the set of points at which the family of functions f^n is not normal.

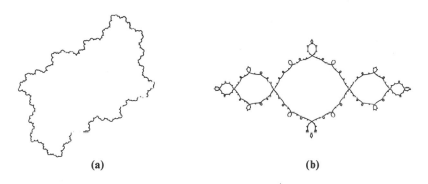

(a)   (b)

FIGURE 6.S.B.2   (a) The Julia set of $f(z) = z^2 + \frac{1}{2}i$, which, is a simple closed curve but is nowhere differentiable. (b) The Julia set of $f(z) = z^2 - 1$, which contains infinitely many closed curves.

This result can be used as an alternative definition for the Julia set, and in fact this was the original definition of Fatou and Julia. We will prove only the following statement here to give a flavor of how the arguments go: *If z is a repelling fixed point of f (and so is in the Julia set), then the family of iterates f^n fails to be normal at z.* Let us assume that this family is normal at z and derive a contradiction. Normal at z means normal in a neighborhood of z, in the same way that we used the terminology "analytic at z." By Definition 6.S.A.1, the family f^n has a subsequence which converges uniformly on a neighborhood of z. Since $f(z) = z$ and $|f'(z)| > 1$, it follows from the chain rule that

$$|f^{n\prime}(z)| = |f'(z)|^n \to \infty$$

that is, that the sequence of derivatives of f^n evaluated at z must tend to infinity as $n \to \infty$. However, the sequence of derivatives must converge to the derivative of the limit function by the analytic convergence theorem (3.1.8), which is finite, giving us the required contradiction.

This discussion represents only the tip of a large collection of very interesting and beautiful results. We hope that the reader will be inspired to look up some of the references on the subject that we have given, and further references that will be found in those sources, and will explore the subject further. We hasten to point out that the iteration of complex mappings is just one part of a larger and growing field called *chaotic dynamics*. For these more general aspects, the reader can consult Devany's book cited above, or the book *Nonlinear Oscillations, Dynamical Systems, and Bifurcation of Vector Fields,* by J. Guckenheimer and P. Holmes (New York: Springer-Verlag, 1983).

REVIEW EXERCISES FOR CHAPTER 6

1.   Let f be analytic on $\{z$ such that $|z| < 1\}$ and let $f(1/n) = 0, n = 1, 2, \ldots$. What can be said about f?

2. Suppose that f and g are analytic on the disk $A = \{z$ such that $|z| < 2\}$ and that neither $f(z)$ nor $g(z)$ is ever 0 for $z \in A$. If

$$\frac{f'(1/n)}{f(1/n)} = \frac{g'(1/n)}{g(1/n)} \qquad \text{for} \qquad n = 1, 2, 3, 4, \ldots$$

show there is a constant c such that $f(z) = cg(z)$ for all $z \in A$. (*Hint:* Consider $(f/g)'(1/n)$.)

3. Suppose f is an entire function and that there is a bounded sequence of distinct real numbers $a_1, a_2, a_3, \ldots$ such that $f(a_k)$ is real for each k.
(a) Show that $f(x)$ is real for all real x.
(b) Suppose $a_1 > a_2 > a_3 > \cdots > 0$ and $\lim_{k \to \infty} a_k = 0$. Show that if $f(a_{2n+1}) = f(a_{2n})$ for all n, then f must be constant.

4. If f is analytic on the set $\{z$ such that $|z| < 1\}$ and $f(1 - 1/n) = 0$, $n = 1, 2, 3, \ldots$, does it follow that $f = 0$?

5. Let f be analytic and bounded on $\{z \mid \text{Im } z < 1\}$ and suppose that f is real on the real axis. Show that f is constant.

6. Let f be analytic and bounded on $|z + i| > \frac{1}{2}$ and real on $]-1, 1[$. Show that f is constant. (*Hint.* Use the Schwarz reflection principle from Sec. 6.1.)

7. Let f be entire and suppose that for $z = x$ real, $f(x + 1) = f(x)$. Show that $f(z + 1) = f(z)$ for all $z \in \mathbf{C}$.

8. Show that for $n > 2$, all the roots of $z^n - (z^2 + z + 1)/4 = 0$ lie inside the unit circle.

9. Suppose that f is analytic in $\mathbf{C}$ except for poles at $n \pm i$, $n = 0, \pm 1, \pm 2, \ldots$. What is the length of the longest interval $]x_0 - R, x_0 + R[$ in $\mathbf{R}$ on which $f(x_0) + f'(x_0)(x - x_0) + f''(x_0)(x - x_0)^2/2 + \cdots + f^{(k)}(x_0)(x - x_0)^k/k! + \cdots$ converges?

10. Let $f : A \to B$ be analytic and onto; assume that $z_1, z_2 \in A$, $z_1 \neq z_2$, implies that $f(z_1) \neq f(z_2)$. Prove that f^{-1} is analytic.

11. Let f be a polynomial. Show that the integral of f'/f around every sufficiently large circle centered at the origin is $2\pi i$ times the degree of f.

12. (a) Prove *Vitali's Convergence Theorem. Let f_n be analytic on a domain A such that*

    (i) *For each closed disk B in A there is a constant M_B such that $|f_n(z)| \leq M_B$ for all $z \in B$ and $n = 1, 2, 3, \ldots$.*

    (ii) *There is a sequence of distinct points z_k of A converging to $z_0 \in A$ such that $\lim_{n \to \infty} f_n(z_k)$ exists for $k = 1, 2, \ldots$.*

Then f_n converges uniformly on every closed disk in A; the limit is an analytic function. (*Hint.* First take the case of a disk B with radius R and $z_k \to z_0 =$ the center of B. Use the Schwarz lemma to show that $|f_n(z) - f_n(z_0)| \leq 2M|z - z_0|/R$. Then show that

$$|f_n(z_0) - f_{n+p}(z_0)| \leq \frac{4M|z - z_0|}{R} + |f_n(z) - f_{n+p}(z)|$$

and deduce that $f_n(z_0)$ converges. Let

$$g_n(z) = \frac{f_n(z) - f_n(z_0)}{z - z_0}$$

and conclude that $g_n(z_0)$ converges. Show that in general, if

$$f_n(z) = \sum_{k=0}^{\infty} a_{n,k}(z - z_0)^k$$

then $a_{n,k} \to a_k$ as $n \to \infty$. Deduce that $f_n(z)$ converges uniformly in $|z - z_0| < R - \epsilon$. Then use connectedness of A to deduce uniform convergence on any closed disk.)

(b) Show that if condition (i) is omitted, the conclusion is false. (Let $f_n(z) = z^n$.)

13. Let f be analytic on a region A and let γ be a closed curve in A homotopic to a point. Show that

$$\mathrm{Re}\left(\int_\gamma \frac{f'}{f}\right) = 0$$

14. Let $f(z)$ be analytic on $\{z \mid 0 < |z| < 2\}$ and suppose that for $n = 0, 1, 2, \ldots$

$$\int_{|z|=1} z^n f(z)\, dz = 0$$

Show that f has a removable singularity at $z = 0$.

15. Let f be analytic and bounded on $A = \{z$ such that $|z| < 1\}$. Show that if f is one-to-one on $\{z \mid 0 < |z| < 1\}$, then f is one-to-one on A.

16. Let $|f(z)| \leq 1$ when $|z| = 1$ and let $f(0) = \frac{1}{2}$ with f analytic. Prove that

$$|f(z)| \leq \begin{cases} 3|z| + 1 & \text{for all } |z| \leq \frac{1}{3} \\ ? \\ 1 & \text{for } \frac{1}{3} \leq |z| \leq 1 \end{cases}$$

17. Let f and g be continuous for $|z| \leq 1$ and analytic for $|z| < 1$. Suppose that $f = g$ on the unit circle. Prove that $f = g$.

18. If $f(z)$ is analytic for $|z| < 1$ and if $|f(z)| \leq 1/(1 - |z|)$, show that the coefficients of the expansion $f(z) = \sum_{n=0}^{\infty} a_n z^n$ are subject to the inequality

$$|a_n| \leq (n + 1) \left(1 + \frac{1}{n}\right)^n < e(n + 1)$$

19. Which of the following statements is/are true?

(a) The radius of convergence of $\sum_{n=0}^{\infty} 2^n z^{2n}$ is $1/\sqrt{2}$.

(b) An entire function that is constant on the unit circle is a constant.

(c) The residue of $1/[z^{10}(z - 2)]$ at the origin is $-(2)^{-10}$.

(d) If f_n is a sequence of entire functions converging to a function f and if the convergence is uniform on the unit circle, then f is analytic in the open unit disk.

(e) $\displaystyle\int_0^\pi \frac{d\theta}{a + \cos\theta} = \frac{2\pi}{a^2 - 1}$.

(f) For sufficiently large r, $\sin z$ maps the exterior of the disk of radius r ($\{z$ such that $|z| > r\}$) into any preassigned neighborhood of ∞.

(g) Let $f : \mathbf{C} \to \mathbf{C}$ be analytic in the open unit disk and let f have a nonremovable singularity at i. Then the radius of convergence of the Taylor series of f at 0 is 1.

(h) Let $f : \mathbf{C} \to \mathbf{C}$ be analytic and nonconstant and let D be a domain in $\mathbf{C}$. Then f maps the boundary of D into the boundary of $f(D)$.

(i) Let f be analytic on $\{z \mid 0 < |z| < 1\}$ and suppose that $|f(z)| \leq \log (1/|z|)$. Then f has a removable singularity at 0.

(j) Suppose that $f : \mathbf{C} \to \mathbf{C}$ is entire and that f has exactly k zeros in the open unit disk but none on the unit circle. Then there exists an $\epsilon > 0$ such that any entire function g that satisfies $|f(z) - g(z)| < \epsilon$ for $|z| = 1$ must also have exactly k zeros in the open unit disk.

20. Prove the *Phragmen-Lindelof Theorem*:

(a) *Suppose that f is analytic in a domain that includes the strip $G = \{z \in \mathbf{C} \mid 0 \leq \operatorname{Re} z \leq 1\}$. If $\lim\limits_{\substack{z \to \infty \\ z \in G}} f(z) = 0$ and if $|f(it)| \leq 1$ and $|f(1 + it)| \leq 1$ for all real t, then $|f(z)| \leq 1$ for all $z \in G$.*

(b) *If g is analytic in a domain containing G, if $\lim\limits_{\substack{z \to \infty \\ z \in G}} g(z) = 0$, and if $|g(it)| \leq M$ and $|g(1 + it)| \leq N$ for all real t, then $|g(z)| \leq M^{1 - \operatorname{Re} z} N^{\operatorname{Re} z}$.* (Hint. Apply the result of (a) to $f(z) = g(z)/M^{1-z}N^z$.)

21. Is it correct to say that $1/\sqrt{z}$ has a pole at $z = 0$?

22. Prove that for the principal value of the logarithm, $|\log z| \leq r/(1 - r)$ if $|1 - z| \leq r < 1$.

23. (a) Let $f : \mathbf{C} \to \mathbf{C}$ be continuous on $\mathbf{C}$ and analytic on $\mathbf{C} \backslash \mathbf{R}$. Is f actually entire?

(b) Let $f = \mathbf{C} \to \mathbf{C}$ be analytic on $\mathbf{C} \backslash \mathbf{R}$. Is f entire?

24. Let $P(z)$ be a polynomial. Prove that

$$\int_{|z|=1} P(z)\, d\bar{z} = -2\pi i P'(0)$$

25. Find the radius of convergence of the series $\sum_0^\infty 2^n z^n$.

26. Show that $f(z) = (z^2 + 1)/(z^2 - 1)$ is one-to-one on $\{z \mid \operatorname{Im} z > 0\}$. Is it one-to-one on any larger set?

CHAPTER 7

ASYMPTOTIC METHODS

This chapter will give an introduction to the theory of asymptotic methods; that is, to the study of functions $f(z)$ as $z \to \infty$. The chapter begins with infinite products and the gamma function. These topics are of interest in their own right but they also provide the student with motivation to study asymptotic expansions, which are analyzed in Sec. 7.2. One of the main techniques used in this analysis, the method of steepest descent, and its variant, the method of stationary phase, are also considered and are applied to Stirling's formula and to Bessel functions in Sec. 7.3.

7.1 INFINITE PRODUCTS AND THE GAMMA FUNCTION

To study the gamma function and subsequent topics, we shall first develop some basic properties of infinite products. They are somewhat analogous to the infinite sums considered in Sec. 3.1. For orientation and motivation, the student should note that any polynomial $p(z)$ can be written in the form

$$p(z) = a_n(z - \alpha_1) \cdots (z - \alpha_n) = a_n \prod_{j=1}^{n} (z - \alpha_j)$$

where $\alpha_1, \ldots, \alpha_n$ are the roots of $p(z) = a_n z^n + \cdots + a_1 z + a_0$ and Π stands for "take the product of" in the same way as Σ stands for "take the sum of." It is natural to attempt to generalize this expression to entire functions, and in doing so we encounter the concept of an infinite product.

Infinite Products

Let $z_1, z_2, \ldots$ be a sequence of complex numbers. We want to consider

$$\prod_{n=1}^{\infty} (1 + z_n) = (1 + z_1)(1 + z_2) \cdots$$

We write $1 + z_n$ because if the product is to converge, the general term should approach

1; that is, $z_n \to 0$. Some technicalities are involved when $z_n = -1$. We want to allow the product to be zero yet be able to impose some convergence condition. The following definition fits our needs.

7.1.1 DEFINITION *The product $\Pi_{n=1}^{\infty} (1 + z_n)$ is said to **converge** iff only a finite number of z_n equal -1 and if $\Pi_{k=m}^{n} (1 + z_k) = (1 + z_m) \cdots (1 + z_n)$ (where $z_k \neq -1$ for $k \geq m$) converges as $n \to \infty$, to a nonzero number.*
    We set

$$\prod_{n=1}^{\infty} (1 + z_n) = \lim_{n \to \infty} \prod_{k=1}^{n} (1 + z_k)$$

(This product will be zero if some $z_k = -1$, and nonzero otherwise.)

For example, consider

$$\prod_{n=2}^{\infty} \left(1 - \frac{1}{n}\right) = \frac{1}{2} \cdot \frac{2}{3} \cdot \frac{3}{4} \cdots$$

The *n*th *partial product* is

$$\frac{1}{2} \cdot \frac{2}{3} \cdots \frac{n-1}{n} = \frac{1}{n} \to 0$$

Thus the product does not converge, because we have demanded convergence to a nonzero number. (We say the product *diverges to zero*.) If we started at $n = 1$, the product would still diverge. A reason for this terminology and convention is that the sequence of logarithms of the partial products diverges to $-\infty$. The relevance of this is made clear by the convergence theorem below.

By starting a given product beyond the point where some $z_n = -1$, we can assume that $z_n \neq -1$ for all n. Such an assumption imposes no real restrictions in the tests for convergence.

7.1.2 CONVERGENCE THEOREM FOR PRODUCTS

(i) *If $\Pi_{n=1}^{\infty} (1 + z_n)$ converges, then $z_n \to 0$.*
(ii) *Suppose that $|z_n| < 1$ for all $n = 1, 2, \ldots$ so that $z_n \neq 1$. Then $\Pi_{n=1}^{\infty} (1 + z_n)$ converges if and only if $\Sigma_{n=1}^{\infty} \log (1 + z_n)$ converges. (log is the principal branch; $|z_n| < 1$ implies that $\log (1 + z_n)$ is defined.)*
(iii) *$\Pi_{n=1}^{\infty} (1 + |z_n|)$ converges iff $\Sigma_{n=1}^{\infty} |z_n|$ converges. (We say that $\Pi_{n=1}^{\infty} (1 + z_n)$ converges absolutely in this case.)*
(iv) *If $\Pi_{n=1}^{\infty} (1 + |z_n|)$ converges, then $\Pi_{n=1}^{\infty} (1 + z_n)$ converges.*

This theorem summarizes the main convergence properties of infinite products. Criteria (iii) and (iv) are particularly important and are easy to apply. Because it is technical, the proof of this theorem appears at the end of this section. Only the plausibility of the theorem will be discussed here. Criterion (i) was explained at the beginning of this section. To explain (ii), note that if we let $S_n = \Sigma_1^n \log (1 + z_k)$ and $P_n = \Pi_1^n (1 + z_k)$, then

$$P_n = e^{S_n}$$

That (ii) is plausible easily follows from this equation. Indeed, if $S_n \to S$ it is clear that $P_n \to e^S$. Once (ii) is shown, (iii) and (iv) follow. The following corollary requires no proof, since it is implicit in the preceding discussion.

7.1.3 COROLLARY *If $|z_n| < 1$ and $\Sigma \log (1 + z_n)$ converges to S, then $\Pi (1 + z_n)$ converges to e^S.*

This corollary is sometimes useful, but when it is applied to concrete problems, the sum of logarithms is often difficult to handle.

Let $f_n(z)$ be a sequence of functions defined on a set $B \subset \mathbf{C}$. The way to define the concept of the uniform convergence of $\Pi_1^\infty (1 + f_n)$ should be fairly clear.

7.1.4 DEFINITION *The product*

$$\prod_{n=1}^{\infty} [1 + f_n(z)]$$

*is said to **converge uniformly** on B iff, for some m, $f_n(z) \neq -1$ for $n \geq m$ and all $z \in B$, if the sequence $P_n(z) = \Pi_{k=m}^n [1 + f_k(z)]$ converges uniformly on B to some P(z), and if $P(z) \neq 0$ for all $z \in B$ (see Sec. 3.1 for the definition of uniform convergence of a sequence of functions).*

The next result follows from the analytic convergence theorem (3.1.8).

7.1.5 ANALYTICITY OF INFINITE PRODUCTS *Suppose that $f_n(z)$ is a sequence of analytic functions on a region A and that $\Pi_{n=1}^\infty [1 + f_n(z)]$ converges uniformly to $f(z)$ on every closed disk in A. Then $f(z)$ is analytic on A. Such uniform convergence holds if $|f_n(z)| < 1$ for $n \geq m$ and if either $\Sigma_{n=m}^\infty \log [(1 + f_n(z)]$ converges uniformly or $\Sigma_{n=1}^\infty |f_n(z)|$ converges uniformly (on closed disks in both cases).*

To check the validity of the last statement, one must check that the proof of the convergence theorem for products works for uniform convergence; this is left as an exercise.

Canonical Products

The following useful theorem is a special case of a theorem of Weierstrass that constructs the most general entire function with a given set of zeros. The special case described here is applicable to many examples, yet it illustrates the main ideas of the general case. (For a statement of the general case, see Exercises 10 and 14 at the end of this section.)

7.1.6 THEOREM ON CANONICAL PRODUCTS *Let $a_1, a_2, \ldots$ be a given sequence (possibly finite) of nonzero complex numbers such that*

$$\sum_{n=1}^{\infty} \frac{1}{|a_n|^2} < \infty$$

Then if $g(z)$ is any entire function, the function

$$f(z) = e^{g(z)} z^k \left[\prod_{n=1}^{\infty} \left(1 - \frac{z}{a_n} \right) e^{z/a_n} \right] \tag{1}$$

*is entire. The product converges uniformly on closed disks, has zeros at $a_1, a_2, \ldots$, and has a zero of order k at $z = 0$, but has no other zeros. Furthermore, if f is any entire function having these properties, it can be written in the same form (Eq. (1)). In particular, f is entire with no zeros if and only if f has the form $f(z) = e^{g(z)}$ for some entire function g. The product $\prod_{n=1}^{\infty} (1 - z/a_n) e^{z/a_n}$ is called a **canonical product**.*

The technical proof of this appears at the end of this section. The result is quite plausible if we note that the product vanishes exactly when z is equal to some a_n and that z^k has a zero of order k at 0. Also, $e^{g(z)}$ vanishes nowhere, since $e^w \neq 0$ for all $w \in \mathbb{C}$. We note that the points $a_1, a_2, \ldots$ need not be distinct; each may be repeated finitely many times. If a_n is repeated l times, f will have a zero of order l at a_n.

The theorem on canonical products will be applied several times in the remainder of this section. One important application of the theorem is found in Worked Example 7.1.10, where it is proved that

$$\sin \pi z = \pi z \prod_{\substack{n=-\infty \\ n \neq 0}}^{\infty} \left(1 - \frac{z}{n} \right) e^{z/n} \tag{2}$$

Another application of the theorem is to the gamma function.

The Gamma Function

The gamma function is a useful solution to an interpolation problem which has been studied since the 1700s. What is the best way to define a continuous function of a real or complex variable which agrees with the factorial function at the integers? The gamma function, $\Gamma(z)$, is one solution. It is analytic on $\mathbf{C}$ except for simple poles at $0, -1, -2, \ldots$, and $\Gamma(n+1) = n!$ for $n = 0, 1, 2, \ldots$. The importance of this function was realized by Euler and Gauss as early as the eighteenth century. Two equivalent definitions of the gamma function are given here; the first will be in terms of infinite products, the second in terms of an integral formula. These two formulas are due to Euler. Significant contributions were also made by Gauss and Legendre. The main facts that are included in the following discussion and in the end-of-section exercises are summarized in Table 7.1.1 at the end of this section.

For the first definition, let us begin with an associated function that is defined by the canonical product

$$G(z) = \prod_{n=1}^{\infty} \left(1 + \frac{z}{n}\right) e^{-z/n} \tag{3}$$

By the theorem on canonical products, this function is entire, with simple zeros at the negative integers $-1, -2, -3, \ldots$. This function satisfies the identity

$$zG(z)G(-z) = \frac{\sin \pi z}{\pi} \tag{4}$$

because of Eq. (2). Now we consider the function

$$H(z) = G(z - 1) \tag{5}$$

This function has zeros at $0, -1, -2, \ldots$. Thus, by the theorem on canonical products, we can write

$$H(z) = e^{g(z)} z \prod_{1}^{\infty} \left(1 + \frac{z}{n}\right) e^{-z/n} = z e^{g(z)} G(z) \tag{6}$$

for some entire function $g(z)$. It will now be shown that $g(z)$ is constant. Using the convergence theorem for products, we get

$$\log H(z) = \log z + g(z) + \sum_{n=1}^{\infty} \left[\log\left(1 + \frac{z}{n}\right) - \frac{z}{n}\right]$$

Since the convergence is uniform on closed disks, we may differentiate term by term:

$$\frac{d}{dz} \log H(z) = \frac{1}{z} + g'(z) + \sum_{n=1}^{\infty} \left(\frac{1}{z+n} - \frac{1}{n}\right) \tag{7}$$

Similarly, by Eq. (3),

$$
\begin{aligned}
\frac{d}{dz} \log G(z-1) &= \sum_{n=1}^{\infty} \left(\frac{1}{z-1+n} - \frac{1}{n} \right) = \frac{1}{z} - 1 + \sum_{n=1}^{\infty} \left(\frac{1}{z+n} - \frac{1}{n+1} \right) \\
&= \frac{1}{z} - 1 + \sum_{n=1}^{\infty} \left(\frac{1}{z+n} - \frac{1}{n} \right) + \sum_{n=1}^{\infty} \left(\frac{1}{n} - \frac{1}{n+1} \right) \\
&= \frac{1}{z} + \sum_{n=1}^{\infty} \left(\frac{1}{z+n} - \frac{1}{n} \right)
\end{aligned} \tag{8}
$$

Comparing Eqs. (7) and (8) and using Eq. (5), we see that $g'(z) = 0$, so $g(z)$ is constant. (Part (ii) of the convergence theorem for products is actually valid only for $|z| < 1$, but this region of validity suffices since two entire functions that agree on $|z| < 1$ are equal by Taylor's theorem or the identity theorem.)

The constant value $g(z) = \gamma$ is called *Euler's constant*. We can determine an expression for it as follows. By Eqs. (3), (5), and (6), we get

$$
G(z-1) = z e^{\gamma} G(z) \tag{9}
$$

and so if we let $z = 1$, then $G(0) = 1 = e^{\gamma} G(1)$. Thus, by Eq. (3),

$$
e^{-\gamma} = \prod_{1}^{\infty} \left(1 + \frac{1}{n} \right) e^{-1/n} = \prod_{1}^{\infty} \left(\frac{n+1}{n} \right) e^{-1/n}
$$

Noting that

$$
\begin{aligned}
\prod_{k=1}^{n} \left(\frac{k+1}{k} \right) e^{-1/k} &= \frac{2}{1} \cdot \frac{3}{2} \cdot \frac{4}{3} \cdots \frac{n+1}{n} e^{-1-1/2-1/3-\cdots-1/n} \\
&= (n+1) e^{-1-1/2-\cdots-1/n} \\
&= n e^{-1-1/2-\cdots-1/n} + e^{-1-1/2-\cdots-1/n}
\end{aligned}
$$

we get $e^{-\gamma} = \lim_{n\to\infty} n e^{-1-1/2-\cdots-1/n}$. Taking logs, we find that

$$
\gamma = \lim_{n\to\infty} \left(1 + \frac{1}{2} + \cdots + \frac{1}{n} - \log n \right) \tag{10}
$$

We do not need a separate proof that the limit in Eq. (10) exists and is finite, because this follows from what we have done. Numerically, we can compute from the limit in Eq. (10) that $\gamma \approx 0.57716 \cdots$.

Now we are ready to define the gamma function. We set

$$\Gamma(z) = [ze^{\gamma z}G(z)]^{-1} = \left[ze^{\gamma z} \prod_{n=1}^{\infty} \left(1 + \frac{z}{n} \right) e^{-z/n} \right]^{-1} \tag{11}$$

From the entireness of G, we can conclude that $\Gamma(z)$ is meromorphic, with simple poles at $0, -1, -2, \ldots$. Since, by Eq. (9), $G(z-1) = ze^{\gamma}G(z)$, we find that

$$\Gamma(z+1) = z\Gamma(z) \quad \text{for} \quad z \neq 0, -1, -2, \ldots \tag{12}$$

which is called the *functional equation for the gamma function* (see Exercise 7). Also, $\Gamma(1) = 1$, since $\Gamma(z) = [ze^{\gamma z}G(z)]^{-1}$ and $G(1) = e^{-\gamma}$ by our construction of γ. Thus, from Eq. (12) we see that $\Gamma(2) = 1 \cdot 1$, $\Gamma(3) = 2 \cdot 1$, $\Gamma(4) = 3 \cdot 2 \cdot 1$, and generally that, as earlier advertised,

$$\Gamma(n+1) = n! \tag{13}$$

This formula will enable us to obtain manageable approximations for $n!$, which are derived in Sec. 7.3. (Fig. 7.1.1 shows a graph of $\Gamma(x)$ for x real.)

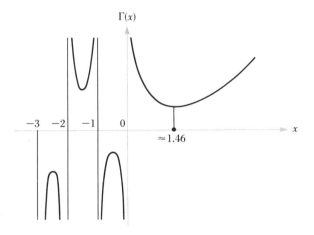

FIGURE 7.1.1   Graph of $\Gamma(x)$ for x real.

From the equation $zG(z)G(-z) = (\sin \pi z)/\pi$ (see Eq. (4)), we get the relation

$$\Gamma(z)\Gamma(1-z) = \frac{\pi}{\sin \pi z} \tag{14}$$

From this it follows that $\Gamma(z) \neq 0$ for all z. (Of course, $z \neq 0, -1, -2, \ldots$.) We know this to be true because if $\Gamma(z) = 0$, we would have $\pi = \Gamma(z)\Gamma(1 - z) \sin \pi z = 0$ as long as $z \neq 0, \pm 1, \pm 2, \ldots$. (These are the points at which $\sin \pi z$ vanishes, so that cross multiplication is invalid at those points.) But we also know that $\Gamma(z) \neq 0$ if $z = 1, 2, 3, \ldots$, since $\Gamma(n + 1) = n!$, $n = 0, 1, 2, \ldots$. Thus we have proved that $\Gamma(z) \neq 0$ if $z \neq 0, -1, -2, \ldots$.

If we let $z = \frac{1}{2}$ in Eq. (14), we get $[\Gamma(\frac{1}{2})]^2 = \pi$. But $\Gamma(\frac{1}{2}) > 0$. To see this, note that Γ is real for real positive z; we have shown that Γ has no zeros and that $\Gamma(n + 1) = n! > 0$. Therefore, since $\Gamma(x)$ is continuous for $x \in {]}0, \infty{[}$ (because Γ is analytic), it follows from the intermediate value theorem that $\Gamma(x) > 0$ for all $x \in {]}0, \infty{[}$ (as in Fig. 7.1.1). Thus $\Gamma(\frac{1}{2}) = \sqrt{\pi}$ (rather than the other possibility, $-\sqrt{\pi}$).

Euler's formula for the gamma function is

$$\Gamma(z) = \frac{1}{z} \prod_{n=1}^{\infty} \left[\left(1 + \frac{1}{n}\right)^{z} \left(1 + \frac{z}{n}\right)^{-1} \right] = \lim_{n \to \infty} \frac{n! n^z}{z(z + 1) \cdots (z + n)} \qquad (15)$$

This formula is proven as follows. By definition,

$$\frac{1}{\Gamma(z)} = z \left(\lim_{n \to \infty} e^{(1 + 1/2 + \cdots + 1/n - \log n)z} \right) \left[\lim_{n \to \infty} \prod_{k=1}^{n} \left(1 + \frac{z}{k}\right) e^{-z/k} \right]$$

$$= z \left[\lim_{n \to \infty} e^{(1 + 1/2 + \cdots + 1/n - \log n)z} \prod_{k=1}^{n} \left(1 + \frac{z}{k}\right) e^{-z/k} \right]$$

$$= z \lim_{n \to \infty} \left[n^{-z} \prod_{k=1}^{n} \left(1 + \frac{z}{k}\right) \right]$$

since $n^{-z} = e^{-\log n \cdot z}$. Thus we get

$$\frac{1}{\Gamma(z)} = z \lim_{n \to \infty} \left[\prod_{k=1}^{n-1} \left(1 + \frac{1}{k}\right)^{-z} \prod_{k=1}^{n} \left(1 + \frac{z}{k}\right) \right]$$

$$= z \lim_{n \to \infty} \left\{ \left(1 + \frac{1}{n}\right)^{z} \left[\prod_{k=1}^{n} \left(1 + \frac{z}{k}\right)\left(1 + \frac{1}{k}\right)^{-z} \right] \right\}$$

The first equality in Eq. (15) now follows. The student is asked to prove the second in Exercise 11.

Another important property of the gamma function is given in the *Gauss formula:* For any fixed positive integer $n \geq 2$,

$$\Gamma(z)\Gamma\left(z + \frac{1}{n}\right) \cdots \Gamma\left(z + \frac{n-1}{n}\right) = (2\pi)^{(n-1)/2} n^{(1/2) - nz} \Gamma(nz) \qquad (16)$$

To prove this formula we first note that we can write Euler's formula (Eq. (15)) as

$$\Gamma(z) = \lim_{m \to \infty} \frac{m!m^z}{z(z+1) \cdots (z+m)} = \lim_{m \to \infty} \frac{(m-1)!m^z}{z(z+1) \cdots (z+m-1)}$$

$$= \lim_{m \to \infty} \frac{(mn-1)!(mn)^z}{z(z+1) \cdots (z+mn-1)}$$

We define $f(z)$ as follows:

$$f(z) = \frac{n^{nz}\Gamma(z)\Gamma\left(z+\frac{1}{n}\right) \cdots \Gamma\left(z+\frac{n-1}{n}\right)}{n\Gamma(nz)}$$

$$= \frac{n^{nz-1} \displaystyle\prod_{k=0}^{n-1} \lim_{m \to \infty} \frac{(m-1)!m^{z+k/n}}{\left(z+\dfrac{k}{n}\right)\left(z+\dfrac{k}{n}+1\right) \cdots \left(z+\dfrac{k}{n}+m-1\right)}}{\displaystyle\lim_{m \to \infty} \frac{(mn-1)!(mn)^{nz}}{nz(nz+1) \cdots (nz+nm-1)}}$$

$$= \lim_{m \to \infty} \frac{[(m-1)!]^n m^{(n-1)/2} n^{mn-1}(nz)(nz+1) \cdots (nz+mn-1)}{(mn-1)! \displaystyle\prod_{k=0}^{n-1} [(nz+k)(nz+k+n) \cdots (nz+k+mn-n)]}$$

$$= \lim_{m \to \infty} \frac{[(m-1)!]^n m^{(n-1)/2} n^{nm-1}}{(nm-1)!}$$

Thus f is constant. Setting $z = 1/n$ gives

$$f(z) = \Gamma\left(\frac{1}{n}\right)\Gamma\left(\frac{2}{n}\right) \cdots \Gamma\left(\frac{n-1}{n}\right) > 0$$

so that

$$[f(z)]^2 = \frac{\pi^{n-1}}{\sin \dfrac{\pi}{n} \sin \dfrac{2\pi}{n} \cdots \sin \dfrac{(n-1)\pi}{n}}$$

using Eq. (14). From the fact that

$$\sin \frac{\pi}{n} \sin \frac{2\pi}{n} \cdots \sin \frac{(n-1)\pi}{n} = \frac{n}{2^{n-1}} \qquad \text{for} \qquad n = 2, 3, \ldots$$

(see Exercise 28, Sec. 1.2), we get

$$[f(z)]^2 = \frac{(2\pi)^{n-1}}{n}$$

Since $f(z) > 0$,

$$f(z) = \frac{(2\pi)^{(n-1)/2}}{\sqrt{n}}$$

The Gauss formula therefore follows.

If we take the special case of Eq. (16), in which $n = 2$, we obtain the *Legendre duplication formula*:

$$2^{2z-1}\Gamma(z)\Gamma\left(z + \tfrac{1}{2}\right) = \sqrt{\pi}\,\Gamma(2z) \tag{17}$$

Let us next show that the residue of $\Gamma(z)$ at $z = -m$, $m = 0, 1, 2, \ldots$ is $(-1)^m/m!$. Indeed,

$$(z + m)\Gamma(z) = (z + m)\frac{\Gamma(z + 1)}{z} = (z + m)\frac{\Gamma(z + 2)}{z(z + 1)}$$

More generally, we find that

$$(z + m)\Gamma(z) = \frac{\Gamma(z + m + 1)}{z(z + 1) \cdots (z + m - 1)}$$

Letting $z \to -m$, we get

$$\frac{\Gamma(1)}{-m(-m + 1) \cdots (-1)} = \frac{(-1)^m}{m!}$$

as required.

There is an important expression for $\Gamma(z)$ as an integral. For Re $z > 0$, we shall establish the following formula known as *Euler's integral for $\Gamma(z)$*:

$$\Gamma(z) = \int_0^\infty t^{z-1}e^{-t}\,dt \tag{18}$$

The student might suspect that this expression can be evaluated by the methods of Sec. 4.3. Unfortunately, the hypotheses of Theorem 4.3.8 do not hold in this case, and so another method is needed to prove Eq. (18). Let us start by defining

$$F_n(z) = \int_0^n \left(1 - \frac{t}{n}\right)^n t^{z-1}\,dt$$

and showing that

$$F_n(z) = \frac{n!n^z}{z(z + 1) \cdots (z + n)} \tag{19}$$

By Euler's formula (Eq. (15)), we will then have proved that $F_n(z) \to \Gamma(z)$ as $n \to \infty$. To prove Eq. (19), we note that by changing variables and letting $t = ns$,

$$F_n(z) = n^z \int_0^1 (1 - s)^n s^{z-1} \, ds$$

Now we integrate this expression successively by parts, the first step being

$$F_n(z) = n^z \left[\frac{1}{z} s^z (1 - s)^n \Big|_0^1 + \frac{n}{z} \int_0^1 (1 - s)^{n-1} s^z \, ds \right] = n^z \frac{n}{z} \int_0^1 (1 - s)^{n-1} s^z \, ds$$

Repeating this procedure, we integrate by parts n times and get

$$F_n(z) = n^z \frac{n \cdot (n - 1) \cdots 1}{z(z + 1) \cdots (z + n - 1)} \int_0^1 s^{z+n-1} \, ds = \frac{n! n^z}{z(z + 1) \cdots (z + n)}$$

which establishes Eq. (19).

From Exercise 15 we obtain a formula that should be well known from calculus:

$$\left(1 - \frac{t}{n} \right)^n \to e^{-t} \qquad \text{as} \qquad n \to \infty \tag{20}$$

If we let $n \to \infty$ in Eq. (19), the validity of Eq. (18) seems assured. However, such a conclusion is not so easily justified.* To do this, we proceed as follows. From Eqs. (15) and (19) we know that

$$\Gamma(z) = \lim_{n \to \infty} \int_0^n \left(1 - \frac{t}{n} \right)^n t^{z-1} \, dt \tag{21}$$

Let $f(z) = \int_0^\infty e^{-t} t^{z-1} \, dt$. This integral converges, since $|e^{-t} t^{z-1}| \le e^{-t} t^{\mathrm{Re}\, z - 1}$ and $\mathrm{Re}\, z > 0$ (compare with $\int_1^\infty e^{-t} t^p \, dt$ and $\int_0^1 t^p \, dt, \ p > -1$). We shall need to know "how fast" $[1 - (t/n)]^n \to e^{-t}$. The following inequalities hold:

$$0 \le e^{-t} - \left(1 - \frac{t}{n} \right)^n \le \frac{t^2 e^{-t}}{n} \qquad \text{for} \qquad 0 \le t \le n \tag{22}$$

(This follows from a calculus lemma whose proof is asked for in Exercise 15.)

---

* The reader who has studied convergence theorems in Lebesque integration theory is urged to apply them to this integral.

From Eq. (21) and the definition of f we have

$$f(z) - \Gamma(z) = \lim_{n \to \infty} \left\{ \int_0^n \left[e^{-t} - \left(1 - \frac{t}{n} \right)^n \right] t^{z-1} \, dt + \int_n^\infty e^{-t} t^{z-1} \, dt \right\} \tag{23}$$

To show that the limit (Eq. (23)) is zero, note that $\int_n^\infty e^{-t} t^{z-1} \, dt \to 0$ as $n \to \infty$. Indeed if $t > 1$, then $|e^{-t} t^{z-1}| \le e^{-t} t^m$ where m is an integer $m \ge \mathrm{Re}\, z > 0$. But from calculus (or directly using integration by parts), we know that $\int_0^\infty e^{-t} t^m \, dt < \infty$, so that $\int_n^\infty e^{-t} t^m \, dt \to 0$ as $n \to \infty$. It remains to be shown that

$$\int_0^n \left[e^{-t} - \left(1 - \frac{t}{n} \right)^n \right] t^{z-1} \, dt \to 0 \qquad \text{as} \qquad n \to \infty$$

By inequality (22),

$$\left| \int_0^n \left[e^{-t} - \left(1 - \frac{t}{n} \right)^n \right] t^{z-1} \, dt \right| \le \int_0^n \frac{e^{-t} t^{\mathrm{Re}\, z+1}}{n} \, dt \le \frac{1}{n} \int_0^\infty e^{-t} t^{\mathrm{Re}\, z+1} \, dt$$

which approaches zero as $n \to \infty$ because the integral converges. This completes the proof of Eq. (18); that is, for $\mathrm{Re}\, z > 0$,

$$\Gamma(z) = \int_0^\infty e^{-t} t^{z-1} \, dt$$

In fact, if we examine that proof, we see that provided $0 < \epsilon < R$, $\epsilon \le |z| \le R$, and $(-\pi/2) + \delta \le \arg z \le (\pi/2) - \delta$, $\delta > 0$, the convergence is uniform in z (see Exercise 18).

Technical Proofs of Theorems 7.1.2 and 7.1.6

7.1.2 CONVERGENCE THEOREM FOR PRODUCTS

(i) *If $\prod_{n=1}^\infty (1 + z_n)$ converges, then $z_n \to 0$.*

(ii) *Suppose that $|z_n| < 1$ for all $n = 1, 2, \ldots$, so that $z_n \ne -1$. Then $\prod_{n=1}^\infty (1 + z_n)$ converges if and only if $\sum_{n=1}^\infty \log (1 + z_n)$ converges. (log is the principal branch; $|z_n| < 1$ implies that $\log (1 + z_n)$ is defined.)*

(iii) *$\prod_{n=1}^\infty (1 + |z_n|)$ converges iff $\sum_{n=1}^\infty |z_n|$ converges. (We say that $\prod_{n=1}^\infty (1 + z_n)$ converges absolutely in this case.)*

(iv) *If $\prod_{n=1}^\infty (1 + |z_n|)$ converges, then $\prod_{n=1}^\infty (1 + z_n)$ converges.*

PROOF

(i) We can assume that $z_n \ne -1$ for all n. Let $P_n = \prod_{k=1}^n (1 + z_k)$; therefore, by assumption, $P_n \to P$ for some $P \ne 0$. Thus $P_n / P_{n-1} \to 1$ by the quotient theorem for limits. But $P_n / P_{n-1} = 1 + z_n$. Thus $z_n \to 0$.

(ii) Let $S_n = \sum_{k=1}^{n} \log(1 + z_k)$ and let $P_n = \prod_{k=1}^{n}(1 + z_k)$. Then $P_n = e^{S_n}$. It is clear that if S_n converges, then P_n also converges because e^z is continuous.

Conversely, suppose that $P_n \to P \neq 0$. To show that S_n converges, it suffices to show that for n sufficiently large, all S_n lie in a period strip (on which e^z has a continuous inverse).

We cannot write $\log \sum_{k=1}^{n}(1 + z_k) = \log P_n$, because P_n could be on the negative real axis. Instead, for purposes of this proof, let us choose the branch of log such that P lies in its domain A. Now $P_n \to P$, so $P_n \in A$ if n is large and therefore we can write $S_n = \log P_n + k_n \cdot 2\pi i$ for an integer k_n. Thus

$$(k_{n+1} - k_n) \cdot 2\pi i = \log(1 + z_{n+1}) - (\log P_{n+1} - \log P_n)$$

Since the left side of the equation is purely imaginary,

$$(k_{n+1} - k_n) \cdot 2\pi i = i[\arg(1 + z_{n+1}) - \arg P_{n+1} + \arg P_n]$$

By (i), $z_{n+1} \to 0$, and so $\arg(1 + z_{n+1}) \to 0$. Also, $\arg P_n \to \arg P$, and therefore $k_{n+1} - k_n \to 0$ as $n \to \infty$. Since the k_n's are integers, they must equal a fixed integer k for n large. Thus $S_n = \log P_n + k \cdot 2\pi i$, so, as $n \to \infty$, $S_n \to S = \log P + k \cdot 2\pi i$.

(iii) By (ii), it suffices to show that for $x_n \geq 0$, $\sum x_n$ converges iff $\sum \log(1 + x_n)$ converges. Indeed, since

$$\log(1 + z) = z - \frac{z^2}{2} + \frac{z^3}{3} - \cdots \qquad \text{for} \qquad |z| < 1$$

we see that

$$\frac{\log(1 + z)}{z} = 1 - \frac{z}{2} + \frac{z^2}{3} - \cdots$$

has a removable singularity at $z = 0$ and that

$$\lim_{z \to 0} \frac{\log(1 + z)}{z} = 1$$

Suppose that $\sum x_n$ converges. Then $x_n \to 0$. Thus, given $\epsilon > 0$, $0 \leq \log(1 + x_n) \leq (1 + \epsilon)x_n$ for sufficiently large n. By the comparison test, $\sum \log(1 + x_n)$ converges. If we use $(1 - \epsilon)x_n \leq \log(1 + x_n)$, we obtain the converse.

(iv) Suppose that $\prod(1 + |z_n|)$ converges. Then by (ii), $\sum \log(1 + |z_n|)$ converges. (We must begin with terms such that the conditions in (ii) hold.) In fact, the argument in (iii) shows that $\sum \log(1 + z_n)$ converges absolutely and hence converges. Thus by (ii), $\prod(1 + z_n)$ converges. ∎

7.1.6 THEOREM ON CANONICAL PRODUCTS *Let $a_1, a_2, \ldots$ be a given sequence (possibly finite) of nonzero complex numbers such that*

$$\sum_{n=1}^{\infty} \frac{1}{|a_n|^2} < \infty$$

Then if $g(z)$ is any entire function, the function

$$f(z) = e^{g(z)} z^k \left[\prod_{n=1}^{\infty} \left(1 - \frac{z}{a_n} \right) e^{z/a_n} \right] \tag{1}$$

is entire. The product converges uniformly on closed disks, has zeros at $a_1, a_2, \ldots$, and has a zero of order k at $z = 0$, but has no other zeros. Furthermore, if f is any entire function having these properties, it can be written in the same form (Eq. (1)). In particular, f is entire with no zeros if and only if f has the form $f(z) = e^{g(z)}$ for some entire function g.

PROOF First we show that $\prod (1 - z/a_n) e^{z/a_n}$ is entire. For each $R > 0$, let $A_R = \{z \text{ such that } |z| < R\}$. Since $a_n \to \infty$, only a finite number of a_n's lie in A_R, say, $a_1, \ldots, a_{N-1}$. Therefore, for $z \in A_R$, only a finite number of terms $(1 - z/a_n)$ vanish.

To effect the proof, we prove the following lemma:

7.1.7 LEMMA *If $1 + w = (1 - a)e^a$ and $|a| < 1$, then*

$$|w| \le \frac{|a|^2}{1 - |a|}$$

PROOF We have

$$(1 - a)e^a = 1 - \frac{a^2}{2} - \cdots - \left(1 - \frac{1}{n} \right) \frac{a^n}{(n-1)!} - \cdots$$

Thus

$$|(1 - a)e^a - 1| = |w| \le \frac{|a|^2}{2} + \cdots + \frac{(n-1)}{n!} |a|^n + \cdots$$

$$\le |a|^2 + |a|^3 + \cdots$$

$$= \frac{|a|^2}{1 - |a|} \qquad \text{since } |a| < 1 \qquad \blacktriangledown$$

The next step in the proof is to show that the series

$$\sum_{n=1}^{\infty} w_n(z) = \sum_{n=1}^{\infty} \left[\left(1 - \frac{z}{a_n} \right) e^{z/a_n} - 1 \right]$$

converges uniformly and absolutely on $A_{R/2}$. This will show that

$$\prod_{n=1}^{\infty} \left(1 - \frac{z}{a_n} \right) e^{z/a_n}$$

is entire (by Theorem 7.1.5).

Indeed, for $n \geq N$, $|z/a_n| < 1$ if $|z| \leq R/2$, and so from the preceding lemma,

$$|w_n(z)| \leq \frac{|z/a_n|^2}{1 - |z/a_n|} \leq \frac{(R/2)^2}{1 - \frac{1}{2}} \cdot \frac{1}{|a_n|^2}$$

since $|z| \leq R/2$ and $|a_n| \geq R$ for $n \geq N$. Thus

$$|w_n(z)| \leq \frac{R^2}{2} \cdot \frac{1}{|a_n|^2} = M_n$$

By assumption, ΣM_n converges, and so by the Weierstrass M test, $\Sigma w_n(z)$ converges uniformly and absolutely.

Thus $f_1(z) = z^k \prod_{n=1}^{\infty} [1 - (z/a_n)] e^{z/a_n}$ is entire. From the definition of the product it is

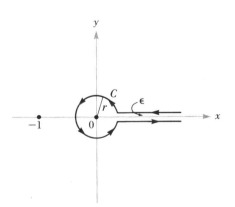

FIGURE 7.1.2   Contour for Hankel's formula.

clear that f_1 has exactly the required number of zeros. Thus so does $e^g f_1$. If f has the given number of zeros, then f / f_1 will be entire and have no zeros (by Proposition 4.1.1). Therefore, we need only prove the following lemma.

Table 7.1.1 Properties of the Gamma Function

Definition:

$$\Gamma(z) = \frac{1}{z e^{\gamma z} \left[\prod_{n=1}^{\infty} \left(1 + \frac{z}{n}\right) e^{-z/n} \right]}, \text{ where } \gamma = \lim_{n \to \infty} \left(1 + \frac{1}{2} + \cdots + \frac{1}{n} - \log n\right) \approx 0.577.$$

1. Γ is meromorphic with simple poles at $0, -1, -2, \ldots$
2. $\Gamma(z + 1) = z\Gamma(z), z \neq 0, -1, -2, \ldots$
3. $\Gamma(n + 1) = n!, n = 0, 1, 2, \ldots$
4. $\Gamma(z)\Gamma(1 - z) = \pi/\sin \pi z.$
5. $\Gamma(z) \neq 0$ for all z.
6. $\Gamma\left(\dfrac{1}{2}\right) = \sqrt{\pi}; \Gamma\left(n + \dfrac{1}{2}\right) = \dfrac{1 \cdot 3 \cdot 5 \cdots (2n - 1)}{2^n} \sqrt{\pi}.$
7. $\Gamma(z) = \dfrac{1}{z} \prod_{n=1}^{\infty} \left[\left(1 + \dfrac{1}{n}\right)^z \left(1 + \dfrac{z}{n}\right)^{-1} \right].$
8. $\Gamma(z) = \lim_{n \to \infty} \dfrac{n! n^z}{z(z + 1) \cdots (z + n)}.$
9. $\Gamma(z)\Gamma\left(z + \dfrac{1}{n}\right) \cdots \Gamma\left(z + \dfrac{n-1}{n}\right) = (2\pi)^{(n-1)/2} n^{(1/2) - nz} \Gamma(nz).$
10. $2^{2z-1}\Gamma(z)\Gamma\left(z + \frac{1}{2}\right) = \sqrt{\pi}\Gamma(2z).$
11. The residue of Γ at $-m$ equals $(-1)^m/m!$.
12. (Euler's integral) $\Gamma(z) = \displaystyle\int_0^{\infty} t^{z-1} e^{-t} \, dt$ for Re $z > 0$.

The convergence is uniform and absolute for $-\pi/2 + \delta \leq \arg z \leq \pi/2 - \delta$, $\delta > 0$, and for $\epsilon \leq |z| \leq R$, where $0 < \epsilon < R$.

13. $\dfrac{\Gamma'(z)}{\Gamma(z)} = -\gamma - \dfrac{1}{z} + \sum_{1}^{\infty} \left(\dfrac{1}{n} - \dfrac{1}{z + n}\right) = \displaystyle\int_0^{\infty} \left(\dfrac{e^{-t}}{t} - \dfrac{e^{-zt}}{1 - e^{-t}}\right) dt.$ (See Worked Example 7.1.11.)

14. (Hankel's formula) $\Gamma(z) = -\dfrac{1}{2i \sin \pi z} \displaystyle\int_C (-t)^{z-1} e^{-t} \, dt.$ (C as in Fig. 7.1.2)

15. $\dfrac{1}{\Gamma(z)} = \dfrac{i}{2\pi} \displaystyle\int_C (-t)^{-z} e^{-t} \, dt.$ (C as in Fig. 7.1.2)

16. $\Gamma(z + 1) \approx \sqrt{2\pi} \, z^{z+1/2} e^{-z}$ for $|z|$ large, Re $z > 0$. (This is Stirling's formula, which will be proved in Sec. 7.3.)

LEMMA *Let $h(z)$ be entire with no zeros. Then there is an entire function $g(z)$ such that* $h = e^g$.

PROOF It is not quite correct to set $g(z) = \log h(z)$, because it is not obvious that $h(\mathbf{C})$ is simply connected, and so we do not know whether log has an analytic branch on $h(\mathbf{C})$. However, we can circumvent this problem taking a clue from logarithmic differentiation. Since h'/h is entire, we can write

$$\frac{h'(z)}{h(z)} = a_0 + a_1 z + \cdots$$

and let $g_1(z) = a_0 z + a_1 z^2/2 + a_2 z^3/3 + \cdots$; that is, $g_1' = h'/h$. (The series for g_1 has an infinite radius of convergence since $\Sigma\, a_n z^n$ does.) Let $f(z) = e^{g_1(z)}$. Then

$$\frac{f'(z)}{f(z)} = g_1'(z) = \frac{h'(z)}{h(z)}$$

and therefore $(d/dz)\,(f/h) = 0$, so $h = K \cdot f$ for a constant K. If we let $z = 0$, we see that $K = h(0) \neq 0$. Since $K \neq 0$, we can let $K = e^c$ for some c, so that $h = e^c \cdot e^{g_1} = e^g$ where $g = g_1 + c$. ∎

Worked Examples

7.1.9 *For what z does $(1 + z) \prod_{n=1}^{\infty} (1 + z^{2^n})$ converge absolutely? Show that the product is* $1/(1 - z)$.

Solution. By Theorem 7.1.2(iii), we have absolute convergence iff $\Sigma_{n=1}^{\infty} z^{2^n}$ converges absolutely. This occurs for $|z| < 1$, since the radius of convergence of the series is 1. Thus the product converges absolutely for $|z| < 1$.

Evaluation of the product requires a trick, which, in this case, is quite simple. Our product is $(1 + z)(1 + z^2)(1 + z^4)(1 + z^8) \cdots$. Notice that $(1 + z)(1 + z^2) = 1 + z + z^2 + z^3$, and so $(1 + z)(1 + z^2)(1 + z^4) = 1 + z + z^2 + z^3 + \cdots + z^7$. Generally, $(1 + z) \prod_{k=1}^{n} (1 + z^{2^k}) = 1 + z + z^2 + \cdots + z^{2^{n+1}-1}$. We know that this series converges to $1/(1 - z)$ as $n \to \infty$ since it is the power series around $z = 0$ for $1/(1 - z)$.

7.1.10 *Prove that*

$$\sin z = z \prod_{n=1}^{\infty} \left(1 - \frac{z^2}{n^2\pi^2}\right)$$

Solution. The zeros of $\sin z$ occur at 0 and $\pm n\pi$; let us define $a_1 = \pi$, $a_2 = -\pi$, $a_3 = 2\pi$, $a_4 = -2\pi$, $\ldots$. All the zeros are simple, and $\Sigma 1/|a_n|^2$ converges. Therefore, by the theorem on canonical products, we can write

$$\sin z = e^{g(z)}z \prod_{n=1}^{\infty} \left(1 - \frac{z}{a_n}\right) e^{z/a_n}$$

$$= e^{g(z)}z \cdot \left[\left(1 - \frac{z}{\pi}\right)e^{z/\pi}\right]\left[\left(1 + \frac{z}{\pi}\right)e^{-z/\pi}\right]\left[\left(1 - \frac{z}{2\pi}\right)e^{z/2\pi}\right]\left[\left(1 + \frac{z}{2\pi}\right)e^{-z/2\pi}\right] \cdots$$

$$= e^{g(z)}z \prod_{n=1}^{\infty} \left(1 - \frac{z^2}{n^2\pi^2}\right)$$

(gathering the terms in pairs). Thus it remains to be shown that $e^{g(z)} = 1$.

This is not so simple, and the student would not be expected routinely to carry out the following argument; it is, admittedly, a trick. Let

$$P_n(z) = e^{g(z)}z \prod_{k=1}^{n} \left(1 - \frac{z^2}{k^2\pi^2}\right)$$

We know that $P_n(z) \to \sin z$ (uniformly on disks), so that $P_n'(z) \to \cos z$. Thus

$$\frac{P_n'(z)}{P_n(z)} \to \cot z \qquad \text{for} \qquad z \neq 0, \pm\pi, \pm 2\pi, \ldots$$

But

$$\frac{P_n'(z)}{P_n(z)} = \frac{d}{dz}\log P_n(z) = \frac{d}{dz}\left[g(z) + \log z + \sum_{k=1}^{n} \log\left(1 - \frac{z^2}{k^2\pi^2}\right)\right]$$

$$= g'(z) + \frac{1}{z} + \sum_{k=1}^{n} \left(\frac{2z}{z^2 - k^2\pi^2}\right)$$

However, we know from Sec. 4.3 that

$$\cot z = \frac{1}{z} + \sum_{n=1}^{\infty} \frac{2z}{z^2 - n^2\pi^2} \qquad \text{for} \qquad z \neq n\pi$$

Thus $g'(z) = 0$, and so $g(z)$ is a constant, say, c. Therefore

$$\frac{\sin z}{z} = e^c \prod_{n=1}^{\infty} \left(1 - \frac{z^2}{n^2\pi^2}\right)$$

Let $z \to 0$. The left side approaches 1 while the right side approaches e^c (Why?). Thus $e^c = 1$ and we have our formula.

7.1.11 *Prove that*

$$\frac{\Gamma'(z)}{\Gamma(z)} = -\gamma - \frac{1}{z} + \lim_{n \to \infty} \sum_{k=1}^{n} \left(\frac{1}{k} - \frac{1}{z+k}\right)$$

Solution. We have

$$\frac{1}{\Gamma(z)} = ze^{\gamma z} \prod_{n=1}^{\infty} \left(1 + \frac{z}{n}\right) e^{-z/n}$$

Taking logs (which we know we can do near each z for which we have convergence), we obtain

$$-\log \Gamma(z) = \gamma z + \log z + \sum_{1}^{\infty} \left[\log\left(1 + \frac{z}{n}\right) - \frac{z}{n}\right]$$

Differentiating, we get

$$-\frac{\Gamma'(z)}{\Gamma(z)} = \gamma + \frac{1}{z} + \sum_{1}^{\infty} \left(\frac{1/n}{1 + z/n} - 1/n\right) = \gamma + \frac{1}{z} + \sum_{1}^{\infty} \left(\frac{1}{n + z} - \frac{1}{n}\right)$$

This is the result claimed.

Exercises

1. Show that $\prod_{n=2}^{\infty} \left(1 - \frac{1}{n^2}\right) = \frac{1}{2}$.

2. Show that $\prod_{n=2}^{\infty} \left(1 - \frac{2}{n(n+1)}\right) = \frac{1}{3}$.

3. Show that $\prod_{n=1}^{\infty} (1 + z_n)$ converges absolutely iff $\sum_{n=1}^{\infty} \log(1 + z_n)$ converges absolutely.

4. Complete the proof of the analyticity of infinite products (7.1.5).

5. Use Worked Example 7.1.10 to establish *Wallis' formula,*

$$\frac{\pi}{2} = \frac{2}{1} \cdot \frac{2}{3} \cdot \frac{4}{3} \cdot \frac{4}{5} \cdot \frac{6}{5} \cdot \frac{6}{7} \cdot \cdots$$

6. Show that $\prod_{n=2}^{\infty} \left(1 - \frac{2}{n^3 + 1}\right) = \frac{2}{3}$.

7. Prove formula 2 of Table 7.1.1.

8. Show that $\prod_{n=1}^{\infty} (1 + z_n)$ converges (assuming that $z_n \neq -1$) iff, for any $\epsilon > 0$, there is an N such that $n \geq N$ implies that

$$|(1 + z_n) \cdots (1 + z_{n+p}) - 1| < \epsilon \qquad \text{for all } p = 0, 1, 2, \ldots$$

(*Hint.* Use the Cauchy criterion for sequences.)

9. Prove formula 4 of Table 7.1.1.

10. Let $a_1, a_2, \ldots$ be points in $\mathbb{C}$, let $a_i \neq 0$, and let

$$\sum_1^\infty \frac{1}{|a_n|^{1+h}} < \infty$$

for a fixed integer $h \geq 0$. Show that the most general entire function having zeros at $a_1, a_2, \ldots$ and a zero of order k at 0 is

$$f(z) = e^{g(z)} z^k \prod_1^m \left[\left(1 - \frac{z}{a_n} \right) e^{[z/a_n + (z/a_n)^2/2 + \cdots + (z/a_n)^h/h]} \right]$$

(Each of the points a_i may be repeated finitely often.) (*Hint.* Prove the following lemma: If $1 + w = (1 - a)e^{a + a^2/2 + \cdots + a^h/h}$ for $|a| < 1$, then $|w| \leq |a|^{h+1}/(1 - |a|)$.)

11. Prove formula 8 of Table 7.1.1

12. Using Euler's formula (formula 7 of Table 7.1.1), prove that $\Gamma(z + 1) = z\Gamma(z)$.

13. Show that in the neighborhood $|z + m| < 1$ for m a fixed positive integer,

$$\Gamma(z) - \frac{(-1)^m}{m!(z + m)}$$

is analytic (that is, has a removable singularity at $z = -m$).

14. (a) Let $E(z, h) = (1 - z) e^{z + z^2/2 + \cdots + z^h/h}$. Show that the most general entire function having zeros at $a_1, a_2, \ldots$, each repeated according to its multiplicity, where $a_n \to \infty$, and having a zero of order k at 0 is

$$f(z) = e^{g(z)} z^k \prod_{n=1}^\infty E\left(\frac{z}{a_n}, n \right)$$

(This is the *Weierstrass factorization theorem.*)
(b) Conclude that every meromorphic function is the quotient of two entire functions.

15. Prove that, for $0 \leq t \leq n$,

$$0 \leq e^{-t} - \left(1 - \frac{t}{n} \right)^n \leq \frac{t^2 e^{-t}}{n}$$

16. Prove that, for Re $z \geq 0$,

$$\frac{\Gamma'(z)}{\Gamma(z)} = \int_0^\infty \left(\frac{e^{-t}}{t} - \frac{e^{-zt}}{1 - e^{-t}} \right) dt$$

(*Hint.* If Re $z \geq 0$, then $1/(z + n) = \int_0^\infty e^{-t(z+n)} dt$. Use $\gamma = \lim_{n \to \infty} (1 + \frac{1}{2} + \cdots + 1/n - \log n)$ and Worked Example 7.1.11.)

17. Let γ be a circle of radius $\frac{1}{2}$ around $z_0 = 0$. Show that $\int_\gamma \Gamma(z) \, dz = 2\pi i$.

18. Establish the uniform convergence in formula 12 of Table 7.1.1.

19. Prove *Hankel's formula* (formula 14 of Table 7.1.1):

$$\Gamma(z) = \frac{-1}{2i \sin \pi z} \int_C (-t)^{z-1} e^{-t} \, dt$$

where C is the contour illustrated in Fig. 7.1.2. For what z is this formula valid? Using $\Gamma(z)\Gamma(1 - z) = \pi/\sin \pi z$, conclude that

$$\frac{1}{\Gamma(z)} = \frac{i}{2\pi} \int_C (-t)^{-z} e^{-t} \, dt$$

20. In answering this question, refer back to Sec. 6.1. Define $\Gamma(z) = \int_0^\infty t^{z-1} e^{-t} \, dt$ for Re $z > 0$.
(a) Show directly that $\Gamma(z)$ is analytic on Re $z > 0$ by showing that $\int_0^n t^{z-1} e^{-t} \, dt$ converges uniformly on closed disks as $n \to \infty$.
(b) Show that $\Gamma(z + 1) = z\Gamma(z)$, Re $z > 0$.
(c) Use (b) and analytic continuation to prove that $\Gamma(z)$ can be extended to a meromorphic function having simple poles at $0, -1, -2, \ldots$. (*Hint.* The procedure used is analogous to that using in proving the Schwarz reflection principle; see Sec. 6.1.)

21. Show that $\int_{-\infty}^\infty e^{-y^2} \, dy = \sqrt{\pi}$ and that $\int_{-\infty}^\infty y^2 e^{-y^2} \, dy = \sqrt{\pi}/2$ by using the gamma function. (*Hint.* Relate these equations by integrating by parts and use $\Gamma(\frac{1}{2}) = \sqrt{\pi}$.)

7.2 ASYMPTOTIC EXPANSIONS AND THE METHOD OF STEEPEST DESCENT

Asymptotic expansions provide a method of using the partial sums of a series to approximate values of a function $f(z)$ for large z. A striking aspect is that the series itself might not converge to the function and might actually diverge. If we use only one term we say that we have an asymptotic approximation or asymptotic formula for f. Stirling's formula for the gamma function is such a formula. This result, proved in Sec. 7.3, states that

$$\Gamma(x) \sim e^{-x} x^{x-1/2} \sqrt{2\pi} \qquad \text{for large } x$$

The expression on the right-hand side may be easier to handle than the Γ function itself and has important applications in fields such as probability and statistical mechanics.

Another famous example is the prime number theorem, which asserts that if $\pi(x)$ is the number of primes less than or equal to the real number x, then

$$\pi(x) \sim \frac{x}{\log x}$$

Exactly what such a formula means and in what sense it is an approximation will be developed in this section. The theory of asymptotic expansions considered in this section will be applied in the next, where Stirling's formula is proved and Bessel functions are studied.

There are methods for studying the asymptotic behavior of functions $f(z)$ other than those we shall develop. For example, if f satisfies a differential equation, then this equation frequently can be used to obtain an asymptotic formula. The reader who wishes to delve more deeply into these topics should consult the references listed in the Preface.

"Big Oh" and "Little oh" Notation

Some notation is useful for keeping track of relationships in behavior between two functions. Suppose $f(z)$ and $g(z)$ are defined for z in some set A. We say $f(z)$ is $O(g(z))$ (pronounced "$f(z)$ is big 'oh' of $g(z)$") for z in A if there is a constant C such that $|f(z)| \leq C|g(z)|$ for all $z \in A$. We will usually write $f(z) = O(g(z))$ although this is somewhat an abuse of notation since the object on the right is a statement of relationship and not a specific quantity to which $f(z)$ is equal. For example, $\sin x = O(x)$ for x in $\mathbf{R}$, since elementary calculus shows that $|\sin x| \leq |x|$ for all x. An easy but useful observation is that $f(z) = O(1)$ just means that $f(z)$ is bounded.

A more useful notation for us will be "little oh." This requires some sort of limiting behavior for its definition. Roughly speaking, the notation $f(z) = o(g(z))$ means that $f(z)/g(z)$ tends to 0 as $z \to z_0$ or $z \to \infty$, etc. (We say "roughly" only because $g(z)$ could vanish.) For example,

$$
\begin{aligned}
1 - \cos x &= o(x) & \text{as } x \to \infty \\
\log x &= o(x) & \text{as } x \to \infty \\
e^{-x} &= o\left(\frac{1}{x^n}\right) & \text{as } x \to \infty \text{ for any } n
\end{aligned}
$$

We will be concerned primarily with $z \to \infty$ in a sector $\alpha \leq \arg(z) \leq \beta$. For the remainder of this section, unless specified otherwise, the symbols will thus be defined as follows:

$f(z) = O(g(z))$ *means:* There are constants R and M such that whenever $|z| \geq R$ and $\alpha \leq \arg(z) \leq \beta$, $|f(z)| \leq M|g(z)|$.

$f(z) = o(g(z))$ *means:* For each $\epsilon > 0$, there is an R such that whenever $|z| \geq R$ and $\alpha \leq \arg(z) \leq \beta$, $|f(z)| \leq \epsilon|g(z)|$.

In some cases we will be interested only in behavior along the positive real axis and will then take $\alpha = \beta = 0$. Notice that if $f(z)$ is $O(1/z^{n+1})$, then it is $o(1/z^n)$, but the converse is not generally true.

Asymptotic Expansions

Ever since Chap. 3, we have been concerned with representing a function by an infinite series which converges to the value of the function and carefully avoiding divergent series. Nonetheless, divergent series can sometimes be useful, though one must be very careful in their interpretation. We will now see that it is sometimes possible to associate with a function an infinite series which may or may not converge, but whose partial sums can be made to yield good approximations to the value of the function.

Consider a series of the form

$$S = a_0 + \frac{a_1}{z} + \frac{a_2}{z^2} + \cdots$$

and let

$$S_n = a_0 + \frac{a_1}{z} + \cdots + \frac{a_n}{z^n}$$

Thus S_n is well defined for $z \neq 0$ but we make no demand that S converge. The correct way to say that S is asymptotic to a given function f is given in the following definition.

7.2.1 DEFINITION *We say that $f \sim S$, or that f is **asymptotic to** S, or that S is an* **asymptotic expansion** *of f, if*

$$f - S_n = o\left(\frac{1}{z^n}\right)$$

for arg z lying in a specified range $[\alpha, \beta]$ (see Fig. 7.2.1).

Although S may be divergent, the partial sums usually result in accurate approximations of f, the error being approximately $1/z^n$. This will be illustrated with an example in the following paragraphs.

If we allowed the full range $[-\pi, \pi]$ for arg z, we might expect $a_0 + a_1/z + a_2/z^2 + \cdots$ to converge if $f(z)$ were analytic outside a large circle, because f has a convergent Laurent series of that form. However, f usually has poles $z_n \to \infty$ (such as $\Gamma(z)$, which has poles at $0, -1, -2, \ldots$), and therefore in many examples, we do not have a Laurent series that is valid on the exterior of any circle. If f has poles $z_n \to \infty$ in the sector

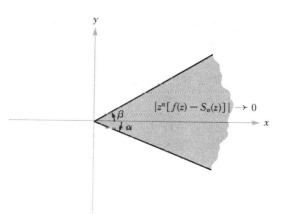

FIGURE 7.2.1   Asymptotic expansion.

arg $z \in [\alpha, \beta]$ and $f \sim S$, then S cannot converge at any z_0. If it did, then S would converge uniformly for all $|z| > |z_0| + 1$. (See Sec. 3.3.) Definition 7.2.1 and the uniform convergence of S_n to S would say that for large enough $|z|$ in that sector, we have $|f(z) - S(z)| < 1$. But this cannot hold near the poles of f.

The following example should help to clarify the concept of asymptotic expansion.

7.2.2 EXAMPLE   *Show that*

$$\int_x^\infty t^{-1} e^{x-t}\, dt \sim \frac{1}{x} - \frac{1}{x^2} + \frac{2!}{x^3} - \frac{3!}{x^4} + \cdots$$

SOLUTION   Let x be real, $x \geq 0$, and set

$$f(x) = \int_x^\infty t^{-1} e^{x-t}\, dt$$

(This is not the gamma function!) Integration by parts gives

$$f(x) = \frac{1}{x} - \frac{1}{x^2} + \frac{2!}{x^3} - \cdots + \frac{(-1)^{n-1}(n-1)!}{x^n} + (-1)^n n! \int_x^\infty \frac{e^{x-t}}{t^{n+1}}\, dt$$

We claim that

$$f(x) \sim S(x) = \frac{1}{x} - \frac{1}{x^2} + \frac{2!}{x^3} - \frac{3!}{x^4} + \cdots$$

Note that the series diverges. Here the sector is $\alpha = \beta = 0$; that is, we are restricting z to the positive real axis.

Indeed, if

$$S_n = \frac{1}{x} - \frac{1}{x^2} + \cdots + \frac{(-1)^{n-1}(n-1)!}{x^n}$$

we have

$$|x^n[f(x) - S_n(x)]| = x^n n! \int_x^\infty \frac{e^{x-t}}{t^{n+1}}\, dt = n! \int_x^\infty \left(\frac{x}{t}\right)^n \frac{e^{x-t}}{t}\, dt$$

$$\leq n! \int_x^\infty \frac{e^{x-t}}{t}\, dt \leq \frac{n!}{x} \int_x^\infty e^{x-t}\, dt = \frac{n!}{x}$$

which approaches zero as $x \to \infty$. Thus $f(x) - S_n(x)$ is $o(1/x^n)$ and so $f \sim S$ as required. Note that even though $n!$ grows quickly, we still have an accurate approximation because

$$|f(x) - S_n(x)| \leq \frac{n!}{x^{n+1}} = O\left(\frac{1}{x^{n+1}}\right)$$

and if x is, say, greater than n, then $n!/x^{n+1}$ is very small. ▲

Some basic properties of asymptotic expansions are summarized next.

7.2.3 PROPOSITION

(i) *If*

$$f(z) \sim S(z) = a_0 + \frac{a_1}{z} + \frac{a_2}{z^2} + \cdots$$

then

$$f(z) - S_n(z) = O\left(\frac{1}{z^{n+1}}\right)$$

and conversely.

(ii) *If*

$$f \sim a_0 + \frac{a_1}{z} + \frac{a_2}{z^2} + \cdots$$

and

$$f \sim \tilde{a}_0 + \frac{\tilde{a}_1}{z} + \frac{\tilde{a}_2}{z^2} + \cdots$$

then $a_i = \tilde{a}_i$. (Asymptotic expansions are unique.)

(iii) *If*

$$f \sim a_0 + \frac{a_1}{z} + \frac{a_2}{z^2} + \cdots$$

and
$$g \sim b_0 + \frac{b_1}{z} + \frac{b_2}{z^2} + \cdots$$

both being valid in the same range of arg z, then in that range

$$f + g \sim (a_0 + b_0) + \frac{(a_1 + b_1)}{z} + \frac{(a_2 + b_2)}{z^2} + \cdots$$

and

$$fg \sim c_0 + \frac{c_1}{z} + \frac{c_2}{z^2} + \cdots \qquad \text{where} \qquad c_n = \sum_{k=0}^{n} a_k b_{n-k}$$

(Asymptotic series may be added and multiplied.)
   (iv) Two different functions can have the same asymptotic expansion.
   (v) Let $\phi: [a, \infty[\to \mathbf{R}$ be continuous and suppose that $\phi(x) = o(1/x^n)$, $n \geq 2$. Then

$$\int_x^{\infty} \phi(t) \, dt = o\left(\frac{1}{x^{n-1}}\right)$$

PROOF

   (i) Since $f \sim S$, we have, by definition, $f - S_{n+1} = o(1/z^{n+1})$. Therefore, we get $f - S_n = f - S_{n+1} + S_{n+1} - S_n = o(1/z^{n+1}) + a_{n+1}/z^{n+1} = O(1/z^{n+1})$.
   (ii) We shall show that $a_n = \tilde{a}_n$ by induction on n. First, by the definition of $f \sim S$, $f(z) - a_0 \to 0$ as $z \to \infty$, and so $a_0 = \lim_{z \to \infty} f(z)$. Thus $a_0 = \tilde{a}_0$. Suppose we have proven that $a_0 = \tilde{a}_0, \ldots, a_n = \tilde{a}_n$. We shall show that $a_{n+1} = \tilde{a}_{n+1}$. Given $\epsilon > 0$, there is an R such that if $|z| \geq R$, we have

$$\left| z^{n+1} \left[f(z) - \left(a_0 + \frac{a_1}{z} + \cdots + \frac{a_{n+1}}{z^{n+1}} \right) \right] \right| < \epsilon$$

and

$$\left| z^{n+1} \left[f(z) - \left(\tilde{a}_0 + \frac{\tilde{a}_1}{z} + \cdots + \frac{\tilde{a}_{n+1}}{z^{n+1}} \right) \right] \right| < \epsilon$$

Therefore, by the triangle inequality,

$$|a_{n+1} - \tilde{a}_{n+1}| = |z^{n+1}| \frac{|a_{n+1} - \tilde{a}_{n+1}|}{|z^{n+1}|}$$

$$= |z^{n+1}| \left| \left[f(z) - \left(a_0 + \cdots + \frac{a_{n+1}}{z^{n+1}} \right) \right] - \left[f(z) - \left(\tilde{a}_0 + \cdots + \frac{\tilde{a}_{n+1}}{z^{n+1}} \right) \right] \right|$$

$$< \epsilon + \epsilon = 2\epsilon$$

Thus $|a_{n+1} - \tilde{a}_{n+1}| < 2\epsilon$ for any $\epsilon > 0$. Hence $a_{n+1} = \tilde{a}_{n+1}$.

(iii) Let $S_n(z) = a_0 + \cdots + a_n/z^n$ and $\tilde{S}_n(z) = b_0 + \cdots + b_n/z^n$. We must show that $f + g - (S_n + \tilde{S}_n) = o(1/z^n)$. To do this, write $f + g - (S_n + \tilde{S}_n) = (f - S_n) + (g - \tilde{S}_n) = o(1/z^n) + o(1/z^n) = o(1/z^n)$ (Why?).

To establish the formula for the product, note that $c_0 + c_1/z + \cdots + c_n/z^n = S_n \tilde{S} + o(1/z^n)$ since $S_n \tilde{S}_n = c_0 + c_1/z + \cdots + c_n/z^n$ plus higher-order terms. Thus $fg - (c_0 + c_1 + \cdots + c_n/z^n) = fg - S_n \tilde{S}_n + o(1/z^n)$. Now write $fg - S_n \tilde{S}_n = (f - S_n)g + S_n(g - \tilde{S}_n)$ and note that both terms are $o(1/z^n)$, since g and S_n are bounded as $z \to \infty$.

(iv) On $\mathbf{R}$, the function e^{-x} is $o(1/x^n)$ as $x \to \infty$ for any n. Thus if $f \sim a_0 + a_1/x + a_2/x^2 + \cdots$, then $f(x) + e^{-x} \sim a_0 + a_1/x + a_2/x^2 + \cdots$ as well.

(v) Since $\phi(t) = o(1/t^n)$, $\lim_{n \to \infty} t^n \phi(t) = 0$. Given $\epsilon > 0$, there is an $x_0 > 0$ such that $t > x_0$ implies $|t^n \phi(t)| < \epsilon$. Thus for $x > x_0$,

$$\left| \int_x^\infty \phi(t) \, dt \right| \le \int_x^\infty \frac{\epsilon}{t^n} \, dt = \frac{\epsilon}{n-1} \cdot \frac{1}{x^{n-1}}$$

and so, for $x > x_0$,

$$\left| x^{n-1} \int_x^\infty \phi(t) \, dt \right| \le \epsilon$$

Therefore $\lim_{x \to \infty} x^{n-1} \int_x^\infty \phi(t) \, dt = 0$, and so $\int_x^\infty \phi(t) \, dt = o(1/x^{n-1})$. ∎

Asymptotic Formulas and Asymptotic Equivalence

If a function has an asymptotic series as just described, then the partial sums of that series can be used to obtain approximations to the function for large z. However, the range of applicability of this method is a bit restricted. If f has the asymptotic series

$$f(z) \sim S(z) = a_0 + \frac{a_1}{z} + \frac{a_2}{z^2} + \cdots$$

then $f(z) - a_0 = o(1)$; that is, $\lim_{x \to \infty} f(x) = a_0$, and so f has a finite limit at infinity in the specified sector. This is too restrictive. We are commonly interested in functions which grow as x grows. The two examples mentioned at the beginning of this section, $\Gamma(z)$ and $\pi(z)$, certainly do this. Thus we shall also write

$$f(z) \sim g(z) \left(a_0 + \frac{a_1}{z} + \frac{a_2}{z^2} + \cdots \right)$$

to mean that

$$f(z) = g(z)\left[a_0 + \frac{a_1}{z} + \cdots + \frac{a_n}{z^n} + o\left(\frac{1}{z^n}\right)\right]$$

In other words, if $g(z) \neq 0$, then

$$\frac{f(z)}{g(z)} \sim a_0 + \frac{a_1}{z} + \frac{a_2}{z^2} + \cdots$$

the hope being that, at least in some sense, $g(z)$ is an easier function to handle for large z than is f. Incorporating the factor a_0 into g, we have

$$f(z) \sim g(z)\left(1 + \frac{b_1}{z} + \frac{b_2}{z^2} + \cdots\right)$$

We will often be interested primarily in obtaining the first term. That would supply a function $g(z)$ with $f(z) = g(z)[1 + O(1/z)]$, or, slightly more generally, $f(z) = g(z)[1 + o(1)]$. In this case we say that f and g are asymptotically equivalent.

7.2.4 DEFINITION *Two functions $f(z)$ and $g(z)$ are **asymptotically equivalent** if $f(z) = g(z)[1 + o(1)]$. In this case we write $f(z) \sim g(z)$.*

Notice that if $g(z) \neq 0$, this says $f(z)/g(z) - 1 = o(1)$ so that $\lim_{z \to \infty} [f(z)/g(z)] = 1$ in the specified sector. The expression $g(z)$ is thought of as giving an asymptotic formula for $f(z)$. It is in this sense that Stirling's formula and the prime number theorem are to be interpreted.

The purpose of all this is to use $g(z)$ to approximate $f(z)$ for large z. But care is required. The approximation need *not* be improving as $z \to \infty$ in the sense we have been using so far in this text. That is, the absolute value of the error $\Delta f = g(z) - f(z)$ need not be shrinking. Instead it is the *relative error* or *percentage error*, the error expressed as a fraction of the true value, which has to be shrinking. The relative error is

$$\frac{\Delta f}{f} = \frac{g(z) - f(z)}{f(z)} = \frac{g(z)}{f(z)} - 1$$

and this goes to 0 as z goes to infinity in the specified sector since it is $o(1)$. The following simple example should make this clear. Let $f(z) = ze^z/(1 + z)$. Then $f(z) \sim e^z$. That is, $g(z) = e^z$ is an asymptotic formula for f. The asymptotic error incurred by using $g(z)$ to approximate $f(z)$ is $\Delta f = e^z - f(z) = e^z/(1 + z)$. This error goes to infinity as z grows along the positive real axis. However, the relative error, the absolute error expressed as a fraction of the true value, is

$$\frac{\Delta f}{f} = \frac{e^z}{1+z} \cdot \frac{1+z}{ze^z} = \frac{1}{z}$$

which does go to 0 as z grows.

It is also important to notice that asymptotic formulas are not unique. The same function might have two different-looking asymptotic formulas although the ratio of the two will tend to 1 as z grows in the specified sector.

Many of the functions one wishes to study either arise as or can be converted to integrals of the form

$$f(z) = \int_\gamma e^{zh(\xi)} g(\xi) \, d\xi$$

The Γ function itself is

$$\Gamma(z) = \int_0^\infty e^{-t} t^z \, dt = \int_0^\infty e^{z \log t} e^{-t} \, dt$$

We will pursue this idea in the next section to obtain Stirling's formula from one of the results at the end of this section. The general plan of attack is to find a point ξ_0 on the curve such that the factor $e^{zh(\xi)}$ is fairly large there but becomes very small away from ξ_0 along the curve for large z. Then most of the contribution to the integral will come from the part of the curve near ξ_0 and we may be able to estimate it in terms of the behavior of h and g near ξ_0. First we turn our attention to some cases in which h is simple enough that we can obtain all terms of the series.

Laplace Transforms

The Laplace transform is a construction very much like the Fourier transform we met earlier. Provided the integral makes sense, the Laplace transform of a function g defined on the positive real axis is

$$\tilde{g}(z) = \int_0^\infty e^{-zt} g(t) \, dt$$

We will devote considerable attention to this construction and some of its applications in Chap. 8. Here we will see how asymptotic series might shed some light on the behavior of $\tilde{g}(z)$ for large z.

7.2.5 PROPOSITION *Suppose g is analytic in a region containing the positive real axis and bounded on the positive real axis. Let the Taylor series for g centered at 0 be*

$\sum_{n=0}^{\infty} a_n z^n$ and let $\tilde{g}(z) = \int_0^{\infty} e^{-zt} g(t) \, dt$. Then,

$$\tilde{g}(z) \sim \frac{a_0}{z} + \frac{a_1}{z^2} + \frac{2a_2}{z^3} + \cdots + \frac{n! a_n}{z^{n+1}} + \cdots$$

as $z \to \infty$, $\arg z = 0$.

PROOF Let $h(z) = [g(z) - (a_0 + a_1 z + \cdots + a_{n-1} z^{n-1})]/z^n$. Then h is bounded on the positive real axis because first, g is bounded, second the polynomial term in the numerator has degree less than n, and third, the limit as z tends to 0 is a_n. Thus there is a constant M with

$$|g(t) - (a_0 + a_1 t + \cdots + a_{n-1} t^{n-1})| < M t^n \qquad \text{for all } t \geq 0$$

and so, for real z,

$$\left| \int_0^{\infty} e^{-zt} [g(t) - (a_0 + a_1 t + \cdots + a_{n-1} t^{n-1})] \, dt \right| \leq M \int_0^{\infty} e^{-zt} t^n \, dt$$

$$\left| \tilde{g}(z) - \sum_{k=0}^{n-1} a_k \int_0^{\infty} e^{-zt} t^k \, dt \right| \leq M \int_0^{\infty} e^{-zt} t^n \, dt$$

Letting $x = zt$, we get

$$\left| \tilde{g}(z) - \sum_{k=0}^{n-1} a_k \frac{1}{z^{k+1}} \int_0^{\infty} e^{-x} x^k \, dx \right| \leq M \frac{1}{z^{n+1}} \int_0^{\infty} e^{-x} x^n \, dx$$

$$\left| \tilde{g}(z) - \sum_{k=0}^{n-1} a_k \frac{\Gamma(k+1)}{z^{k+1}} \right| \leq M \frac{\Gamma(n+1)}{z^{n+1}} = \frac{M n!}{z^{n+1}} = o\left(\frac{1}{z^n}\right)$$

which is exactly what we wanted. ∎

The assumption of analyticity for the function g fits nicely into the theme of this text and makes the simple proof just given possible. It is worth noting, and important for many applications, that the same result holds with slightly different assumptions on g. Analyticity is not so essential as that g be infinitely differentiable.

7.2.6 PROPOSITION *Suppose g is infinitely differentiable on the positive real axis and that g and each of its derivatives are of exponential order. That is, there are constants*

A_n and B_n such that $|g^{(n)}(t)| \leq A_n e^{B_n t}$ for $t \geq 0$. Let $\tilde{g}(z) = \int_0^\infty e^{-zt} g(t)\, dt$. Then

$$\tilde{g}(z) \sim \frac{g(0)}{z} + \frac{g'(0)}{z^2} + \frac{g''(0)}{z^3} + \cdots + \frac{g^{(n)}(0)}{z^{n+1}} + \cdots$$

as $z \to \infty$, $\arg z = 0$.

PROOF   Fix $n \geq 0$ and suppose $z > \max (B_0, B_1, \ldots, B_n)$. Then repeated integration by parts gives

$$\tilde{g}(z) = -\sum_{k=0}^{n-1} \left(\lim_{T \to \infty} \frac{e^{-zt} g^{(k)}(t)}{z^{k+1}} \Bigg|_{t=0}^{T} \right) + \frac{1}{z^n} \int_0^\infty e^{-zt} g^{(n)}(t)\, dt$$

$$= \sum_{k=0}^{n-1} \frac{g^{(k)}(0)}{z^{k+1}} + \frac{1}{z^n} \int_0^\infty e^{-zt} g^{(n)}(t)\, dt$$

since $|e^{-zt} g^{(k)}(T)| \leq A_k e^{(B_k - z)T}$, and this last term goes to 0 as T grows. Therefore

$$z^n |\tilde{g}(z) - S_n(z)| \leq \int_0^\infty |e^{-zt} g^{(n)}(t)|\, dt \leq \int_0^\infty A_n e^{(B_n - z)t}\, dt = \frac{A_n}{z - B_n}$$

and this goes to 0 as z grows, as we need. ∎

The second result (7.2.6) applies to infinitely differentiable functions which are not analytic, as well as functions such as polynomials which are not bounded. The first (7.2.5) includes functions such as $g(t) = \sin e^{t^2}$ whose derivative is not of exponential order.

Watson's Theorem

Almost the same argument as used for Proposition 7.2.5 will establish an expansion with a slightly more complicated function for h.

7.2.7 WATSON'S THEOREM   *Let $g(z)$ be analytic and bounded on a domain containing the real axis. Set*

$$f(z) = \int_{-\infty}^\infty e^{-zy^2/2} g(y)\, dy$$

for z real. Then

$$f(z) \sim \frac{\sqrt{2\pi}}{\sqrt{z}} \left(a_0 + \frac{a_2}{z} + \frac{a_4 \cdot 1 \cdot 3}{z^2} + \frac{a_6 \cdot 1 \cdot 3 \cdot 5}{z^3} + \cdots \right)$$

as $z \to \infty$, $\arg z = 0$, where $g(z) = \sum_{n=0}^{\infty} a_n z^n$ near zero.

PROOF We first observe that

$$h(z) = \frac{g(z) - (a_0 + a_1 z + \cdots + a_{2n-1} z^{2n-1})}{z^{2n}}$$

is bounded on the real axis since g is bounded and since $h(z) \to a_{2n}$ as $z \to 0$. Therefore, we obtain

$$\left| \int_{-\infty}^{\infty} e^{-zy^2/2} [g(y) - (a_0 + a_1 y + \cdots + a_{2n-1} y^{2n-1})] \, dy \right| \le M \int_{-\infty}^{\infty} e^{-zy^2/2} y^{2n} \, dy$$

Now we use the fact that for $z > 0$,

$$\int_{-\infty}^{\infty} e^{-zy^2/2} y^{2k} \, dy = \sqrt{2\pi} \cdot 1 \cdot 3 \cdots \frac{2k-1}{z^{k+1/2}} \qquad \text{and} \qquad \int_{-\infty}^{\infty} e^{-zy^2} y^{2k+1} \, dy = 0$$

(see Exercise 7), to obtain

$$\left| \int_{-\infty}^{\infty} e^{-zy^2} g(y) \, dy - \left(\frac{a_0 \sqrt{2\pi}}{z^{1/2}} + \frac{a_2 \sqrt{2\pi}}{z^{1+1/2}} + \frac{a_4 \sqrt{2\pi} \cdot 1 \cdot 3}{z^{2+1/2}} + \cdots \right. \right.$$
$$\left. \left. + \frac{a_{2n-2} \cdot \sqrt{2\pi} \cdot 1 \cdot 3 \cdots (2n-3)}{z^{n-1+1/2}} \right) \right| \le M\sqrt{2\pi} \frac{1 \cdot 3 \cdots (2n-1)}{z^{n+1/2}}$$

from which the theorem follows. ∎

Method of Steepest Descent

Finally we turn to situations in which h may be more complicated and in which more sophisticated techniques are needed. One of these is called the method of steepest descent or the saddle-point method. It was discovered by P. Debye in approximately 1909. We shall seek an expansion of the form

$$f(z) \sim g(z) \left(1 + \frac{a_1}{z} + \frac{a_2}{z^2} + \cdots \right)$$

for a particularly simple function $g(z)$, and shall be mainly interested in obtaining the first term. The method described here works well if f has the special form $f(z) = \int_\gamma e^{zh(\xi)} d\xi$. We shall use contours $\gamma(t)$ defined for all $t \in \mathbf{R}$. We can integrate over such infinite contours in the same manner as we would integrate over ordinary ones, as long as we check convergence of the integrals.

7.2.8 STEEPEST DESCENT THEOREM   *Let $\gamma:]-\infty, \infty[\to \mathbf{C}$ be a C^1 curve in $\mathbf{C}$. (γ may also be defined only on a finite interval.) Let $\zeta_0 = \gamma(t_0)$ be a point on γ and let $h(\zeta)$ be a function continuous along γ and analytic at ζ_0. Make the following hypotheses: For $|z| \geq R$ and $\arg z$ fixed,*

(i) *$f(z) = \int_\gamma e^{zh(\zeta)} d\zeta$ converges absolutely.*
(ii) *$h'(\zeta_0) = 0$; $h''(\zeta_0) \neq 0$.*
(iii) *Im $zh(\zeta)$ is constant for ζ on γ in some neighborhood of ζ_0.*
(iv) *Re $zh(\zeta)$ has a strict maximum at ζ_0 along the entire curve γ.*

Then

$$f(z) \sim \frac{e^{zh(\zeta_0)}\sqrt{2\pi}}{\sqrt{z}\sqrt{-h''(\zeta_0)}} \tag{1}$$

as $z \to \infty$, $\arg z$ fixed. The sign of the square root is chosen such that $\sqrt{z}\sqrt{-h''(\zeta_0)} \cdot \gamma'(t_0) > 0$.

REMARKS

(i) To achieve conditions (i) to (iv) it may be necessary to deform γ by applying Cauchy's theorem. A path γ verifying these conditions is called a *path of steepest descent*.
(ii) The asymptotic expansion in the conclusion of the theorem depends only on $h(\zeta_0)$ and $h''(\zeta_0)$ and not on the behavior of h elsewhere on γ (except, of course, that h must satisfy the hypotheses of the theorem). Higher-order derivatives would be used if further terms in the expansion were needed.
(iii) The origin of the term "steepest descent" can be traced to conditions (iii) and (iv) in the following way. Recall that Im $zh(\zeta) = v(\zeta)$ and Re $zh(\zeta) = u(\zeta)$ are harmonic conjugates, and the fact that v is constant on γ means that u is changing fastest in the direction of γ. Since ζ_0 is a maximum, $u(\zeta) = $ Re $zh(\zeta)$ is decreasing fastest when moving away from ζ_0 in the direction of γ. Hence the curve γ is called the path of steepest descent. The term "saddle-point method" originated as follows. The function $u(\zeta) = $ Re $zh(\zeta)$ has a maximum on γ at ζ_0. But $h''(\zeta_0) \neq 0$ implies that u is not constant, and so ζ_0 must be a saddle point of u since harmonic functions never have local maxima or minima (see Fig. 7.2.2).
(iv) Often the correct sign for the square root may be determined by examining the sign of the integral defining $f(z)$.

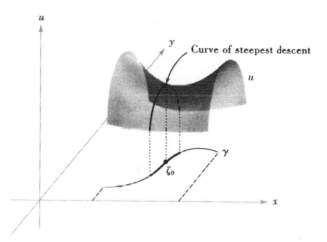

FIGURE 7.2.2   Saddle-point method.

PROOF   Break up γ into three portions, γ_1, C, and γ_2, as illustrated in Fig. 7.2.3. Choose C so that it lies in a neighborhood of ζ_0 small enough that (1) $h(\zeta)$ is analytic and (2) condition (iii) holds. Clearly, $f(z) = I_1(z) + I_2(z) + J(z)$ where we use the notations $J(z) = \int_C e^{zh(\zeta)}\, d\zeta$ and $I_k(z) = \int_{\gamma_k} e^{zh(\zeta)}\, d\zeta$, $k = 1, 2$.

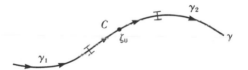

FIGURE 7.2.3   Method of steepest descent.

We will show that for large z the part of the integral which really matters is $J(z)$ so that an asymptotic approximation for $J(z)$ will also give one for $f(z)$. Worked Example 7.2.11 says that in order to do this it is enough to show that $I_k(z)/J(z) = o(1/z^n)$ for all positive n. To prove this, note that

$$\left| |I_k(z)| - \left| \int_{\gamma_k} e^{zh(\zeta)}\, d\zeta \right| \right| \le \int_{\gamma_k} e^{\operatorname{Re} zh(\zeta)}\, |d\zeta|$$

But

$$J(z) = \int_C e^{zh(\zeta)}\, d\zeta = \int_C e^{\operatorname{Re} zh(\zeta)} e^{i\operatorname{Im} zh(\zeta)}\, d\zeta$$

Since Im $zh(\zeta)$ is constant on C, we get

$$|J(z)| = \left| \int_C e^{\operatorname{Re} zh(\zeta)} \, d\zeta \right|$$

If C is short enough that arg (γ') changes by less than $\pi/4$ along C, then we obtain $|J(z)| > (1/\sqrt{2}) \int_C e^{\operatorname{Re} zh(\zeta)} |d\zeta|$. Thus

$$\left| \frac{I_k(z)}{J(z)} \right| \leq \frac{\displaystyle\int_{\gamma_k} e^{\operatorname{Re} zh(\zeta)} |d\zeta|}{\displaystyle\int_C e^{\operatorname{Re} zh(\zeta)} |d\zeta|} \sqrt{2}$$

Now let $\tilde{C}$ be a strictly smaller subinterval of C, centered at ζ_0. Then

$$\left| \frac{I_k(z)}{J(z)} \right| \leq \frac{\displaystyle\int_{\gamma_k} e^{\operatorname{Re} zh(\zeta)} |d\zeta|}{\displaystyle\int_{\tilde{C}} e^{\operatorname{Re} zh(\zeta)} |d\zeta|} \sqrt{2}$$

Fix z_0 and let α be the minimum of Re $(zh(\zeta))$ on $\tilde{C}$. There is an $\epsilon > 0$ such that Re $(z_0 h(\zeta)) \leq \alpha - \epsilon$ for all $\zeta \in \gamma_k$. Thus, using the fact that z lies on the same ray as z_0,

$$\frac{\displaystyle\int_{\gamma_k} e^{\operatorname{Re} zh(\zeta)} |d\zeta|}{\displaystyle\int_{\tilde{C}} e^{\operatorname{Re} zh(\zeta)} |d\zeta|} = \frac{\displaystyle\int_{\gamma_k} e^{\operatorname{Re} z_0 h(\zeta)} e^{\operatorname{Re} (z - z_0) h(\zeta)} |d\zeta|}{\displaystyle\int_{\tilde{C}} e^{\operatorname{Re} z_0 h(\zeta)} e^{\operatorname{Re} (z - z_0) h(\zeta)} |d\zeta|} \leq \frac{\left(\displaystyle\int_{\gamma_k} e^{\operatorname{Re} z_0 h(\zeta)} |d\zeta| \right) e^{|z - z_0|(\alpha - \epsilon)/|z_0|}}{\left(\displaystyle\int_{\tilde{C}} e^{\operatorname{Re} z_0 h(\zeta)} |d\zeta| \right) e^{|z - z_0|\alpha/|z_0|}}$$

This expression is a constant factor, say M, times $e^{-|z - z_0|\epsilon/|z_0|}$. The latter is certainly $O(1/z)$ (and in fact is $O(1/z^n)$ for all $n \geq 1$), so we have proved that $I_k(z)/J(z) = O(1/z^n)$ for all $n \geq 1$. This localizes the problem to a neighborhood around ζ_0 where the bulk of the contribution to the integral is made. Also, we can shrink the length of C without affecting the conclusion that $f(z) \sim J(z)$ as $z \to \infty$.

Next, we write

$$h(\zeta) = h(\zeta_0) - w(\zeta)^2$$

where $w(\zeta)$ is analytic and invertible (abusing notation, we denote the inverse by $\zeta(w)$), where $w(\zeta_0) = 0$, and where

$$[w'(\zeta_0)]^2 = \frac{-h''(\zeta_0)}{2}$$

(see Worked Example 6.3.7). Since Im $(zh(\zeta)) =$ Im $(zh(\zeta_0))$ on C and Re $(zh(\zeta)) <$ Re $(zh(\zeta_0))$, we see that $z[w(\zeta)]^2$ is real and greater than zero on C; also, by our choice of branch for $\sqrt{}$, $\sqrt{z}w(\zeta)$ is real and, as a function of the curve parameter t, has positive derivative at t_0. Thus, by shrinking C if necessary, we can assume that $\sqrt{z}w(\zeta)$ is increasing along C.

Note that

$$J(z) = \int_C e^{zh(\zeta)}\, d\zeta = \int_C e^{zh(\zeta_0)} \cdot e^{-zw(\zeta)^2}\, d\zeta = e^{zh(\zeta_0)} \int_C e^{-z[w(\zeta)]^2}\, d\zeta$$

We can change variables by setting $\sqrt{z}w(\zeta) = y$, and we get

$$J(z) = e^{zh(\zeta_0)} \int_{-\sqrt{|z|}\epsilon_1}^{\sqrt{|z|}\epsilon_2} e^{-y^2}\, \frac{d\zeta}{dw}\, \frac{dy}{\sqrt{z}} = \frac{e^{zh(\zeta_0)}}{\sqrt{z}} \int_{-\sqrt{|z|}\epsilon_1}^{\sqrt{|z|}\epsilon_2} e^{-y^2}\, \frac{d\zeta}{dw}\, dy$$

since y is real on C; we choose positive numbers ϵ_1 and ϵ_2 such that $[-\sqrt{|z|}\epsilon_1, \sqrt{|z|}\epsilon_2]$ is in the range of y corresponding to ζ on C. Next we write

$$\zeta = \zeta_0 + a_1 w + a_2 w^2 + \cdots$$

so that

$$\frac{d\zeta}{dw} = a_1 + 2a_2 w + 3a_3 w^2 + \cdots$$

where $w = y/\sqrt{z}$. Thus,

$$\begin{aligned}
\frac{J(z)}{e^{zh(\zeta_0)}\sqrt{\pi}} &= \int_{-\sqrt{|z|}\epsilon_1}^{\sqrt{|z|}\epsilon_2} e^{-y^2} \left[\sum_{k=1}^{\infty} k a_k \left(\frac{y}{\sqrt{z}}\right)^{k-1} \right] dy \\
&= \int_{-\sqrt{|z|}\epsilon_1}^{\sqrt{|z|}\epsilon_2} e^{-y^2} \left[\sum_{k=0}^{N} (k+1)a_{k+1} \left(\frac{y}{\sqrt{z}}\right)^{k} \right] dy \\
&\quad + \int_{-\sqrt{|z|}\epsilon_1}^{\sqrt{|z|}\epsilon_2} e^{-y^2} O\left(\left(\frac{y}{\sqrt{z}}\right)^{N+1} \right) dy \\
&= \sum_{k=0}^{N} \frac{(k+1)a_{k+1}}{(\sqrt{z})^k} \int_{-\sqrt{|z|}\epsilon_1}^{\sqrt{|z|}\epsilon_2} e^{-y^2} y^k\, dy \\
&\quad + \int_{-\sqrt{|z|}\epsilon_1}^{\sqrt{|z|}\epsilon_2} e^{-y^2} O\left(\left(\frac{y}{\sqrt{z}}\right)^{N+1} \right) dy
\end{aligned}$$

By Exercise 7 we know that

$$\int_{-\infty}^{\infty} e^{-y^2} y^k\, dy = \begin{cases} \dfrac{(2m)!\sqrt{\pi}}{m!2^{2m}} & \text{if } k = 2m \text{ is even} \\[2mm] 0 & \text{if } k = 2m+1 \text{ is odd} \end{cases}$$

and so we are led to the series

$$S \equiv \sum \frac{(k+1)a_{k+1}}{(\sqrt{z})^k} \int_{-\infty}^{\infty} e^{-y^2} y^k \, dy = \sum_{m=0}^{\infty} \frac{(2m)! \sqrt{\pi}}{m! 2^{2m}} \frac{(2m+1)a_{2m+1}}{z^m}$$

This gives the desired series:

$$\frac{J(z)}{e^{zh(\zeta_0)}/\sqrt{z}} - S_M = -\sum_{k=0}^{2M} \frac{(k+1)a_{k+1}}{(\sqrt{z})^k} \left(\int_{-\infty}^{-\sqrt{|z|}\,\epsilon_1} e^{-y^2} y^k \, dy + \int_{\sqrt{|z|}\,\epsilon_2}^{\infty} e^{-y^2} y^k \, dy \right)$$

$$+ \int_{-\sqrt{|z|}\,\epsilon_1}^{\sqrt{|z|}\,\epsilon_2} e^{-y^2} O\left(\left(\frac{y}{\sqrt{z}} \right)^{2M+1} \right) dy$$

The first two integrals are $o(1/(\sqrt{z})^{2M})$ by Proposition 7.2.3(v), since $e^{-y^2} y^k = o(1/y^{2M+1})$. In the third, there is a constant B_M such that the integrand is bounded by

$$\frac{B_M e^{-y^2} |y|^{2M+1}}{|\sqrt{z}|^{2M+1}} = \left(\frac{B_M}{|z|^M} \right) \left(\frac{1}{\sqrt{|z|}} \right) e^{-y^2} |y|^{2M+1}$$

Since

$$\int_{-\infty}^{\infty} e^{-y^2} |y|^{2M+1} \, dy < \infty$$

this term is also $o(1/|z|^M)$. Thus,

$$J(z) \sim \frac{e^{zh(\zeta_0)} S}{\sqrt{z}}$$

and by Worked Example 7.2.11, so is $f(z)$. Thus,

$$f(z) \sim e^{zh(\zeta_0)} \sqrt{\frac{\pi}{z}} \left(a_1 + \frac{1 \cdot 3a_3}{z} + \frac{1 \cdot 3 \cdot 5a_5}{z^2} + \cdots \right)$$

To complete the proof, note that

$$a_1 = \frac{d\zeta}{dw}(0) = \frac{1}{\dfrac{dw}{d\zeta}(\zeta_0)} = \frac{\sqrt{2}}{\sqrt{-h''(\zeta_0)}}$$

so that

$$f(z) \sim \frac{e^{zh(\zeta_0)} \sqrt{2\pi}}{\sqrt{z}\sqrt{-h''(\zeta_0)}}$$

as desired. ∎

Note that to obtain the higher-order terms in the expansion

$$f(z) \sim \frac{e^{zh(\zeta_0)}}{\sqrt{z}} \cdot \frac{\sqrt{2\pi}}{\sqrt{-h''(\zeta_0)}} \left(1 + \frac{A_1}{z} + \frac{A_2}{z^2} + \cdots \right) \qquad (2)$$

one must be able to compute more terms in the series $\zeta = \zeta_0 + a_1 w + a_2 w^2 + \cdots$ in the preceding proof. In simple cases, these higher-order terms can be evaluated explicitly; see Watson's theorem (7.2.7). The details of the method of obtaining higher-order terms will not be given here, because such terms are needed only in very refined calculations. The leading term given in the steepest descent theorem is the important one.

The applications of this theorem given in the next section deal primarily with the case in which z is real and positive. Clearly, in that case conditions (iii) and (iv) of the theorem can be written equivalently with or without the z.

A proof similar to that given above leads to the following. (See Exercise 8.)

7.2.9 GENERALIZED STEEPEST DESCENT THEOREM *Let the conditions of the steepest descent theorem (7.2.8) hold but let f have the form $f(z) = \int_\gamma e^{zh(\zeta)} g(\zeta)\, d\zeta$ where $g(\zeta)$ is a bounded continuous function on γ with $g(\zeta_0) \neq 0$. Then*

$$f(z) \sim \frac{e^{zh(\zeta_0)} \sqrt{2\pi} g(\zeta_0)}{\sqrt{z}\sqrt{-h''(\zeta_0)}}$$

Method of Stationary Phase

If the exponent in the integrand of Theorem 7.2.8 is purely imaginary we can obtain a related result known as the *method of stationary phase*. This method was developed in part by Lord Kelvin in 1887 and will be applied to the study of Bessel functions in the next section.

7.2.10 STATIONARY PHASE THEOREM *Let $[a, b]$ be a bounded interval on the real axis. Let $h(t)$ be analytic in a neighborhood of $[a, b]$ and be real for real t. Let $g(t)$ be a real- or complex-valued function on $[a, b]$ with continuous derivative. Suppose*

$$f(z) = \int_a^b e^{izh(t)} g(t)\, dt$$

If $h'(t) = 0$ at exactly one point t_0 in $]a, b[$ and $h''(t_0) \neq 0$, then as $z \to \infty$ on the positive real axis we have

$$f(z) \sim \frac{e^{izh(t_0)} \sqrt{2\pi}}{\sqrt{z}\sqrt{\pm h''(t_0)}} e^{\pm \pi i/4} g(t_0)$$

The plus signs are used if $h''(t_0) > 0$, and the minus signs are used if $h''(t_0) < 0$.

The asymptotic formula for f can also be written as

$$\lim_{z \to +\infty} \sqrt{z} e^{-izh(t_0)} \int_a^b e^{izh(t)} g(t)\, dt = \frac{\sqrt{2\pi} e^{\pm \pi i/4} g(t_0)}{\sqrt{\pm h''(t_0)}}$$

In this form it also makes sense when $g(t_0) = 0$, and we will establish it in this form. We note that by breaking g into its real and imaginary parts, it is sufficient to prove the theorem for g real-valued.

The name "stationary phase" comes from the interpretation that the integrand is a complex quantity with amplitude (magnitude) $g(t)$ and phase angle $zh(t)$. The intuition behind the formula is that the main contribution to the integral should come from the neighborhood of t_0, where the phase angle is varying as slowly as possible. To see why, think of the integral in terms of its real and imaginary parts:

$$f(z) = \int_a^b g(t) \cos (zh(t))\, dt + i \int_a^b g(t) \sin (zh(t))\, dt$$

If z is very large, then $zh(t)$ is changing quite rapidly in regions where $h'(t)$ is not zero. Thus $\cos (zh(t))$ and $\sin (zh(t))$ are oscillating rapidly. Figure 7.2.4 illustrates this with

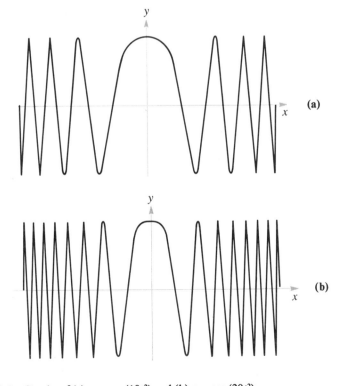

FIGURE 7.2.4   Graphs of (a) $y = \cos (10t^2)$ and (b) $y = \cos (20t^2)$

$h(t) = t^2$ by the graphs of cos $(10t^2)$ and cos $(20t^2)$. If g is at all reasonable, the resulting oscillations of the integral should tend to cancel out *except* near the points where $h'(t) = 0$.

The endpoints of the interval of integration might also be expected to contribute, but it turns out that this contribution is at worst proportional to $1/z$ and so will not interfere with the result we hope to prove: that the integral behaves about like $1/\sqrt{z}$.

We should be able to estimate the integral for large z by using only a portion of the path very near to t_0. In this short interval we then approximate $g(t)$ by the constant $g(t_0)$ and $h(t)$ by its second-order Taylor approximation $h(t_0) + [h''(t_0)/2](t - t_0)^2$ to obtain

$$f(z) \approx e^{izh(t_0)}g(t_0) \int e^{izh''(t_0)(t-t_0)^2/2} \, dt$$

where the integral is over some short interval centered at t_0. Changing variables produces

$$f(z) \approx \frac{e^{izh(t_0)}g(t_0)\sqrt{2}}{\sqrt{z}\sqrt{h''(t_0)}} \int e^{ix^2} \, dx$$

Now the integral is over some interval of the form $[-A\sqrt{z}, A\sqrt{z}]$. We know from Worked Example 4.3.17 on the Fresnel integrals that as z goes to infinity this integral converges to $\sqrt{\pi/2}(1 + i) = \sqrt{\pi}e^{\pi i/4}$. This leaves us with exactly the result we wish with obvious modifications for the case in which $h''(t_0) < 0$.

We will see something of the applicability of this formula in the next section when we study the Bessel functions. Kelvin used it in 1891 to study the pattern of bow and stern waves from a moving ship. In any particular application the amplitude $g(t)$ is usually quite well behaved. But actually turning the intuitive derivation just given into a proof is a bit tricky. The very first step requires that the function g be smooth enough so that when multiplied by the rapidly oscillating cos $(zh(t))$ and sin $(zh(t))$ it gives something for which the integral effectively cancels out away from t_0 and for which any cancellation is not so effective near t_0. This may not happen if g itself has a lot of oscillation at very high frequencies. Continuity alone is not enough to prevent this, as can be seen from the following example.

7.2.11 EXAMPLE *Find a continuous function g for which the conclusion of the stationary phase theorem (7.2.10) is false.*

SOLUTION Let $\phi(t) - \sum_{k=1}^{\infty} (1/k^7) \cos(k^6 t)$. This series converges uniformly and absolutely for $t \in \mathbf{R}$ since the kth term is dominated by $1/k^2$. Thus ϕ is continuous. Define $g(t)$ on the interval $\mathbf{I} = [-\sqrt{2\pi}, \sqrt{2\pi}]$ by $g(t) = 2t\phi(t^2)$ when $t \geq 0$ and by

$g(t) = 0$ when $t < 0$ and consider the integral

$$f(z) = \int_I e^{-izt^2} g(t)\, dt$$

This fits the pattern of (7.2.10) with $h(t) = -t^2$, $t_0 = 0$, and $h''(0) < 0$. However, for a positive integer n we have

$$
\begin{aligned}
f(n) &= \int_I e^{-int^2} g(t)\, dt = \int_0^{\sqrt{2\pi}} e^{-int^2} \phi(t^2) 2t\, dt \\
&= \int_0^{2\pi} e^{-inx} \phi(x)\, dx = \int_0^{2\pi} e^{-inx} \left[\sum_{k=1}^{\infty} \frac{1}{k^2} \cos(k^6 x) \right] dx \\
&= \sum_{k=1}^{\infty} \frac{1}{k^2} \int_0^{2\pi} e^{-inx} \cos(k^6 x)\, dx \\
&= \sum_{k=1}^{\infty} \frac{1}{k^2} \left[\int_0^{2\pi} \cos(nx) \cos(k^6 x)\, dx - i \int_0^{2\pi} \sin(nx) \cos(k^6 x)\, dx \right]
\end{aligned}
$$

These integrals are all 0 except the first in the single case $k^6 = n$, and that one is π. Thus $f(k^6) = \pi/k^2$. In other words, for positive integer z we have $f(z) = 0$ unless z is a sixth power, in which case $f(z) = \pi/\sqrt[3]{z}$. Thus $\sqrt{z} f(z)$ does not remain bounded and the conclusion of Theorem 7.2.10 cannot possibly hold. ▲

The function g in this example is not smooth enough to make Theorem 7.2.10 work. It has too much influence from its high-frequency components, and some condition is needed to prevent this. The requirement of a continuous first derivative (in the sense of one real variable) specified in Theorem 7.2.10 is one such condition. It implies a property which is phrased specifically in terms of the oscillations of g and which is of central importance in the theory of integration called *bounded variation*. A few of the ideas about this property and a proof of Theorem 7.2.10 are outlined in the supplement to this section, which may be regarded as optional.

SUPPLEMENT TO SECTION 7.2: BOUNDED VARIATION AND THE PROOF OF THE STATIONARY PHASE FORMULA

We saw in the method of stationary phase that we needed to impose on the amplitude some condition which would limit the amount of high-frequency oscillation it could have. This sort of thing is often needed in theory involving integrals; the notion of bounded variation provides the appropriate tools to deal with the situation.

7.2.12 DEFINITION   *Suppose f: [a, b] → **R**.*

(i) *If P is a partition of [a, b] given by a = $t_0 < t_1 < \cdots < t_n = b$, then the **variation of** f **on** [a, b] **relative to** P is $V_P f = \sum_{k=1}^{n} |f(t_k) - f(t_{k-1})|$.*

(ii) *The **total variation of** f **on** [a, b] is $V_{[a,b]} f = \sup \{V_P f\}$, where the least upper bound is taken over all possible partitions. (It might be $+\infty$.)*

(iii) *If $V_{[a,b]} f < \infty$ we say that f is of **bounded variation** and write $f \in$ BV ([a, b]).*

Some important examples of such functions are included in the following:

7.2.13 PROPOSITION

(i) *If f is monotone and bounded on [a, b], then $f \in$ BV ([a, b]) and $V_{[a,b]} f = |f(b) - f(a)|$.*

(ii) *If f is differentiable on a bounded interval [a, b] and $|f'(x)| < M$ for all $x \in$ [a, b], then $f \in$ BV ([a, b]) and $V_{[a,b]} f \leq |b - a|M$.*

(iii) *If f has a continuous derivative on the bounded interval [a, b]—that is, if $f \in C^1([a, b])$—then $f \in$ BV ([a, b]).*

PROOF   The first result is immediate, since the succeeding differences from point to point along any partition are all of the same sign and values at intermediate points cancel out. The second is shown by applying the mean value theorem to each subinterval of any partition, and the third follows from it since if f' is continuous on the compact interval [a, b] then it is bounded. ∎

It is possible for a continuous function not to have bounded variation. On [−1, 1] set $f(0) = 0$ and $f(x) = x \cos (1/x)$ for $x \neq 0$. (See Fig. 7.2.5.) Then we have

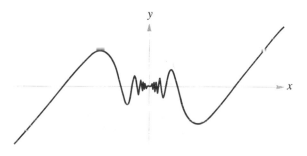

FIGURE 7.2.5   The continuous function $x \cos (1/x)$ has unbounded variation.

$|f(1/n\pi) - f(1/((n+1)\pi)| = (2n+1)/n(n+1)\pi > 1/n\pi$. Since the harmonic series diverges, partitions may be created using these points which give arbitrarily large variation.

Some of the important properties of functions of bounded variation are outlined in the following proposition.

7.2.14 PROPOSITION   *Suppose* $f \in$ BV ([a, b]).

(i) *If* $[c, d] \subset [a, b]$, *then* $f \in$ BV ([c, d]) *and* $V_{[c,d]}f \le V_{[a,b]}f$.
(ii) $V_{[a,c]}f + V_{[c,b]}f = V_{[a,b]}f$ *if* $a < c < b$.
(iii) $(Vf)(x) = V_{[a,x]}f$ *is a bounded increasing function on* $[a, b]$ *with* $(Vf)(a) = 0$ *and* $(Vf)(b) = V_{[a,b]}f$.
(iv) *If* $a \le x \le y \le b$, *then* $(Vf)(y) - (Vf)(x) = V_{[x,y]}f$.
(v) f *is the difference of two bounded increasing functions:* $f = f_1 - f_2$ *with* $f_1 = (Vf + f)/2$ *and* $f_2 = (Vf - f)/2$.

PROOF   The first assertion follows since any partition of $[c, d]$ can be extended by the intervals $[a, c]$ and $[d, b]$ to obtain a partition of $[a, b]$ offering a larger candidate for $V_{[a,b]}f$. For the second, adjoin partitions of $[a, c]$ and $[c, b]$ to get a partition of $[a, b]$ and show $V_{[a,c]}f + V_{[c,b]}f \le V_{[a,b]}f$. For the opposite inequality let $a = t_0 < t_1 < \cdots < t_n = b$ be any partition of $[a, b]$ with $\sum_{k=1}^{n} |f(t_k) - f(t_{k-1})| > V_{[a,b]} - \epsilon$. Pick N with $t_N \le c \le t_{N+1}$. Then

$$V_{[a,b]}f < \sum_{0}^{N} |f(t_k) - f(t_{k-1})| + \sum_{N}^{N+1} |f(t_k) - f(t_{k-1})| + \epsilon$$

$$< \sum_{0}^{N} |f(t_k) - f(t_{k-1})| + |f(c) - f(t_N)| + |f(t_{N+1}) - f(c)| + \sum_{N+2}^{n} |f(t_k) - f(t_{k-1})| + \epsilon$$

$$\le V_{[a,c]}f + V_{[c,b]}f + \epsilon$$

Since this holds for any $\epsilon \ge 0$ we have the desired inequality. The third assertion is clear and the fourth follows from it and the second. For the last assertion, use (iv) to show that the functions indicated are increasing. ■

The last property will be the one directly utilized in the proof of Theorem 7.2.10. The tool by which we will use it is the second mean value theorem for integrals.

7.2.15 SECOND MEAN VALUE THEOREM FOR INTEGRALS   *If* f *is bounded and increasing on* $[a, b]$ *and* g *is integrable, then there is a point* c *in* $[a, b]$ *such that*

$$\int_{a}^{b} f(t)g(t)\, dt = f(a) \int_{a}^{c} g(t)\, dt + f(b) \int_{c}^{b} g(t)\, dt$$

PROOF Let

$$F(x) = f(a) \int_a^x g(t) \, dt + f(b) \int_x^b g(t) \, dt$$

Then $F(a) = f(b) \int_a^b g(t) \, dt$ and $F(b) = f(a) \int_a^b g(t) \, dt$ and

$$f(a) \int_a^b g(t) \, dt \le \int_a^b f(t) g(t) \, dt \le f(b) \int_a^b g(t) \, dt$$

Since F is continuous on $[a, b]$, the conclusion follows from the intermediate value theorem. ∎

We will also need the following estimates.

7.2.16 LEMMA *If f is differentiable on $[a, b]$ and $|f'(x)| \le M$ for all x in $[a, b]$, then $|f(y) - f(x)|$, $|(Vf)(y) - (Vf)(x)|$, $|f_1(y) - f_1(x)|$, and $|f_2(y) - f_2(x)|$ are each bounded above by $M|y - x|$.*

PROOF The first conclusion follows from the mean value theorem and the second from Proposition 7.2.13(ii) and Proposition 7.2.14(iv). The last two follow from the first two and the formulas for f_1 and f_2 given in Proposition 7.2.14(v). ▼

The first step in the intuitive derivation of Theorem 7.2.10 was that contributions to the integral from parts of the interval away from t_0 tended to cancel out and could be neglected by comparison with the contribution from a short interval near t_0. The only critical point of h was at t_0 and h' and h'' were continuous, so that away from t_0 the derivative stays away from 0 and we can apply the following lemma.

7.2.17 LEMMA *Suppose h has a continuous second derivative on $[a, b]$, that $h'(x)$ is never 0 in $[a, b]$, and that g has a continuous derivative on $[a, b]$. Then $\int_a^b e^{izh(t)} g(t) \, dt = O(1/z)$.*

PROOF $\psi(x) = g(x)/h'(x)$ has a continuous derivative on $[a, b]$ and so has bounded variation and may be written as a difference of two increasing functions, $\psi = \psi_1 - \psi_2$. Then

$$\int_a^b e^{izh(t)} g(t) \, dt - \frac{1}{z} \int_a^b \psi_1(t) \cos(zh(t)) zh'(t) \, dt + \frac{i}{z} \int_a^b \psi_1(t) \sin(zh(t)) zh'(t) \, dt$$

$$- \frac{1}{z} \int_a^b \psi_2(t) \cos(zh(t)) zh'(t) \, dt - \frac{i}{z} \int_a^b \psi_2(t) \cos(zh(t)) zh'(t) \, dt$$

Each of these integrals may be handled by the second mean value theorem for integrals. The first is typical. There is a point x between a and b with

$$\left| \int_a^b \psi_1(t) \cos(zh(t)) zh'(t)\, dt \right| = \left| \psi_1(a) \int_a^x \cos(zh(t)) zh'(t)\, dt \right.$$
$$\left. + \psi_1(b) \int_x^b \cos(zh(t)) zh'(t)\, dt \right|$$
$$= |\psi_1(a)[\sin(zh(x)) - \sin(zh(c))]$$
$$+ \psi_1(b)[\sin(zh(b)) - \sin(zh(x))]|$$
$$\leq 2|\psi_1(a)| + 2|\psi_1(b)|$$

The others are treated similarly to obtain

$$\left| \int_a^b e^{izh(t)} g(t)\, dt \right| \leq \frac{4}{z} [|\psi_1(a)| + |\psi_1(b)| + |\psi_2(a)| + |\psi_2(b)|]$$

as needed. ▼

We are now ready to complete the proof of Theorem 7.2.10. Since t_0 is the only critical point of h in $[a, b]$, we know that for any $\delta > 0$, $h'(t)$ is never 0 on $[a, t_0 - \delta]$ or $[t_0 + \delta, b]$, and so by the last lemma the integrals of $e^{izh(t)} g(t)$ over each are $O(1/z)$ and so $o(1/\sqrt{z})$. Thus to establish Theorem 7.2.10 it is enough to show that

$$\lim_{z \to \infty} \sqrt{z}\, e^{-ih(t_0)} \int_J e^{izh(t)} g(t)\, dt = \frac{\sqrt{2\pi}}{\sqrt{\pm h''(t_0)}}\, e^{\pm \pi i/4} g(t_0)$$

where $J = [t_0 - \delta, t_0 + \delta]$. We may fix δ as small as we please so long as its choice does not depend on z. In the course of the proof we shall find conditions for that choice.

We know h is analytic in a neighborhood of t_0. By Worked Example 6.3.7 there is an analytic function $w(t)$ such that $h(t) = h(t_0) \pm [w(t)]^2$ for t near t_0 and w is locally one-to-one. We may choose w to be real and strictly increasing on J if δ is selected small enough. This is our first criterion for δ. We choose the plus sign if $h''(t_0) > 0$ and the minus sign if $h''(t_0) < 0$. Since $w(t_0) = 0$ and w is continuous, $w(t_0 + \delta) = c$ and $w(t_0 - \delta) = d$, where $c < 0 < d$. The change of variables $x = w(t)$ gives

$$\int_J e^{izh(t)} g(t)\, dt = e^{izh(t_0)} \int_c^d e^{\pm izx^2} \psi(x)\, dx$$

where $\psi(x) = g(w^{-1}(x))/(w^{-1})'(x)$. The function ψ has a continuous derivative on $[c, d]$. The point $x = 0$ corresponds to $t = t_0$, and $h''(t_0) = \pm 2w(t_0)w''(t_0) \pm 2[w'(t_0)]^2$. Thus $(w^{-1})'(0) = 1/w'(t_0) = \sqrt{\pm h''(t_0)/2}$. Since ψ' is continuous, ψ has bounded variation and can be written as a difference $\psi_1 - \psi_2$ of two increasing functions. Let $\epsilon > 0$. Since c and

d go to 0 as $\delta \to 0$, we can use Lemma 7.2.16 to select δ small enough so that the quantities $|\psi_1(c) - \psi_1(0)|$, $|\psi_1(d) - \psi_1(0)|$, $|\psi_2(c) - \psi_2(0)|$, and $|\psi_2(d) - \psi_2(0)|$ are all smaller than ϵ. Thus,

$$\sqrt{z}e^{-izh(t_0)} \int_J e^{izh(t)}g(t)\,dt = \sqrt{z} \int_c^d e^{\pm izx^2}\psi(x)\,dx$$

$$= \int_c^d \cos(zx^2)\,\psi_1(x)\sqrt{z}\,dx \pm i \int_c^d \sin(zx^2)\,\psi_1(x)\sqrt{z}\,dx$$

$$- \int_c^d \cos(zx^2)\,\psi_2(x)\sqrt{z}\,dx \mp i \int_c^d \sin(zx^2)\,\psi_2(x)\sqrt{z}\,dx$$

As in the proof of Lemma 7.2.17, each integral may be handled by the second mean value theorem for integrals and the first is typical. There is a point y between c and d with

$$\int_c^d \cos(zx^2)\,\psi_1(x)\sqrt{z}\,dx = \psi_1(c) \int_c^y \cos(zx^2)\,\sqrt{z}\,dx + \psi_1(d) \int_y^d \cos(zx^2)\,\sqrt{z}\,dx$$

$$= \psi_1(c) \int_{c\sqrt{z}}^{y\sqrt{z}} \cos(u^2)\,du + \psi_1(d) \int_{y\sqrt{z}}^{d\sqrt{z}} \cos(u^2)\,du$$

Using the Fresnel integrals of Worked Example 4.3.17, these converge as z goes to $+\infty$. Since $c < 0 < d$, the limit is $\psi_1(d)\sqrt{\pi/2}$ if $y < 0$, is $\psi_1(c)\sqrt{\pi/2}$ if $y > 0$, and is $\{[\psi_1(c) + \psi_1(d)]/2\}\sqrt{\pi/2}$ if $y = 0$. But each of these is within $\epsilon\sqrt{\pi/2}$ of $\psi_1(0)\sqrt{\pi/2}$. Similar arguments for the other three integrals show that the whole sum converges to a limit which is

$$\psi_1(0)\sqrt{\frac{\pi}{2}} \pm i\psi_1(0)\sqrt{\frac{\pi}{2}} - \psi_2(0)\sqrt{\frac{\pi}{2}} + i\psi_2(0)\sqrt{\frac{\pi}{2}}$$

with an error of no more than $\epsilon\sqrt{\pi/2}$ in each term. Thus, we do get a limit which is no more than $2\epsilon\sqrt{2\pi}$ away from the point $[\psi_1(0) - \psi_2(0)](1 \pm i)\sqrt{\pi/2} = \psi(0)\sqrt{\pi}e^{\pm\pi i/4} = \sqrt{2\pi}g(t_0)e^{\pm\pi i/4}/\sqrt{\pm h''(t_0)}$, just as desired. This completes the proof of Theorem 7.2.10. ∎

Worked Examples

7.2.18   *Suppose that $f(z) = I(z) + J(z)$, that $I(z)/J(z) = O(1/z^M)$ for every positive integer M, and that*

$$J(z) \sim g(z)\left(a_0 + \frac{a_1}{z} + \frac{a_2}{z^2} + \cdots\right)$$

Show that

$$f(z) \sim g(z)\left(a_0 + \frac{a_1}{z} + \frac{a_2}{z^2} + \cdots\right)$$

Solution. Since $I(z)/J(z) = O(1/z^M)$, we know that $z^M I(z)/J(z)$ stays bounded, and therefore $z^{M-1}I(z)/J(z) \to 0$. Thus $I(z)/J(z) = o(1/z^N)$ for every integer $N \geq 0$. There is a function $B_N(z)$ such that $z^N I(z) = B_N(z)J(z)$ and $B_N(z) \to 0$ as $z \to \infty$. Now compute

$$\left| z^N \left[\frac{f(z)}{g(z)} - S_N(z) \right] \right| \leq \left| \frac{z^N I(z)}{g(z)} \right| + \left| z^N \left[\frac{J(z)}{g(z)} - S_N(z) \right] \right|$$

$$\leq \left| B_N(z) \frac{J(z)}{g(z)} \right| + \left| z^N \left[\frac{J(z)}{g(z)} - S_N(z) \right] \right|$$

The first term goes to 0, since $B_N(z) \to 0$ and $J(z)/g(z) \to a_0$. The second term goes to 0, since $J(z) \sim g(z)(a_0 + a_1/z + a_2/z^2 + \cdots)$. This completes the proof.

7.2.19 *Let $h(\zeta) = \zeta^2$, $\zeta_0 = 0$. Find a curve γ satisfying the hypotheses of the steepest descent theorem. (In other words, find a path of steepest descent.) Take $\arg z = 0$; that is, z is real, $z > 0$.*

Solution. Let $h(\zeta) = u + iv$, so that if $\zeta = \xi + i\eta$, $u = \xi^2 - \eta^2$ and $v = 2\xi\eta$. The discussion following the steepest descent theorem indicated that the path of steepest descent is defined by $v = $ constant (since in our case z is real, $z > 0$). Thus the line of steepest descent through $\zeta_0 = 0$ is either $\xi = 0$ or $\eta = 0$. Since u must have a maximum at $\zeta_0 = 0$, the curve γ is defined by $\xi = 0$.

7.2.20 *Prove that*

$$f(z) = \int_{-\infty}^{\infty} e^{-zy^2/2} \cos y \, dy \sim \frac{\sqrt{2\pi}}{\sqrt{z}} \left(1 - \frac{1}{2z} + \frac{1}{2^2 2! z^2} - \cdots \right)$$

as $z \to \infty$; $\arg z = 0$.

Solution. We apply Watson's theorem. Here

$$\cos y = 1 - \frac{y^2}{2!} + \frac{y^4}{5!} - \cdots$$

Therefore, $a_0 = 1$, $a_2 = \dfrac{-1}{2!}$, $a_4 = \dfrac{1}{4!}, \cdots$

and thus

$$f(z) \sim \sqrt{\frac{2\pi}{z}} \left(1 - \frac{1}{2z} + \frac{1 \cdot 3}{4! z^2} - \cdots \right) = \sqrt{\frac{2\pi}{z}} \left(1 - \frac{1}{2z} + \frac{1}{2^2 2! z^2} - \cdots \right)$$

Exercises

1. Show that if $f(z) = O(h(z))$ and $g(z) = O(h(z))$ and a and b are constants, then $af(z) + bg(z) = O(h(z))$.

2. Show that asymptotic equivalence is an equivalence relation in the sense that the following three properties hold:

(a) *Reflexive:* $f \sim f$.

(b) *Symmetric:* If $f \sim g$, then $g \sim f$.

(c) *Transitive:* If $f \sim g$ and $g \sim h$, then $f \sim h$.

3. If $f(x) \sim a_2/x^2 + a_3/x^3 + \cdots$ for $x \in [0, \infty[$, show that

$$g(x) = \int_x^\infty f(t)\, dt \sim \frac{a_2}{x} + \frac{a_3}{2x^2} + \frac{a_4}{3x^3} + \cdots$$

4. Let $f(x) = \int_x^\infty e^{-t}/t\, dt$. Use integration by parts to show

$$f(x) \sim e^{-x}\left(\frac{1}{x} - \frac{1}{x^2} + \frac{1 \cdot 2}{x^3} - \cdots\right)$$

5. Show that

$$f(x) = \int_0^\infty \frac{e^{-xt}}{1+t^2}\, dt \sim \frac{1}{x} - \frac{2!}{x^3} + \frac{4!}{x^5} - \cdots$$

6. Let $g(z)$ be analytic at z_0 and let $g'(z_0) = 0$ and $g''(z_0) \neq 0$, so that near z_0, $g(z) - g(z_0) = [w(z)]^2$ for w analytic, $w'(z_0) \neq 0$. Prove that there are exactly two perpendicular curves on which Re g (alternatively, Im g) are constant through z_0. (Recall that Proposition 1.5.12 shows that if $f'(z_0) \neq 0$, Re f has exactly one level curve through z_0.) Show also that lines of constant Re g and Im g intersect at $45°$.

7. (a) (See Exercise 21, Sec. 7.1.) Show that if $z > 0$, then for integers $k \geq 0$,

$$\int_{-\infty}^\infty e^{-zy^2/2} y^{2k}\, dy = \sqrt{2\pi}\, \frac{1 \cdot 3 \cdot 5 \cdots (2k-1)}{z^{k+1/2}}$$

and

$$\int_{-\infty}^\infty e^{-zy^2/2} y^{2k+1}\, dy = 0$$

(b) Show that for integers $m \geq 0$,

$$\int_{-\infty}^\infty e^{-y^2} y^{2m}\, dy = \sqrt{\pi}\, \frac{1 \cdot 3 \cdot 5 \cdots (2m-1)}{2^m} = \frac{(2m)!\sqrt{\pi}}{m! 2^{2m}}$$

and

$$\int_{-\infty}^\infty e^{-y^2} y^{2m+1}\, dy = 0$$

8. Let $h_n(t) = \sqrt{n}\,e^{-nt^2}$. The area under the graph of $h_n(t)$ is $\sqrt{\pi}$ and for any $\epsilon > 0$, $h_n(t) \to 0$ uniformly outside $]-\epsilon, \epsilon[$. Such a sequence is called an *approximating δ sequence*. See Fig. 7.2.6.
(a) Show that if $g(t)$ is continuous and $0 < N < \infty$, then $\int_{-N}^{N} g(t)h_n(t)\,dt \to g(0)\sqrt{\pi}$ as $n \to \infty$.
(b) Show that if $g(t)$ is continuous and bounded, then $\int_{-\infty}^{\infty} g(t)h_n(t)\,dt \to g(0)\sqrt{\pi}$ as $n \to \infty$.

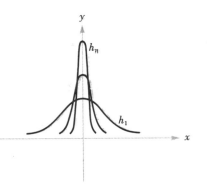

FIGURE 7.2.6 Approximating δ sequence.

9. The expansion

$$\int_x^\infty t^{-1}e^{x-t}\,dt \sim \frac{1}{x} - \frac{1}{x^2} + \cdots$$

was discussed in Example 7.2.2. Compute $S_4(10)$ and $S_5(10)$ numerically and find an upper bound for the respective errors. Discuss how the errors change in $S_n(x)$ as n and x increase. For example, for a given x, are errors reduced if we take n very large?

10. Sketch the proof of the generalized steepest descent theorem (7.2.9) using Exercise 8.

11. Find an asymptotic expansion for

$$f(z) = \int_{-\infty}^{\infty} e^{-zy^2/2} \sin y^2 \, dy$$

(Assume that $z \to \infty$, $z > 0$.)

12. Show that if $f(z) = O(\phi(z))$ and $g(z) = o(h(z))$, then $f(z)g(z) = o(\phi(z)h(z))$.

13. Find the path of steepest descent through $t_0 = 0$ if $h(t) = \cos t$. (Take z real, $z > 0$.)

14. Prove that $\int_C e^{-y^2}\,dy = \sqrt{\pi}$, where C is the 45° line $z = t + it$ with $-\infty < t < \infty$, by showing that

$$\int_C e^{-y^2}\,dy = \int_{-\infty}^{\infty} e^{-y^2}\,dy$$

(*Hint.* Show that $\int_{\gamma_x} e^{-\zeta^2}\,d\zeta \to 0$ as $x \to \infty$ where γ_x is the vertical line joining x to $x + ix$ or use Worked Example 4.3.17.)

15. Repeat Exercise 13 but assume that z lies on the positive imaginary axis.

16. Show that the first term in Watson's theorem may be obtained as a special case of the generalized steepest descent theorem if $g \geq 0$ on the real axis.

17. Find the asymptotic formula for f when the path found in Exercise 13 is used in the steepest descent theorem.

18. Use the steepest descent theorem to obtain the asymptotic formula for f using the path γ that was obtained in Worked Example 7.2.19.

19. Find the asymptotic formula for f when the path γ found in Exercise 15 is used in the steepest descent theorem.

7.3 STIRLING'S FORMULA AND BESSEL FUNCTIONS

In this section the method of steepest descent will be applied to prove Stirling's formula for the gamma function $\Gamma(z)$. Some properties of Bessel functions $J_n(z)$, which are defined for $n = \ldots, -1, 0, 1, \ldots$, will also be developed and the method of stationary phase will be used to obtain an asymptotic formula for these functions.

Stirling's Formula

7.3.1 STIRLING'S FORMULA

$$\Gamma(z + 1) \sim \sqrt{2\pi}\, z^{z+1/2} e^{-z} \tag{1}$$

as $z \to \infty$ on the positive real axis.

REMARK An extension of the proof given here shows that this result also holds for $-\pi/2 + \delta \leq \arg z \leq \pi/2 - \delta$, for any $\delta > 0$ (see Fig. 7.3.1).

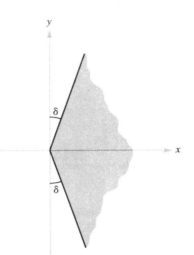

FIGURE 7.3.1   Region of validity for Stirling's formula.

PROOF   Recall from formula 12 in Table 7.1.1 that for Re $z > 0$, we have Euler's integral

$$\Gamma(z) = \int_0^\infty e^{-t} t^{z-1}\, dt$$

We are concerned with the case in which z is real and positive. We want to rewrite the integral so that the steepest descent theorem (7.2.8) will apply. To do this, we make the change of variables $t = z\tau$. We get

$$\Gamma(z+1) = \int_0^\infty e^{-t} t^z\, dt = z^{z+1} \int_0^\infty e^{z(\log \tau - \tau)}\, d\tau$$

Thus $\Gamma(z+1)/z^{z+1}$ has the form

$$\int_\gamma e^{zh(\zeta)}\, d\zeta$$

where $h(\zeta) = \log \zeta - \zeta$ and γ is the positive real axis, $[0, \infty[$. We must check hypotheses (i) to (iv) of the method of steepest descent (7.2.8). Let $\zeta_0 = 1$. Clearly, $h(\zeta_0) = -1$, $h'(\zeta_0) = 0$, and $h''(\zeta_0) \neq 0$; thus hypotheses (i) and (ii) of the method hold. Also, $h(\zeta)$ is real on γ, and so (iii) is valid. To prove (iv), we know that Re $(zh(t)) = xh(t)$ has a maximum iff $h(t)$ does. But $h(t)$ has a maximum of -1 at $\zeta_0 = 1$ on γ (see Fig. 7.3.2).

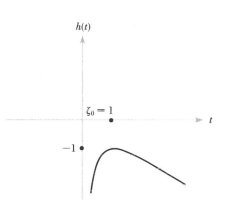

FIGURE 7.3.2 Graph of $h(t) = \log t - t$.

Thus (iv) holds. Therefore,

$$\frac{\Gamma(z+1)}{z^{z+1}} \sim \frac{e^{zh(\zeta_0)}}{\sqrt{z}\sqrt{-h''(\zeta_0)}} \cdot \sqrt{2\pi} = \frac{e^{-z}}{\sqrt{z}} \cdot \sqrt{2\pi}$$

Hence $\Gamma(z+1) \sim z^{z+1/2}e^{-z}\sqrt{2\pi}$, as required. ∎

If $z = re^{i\theta}$ were not real, the x axis would no longer be the path of steepest descent and the path of integration would have to be deformed into such a path using Cauchy's theorem.

If one examines this method more carefully, one finds that the first few terms are

$$\Gamma(z+1) \sim \sqrt{2\pi}\, z^{z+1/2} e^{-z} \left(1 + \frac{1}{12z} + \frac{1}{288z^2} + \cdots \right) \tag{2}$$

that is,

$$\Gamma(z+1) = \sqrt{2\pi}\, z^{z+1/2} e^{-z} \left[1 + \frac{1}{12z} + \frac{1}{288z^2} + O\left(\frac{1}{z^3}\right) \right]$$

However, when solving particular problems, we usually find that the first term is the most important one.

Since $\Gamma(z+1) = z\Gamma(z)$, we obtain $\Gamma(x) \sim e^{-x}x^{x-1/2}(2\pi)^{1/2}$, mentioned earlier.

Bessel Functions

The remainder of this section will include a brief discussion of some basic properties of Bessel functions and will describe a way in which the method of stationary phase can be applied to obtain an asymptotic formula. Bessel functions (the main properties of which are listed in Table 7.3.1) are studied because they arise naturally in solutions to certain

Table 7.3.1   Summary of Properties of Bessel Functions

1. $J_n(z) = \dfrac{1}{\pi} \displaystyle\int_0^\pi \cos(n\theta - z\sin\theta)\, d\theta$, where n is an integer.

2. $|J_n(z)| \le 1$ for z real.

3. $J_n(z) = \displaystyle\sum_{k=0}^\infty \dfrac{(-1)^k z^{n+2k}}{2^{n+2k} k!(n+k)!}$, $n \ge 0$.

4. J_n is entire and has a zero of order n at $z = 0$; $J_0(0) = 1$.

5. $J_n(z) = (-1)^n J_{-n}(z)$.

6. Bessel's equation:

$$\frac{d^2 J_n}{dz^2} + \frac{1}{z}\frac{dJ_n}{dz} + \left(1 - \frac{n^2}{z^2}\right) J_n = 0$$

7. $\dfrac{d}{dz}[z^{-n} J_n(z)] = -z^{-n} J_{n+1}(z)$.

8. $\dfrac{d}{dz} J_n(z) = \dfrac{n}{z} J_n(z) - J_{n+1}(z)$.

9. $\dfrac{d}{dz} J_n(z) = \dfrac{J_{n-1}(z) - J_{n+1}(z)}{2}$

10. $J_n(z) \sim \sqrt{\dfrac{2}{\pi z}} \cos\left(z - \dfrac{n\pi}{2} - \dfrac{\pi}{4}\right)$ as $z \to \infty$, z real and greater than zero.

partial differential equations, such as Laplace's equation, when these equations are expressed in terms of cylindrical coordinates. Bessel functions can be defined in several different ways. We will find the following definition convenient.

7.3.2 DEFINITION   *Let $z \in \mathbf{C}$ be fixed and consider the function*

$$f(\zeta) = e^{z(\zeta - 1/\zeta)/2}$$

*Expand $f(\zeta)$ in a Laurent series around 0. The coefficient of ζ^n where n is positive or negative is denoted $J_n(z)$ and is called the **Bessel function** of order n. We call $e^{z(\zeta - 1/\zeta)/2}$ the **generating function**.*

To begin the development of these properties we rewrite the definition as,

$$e^{z(\zeta - 1/\zeta)/2} = \sum_{n=-\infty}^{\infty} J_n(z)\zeta^n \tag{3}$$

From the formula for the coefficients of a Laurent expansion (see Theorem 3.3.1), we see that

$$J_n(z) = \frac{1}{2\pi i} \int_\gamma \zeta^{-n-1} e^{z(\zeta - 1/\zeta)/2} \, d\zeta$$

where γ is any circle around 0. If we use the unit circle $\zeta = e^{i\theta}$ and write out the integral explicitly, we get

$$J_n(z) = \frac{1}{2\pi} \int_0^{2\pi} e^{-(n+1)i\theta} e^{iz\sin\theta} e^{i\theta} \, d\theta$$

$$= \frac{1}{2\pi} \int_0^{\pi} e^{iz\sin\theta - ni\theta} \, d\theta + \frac{1}{2\pi} \int_0^{\pi} e^{-iz\sin\theta + ni\theta} \, d\theta$$

$$= \frac{1}{\pi} \int_0^{\pi} \cos(n\theta - z\sin\theta) \, d\theta \tag{4}$$

Although $J_n(z)$ will be defined for noninteger values of n in Eq. (10), Eq. (4) is valid only if n is an integer. Equation (4) shows that $|J_n(z)| \le 1$ for z real. The graphs of $J_0(x)$ and $J_1(x)$ are shown in Fig. 7.3.3.

Next it will be shown that $J_n(z)$ is entire and a method will be described for finding its power series. To carry out these tasks, it is convenient to change variables by $\zeta \mapsto 2\zeta/z$

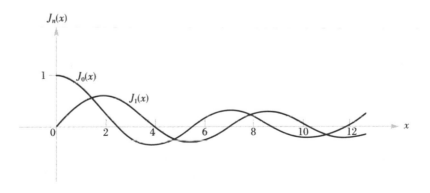

FIGURE 7.3.3 Bessel functions $J_0(x)$, $J_1(x)$.

and obtain, for each fixed z,

$$J_n(z) = \frac{1}{2\pi i} \left(\frac{z}{2}\right)^n \int_\gamma \zeta^{-n-1} \exp\left(\zeta - \frac{z^2}{4\zeta}\right) d\zeta \tag{5}$$

Writing the exponential as a power series gives

$$\exp\left(\frac{-z^2}{4\zeta}\right) = 1 - \frac{z^2}{4\zeta} + \frac{z^4}{2 \cdot (4\zeta)^2} - \cdots$$

This series converges uniformly in ζ on γ (Why?), so we can integrate Eq. (5) term by term (see Theorem 3.1.9) and obtain

$$J_n(z) = \frac{1}{2\pi i} \sum_{k=0}^\infty \frac{(-1)^k}{k!} \left(\frac{z}{2}\right)^{n+2k} \int_\gamma \frac{e^\zeta}{\zeta^{n+k+1}} d\zeta$$

If $n \geq 0$, the residue of e^ζ/ζ^{n+k+1} at $\zeta = 0$ is $1/(n+k)!$ (Why?), and so

$$J_n(z) = \sum_{k=0}^\infty \frac{(-1)^k (z)^{n+2k}}{2^{n+2k} k!(n+k)!}$$

$$= \frac{z^n}{2^n n!} \left[1 - \frac{z^2}{2^2 \cdot 1(n+1)} + \frac{z^4}{2^4 \cdot 1 \cdot 2(n+1)(n+2)} - \cdots \right] \tag{6}$$

Thus $J_n(z)$ is entire for $n \geq 0$ and has a zero of order n at $z = 0$.
Similarly, for $n \leq 0$, one finds that

$$J_n(z) = \sum_{k=0}^\infty \frac{(-1)^{k-n} z^{-n+2k}}{2^{-n+2k}(k-n)!k!} \tag{7}$$

(see Exercise 11). It follows that for $n \leq 0$,

$$J_n(z) = (-1)^n J_{-n}(z) \tag{8}$$

The relationship of Bessel functions to differential equations is as follows: $J_n(z)$ is a solution of *Bessel's equation:*

$$\frac{d^2 J_n}{dz^2} + \frac{1}{z} \frac{dJ_n}{dz} + \left(1 - \frac{n^2}{z^2}\right) J_n = 0 \tag{9}$$

Equation (9) is obtained by differentiating either Eq. (4) or Eq. (5) for $J_n(z)$ and inserting the result into Bessel's equation (see Exercise 1). Note that both J_n and J_{-n} satisfy Eq. (9).

If $n \geq 0$ but n is not an integer, we can still make sense of $J_n(z)$ in Eq. (6) by setting

$$J_n(z) = \sum_{k=0}^{\infty} \frac{(-1)^k z^{n+2k}}{2^{n+2k} k! \Gamma(n+k+1)} \tag{10}$$

Some basic identities can be obtained from the following relations:

$$\frac{d}{dz}[z^{-n}J_n(z)] = -z^{-n}J_{n+1}(z) \qquad \text{if} \qquad z \neq 0 \tag{11}$$

This can be proven directly by differentiating the power series. (Term-by-term differentiation is, of course, valid—see Sec. 3.1 and 3.2.) The student is requested to establish such a proof in Exercise 4.

In identity (11) we differentiate $z^{-n}J_n(z)$ to obtain

$$\frac{d}{dz}J_n(z) = \frac{n}{z}[J_n(z)] - J_{n+1}(z) \tag{12}$$

Writing $-n$ for n in Eq. (11), we get

$$\frac{d}{dz}[z^n J_{-n}(z)] = -z^n J_{-n+1}(z)$$

But $J_{-n}(z) = (-1)^n J_n(z)$, and so

$$\frac{d}{dz}[z^n J_n(z)] = z^n J_{n-1}(z)$$

that is,

$$\frac{d}{dz}J_n(z) = J_{n-1}(z) - \frac{n}{z}J_n(z) \tag{13}$$

Combining Eqs. (12) and (13), we get

$$\frac{d}{dz}J_n(z) = \frac{1}{2}[J_{n-1}(z) - J_{n+1}(z)] \tag{14}$$

Relations (11), (12), and (14) are called the *recurrence relations* for Bessel functions. For example, if we know J_n and J_{n-1}, Eq. (14) recursively determines J_{n+1}.

Our study is concluded with the asymptotic formula for $J_n(z)$.

7.3.3 ASYMPTOTIC FORMULA FOR BESSEL FUNCTIONS   *The following formula holds for any integer n:*

$$J_n(z) \sim \sqrt{\frac{2}{\pi z}} \left[\cos\left(z - \frac{n\pi}{2} - \frac{\pi}{4} \right) \right] \tag{15}$$

as $z \to \infty$, z real and greater than zero. (Equation (15) is also valid for $|\arg z| < \pi$.)

PROOF   We use the stationary phase theorem (7.2.10) and the representation

$$J_n(z) = \frac{1}{2\pi} \left(\int_0^\pi e^{iz\sin\theta - ni\theta} \, d\theta + \int_0^\pi e^{-iz\sin\theta + ni\theta} \, d\theta \right) \tag{16}$$

obtained in Eq. (4). First, let us consider the function

$$f(z) = \int_0^\pi e^{iz\sin\theta - ni\theta} \, d\theta$$

In the notation of the stationary phase theorem, $h(t) = \sin t$ and $g(t) = e^{-int}$. Clearly h is analytic and real for real t, and g is C^1. The interval $[a, b]$ is $[0, \pi]$, and $h'(t) = \cos t$ vanishes only at $t_0 = \pi/2$. At this point, $h''(t_0) = -\sin(\pi/2) = -1 < 0$. Thus, we use the minus sign in the asymptotic formula for f, giving

$$f(z) \sim \frac{e^{iz}\sqrt{2\pi}e^{-\pi i/4}}{\sqrt{z}} \cdot e^{-ni\pi/2} = \sqrt{\frac{2\pi}{z}} \, e^{i(z - n\pi/2 - \pi/4)} \tag{17}$$

Similarly, if we set $g(z) = \int_0^\pi e^{-iz\sin\theta + in\theta} \, d\theta$, we get

$$g(z) \sim \sqrt{\frac{2\pi}{z}} \, e^{-i(z - n\pi/2 - \pi/4)} \tag{18}$$

(Proof of this is requested in Exercise 9.) Adding Eqs. (17) and (18), we obtain, from Eq. (16),

$$J_n(z) \sim \left(\frac{1}{2\pi} \right) \sqrt{\frac{2\pi}{z}} \cdot 2 \cos\left(z - \frac{n\pi}{2} - \frac{\pi}{4} \right)$$

which is the result claimed.   ∎

Thus, for large x, $J_n(x)$ behaves like $\sqrt{2/\pi x} \, [\cos(x - \theta)]$, where θ is a certain angle called the *phase shift*.

Exercises

1. Prove that $J_n(z)$ satisfies Bessel's equation (formula 6 of Table 7.3.1).

2. Show that

$$\Gamma(z+1) = \sqrt{2\pi} z^{z+1/2} e^{-z} \left[1 + O\left(\frac{1}{z}\right) \right]$$

and that

$$\Gamma(z+1) = \sqrt{2\pi} z^{z+1/2} e^{-z} \left[1 + \frac{1}{12z} + O\left(\frac{1}{z^2}\right) \right]$$

as $z \to \infty$. (z is real and greater than zero.)

3. Prove that $J_0'(z) = -J_1(z)$ by using Eq. (4).

4. Prove that $d[z^{-n} J_n(z)]/dz = -z^{-n} J_{n+1}(z)$, for all n.

5. Prove that $J_2(z) - J_0''(z) = J_0'(z)/z$.

6. Use the recurrence relations for Bessel functions and Rolle's theorem from calculus to show that between two consecutive real positive zeros of $J_n(x)$, there is exactly one zero of $J_{n+1}(x)$. Show that $J_n(x)$ and $J_{n+1}(x)$ have no common roots.

7. Prove that $J_{1/2}(z) = \sqrt{2/\pi z} \,(\sin z)$, using the definition of $J_n(z)$ for nonintegral n given by Eq. (10).

8. Verify that the asymptotic expansion for $J_n(z)$ is consistent with Bessel's equation.

9. Complete the proof that

$$J_n(z) \sim \sqrt{\frac{2}{\pi z}} \cos\left(z - \frac{n\pi}{2} - \frac{\pi}{4}\right)$$

by showing that

$$\int_0^\pi e^{-iz\sin\theta + ni\theta} \, d\theta \sim \sqrt{\frac{2\pi}{z}} \, e^{-i(z - n\pi/2 - \pi/4)}$$

(The function on the left side of the expression is called a *Hankel function*.)

10. Verify that $\phi(x) = J_n(kx)$ is a solution of

$$\frac{d}{dx}(x\phi') + \left(k^2 x - \frac{n^2}{x}\right)\phi(x) = 0$$

with $\phi(0) = 0$, $\phi(a) = 0$, where ka is any of the zeros of J_n, $n \neq 0$.

11. Establish Eqs. (7) and (8).

REVIEW EXERCISES FOR CHAPTER 7

1. Establish the convergence of and evaluate the infinite product

$$\prod_1^\infty \left(1 + \frac{1}{n(n+2)}\right)$$

2. Establish the convergence of and evaluate the infinite product

$$\prod_1^\infty \left(\frac{n^2 + 3n + 2}{n^2 + 3n}\right)$$

3. Use Worked Example 7.1.10 to show that

$$\sqrt{2} = (\tfrac{3}{2})(\tfrac{5}{6})(\tfrac{7}{6})(\tfrac{9}{10})(\tfrac{11}{10})(\tfrac{13}{14}) \cdots$$

4. Use Worked Example 7.1.10 to show that

$$\sqrt{3} = 2(\tfrac{4}{5})(\tfrac{8}{7})(\tfrac{10}{11})(\tfrac{14}{13})(\tfrac{16}{17})(\tfrac{20}{19}) \cdots$$

5. On what region is each of the following absolutely convergent?

(a) $\displaystyle\prod_1^\infty (1 - z^n)$ (b) $\displaystyle\prod_1^\infty (1 - n^{-z})$

6. Let $H_m = \sum_{n=1}^\infty 1/n^m$. Prove that

$$\log \Gamma(1 + z) = -\gamma z + \sum_{n=2}^\infty \frac{(-1)^n}{n} H_n z^n \qquad \text{for} \qquad |z| < 1$$

where γ is Euler's constant.

7. Let $f(z) = \int_0^\pi e^{iz\sin t} \sin^2 t \, dt$. Use the method of stationary phase to give an asymptotic formula for $f(z)$ as $z \to \infty$, z real and positive.

8. Prove that

$$\Gamma(z) = \sum_{n=0}^{\infty} \frac{(-1)^n}{n!(z+n)} + Q(z)$$

where $Q(z)$ is entire. In addition, show that

$$Q(z) = \int_{1}^{\infty} t^{z-1} e^{-t}\, dt$$

9. Write out $\int_0^{\infty} e^{-t^2}\, dt$ in terms of the gamma function. (*Hint.* Change to the variable $y = t^3$.)

10. Show that $|d^k J_n(z)/dz^k| \leq 1$, $k = 0, 1, 2, \ldots$, for any n, z real.

11. Prove that $\Gamma|\tfrac{1}{2} + iy| \to 0$ as $y \to \infty$.

12. Show that $\displaystyle\lim_{x \to \infty} J_n(x) = 0$.

13. Obtain an asymptotic expansion for $\int_{-\infty}^{\infty} e^{-zy^2/2} \cos y^2\, dy$ (as $z \to \infty$, $z > 0$).

14. Prove that $x^n J_n(x) - \int_0^x t^n J_{n-1}(t)\, dt$, $n = 1, 2, \ldots$.

15. Prove that $J_n(iy) \sim i^n e^y/\sqrt{2\pi y}$ (as $y \to \infty$, $y > 0$).

16. In this exercise you are asked to develop some properties of the *Legendre functions* (see Review Exercise 34, Chap. 3). These functions are encountered in the study of differential equations (specifically Laplace's equation in three dimensions, which describes a wide range of physical phenomena) when spherical coordinates arc used.[*]

(a) For $-1 < x < 1$, set

$$P_n(x) = \frac{1}{2^{n+1}\pi i} \int_{\gamma} \frac{(t^2 - 1)^n}{(t - x)^{n+1}}\, dt$$

where γ is the contour as shown in Fig. 7.R.1. By differentiating under the integral sign (see Worked Example 2.4.15), show that $P_n(x)$ solves *Legendre's equation:*

$$(1 - x^2)y'' - 2xy' + n(n + 1)y = 0$$

---

[*] Consult, for example, G. F. D. Duff and D. Naylor, *Differential Equations of Applied Mathematics* (New York: Wiley, 1965).

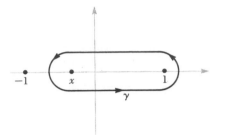

FIGURE 7.R.1   Contour for the Legendre function.

(b) For an integer n, derive *Rodrigues' formula:*

$$P_n(x) = \frac{1}{2^n n!} \frac{d^n}{dx^n} (x^2 - 1)^n$$

This formula yields an obvious analytic extension $P_n(z)$.

(c) Show that

$$P_n(x) = \frac{1}{n!} \frac{d^n}{dt^n} \frac{1}{(t^2 - 2tx + 1)^{1/2}} \Big|_{t=0}$$

and deduce from Taylor's theorem that

$$\frac{1}{(1 - 2tz + t^2)^{1/2}} = \sum_{n=0}^{\infty} P_n(z) t^n$$

(d) Develop recurrence relations for the coefficients of solutions of Legendre's equation. Then use these relations to show that entire solutions must be of the form

$$\psi(x) = \sum_{k=0}^{\infty} a_k x^{2k}, \quad n \text{ even}$$

or of the form

$$\rho(x) = \sum_{k=0}^{\infty} b_k x^{2k+1}, \quad n \text{ odd}$$

Show that these are actually polynomials, that is, that a_k and b_k vanish for large k.

(e) Using (c), show that $nP_n(x) = (2n - 1)xP_{n-1}(x) - (n - 1)P_{n-2}(x)$.

(f) Prove that $P_0(x) = 1$, $P_1(x) = x$, $P_2(x) = (3x^2 - 1)/2$.

(g) Show that

$$\int_{-1}^{1} P_n(x)P_m(x)\, dx = \begin{cases} 0 & n \neq m \\ \dfrac{2}{(2n+1)} & n - m \end{cases}$$

(*Hint.* Use (b) to prove the case where $n \neq m$; use (c) to prove the case where $n = m$.)

17. Obtain the asymptotic formula $P_n(z) \sim [(2n)!/2^n(n!)^2]z^n$ as $z \to \infty$, using part (b) of Exercise 16.

CHAPTER 8

THE LAPLACE TRANSFORM
AND APPLICATIONS

This final chapter gives an introduction to the Laplace transform and some of its applications. The first section introduces two key properties that make the Laplace transform useful for differential equations: First, it behaves well with respect to differentiation, and second, a function can be recovered if its Laplace transform is known. The closely related Fourier transform also enjoys these properties. It was discussed in Sec. 4.3; see also the supplement to Sec. 8.3.

Section 8.2 develops techniques for inverting Laplace transforms; Sec. 8.3 considers some applications of Laplace transforms to ordinary differential equations.

8.1 BASIC PROPERTIES OF LAPLACE TRANSFORMS

The Laplace transform is a powerful tool that is used in both pure and applied mathematics. It is important, therefore, to have a good grasp of both its basic theory and its usefulness. Consider a (real- or complex-valued) function $f(t)$ defined on $[0, \infty[$. The *Laplace transform* of f is defined to be the function $\tilde{f}$ of a complex variable z given by

$$\tilde{f}(z) = \int_0^\infty e^{-zt}f(t)\, dt \qquad (1)$$

$\tilde{f}$ will be defined for those $z \in \mathbf{C}$ for which the integral converges. Other common notations for $\tilde{f}$ are $\mathcal{L}(f)$ or simply F.

For technical reasons, it will be convenient to impose a mild restriction on the functions we consider. We will want $f: [0, \infty[\to \mathbf{C}$ (or $\mathbf{R}$) to be of *exponential order.* This means that there should be constants $A > 0$, $B \in \mathbf{R}$, such that

$$|f(t)| \leq Ae^{tB} \qquad (2)$$

for all $t \geq 0$. In other words, f should not grow too fast; for example, any polynomial satisfies this condition (Why?). *All functions considered in the remainder of this chapter will be assumed to be of exponential order.* It will also be assumed that on any finite interval $[0, a]$, f is bounded and integrable. (If, for example, we assume that f is piecewise continuous, this last condition will hold.)

Abscissa of Convergence

The first important result in this chapter concerns the nature of the set on which $\tilde{f}(z)$ is defined and is analytic.

8.1.1 CONVERGENCE THEOREM FOR LAPLACE TRANSFORMS *Let $f: [0, \infty[\to \mathbf{C}$ (or $\mathbf{R}$) be of exponential order and let*

$$\tilde{f}(z) = \int_0^\infty e^{-zt}f(t)\, dt$$

There exists a unique number σ, $-\infty \le \sigma < \infty$, such that

$$\int_0^\infty e^{-zt}f(t)\, dt \begin{cases} converges\ if\ \mathrm{Re}\ z > \sigma \\ diverges\ if\ \mathrm{Re}\ z < \sigma \end{cases} \tag{3}$$

Furthermore, $\tilde{f}$ is analytic on the set $A = \{z \mid \mathrm{Re}\ z > \sigma\}$ and we have

$$\frac{d}{dz}\tilde{f}(z) = -\int_0^\infty te^{-zt}f(t)\, dt \tag{4}$$

*for $\mathrm{Re}\ z > \sigma$. The number σ is called the **abscissa of convergence**, and if we define the number ρ by*

$$\rho = \inf\{B \in \mathbf{R} \mid there\ exists\ an\ A > 0\ such\ that\ |f(t)| \le Ae^{Bt}\} \tag{5}$$

then $\sigma \le \rho$.

The set $\{z \mid \mathrm{Re}\ z > \sigma\}$ is called the *half plane of convergence*. (If $\sigma = -\infty$, it is all of $\mathbf{C}$.) See Fig. 8.1.1. In general, it is difficult to tell whether $\tilde{f}(z)$ will converge on the vertical line $\mathrm{Re}\ z = \sigma$.

The proof of this theorem and more detailed convergence results are given in the supplement at the end of the section.

If there is any danger of confusion we can write $\sigma(f)$ for σ or $\rho(f)$ for ρ. A convenient way to compute $\sigma(f)$ is described in Worked Examples 8.1.12 and 8.1.13. The map $f \mapsto \tilde{f}$ is clearly linear in the sense that $(af + bg)\tilde{} = a\tilde{f} + b\tilde{g}$, valid for $\mathrm{Re}\ z > \max[\sigma(f), \sigma(g)]$. It is also true that the map is one-to-one; that is, $\tilde{f} = \tilde{g}$ implies that $f = g$; a function $\phi(z)$ is the Laplace transform of at most one function. Precisely:

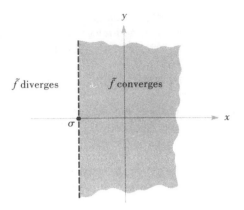

FIGURE 8.1.1   Half plane of convergence of the Laplace transform.

8.1.2 UNIQUENESS THEOREM FOR LAPLACE TRANSFORMS   *Suppose that f and h are continuous and that $\tilde{f}(z) = \tilde{h}(z)$ for Re $z > y_0$ for some y_0. Then $f(t) = h(t)$ for all $t \in [0, \infty[$.*

This theorem is not as simple as it seems. There is not enough machinery to give a complete proof, but the main ideas of a proof are given at the end of the section. Using ideas from integration theory, we could extend the result of the uniqueness theorem to discontinuous functions as well, but we would have to modify what we mean by "equality of functions." For example, if $f(t)$ is changed at a single value of t, $\tilde{f}$ is unchanged.

The uniqueness theorem enables us to give a meaningful answer to the problem, "Given $g(z)$, find $f(t)$ such that $\tilde{f} = g$," because it makes clear that there can be at most one such (continuous) f. We call f the *inverse Laplace transform* of g; methods for finding f when g is given are considered in the next section.

Laplace Transforms of Derivatives

The main utility of Laplace transforms is that they enable us to transform differential problems into algebraic problems. When the latter are solved, the answers to the original problems are obtained by using the inverse Laplace transform. The procedure is based on the following theorem.

8.1.3 PROPOSITION   *Let $f(t)$ be continuous on $[0, \infty[$ and piecewise C^1, that is, piecewise continuously differentiable. Then for Re $z > \rho$ (as defined in the convergence theorem (8.1.1)),*

$$\left(\frac{df}{dt}\right)^{\tilde{}}(z) = z\tilde{f}(z) - f(0) \tag{6}$$

PROOF By definition,

$$\left(\frac{df}{dt}\right)^{\sim}(z) = \int_0^{\infty} e^{-zt} \frac{df}{dt}(t)\, dt$$

Integrating by parts, we get

$$\lim_{t_0 \to \infty}\left(e^{-zt}f(t)\Big|_0^{t_0}\right) + \int_0^{\infty} ze^{-zt}f(t)\, dt$$

By definition of ρ, $|e^{-Bt_0} \cdot f(t_0)| \le A$ for some $B < \mathrm{Re}\, z$. Thus we get $|e^{-zt_0} \cdot f(t_0)| = |e^{-(z-B)t_0}||e^{-Bt_0} \cdot f(t_0)| \le e^{-(\mathrm{Re}\, z - B)t_0}A$, which approaches 0 as $t_0 \to \infty$. Therefore, we obtain $-f(0) + z\tilde{f}(z)$, as asserted. ∎

Note that we have proved that $(df/dt)^{\sim}(z)$ exists for $\mathrm{Re}\, z > \rho$, although its abscissa of convergence might be smaller than ρ.

If we apply Eq. (6) to d^2f/dt^2, we obtain

$$\left(\frac{d^2f}{dt^2}\right)^{\sim}(z) = z^2\tilde{f}(z) - zf(0) - \frac{df}{dt}(0) \qquad (7)$$

Equation (4) of the convergence theorem (8.1.1) is the related formula $\tilde{g}(z) = d\tilde{f}(z)/dz$, where $g(t) = -tf(t)$. In Exercise 19 the student is asked to prove the next proposition, which contains a similar formula for integrals.

8.1.4 PROPOSITION *Let* $g(t) = \displaystyle\int_0^t f(\tau)\, d\tau$. *Then for* $\mathrm{Re}\, z > \max[0, \rho(f)]$,

$$\tilde{g}(z) = \frac{\tilde{f}(z)}{z} \qquad (8)$$

Shifting and Convolution Theorems

Table 8.1.1 at the end of this section lists some formulas that are useful for computing $\tilde{f}(z)$. The proofs of these formulas are straightforward and are included in the exercises and examples. However, three of the formulas are sufficiently important to be given separate explanation, which is done in the following three theorems.

8.1.5 FIRST SHIFTING THEOREM *Fix* $a \in \mathbf{C}$ *and let* $g(t) = e^{-at}f(t)$. *Then for* $\mathrm{Re}\, z > \sigma(f) - \mathrm{Re}\, a$, *we have*

$$\tilde{g}(z) = \tilde{f}(z + a) \qquad (9)$$

PROOF   By definition

$$\tilde{g}(z) = \int_0^\infty e^{-zt}e^{-at}f(t)\, dt = \int_0^\infty e^{-(z+a)t}f(t)\, dt = \tilde{f}(z+a)$$

which is valid if Re $(z + a) > \sigma$. ∎

8.1.6 SECOND SHIFTING THEOREM   *Let*

$$H(t) = \begin{cases} 0 & \text{if } t < 0 \\ 1 & \text{if } t \ge 0 \end{cases}$$

*(This is called the **Heaviside**, or **unit step function**.) Also, let $a \ge 0$ and let $g(t) = f(t-a)H(t-a)$; that is,*

$$g(t) = \begin{cases} 0 & \text{if } t < a \\ f(t-a) & \text{if } t \ge a \end{cases}$$

(see Fig. 8.1.2). Then for Re $z > \sigma$, we have

$$\tilde{g}(z) = e^{-az}\tilde{f}(z) \tag{10}$$

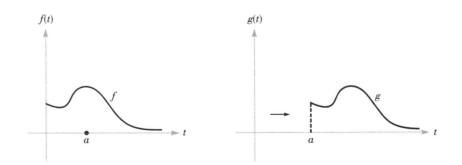

FIGURE 8.1.2   The function g in the second shifting theorem.

PROOF   By definition and because $g = 0$ for $0 \le t < a$,

$$\tilde{g}(z) = \int_0^\infty e^{-zt}g(t)\, dt = \int_a^\infty e^{-zt}f(t-a)\, dt$$

Letting $\tau = t - a$, we get

$$\tilde{g}(z) = \int_0^\infty e^{-z(\tau+a)}f(\tau)\,d\tau = e^{-za}\tilde{f}(z) \qquad \blacksquare$$

From Eq. (10) we can deduce that if $a \geq 0$ and $g(t) = f(t)H(t - a)$, then $\tilde{g}(z) = e^{-az}\tilde{F}(z)$ where $F(t) = f(t + a)$, $t \geq 0$ (see Fig. 8.1.3).

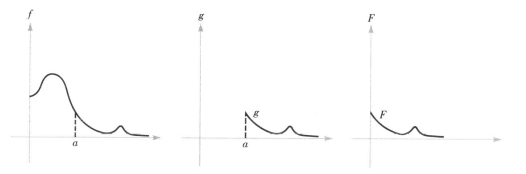

FIGURE 8.1.3 F is obtained from f by shifting and truncating.

The *convolution* of functions $f(t)$ and $g(t)$ is defined by

$$(f * g)(t) = \int_0^\infty f(t - \tau) \cdot g(\tau)\,d\tau \qquad \text{for} \qquad t \geq 0 \tag{11}$$

where we set $f(t) = 0$ if $t < 0$. (Thus the integration is really only from 0 to t.) The convolution operation is related to Laplace transforms in the following way.

8.1.7 CONVOLUTION THEOREM   *We have $f * g = g * f$ and*

$$(f * g)^{\sim}(z) = \tilde{f}(z) \cdot \tilde{g}(z) \tag{12}$$

whenever Re $z > $ max $[\rho(f), \rho(g)]$.

In brief, Eq. (12) states that the Laplace transform of a convolution of two functions is the product of their Laplace transforms. It is precisely this property that makes the convolution an operation of interest to us.

PROOF   We have

$$(f * g)^{\sim}(z) = \int_0^\infty e^{-zt} \left[\int_0^\infty f(t - \tau) \cdot g(\tau) \, d\tau \right] dt$$

$$= \int_0^\infty e^{-z\tau} e^{-z(t-\tau)} \int_0^\infty f(t - \tau) g(\tau) \, d\tau \, dt$$

For Re $z > \max [\rho(f), \rho(g)]$ the integrals for $\tilde{f}(z)$ and $\tilde{g}(z)$ converge absolutely, so we can interchange the order of integration† and thus obtain

$$\int_0^\infty e^{-z\tau} \left[\int_0^\infty e^{-z(t-\tau)} f(t - \tau) \, dt \right] g(\tau) \, d\tau$$

Letting $s = t - \tau$ and remembering that $f(s) = 0$ if $s < 0$, we get

$$\int_0^\infty e^{-z\tau} \tilde{f}(z) g(\tau) \, d\tau = \tilde{f}(z) \cdot \tilde{g}(z)$$

By changing variables, it is not difficult to verify that $f * g = g * f$, but such verification also follows from what we have done if f and g are continuous. We have $(f * g)^{\sim} = \tilde{f} \cdot \tilde{g} = \tilde{g} \cdot \tilde{f} = (g * f)^{\sim}$. Thus $(f * g - g * f)^{\sim} = 0$, so by the uniqueness theorem (8.1.2), $f * g - g * f = 0$. ∎

TECHNICAL PROOFS OF THEOREMS

To prove the convergence theorem (8.1.1), we shall use the following important result.

8.1.8 LEMMA   *Suppose that $\tilde{f}(z) = \int_0^\infty e^{-zt} f(t) \, dt$ converges for $z = z_0$. Assume that $0 \le \theta < \pi/2$ and define the set*

$$S_\theta = \{z \text{ such that } |\arg(z - z_0)| \le \theta\}$$

(see Fig. 8.1.4). Then $\tilde{f}$ converges uniformly on S_θ

---

† This is a theorem concerning integration theory from advanced calculus. See, for instance, J. Marsden, *Elementary Classical Analysis* (New York: W. H. Freeman and Company, 1974), Chap. 9.

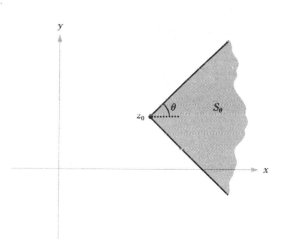

FIGURE 8.1.4 Sector of uniform convergence.

PROOF Let $h(x) = \int_0^x e^{-z_0 t} f(t)\, dt - \int_0^\infty e^{-z_0 t} f(t)\, dt$, so that $h \to 0$ as $x \to \infty$. We must show that for every $\epsilon > 0$, there is a t_0 such that $t_1, t_2 \geq t_0$ implies that

$$\left| \int_{t_1}^{t_2} e^{-zt} f(t)\, dt \right| < \epsilon$$

for all $z \in S_\theta$. It follows that $\int_0^x e^{-zt} f(t)\, dt$ converges uniformly on S_θ as $x \to \infty$, by the Cauchy criterion. We will make use of the function $h(x)$ as follows. Write

$$\int_{t_1}^{t_2} e^{-zt} f(t)\, dt = \int_{t_1}^{t_2} e^{-(z - z_0)t} [e^{-z_0 t} f(t)]\, dt$$

If we integrate by parts, we get (as the student can easily check)

$$e^{-(z - z_0)t_2} h(t_2) - e^{-(z - z_0)t_1} h(t_1) + (z - z_0) \int_{t_1}^{t_2} e^{-(z - z_0)t} h(t)\, dt$$

Given $\epsilon > 0$, choose t_0 such that $|h(t)| < \epsilon/3$ and $|h(t)| < \epsilon' = \epsilon/(6 \sec \theta)$ if $t \geq t_0$. Then for $t_2 > t_0$,

$$|e^{-(z - z_0)t_2} h(t_2)| \leq |h(t_2)| < \frac{\epsilon}{3}$$

since $|e^{-(z - z_0)t_2}| = e^{-(\operatorname{Re} z - \operatorname{Re} z_0)t_2} \leq 1$ because $\operatorname{Re} z \geq \operatorname{Re} z_0$. Similarly, for $t_1 > t_0$,

$$|e^{-(z - z_0)t_1} h(t_1)| < \frac{\epsilon}{3}$$

We must still estimate the last term:

$$\left| (z - z_0) \int_{t_1}^{t_2} e^{-(z-z_0)t} h(t) \, dt \right| \leq |z - z_0| \epsilon' \int_{t_1}^{t_2} e^{-(x-x_0)t} \, dt$$

where $x = \text{Re } z$ and $x_0 = \text{Re } z_0$. If $z = z_0$, this term is zero. If $z \neq z_0$, then $x \neq x_0$, and we get

$$\epsilon' \frac{|z - z_0|}{x - x_0} (e^{-(x-x_0)t_1} - e^{-(x-x_0)t_2}) < 2\epsilon' \frac{|z - z_0|}{x - x_0} \leq 2\epsilon' \sec \theta = \frac{\epsilon}{3}$$

(see Fig. 8.1.5). Note that the restriction $0 \leq \theta < \pi/2$ is necessary for $\sec \theta = 1/\cos \theta$ to be finite.

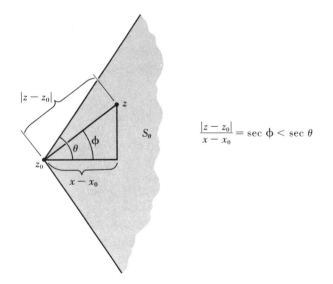

$$\frac{|z - z_0|}{x - x_0} = \sec \phi < \sec \theta$$

FIGURE 8.1.5   Some geometry in the region S_θ.

Combining the preceding inequalities, we get

$$\left| \int_{t_1}^{t_2} e^{-zt} f(t) \, dt \right| < \epsilon$$

if $t_1, t_2 \geq t_0$ for all $z \in S_\theta$, thus completing the proof of the lemma. ▼

PROOF OF THE CONVERGENCE THEOREM Let $\sigma = \inf\{x \in \mathbf{R} \mid \int_0^\infty e^{-xt}f(t)\, dt$ converges$\}$, where inf stands for "greatest lower bound." We note from Lemma 8.1.8 that if $\tilde{f}(z_0)$ converges, then, more specifically, $\tilde{f}(z)$ converges if Re $z >$ Re z_0 because z lies in some S_θ for z_0 (Why?).

Let Re $z > \sigma$. By the definition of σ there is an $x_0 <$ Re z such that $\int_0^\infty e^{-x_0 t}f(t)\, dt$ converges. Hence $\tilde{f}(z)$ converges by Lemma 8.1.8. Conversely, assume Re $z < \sigma$ and Re $z < x < \sigma$. If $\tilde{f}(z)$ converges, then so does $\tilde{f}(x)$, and therefore to say that $\sigma \leq x$ is a contradiction. Thus $\tilde{f}(z)$ does not converge if Re $z < \sigma$.

Next, using the analytic convergence theorem (Sec. 3.1), we prove that $\tilde{f}$ is analytic on $\{z \mid$ Re $z > \sigma\}$. Let $g_n(z) = \int_0^n e^{-zt}f(t)\, dt$. Then $g_n(z) \to \tilde{f}(z)$. By Worked Example 2.4.15, g_n is analytic with $g_n'(z) = -\int_0^n te^{-zt}f(t)\, dt$. We must show that $g_n \to \tilde{f}$ uniformly on closed disks in $\{z \mid$ Re $z > \sigma\}$. But each disk lies in some S_θ relative to some z_0 with Re $z_0 > \sigma$ (Fig. 8.1.6). This statement is geometrically clear and can also be checked analytically.

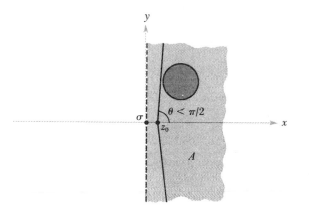

FIGURE 8.1.6 Each disk lies in S_θ for some θ, $0 \leq \theta < \pi/2$.

Thus, by the analytic convergence theorem (3.1.8), $\tilde{f}$ is analytic on $\{z \mid$ Re $z > \sigma\}$ and

$$(\tilde{f})'(z) = -\int_0^\infty te^{-zt}f(t)\, dt$$

We know automatically that this integral representation for the derivative of $\tilde{f}$ will converge for Re $z > \sigma$, as will all the iterated derivatives.

It remains to be shown that $\sigma \leq \rho$. To prove this we merely need to show that $\sigma \leq B$ if $|f(t)| \leq Ae^{Bt}$. This will hold, by what we have proven, if $\tilde{f}(z)$ converges whenever Re $z > B$. Indeed we shall show absolute convergence. Note that $|e^{-zt}f(t)| -$

$|e^{-(z-B)t}e^{-Bt}f(t)| \le e^{-(\text{Re} z - B)t}A$. Since $\int_0^\infty e^{-\alpha t}\, dt = 1/\alpha$ converges for $\alpha > 0$, $\int_0^\infty e^{-zt}f(t)\, dt$ converges absolutely. ∎

To prove that $\tilde{f} = \tilde{h}$ implies that $f = h$ for continuous functions f and h, it suffices, by considering $f - h$, to prove the following special case of theorem 8.1.2.

8.1.9 PROPOSITION   *Suppose that f is continuous and that $\tilde{f}(z) = 0$ whenever Re $z > y_0$ for some y_0. Then $f(t) = 0$ for all $t \in [0, \infty[$.*

The crucial lemma we use to prove this is the following.

8.1.10 LEMMA   *Let f be continuous on $[0, 1]$ and suppose that $\int_0^1 t^n f(t)\, dt = 0$ for all $n = 0, 1, 2, \ldots$. Then $f = 0$.*

This assertion is reasonable since it follows that $\int_0^1 P(t)f(t)\, dt = 0$ for any polynomial P.

PROOF   The precise proof depends on the *Weierstrass approximation theorem*, which states that any continuous function is the uniform limit of polynomials; see, for example, J. Marsden, *Elementary Classical Analysis* (New York: W. H. Freeman and Company, 1974), Chap. 5. By this theorem we get $\int_0^1 g(t)f(t)\, dt = 0$ for any continuous g. The result follows by taking $g = f$ and applying the fact that if the integral of a nonnegative continuous function is zero, then the function is zero. ▼

PROOF OF THEOREM 8.1.9   Suppose that

$$\tilde{f}(z) = \int_0^\infty e^{-zt}f(t)\, dt = 0$$

whenever Re $z > \sigma$. Fix $x_0 > y_0$ real and let $s = e^{-t}$. By changing variables to express the integrals in terms of s, we get, at $z = x_0 + n$ for $n = 0, 1, 2, \ldots,$

$$\int_0^\infty e^{-nt}e^{-x_0 t}f(t)\, dt = \int_0^\infty s^n h(s)\, ds = 0 \qquad n = 0, 1, 2, \ldots$$

where $h(s) = e^{-x_0 t + t}f(t)$. Thus, by Lemma 8.1.10, h, and hence f, is zero. ∎

REMARK   It is useful to note that $\tilde{f}(z) \to 0$ as Re $z \to \infty$. This follows from the arguments used to prove Theorem 8.1.1 (see Review Exercise 10).

Table 8.1.1   Some Common Laplace Transforms

Definition. $\tilde{f}(z) = \int_0^\infty e^{-zt} f(t)\, dt$

1. $\tilde{g}(z) = -\dfrac{d}{dz} \tilde{f}(z)$ where $g(t) = tf(t)$.

2. $(af + bg)\tilde{} = a\tilde{f} + b\tilde{g}$.

3. $\left(\dfrac{df}{dt}\right)\tilde{}\,(z) = z\tilde{f}(z) - f(0)$. (Assume that f is piecewise C^1.)

4. $\tilde{g}(z) = \dfrac{1}{z} \tilde{f}(z)$ where $g(t) = \int_0^t f(\tau)\, d\tau$.

5. $\tilde{g}(z) = \tilde{f}(z + a)$ where $g(t) = e^{-at} f(t)$.

6. $\tilde{g}(z) = e^{-az} \tilde{f}(z)$ where $a > 0$ and $g(t) = \begin{cases} 0 & t < a \\ f(t - a) & t \geq a \end{cases}$

7. $\tilde{g}(z) = e^{-az} \tilde{F}(z)$, where $a \geq 0$, $F(t) = f(t + a)$, and $g(t) = \begin{cases} 0 & 0 \leq t < a \\ f(t) & t \geq a \end{cases}$

8. $(f * g)\tilde{}\,(z) = \tilde{f}(z) \cdot \tilde{g}(z)$, where $(f * g)(t) = \int_0^\infty f(t - \tau) g(\tau)\, d\tau$.

9. If $f(t) = e^{-at}$, then $\tilde{f}(z) = \dfrac{1}{z + a}$ and $\sigma(f) = -\,\mathrm{Re}\ a$.

10. For $f(t) = \cos at$, $\tilde{f}(z) = \dfrac{z}{z^2 + a^2}$ and $\sigma(f) = |\mathrm{Im}\ a|$.

11. If $f(t) = \sin at$, $\tilde{f}(z) = \dfrac{a}{z^2 + a^2}$ and $\sigma(f) = |\mathrm{Im}\ a|$.

12. If $f(t) = t^a$, $a > -1$, $\tilde{f}(z) = \dfrac{\Gamma(a + 1)}{z^{a+1}}$ and $\sigma(f) = 0$.

13. If $f(t) = 1$, $\tilde{f}(z) = \dfrac{1}{z}$ and $\sigma(f) = 0$.

Worked Examples

8.1.11   *Prove formula 9 in Table 8.1.1 and find $\sigma(f)$ in that case.*

Solution. By definition,

$$\tilde{f}(z) = \int_0^\infty e^{-at} e^{-zt}\, dt = \int_0^\infty e^{-(a+z)t}\, dt = -\frac{e^{-(a+z)t}}{a + z}\bigg|_0^\infty = \frac{1}{z + a}$$

The evaluation at $t = \infty$ is justified by noting that $\lim_{t \to \infty} e^{-(a+z)t} = 0$ if Re $(a + z) > 0$, since $|e^{-(a+z)t}| = e^{-\mathrm{Re}(a+z)t} \to 0$ as $t \to \infty$. Thus the formula is valid if Re $z > -$Re a.

Note. The formula for $\tilde{f}$ is valid only for Re $z > -$Re a, although $\tilde{f}$ coincides there with a function that is analytic except at $z = -a$. This situation is similar to that for the gamma function (see formula 12 of Table 7.1.1).

Finally, we show that for $f(t) = e^{-at}$, $\sigma(f) = -$Re a. We have already shown that $\sigma(f) \le -$Re a. But the integral diverges at $z = a$, so $\sigma(f) \ge -$Re a, and thus $\sigma(f) = -$Re a.

In the case where $a = 0$, this example specializes to formula 13 of Table 8.1.1.

8.1.12 *Suppose that we have computed $\tilde{f}(z)$ and found it to converge for Re $z > \gamma$. Suppose also that $\tilde{f}$ coincides with an analytic function that has a pole on the line Re $z = \gamma$. Show that $\sigma(f) = \gamma$.*

Solution. We know that $\sigma(f) \le \gamma$ by the basic property of σ in the convergence theorem. Also, since $\tilde{f}$ is analytic for Re $z > \sigma$, there can be no poles in the region $\{z \mid$ Re $z > \sigma\}$. If $\sigma(f)$ were $< \gamma$, there would be a pole in this region. Hence $\sigma(f) = \gamma$ (see Fig. 8.1.7).

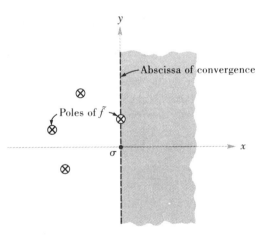

FIGURE 8.1.7   Location of poles of $\tilde{f}$.

8.1.13 *Let $f(t) = \cosh t$. Compute $\tilde{f}$ and $\sigma(f)$.*

Solution. $f(t) = \cosh t = (e^t + e^{-t})/2$. Thus, by formulas 2 and 9 of Table 8.1.1,

$$\tilde{f}(z) = \frac{1}{2}\left(\frac{1}{z-1} + \frac{1}{z+1}\right) = \frac{z}{z^2 - 1}$$

Here $\sigma(f) = 1$ by Worked Example 8.1.12; $\sigma(e^t) = 1$ and $\sigma(e^{-t}) = -1$, so $\sigma(f) \le 1$ but it cannot be < 1 since $\tilde{f}$ has a pole at $z = 1$.

Exercises

In Exercises 1 through 9 compute the Laplace transform of $f(t)$ and find the abscissa of convergence.

1. $f(t) = t^2 + 2$

2. $f(t) = \sinh t$

3. $f(t) = t + e^{-t} + \sin t$

4. $f(t) = \begin{cases} 0 & 0 \le t \le 1 \\ 1 & 1 < t < 2 \\ 0 & t \ge 2 \end{cases}$

5. $f(t) = (t + 1)^n$, n a positive integer

6. $f(t) = \begin{cases} \sin t & 0 \le t \le \pi \\ 0 & t > \pi \end{cases}$

7. $f(t) = t \sin at$

8. $f(t) = t \sinh at$

9. $f(t) = t \cos at$

10. Use the shifting theorems to show the following:
(a) If $f(t) = e^{-at} \cos bt$, then

$$\tilde{f}(z) = \frac{z + a}{(z + a)^2 + b^2}$$

(b) If $f(t) = e^{-at} t^n$, then

$$\tilde{f}(z) = \frac{\Gamma(n + 1)}{(z + a)^{n+1}}$$

What is $\sigma(f)$ in each case?

11. Prove formula 10 of Table 8.1.1.

12. Prove formula 11 of Table 8.1.1.

13. Prove formula 12 of Table 8.1.1.

14. Prove formula 13 of Table 8.1.1.

15. Suppose that f is periodic with period p (that is, $f(t + p) = f(t)$ for all $t \geq 0$). Prove that

$$\tilde{f}(z) = \frac{\displaystyle\int_0^p e^{-zt}f(t)\, dt}{1 - e^{-pz}}$$

is valid if Re $z > 0$. (*Hint.* Write out $\tilde{f}(z)$ as an infinite sum.)

16. Use Exercise 15 to prove that

$$\tilde{f}(z) = \frac{1}{z} \cdot \frac{1 - e^{-z}}{1 - e^{-2z}}$$

where $f(t)$ is as illustrated in Fig. 8.1.8.

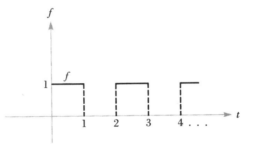

FIGURE 8.1.8 The unit pulse function.

17. Let $g(t) = \displaystyle\int_0^t e^{-s} \sin s\, ds$. Compute $\tilde{g}(z)$. Compute $\tilde{f}(z)$ if $f(t) = tg(t)$.

18. Let $f(t) = (\sin at)/t$. Show that $\tilde{f}(z) = \tan^{-1}(a/z)$.

19. Prove Proposition 8.1.4. First establish that $\rho(g) \leq \max\,[0, \rho(f)]$.

20. Give a direct proof that $f * g = g * f$ (see the convolution theorem (8.1.7)).

21. Let $f(t) = e^{-e^t}$, $t \geq 0$. Show that $\sigma(f) = -\infty$.

22. Referring to the convergence theorem (8.1.1), show that, in general, $\sigma \neq \rho$. (*Hint.* Consider $f(t) = e^t \sin e^t$ and show that $\sigma = 0$, $\rho = 1$.)

8.2 THE COMPLEX INVERSION FORMULA

It is important to be able to compute $f(t)$ when $\tilde{f}(z)$ is known. A general formula for such a computation, called the *complex inversion formula*, will be established in this section. Also, by using the formulas of Table 8.1.1 in reverse, we can obtain a number of useful alternative techniques. (See Worked Examples 8.2.4 and 8.2.5.)

Complex Inversion Formula

The proof of the complex inversion formula draws on many of the main points developed in the first four chapters of this book. It should be regarded as one of the key results of our analysis of the Laplace transform.

8.2.1 COMPLEX INVERSION FORMULA *Suppose that $F(z)$ is analytic on $\mathbb{C}$ except for a finite number of isolated singularities and that F is analytic on the half plane $\{z \mid \text{Re } z > \sigma\}$. Suppose also that there are positive constants M, R, and β such that $|F(z)| \leq M/|z|^\beta$ whenever $|z| \geq R$; this is true, for example, if $F(z) = P(z)/Q(z)$ for polynomials P and Q with $\deg(Q) \geq 1 + \deg(P)$. For $t \geq 0$, let*

$$f(t) - \sum \{\text{residues of } e^{zt}F(z) \text{ at each of its singularities in } \mathbb{C}\} \tag{1}$$

Then $\tilde{f}(z) = F(z)$ for Re $z > \sigma$.

PROOF Let $\alpha > \sigma$ and consider a large rectangle Γ with sides along Re $z = \pm x_1$, Im $z = y_2$, and Im $z = -y_1$ selected large enough so that all the singularities of F are inside Γ and $|z| > R$ everywhere on Γ. Split Γ into a sum of two rectangular paths γ and $\tilde{\gamma}$ by a vertical line through Re $z = \alpha$. (See Fig. 8.2.1.) The proof of the complex inversion formula (1) could just as well be carried out using a large circle instead of the rectangle Γ. In fact, in the last paragraph of the proof, Γ is briefly deformed to such a circle. However, the rectangular path will be useful in the corollary, 8.2.2, in which it plays a role like that of the rectangular path in the proof of Proposition 4.3.4 concerning the evaluation of Fourier transforms.

Since all singularities of F are inside γ, we have

$$\int_\gamma e^{zt}F(z)\, dz = 2\pi i \sum \{\text{residues of } e^{zt}F(z)\} = 2\pi i f(t)$$

so that

$$2\pi i \tilde{f}(z) = \lim_{r \to \infty} \int_0^r e^{-zt}\left[\int_\gamma e^{\zeta t}F(\zeta)\, d\zeta\right] dt = \lim_{r \to \infty} \int_\gamma \int_0^r e^{(\zeta - z)t}F(\zeta)\, dt\, d\zeta$$

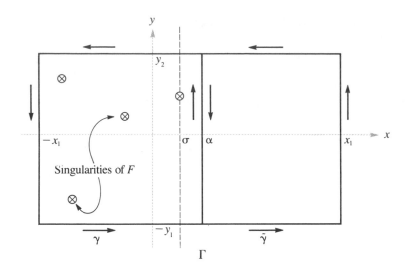

FIGURE 8.2.1 $\Gamma = \gamma + \tilde{\gamma}$.

We were able to interchange the order of integration, because both integrals are over finite intervals. Therefore,

$$2\pi i \tilde{f}(z) = \lim_{r \to \infty} \int_{\gamma} (e^{(\zeta - z)r} - 1) \frac{F(\zeta)}{\zeta - z} d\zeta$$

With z fixed in the half plane Re $z > \alpha$, the term $e^{(\zeta - z)r}$ approaches 0 and the integrand converges uniformly to $-F(\zeta)/(\zeta - z)$ on γ. We obtain

$$2\pi i \tilde{f}(z) = -\int_{\gamma} \frac{F(\zeta)}{\zeta - z} d\zeta = \int_{\tilde{\gamma}} \frac{F(\zeta)}{\zeta - z} d\zeta - \int_{\Gamma} \frac{F(\zeta)}{\zeta - z} d\zeta$$

$$= 2\pi i F(z) - \int_{\Gamma} \frac{F(\zeta)}{\zeta - z} d\zeta$$

provided Γ is large enough so that z is inside $\tilde{\gamma}$. Finally,

$$\left| \int_{\Gamma} \frac{F(\zeta)}{\zeta - z} d\zeta \right| \leq \int_{\Gamma} \frac{M}{|z|^{\beta}|\zeta - z|} |d\zeta| \leq \frac{2\pi M \rho}{\rho^{\beta}(\rho - R)}$$

by choosing Γ large enough so that it lies outside the circle $|\zeta| = \rho > R$ with all the singularities of $F(\zeta)/(\zeta - z)$ inside this circle, and then deforming Γ to the circle. This last expression goes to 0 as $\rho \to \infty$. Thus, letting Γ expand outward toward ∞, we obtain

$\tilde{f}(z) = F(z)$. With the proper choice of α, this can be done for any z in the half plane $\operatorname{Re} z > \sigma$. ∎

The following corollary of the theorem and of its proof should be noted.

8.2.2 COROLLARY *Let the conditions of the complex inversion formula hold. If $F(z)$ is analytic for $\operatorname{Re} z > \sigma$ and has a singularity on the line $\operatorname{Re} z = \sigma$, then* (i) *the abscissa of convergence of f is σ, and* (ii)

$$f(t) = \frac{1}{2\pi i} \int_{\alpha - i\infty}^{\alpha + i\infty} e^{zt} F(z) \, dz = \frac{1}{2\pi} \int_{-\infty}^{\infty} e^{(\alpha + iy)t} F(\alpha + iy) \, dy \qquad (2)$$

for any constant $\alpha > \sigma$. The first integral is taken along the vertical line $\operatorname{Re} z = \alpha$ and converges as an improper Riemann integral; the second integral is used as alternative notation for the first.

PROOF

(i) The complex inversion formula (8.2.1) shows that $\sigma(f) \leq \sigma$ since $\tilde{f}(z)$ converges for $\operatorname{Re} z > \sigma$. If $\sigma(f)$ were $< \sigma$, then $\tilde{f}(z)$ would be analytic for $\operatorname{Re} z > \sigma(f)$ by the convergence theorem (8.1.1). But F has a singularity at a point z_0 on the line $\operatorname{Re} z = \sigma$, and so there is a sequence of points $z_1, z_2, z_3, \ldots$ converging to z_0 with $F(z_n) \to \infty$. Since $\tilde{f}(z) = F(z)$ for $\operatorname{Re} z > \sigma$, and since both are analytic in a deleted neighborhood of z_0, they would be equal in that deleted neighborhood by the principle of analytic continuation. This would mean that $\tilde{f}(z_n) \to \infty$. But that is impossible, since $\tilde{f}(z)$ is analytic on $\operatorname{Re} z > \sigma(f)$. Thus $\sigma(f) < \sigma$ is not possible. We must have $\sigma(f) = \sigma$.

(ii) From the complex inversion formula (8.2.1), $2\pi i f(t) = \int_\gamma e^{zt} F(z) \, dz$. This integral converges to the integral of Eq. (2), exactly as in the proof of Proposition 4.3.4, as x_1, y_1, and $y_2 \to \infty$. Since y_1 and y_2 go independently to ∞, this establishes convergence of the improper integral. (The situation here is rotated by 90° from that of Proposition 4.3.4.) ∎

In working examples, all conditions of the theorem must be checked. If they do not hold, Eq. (2) for $f(t)$ may not be valid. Equation (1) is sometimes more convenient than Table 8.1.1 for computing inverse Laplace transforms since it is systematic and requires no guesswork as to which formula is appropriate. However, the table may be useful in cases in which hypotheses of the theorem do not apply, or are inconvenient to check.

Heaviside Expansion Theorem

Now we apply the complex inversion formula to the case in which $F(z) = P(z)/Q(z)$ where P and Q are polynomials. We give a simple case here.

8.2.3 HEAVISIDE EXPANSION THEOREM *Let $P(z)$ and $Q(z)$ be polynomials with $\deg Q \geq \deg P + 1$. Suppose that the zeros of Q are located at the points $z_1, \ldots, z_m$ and are simple zeros. Then the inverse Laplace transform of $F(z) = P(z)/Q(z)$ is given by*

$$f(t) = \sum_{i=1}^{n} e^{z_i t} \frac{P(z_i)}{Q'(z_i)} \tag{3}$$

Furthermore, $\sigma(f) = \max \{\operatorname{Re} z_i \mid i = 1, 2, \ldots, m\}$.

PROOF Since $\deg Q \geq \deg P + 1$, the conditions of the complex inversion formula (8.2.1) are met (compare Proposition 4.3.4). Thus $f(t) = \Sigma$ {residues of $e^{zt}[P(z)/Q(z)]$}. But the poles are all simple and so, by formula 4 of Table 4.1.1, we have

$$\operatorname{Res}\left(e^{zt} \frac{P(z)}{Q(z)}, z_i\right) = e^{z_i t} \frac{P(z_i)}{Q'(z_i)}$$

The formula for $\sigma(f)$ is a consequence of Corollary 8.2.2. ∎

Worked Examples

8.2.4   *If $\tilde{f}(z) = 1/(z - 3)$, find $f(t)$.*

Solution. Refer to formula 9 of Table 8.1.1. Let $a = -3$; then we get $f(t) = e^{3t}$. Alternatively, we could get the same result by using the Heaviside expansion theorem. In this example, $\sigma(f) = 3$.

8.2.5   *If $\tilde{f}(z) = \log(z^2 + z)$, what is $f(t)$?*

Solution. Note that if $g(t) = tf(t)$, then, by formula 1 of Table 8.1.1,

$$\tilde{g}(z) = -\frac{d}{dz}\tilde{f}(z) = -\frac{d}{dz}\log(z^2 + z) = -\frac{2z + 1}{z^2 + z}$$

To find $g(t)$ we use partial fractions:

$$\tilde{g}(z) = -\frac{2z + 1}{z^2 + z} = -\frac{1}{z} - \frac{1}{z + 1}$$

Therefore, $g(t) = -1 - e^{-t}$, and thus

$$f(t) = -\frac{1}{t}(1 + e^{-t})$$

Although this argument seems satisfactory, it is deceptive because *there is in fact no f(t) whose Laplace transform is* $\log(z^2 + z)$. If there were, then this procedure would yield $f(t) = -(1 + e^{-t})/t$. For any real x,

$$\int_0^\infty e^{-xt} f(t)\, dt$$

cannot converge, because near 0, e^{-xt} is $\geq \frac{1}{2}$, and $f(t) \geq 1/t$ but $1/t$ is not integrable. Thus $\tilde{f}$ does not exist in any sense we have discussed. The previous argument obtaining $f(t)$ is specious because it assumes the existence of an $f(t)$; see also the remark at the end of Sec. 8.1.

8.2.6 *Compute the inverse Laplace transform of*

$$F(z) = \frac{z}{(z+1)^2(z^2 + 3z - 10)}$$

Then compute $\sigma(f)$, the abscissa of convergence of f.

Solution. In this case Eq. (1) is convenient because the hypotheses of the complex inversion formula (8.2.1) clearly hold. Thus

$$f(t) = \sum \left\{ \text{residues of } \frac{e^{zt}z}{(z+1)^2(z^2 + 3z - 10)} = \frac{e^{zt}z}{(z+1)^2(z+5)(z-2)} \right\}$$

The poles are at $z = -1$, $z = -5$, and $z = 2$. The pole at -1 is double, whereas the others are simple. The residue at -1 is, by formula 7 of Table 4.1.1, $g'(-1)$ where $g(z) = (e^{zt}z)/(z^2 + 3z - 10)$. Thus we obtain

$$\frac{-te^{-t}}{-12} + \frac{e^{-t}}{-12} - \frac{(-e^{-t}) \cdot [2 \cdot (-1) + 3]}{144} = \frac{1}{12}\left(te^{-t} - e^{-t} + \frac{e^{-t}}{12} \right)$$

The residue at -5 is $e^{-5t} \cdot 5/16 \cdot 7$; the residue at 2 is $e^{2t} \cdot 2/9 \cdot 7$. Thus

$$f(t) = \frac{1}{12}\left(te^{-t} - e^{-t} + \frac{e^{-t}}{12} \right) + \frac{5e^{-5t}}{16 \cdot 7} + \frac{2e^{2t}}{63}$$

By Corollary 8.2.2, $\sigma(f) = 2$.

8.2.7 *Check formula 9 of Table 8.1.1 using the complex inversion formula.*

Solution. We can find the inverse Laplace transform of $1/(z + a)$ by using Eq. (1). The only pole, which is simple, is at $z = -a$. The residue of $e^{zt}/(z + a)$ at $z = -a$ is clearly e^{-at}, which agrees with Table 8.1.1. Also, $\sigma(f) = -\text{Re } a$ because the pole of F lies on the line Re $z = -\text{Re } a$.

Exercises

1. Compute the inverse Laplace transform of each of the following.

(a) $F(z) = \dfrac{z}{z^2 + 1}$
(b) $F(z) = \dfrac{1}{(z + 1)^2}$

(c) $F(z) = \dfrac{z^2}{z^3 - 1}$

2. Check formulas 10 and 11 of Table 8.1.1 by using Theorem 8.2.1.

3. Explain what is wrong with the following reasoning. Let

$$g(t) = \begin{cases} 0 & 0 \le t < 1 \\ 1 & t \ge 1 \end{cases}$$

Then, by formulas 6 and 13 of Table 8.1.1, $\tilde{g}(z) = e^{-z}/z$. By Eq. (1), $g(t) = \text{Res } (e^{z(t-1)}/z, 0) = 1$. Therefore, $1 = 0$.

4. Prove a Heaviside expansion theorem for P/Q when Q has double zeros.

5. Compute the inverse Laplace transform of each of the following:

(a) $\dfrac{z}{(z + 1)(z + 2)}$
(b) $\sinh z$

(c) $\dfrac{1}{(z + 1)^3}$

6. Use a shifting theorem (formula 7 of Table 8.1.1) to find a formula for the inverse Laplace transform of $e^{-z}/(z^2 + 1)$.

7. Use the convolution theorem to find a formula for the inverse Laplace transform of the function $\tilde{f}(z)/(z^2 + 1)$ for given $f(t)$.

8. Find the inverse Laplace transform of $1/\sqrt{z}$ by suitably modifying the proof of the complex inversion formula (8.2.1).

9. Find the inverse Laplace transform of $(z + 1)/[z(z + 3)^2]$.

10. Let $f(t) = J_1(kt)/t$ where J_1 is the Bessel function. Show that

$$\tilde{f}(z) = \frac{\sqrt{k^2 + z^2} - z}{k}$$

8.3 APPLICATION OF LAPLACE TRANSFORMS TO ORDINARY DIFFERENTIAL EQUATIONS

This section will present a brief introduction to one of the many applications of Laplace transforms. The method is based on the formula

$$\left(\frac{df}{dt}\right)^{\tilde{}}(z) = z\tilde{f}(z) - f(0)$$

and the techniques for finding inverse Laplace transforms that were developed in the preceding section.

We shall assume that solutions exist and attempt to find formulas for them. When the formula is found, we may verify that it is indeed a solution. However, sometimes the solutions are not differentiable and thus do not strictly satisfy the equation. Such solutions should be regarded as *generalized solutions*. Further analysis of these solutions would lead to the subject of distribution theory. It should be remarked that even if f is not differentiable but is continuous at 0, $z\tilde{f}(z) - f(0)$ is still defined and can be regarded as the *generalized definition of* $(df/dt)^{\tilde{}}$. This is illustrated in Example 8.3.2.

8.3.1 EXAMPLE *Solve the equation* $y'' + 4y' + 3y = 0$ *for* $y(t)$, $t \geq 0$, *subject to the conditions that* $y(0) = 0$ *and* $y'(0) = 1$. *Here* y' *stands for* dy/dt.

SOLUTION We take the Laplace transform on each side of the equation using formula 3 of Table 8.1.1:

$$\left(\frac{dy}{dt}\right)^{\tilde{}}(z) = z\tilde{y}(z) - y(0) = z\tilde{y}(z)$$

Applying this again gives

$$\left(\frac{d^2y}{dt^2}\right)^{\tilde{}}(z) - z^2\tilde{y}(z) - zy(0) - y'(0) = z^2\tilde{y}(z) - 1$$

Therefore our equation becomes $z^2\tilde{y}(z) - 1 + 4z\tilde{y}(z) + 3\tilde{y}(z) = 0$, and so

$$\tilde{y}(z) = \frac{1}{z^2 + 4z + 3} = \frac{1}{(z+1)(z+3)}$$

The inverse Laplace transform of this function is, by the inversion formula,

$$y(t) = \sum \left\{ \text{residues of} \frac{e^{zt}}{(z+1)(z+3)} \text{ at } -1, -3 \right\}$$

Thus

$$y(t) = \frac{e^{-t} - e^{-3t}}{2}$$

This is the desired solution, as can be checked directly by substitution into the differential equation. ▲

8.3.2 EXAMPLE *Solve the equation* $y'(t) - y(t) = H(t - 1)$, $t \geq 0$, $y(0) = 0$, *where H is the Heaviside function.*

SOLUTION Again we take the Laplace transforms of both sides of the equation. We get $z\tilde{y}(z) - y(0) - \tilde{y}(z) = e^{-z}/z$. Therefore, $\tilde{y}(z) = e^{-z}/z(z - 1)$. The inverse Laplace transform of $1/[z(z - 1)]$ is $1 - e^{-t}$, and so that of $e^{-z}/[z(z - 1)]$ is, by formula 6 of Table 8.1.1,

$$y(t) = \begin{cases} 0 & 0 \leq t < 1 \\ -1 + e^{t-1} & t \geq 1 \end{cases}$$

(Note that the complex inversion formula (8.2.1) does not apply as stated). This solution (see Fig. 8.3.1) is not differentiable and thus cannot be considered a solution in the strict

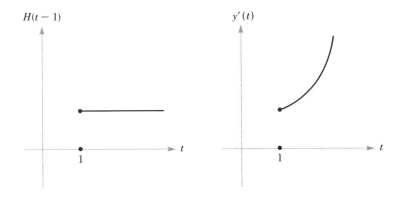

FIGURE 8.3.1 At $t = 1$, y receives an impulse.

sense. However, it is a solution in a generalized sense, as previously explained. In Fig. 8.3.1, the discontinuity in $H(t - 1)$ causes the sudden jump in $y'(t)$. We say that $y(t)$ receives an "impulse" at $t = 1$. ▲

8.3.3 EXAMPLE   *Find a particular solution of $y''(t) + 2y'(t) + 2y(t) = f(t)$.*

SOLUTION   Let us find the solution with $y(0) = 0$, $y'(0) = 0$. Taking Laplace transforms, $z^2 \tilde{y}(z) + 2z\tilde{y}(z) + 2\tilde{y}(z) = \tilde{f}(z)$, and so $\tilde{y}(z) = \tilde{f}(z)/(z^2 + 2z + 2)$. The inverse Laplace transform of $1/(z^2 + 2z + 2)$ is

$$g(t) = \frac{e^{z_1 t}}{2(z_1 + 1)} + \frac{e^{z_2 t}}{2(z_2 + 1)}$$

where z_1, z_2 are the two roots of $z^2 + 2z + 2$, namely, $-1 \pm i$. By simplifying, we get $g(t) = e^{-t} \sin t$. Thus, by formula 8 of Table 8.1.1,

$$y(t) = (g * f)(t) = \int_0^\infty f(t - \tau) g(\tau) \, d\tau$$

$$= \int_0^\infty f(t - \tau) e^{-\tau} \sin \tau \, d\tau$$

This is the particular solution we sought. Generally such particular solutions to equations of the form

$$a_n y^{(n)} + \cdots + a_1 y = f \qquad a_1, \ldots, a_n \text{ are constants}$$

will be expressed in the form of a convolution. To obtain a solution with $y(0)$, $y'(0), \ldots, y^{(n-1)}(0)$ prescribed, we can add a particular solution y_p satisfying $y_p(0) = 0$, $y_p'(0) = 0, \ldots, y_p^{(n-1)}(0) = 0$ to a solution y_c of the homogeneous equation in which f is set equal to zero and with $y_c(0)$, $y_c'(0), \ldots, y_c^{(n-1)}(0)$ prescribed. The sum $y_p + y_c$ is the solution sought. (These statements are easily checked.) ▲

The method of Laplace transforms is a systematic method for handling constant coefficient differential equations. (Of course, these equations can be handled by other means as well.) If the coefficients are not constant, the method fails, because transformation of a product then involves a convolution, and finding a solution for $\tilde{y}(z)$ becomes difficult.

SUPPLEMENT TO SECTION 8.3
THE FOURIER TRANSFORM AND THE WAVE EQUATION

The Fourier transform, which was introduced in Sec. 4.3, provides a second major tool for solving differential equations. We illustrate that use and the role of complex variables by focusing on the wave equation. Our discussion will be somewhat informal and we shall forgo the rigorous formulation of theorems.

The Wave Equation

The wave equation is the equation of motion that describes the development of a wave disturbance propagating in a medium. It describes, for example, the vertical displacement of a vibrating string (see Fig. 8.3.2), the propagation of an electromagnetic wave through space and of a sound wave in a concert hall, and some types of water wave motion.

FIGURE 8.3.2 ϕ is the wave amplitude.

Let us first consider the homogeneous problem, the simplest case of which is a wave traveling down a string of constant density ρ and under constant tension T. The vertical displacement $\phi(x, t)$ at position x and time t satisfies the partial differential equation

$$\frac{1}{c^2} \cdot \frac{\partial^2 \phi}{\partial t^2} = \frac{\partial^2 \phi}{\partial x^2} \tag{1}$$

where $c = \sqrt{T/\rho}$ is the velocity of propagation, a constant. We accept this fact from elementary physics. (The derivation assumes the amplitude is small.)

Note that if we were to have $c = \sqrt{-1}$ in Eq. (1), the equation of motion, we would recover the Laplace equation (see Secs. 2.5 and 5.3). Indeed, just as that equation admitted solutions of the form $f(x \pm \sqrt{-1}y)$, the solutions to the wave equation take the form $f(x \pm ct)$. The fact that Eq. (1) is of second order in the t variable suggests that a solution is uniquely given when two pieces of initial data at $t = 0$ are specified. These data consist of $\phi(x, 0)$ and $\partial\phi/\partial t$ at $(x, 0)$; Eq. (1) then gives the development of $\phi(x, t)$ for all subsequent t.

To solve Eq. (1) we perform a transform on the x variable to obtain a simpler equation involving the transform variable k. However, here x runs from $-\infty$ to $+\infty$, and so instead of using the Laplace transform we use the *Fourier transform*. Let $f: \mathbf{R} \to \mathbf{C}$; the Fourier transform function $\hat{f}$ of f is then defined by

$$\hat{f}(k) = \int_{-\infty}^{+\infty} e^{-ikx} f(x) \, dx$$

There is an inversion formula that is analogous to the Laplace inversion formula. It is:

$$f(x) = \frac{1}{2\pi} \int_{-\infty}^{+\infty} e^{ikx} \hat{f}(k) \, dk$$

The Fourier transform of the function $\phi(x, t)$ is defined by

$$\hat{\phi}(k, t) = \int_{-\infty}^{+\infty} e^{-ikx}\phi(x, t)\, dx \tag{2}$$

Here we perform the integral with respect to the x variable, regarding t as a fixed parameter. The Fourier inversion formula now reads

$$\phi(x, t) = \frac{1}{2\pi} \int_{-\infty}^{+\infty} e^{ikx}\hat{\phi}(k, t)\, dk \tag{2'}$$

We are now ready to solve Eq. (1). By applying Eq. (1) to Eq. (2') and differentiating under the integral, we obtain*

$$\frac{1}{c^2} \cdot \frac{\partial^2 \hat{\phi}}{\partial t^2}(k, t) + k^2\hat{\phi}(k, t) = 0 \tag{3}$$

In other words, our transformation technique has replaced the partial differential equation for $\phi(x, t)$ with a simple ordinary differential equation for $\hat{\phi}(k, t)$. Equation (3) is easily solved. The solution is

$$\hat{\phi}(k, t) = A(k)e^{ikct} + B(k)e^{-ikct} \tag{4}$$

where $A(k)$, $B(k)$ are two constants of integration that may depend on the parameter k. Applying the inversion formula (2'), we have

$$\phi(x, t) = \frac{1}{2\pi} \int_{-\infty}^{+\infty} [A(k)e^{ik(x+ct)} + B(k)e^{ik(x-ct)}]\, dk \tag{4'}$$

This is our solution to Eq. (1). The functions $A(k)$, $B(k)$ are determined by the initial data $\phi(x, 0)$ and $\partial\phi(x, 0)/\partial t$.

Also note that the first integral in Eq. (4') depends only on the variable $x + ct$, whereas the second depends only on $x - ct$; that is, $\phi(x, t)$ takes the form

$$\phi(x, t) = f(x + ct) + g(x - ct)$$

where the functions f, g are again determined by ϕ and $\partial\phi/\partial t$ at $t = 0$. We can verify by substitution into Eq. (1) that this formula for ϕ does indeed give a solution of Eq. (1).

---

* Equation (3) can also be obtained by taking the Fourier transform of Eq. (1) and using the fact that $\phi(x, t)$ and $\partial\phi(x, t)/\partial x$ converge to zero as $x \to \infty$.

Some special solutions of Eq. (1) deserve separate attention. These are monochromatic (single-frequency) waves and are of the form

$$\phi(x,\,t) = e^{i(x/c - t)\omega}$$

where ω is the frequency. This ϕ represents a wave of frequency ω traveling to the right down the string. Generally, $f(x + ct)$ is a wave moving to the left whose shape is that of the graph of f, with velocity c. Similarly, $g(x - ct)$ is a wave moving to the right.

Next we shall deal with the inhomogeneous problem, which occurs when an external force is applied to the wave. For example, suppose that the string illustrated in Fig. 8.3.2 were given a constant charge density q and then placed in an external electric field $E(x, t)$ pointing in the y direction. This would result in the application to the string of a force $F(x, t)$, proportional to $qE(x, t)$. We must then solve the following equation of motion for the displacement $\phi(x, t)$:

$$\frac{1}{c^2} \cdot \frac{\partial^2 \phi}{\partial t^2} = \frac{\partial^2 \phi}{\partial x^2} + F(x,\,t) \tag{5}$$

For simplicity, we take $F(x, t)$ in Eq. (5) to be periodic with frequency ω:

$$F = f(x,\,\omega)e^{i\omega t}$$

This allows us to consider the simpler problem

$$\frac{1}{c^2} \cdot \frac{\partial^2 \phi}{\partial t^2} = \frac{\partial^2 \phi}{\partial x^2} + f(x,\,\omega)e^{i\omega t} \tag{6}$$

We write the solution of Eq. (6) as $\phi(x, t, \omega)$. Once we have solved Eq. (6), we can deal with the problem of a general force $F(x, t)$. First we "Fourier-analyze" it; that is, we write F as follows:

$$F(x,\,t) = \frac{1}{2\pi} \int_{-\infty}^{+\infty} f(x,\,\omega)e^{-i\omega t}\,d\omega$$

where

$$f(x,\,\omega) = \int_{-\infty}^{+\infty} F(x,\,t)e^{it\omega}\,dt$$

We then superimpose the solutions of Eq. (6) to obtain

$$\phi(x,\,t) = \frac{1}{2\pi} \int_{-\infty}^{+\infty} \phi(x,\,t,\,\omega)\,d\omega$$

We solve Eq. (6) by again taking the Fourier transform with respect to the variable x, to get

$$\frac{1}{c^2} \cdot \frac{\partial^2 \hat{\phi}}{\partial t^2} + k^2 \hat{\phi} = \hat{f}(k, \omega) e^{i\omega t} \tag{7}$$

where

$$\hat{f}(k, \omega) = \int_{-\infty}^{+\infty} e^{ikx} f(x, \omega) \, dx$$

Equation (7) is a simple inhomogeneous second-order differential equation. Adding the particular solution

$$\frac{\hat{f}(k, \omega) e^{i\omega t}}{k^2 - (\omega/c)^2}$$

of Eq. (7) to the homogeneous solutions in Eq. (4), we obtain the general solution to Eq. (7):

$$\hat{\phi}(t, \omega) = A(k) e^{ikct} + B(k) e^{-ikct} + \frac{\hat{f}(k, \omega)}{k^2 - (\omega/c)^2} e^{i\omega t} \tag{8}$$

The solution to Eq. (6) is thus

$$\phi(x, t, \omega) = h(x + ct) + g(x - ct) + (G * f) e^{i\omega t} \tag{8'}$$

The terms in Eq. (8') are explained as follows. The first two terms are solutions to the homogeneous equation (1), and again they are to be chosen so that the initial data at $t = 0$ are satisfied. The last term, a particular solution to the inhomogeneous equation (6), is given by taking the inverse Fourier transform of the last term in Eq. (8):

$$\frac{1}{2\pi} \int_{-\infty}^{+\infty} e^{ikx} \left[\frac{\hat{f}(k, \omega)}{k^2 - (\omega/c)^2} \right] dk$$

As with the Laplace transform, this term is the convolution of G and f where $\hat{G} = 1/[k^2 - (\omega/c)^2]$. This function $\hat{G}$ plays a central role in the theory of partial differential equations. Its transform,

$$G(x, \omega) = \frac{1}{2\pi} \int_{-\infty}^{+\infty} \left[\frac{e^{ikx}}{k^2 - (\omega/c)^2} \right] dk \tag{9}$$

is called "Green's function," and we can use contour integration to evaluate it in closed form as follows.

The integrand of Eq. (9) has simple poles at $k = \pm(\omega/c)$. In its present form, the integral in Eq. (9) is not convergent. To specify its value we must use the Cauchy principal value. Several possible values may be obtained depending on how we interpret our integrals. To select the value we want, we shall evaluate Eq. (9) by closing the contour of integration in the upper half of the complex k plane for $x > 0$ and in the lower half of the plane for $x < 0$. This is necessary if the integral over the semicircle is to approach zero as the radius approaches infinity. By Cauchy's theorem, we pick up the residues of the enclosed poles. We still must specify how we are to go around the singularities at $k = \pm\omega/c$. Different choices will lead to different, but still mathematically acceptable, values of G. Our final choice is determined by the asymptotic behavior that we want G to have as $x \to \infty$. The homogeneous solutions to Eq. (1) in which we are interested behave like $\exp(\pm ikx)$ as a function of x, and we will require the same behavior of G. This can be specified by the "$i\epsilon$ prescription":

$$G(x, \omega) = \lim_{\substack{\epsilon \to 0 \\ \epsilon > 0}} \frac{1}{2\pi} \int_{-\infty}^{+\infty} \left[\frac{e^{ikx}}{k^2 - (\omega/c - i\epsilon)^2} \right] dk \qquad (9')$$

in which we still close the contour (as shown in Fig. 8.3.3) according to the sign of x.

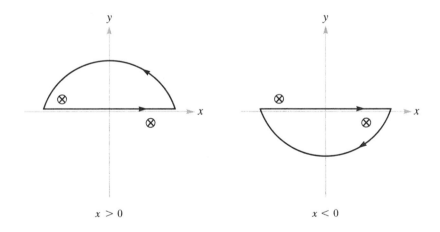

FIGURE 8.3.3 Contours for $G(x, \omega)$.

We can now evaluate $G(x, \omega)$. From Eq. (9') and the residue theorem we obtain, for $x > 0$,

$$G(x, \omega) = \lim_{\substack{\epsilon \to 0 \\ \epsilon > 0}} \frac{1}{2\pi} \left[2\pi i \cdot \frac{e^{ix(i\epsilon + \omega/c)}}{-2(\omega/c - i\epsilon)} \right] = -\frac{ic}{2\omega} e^{i\omega x/c}$$

Making a similar computation for $x < 0$ and using the right-hand contour in Fig. 8.3.3, we obtain

$$G(x, \omega) = \begin{cases} \dfrac{-ic}{2\omega} e^{i\omega x/c} & x > 0 \\[2ex] \dfrac{-ic}{2\omega} e^{-i\omega x/c} & x < 0 \end{cases}$$

Equivalently,†

$$G(x, \omega) = \frac{c}{2i\omega} e^{i\omega |x|/c} \tag{9''}$$

The Scattering Problem

When the medium through which the wave propagates is not homogeneous, we encounter the scattering problem. For example, suppose that the vibrating string of Fig. 8.3.2 now consists of three pieces smoothly joined together, with one piece, of length a, having a density of ρ_2 (region II in Fig. 8.3.4) and the other two pieces each having a

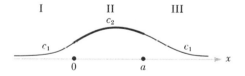

FIGURE 8.3.4   One-dimensional scattering.

---

† In textbooks on the subject of differential equations, G is obtained as the solution to

$$\frac{d^2 G}{dx^2} + \omega^2 G = \delta(x - y)$$

where δ is the "Dirac δ function." The solution is found by the general formula

$$G(x, y, \omega) = \begin{cases} -u(x)v(y)/w & x > y \\ -u(y)v(x)/w & x - y \end{cases}$$

Here u, v are solutions of the corresponding homogeneous equation and w is their Wronskian, $w = uv' - vu'$. In this case, we have $u = e^{i\omega x}$ and $v = e^{-i\omega x}$. We recover Eq. (9'') by setting $y = 0$.

density of ρ_1 (regions I, III in Fig. 8.3.4). Let c_1 and c_2 denote the corresponding velocities of propagation. Assume that $\rho_2 > \rho_1$. Imagine that an incident wave $e^{i(x/c_1 - t)\omega}$ from the left travels down the string. As the wave moves onto the denser material at $x = 0$, part of it will be reflected backward, while part will be transmitted onward. At $x = a$, some of the wave will again be reflected backward while the rest travels forward (see Fig. 8.3.5).

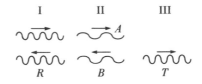

FIGURE 8.3.5 Reflection in one-dimensional scattering.

We must solve the wave equation in each region:

$$\frac{\partial^2 \phi}{\partial x^2} = \frac{1}{c_1^2} \cdot \frac{\partial^2 \phi}{\partial t^2} \qquad \text{for regions I, III}$$

$$\frac{\partial^2 \phi}{\partial x^2} = \frac{1}{c_2^2} \cdot \frac{\partial^2 \phi}{\partial t^2} \qquad \text{for region II}$$

It is not unreasonable to expect solutions of the following forms:

$$\phi_I(x, t) = e^{i(x/c_1 - t)\omega} + Re^{i(-x/c_1 - t)\omega}$$
$$\phi_{II}(x, t) = Ae^{i(x/c_2 - t)\omega} + Be^{i(-x/c_2 - t)\omega}$$
$$\phi_{III}(x, t) = Te^{i(x/c_1 - t)\omega}$$

We shall require that, at $x = 0$ and $x = a$, the solutions join onto each other and have no sharp bend. Mathematically, this means that we impose the boundary conditions that ϕ and $\partial \phi/\partial x$ be continuous at $x = 0, a$. In other words,

$$\phi_I(0, t) = \phi_{II}(0, t) \qquad \phi_{II}(a, t) = \phi_{III}(a, t)$$
$$\frac{\partial \phi_I}{\partial x}(0, t) = \frac{\partial \phi_{II}}{\partial x}(0, t) \qquad \frac{\partial \phi_{II}}{\partial x}(a, t) = \frac{\partial \phi_{III}}{\partial x}(a, t)$$

These four equations allow us to solve for the coefficients R, A, B, T. We are particularly interested in T. After performing some algebraic manipulations we find that

$$T = \frac{4c_1 c_2 e^{it[(1/c_2) - (1/c_1)]\omega a}}{(c_1 + c_2)^2 - (c_1 - c_2)^2 e^{2ia\omega/c_2}} \tag{10}$$

(see Exercise 11). T is called the *scattering amplitude,* and the square of its absolute value represents the intensity of the wave transmitted into region III.

We now allow ω in Eq. (10) to become a complex variable, and we see that $T(\omega)$ has the following properties:

(i) T is meromorphic in ω and has poles in the lower half plane at $\omega = (c_2/a)(n\pi - ip)$, where p is determined by $e^{2p} = (c_1 + c_2)^2/(c_1 - c_2)^2$.
(ii) T has absolute value 1 at those values of ω for which $e^{2i\omega/c_2} = 1$.
(iii) As $\omega \rightarrow + i\infty$, $T \rightarrow 0$.
(iv) As $\omega \rightarrow - i\infty$, $T \rightarrow \infty$.

Dispersion Relations

When a function of a complex variable $f(z)$ shares the same four properties as $T(\omega)$, the Cauchy theorem can be used to obtain an interesting and useful representation for $f(z)$, as shown in the following.

8.3.4 HILBERT TRANSFORM THEOREM *If $f(z)$ is analytic for Im $(z) \geq 0$ and $f(z) \rightarrow 0$ uniformly as $z \rightarrow \infty$ in the half plane $0 < \arg z < \pi$, then $f(z)$ satisfies the following integral relationships:*

(i) *If $z_0 = x_0 + iy_0$ with $y_0 > 0$, then*

$$f(z_0) = \frac{y_0}{\pi} \int_{-\infty}^{\infty} \frac{f(x, 0)}{(x_0 - x)^2 + y_0^2} \, dx \tag{11}$$

and

$$f(z_0) = \frac{1}{\pi i} \int_{-\infty}^{\infty} \frac{f(x, 0)(x - x_0)}{(x - x_0)^2 + y_0^2} \, dx \tag{12}$$

(ii) *If $z_0 = x_0$ is real, then*

$$f(z_0) = \frac{1}{\pi i} \int_{-\infty}^{\infty} \frac{f(x, 0)}{x - x_0} \, dx \tag{13}$$

PROOF Because of the assumptions, we can apply Cauchy's theorem, using a large semicircle in the upper half plane, to give

$$f(z_0) = \frac{1}{2\pi i} \int_{-\infty}^{+\infty} \frac{f(x, 0)}{x - z_0} \, dx$$

(see Sec. 4.3). We also have

$$0 = \frac{1}{2\pi i} \int_{-\infty}^{+\infty} \frac{f(x, 0)}{x - \bar{z}_0} \, dx$$

where $\bar{z}_0$ lies in the lower half plane. If we subtract these last two equations we obtain Eq. (11); if we add them we get Eq. (12). Equation (13) follows from formula 6 of Table 4.2.1. ∎

As a corollary, by taking real and imaginary parts of each side of Eq. (11), we get

$$u(x_0, y_0) = \frac{y_0}{\pi} \int_{-\infty}^{+\infty} \frac{u(x, 0)}{(x - x_0)^2 + y_0^2} \, dx \tag{11'}$$

where $f = u + iv$; a similar equation holds for $v(x, y)$. From Eq. (12) we have

$$u(x_0, y_0) = \frac{1}{\pi} \int_{-\infty}^{+\infty} \frac{(x - x_0)v(x, 0)}{(x - x_0)^2 + y_0^2} \, dx$$
$$v(x_0, y_0) = \frac{-1}{\pi} \int_{-\infty}^{+\infty} \frac{(x - x_0)u(x, 0)}{(x - x_0)^2 + y_0^2} \, dx \tag{12'}$$

Finally, from Eq. (13), we obtain

$$u(x_0, 0) = \frac{1}{\pi} \text{P.V.} \int_{-\infty}^{+\infty} \frac{v(x, 0)}{x - x_0} \, dx$$
$$v(x_0, y_0) = \frac{-1}{\pi} \text{P.V.} \int_{-\infty}^{+\infty} \frac{u(x, 0)}{x - x_0} \, dx \tag{13'}$$

Equation (11') gives us the values of a harmonic function in the upper half plane, in terms of its boundary values on the real axis, and thus provides a solution to the Laplace equation in the upper half plane (see Exercise 12).

Note that if $f(z)$ satisfies the symmetry property $\overline{f(-x)} = f(x)$, then we can write

$$\text{Re} \, f(x_0) = \frac{2}{\pi} \text{P.V.} \int_0^{\infty} \frac{\text{Im} \, f(x)}{x^2 - x_0^2} \, x \, dx$$

whereas if $f(z)$ satisfies $\overline{f(-x)} = -f(x)$, then

$$\text{Re} \, f(x_0) = \frac{2x_0}{\pi} \text{P.V.} \int_0^{\infty} \frac{\text{Im} \, f(x)}{x^2 - x_0^2} \, dx$$

Equations (12') and (13') can be regarded as integral versions of the Cauchy-Riemann equations; they simply tell us, for example, the values that the real part of an analytic function must take when the imaginary part is specified. When functions u, v satisfy equation set (13'), we say that u and v are "Hilbert transforms" of each other. Historically, the Hilbert transforms were the forerunners of a series of such relations called dispersion relations. They were first observed to hold for the complex dielectric constant as a function of incident frequency by H. A. Kramers and R. de L. Kronig in 1924. Since approximately 1950, they have been systematically studied and applied to the scattering amplitude $T(\omega)$ and to quite general classes of scattering problems for which this amplitude is defined. The extension of these relations to three-dimensional scattering problems will be considered later in this supplement.

The relations derived assumed that $f(z)$ is analytic only for Im $z \geq 0$. However, there is a second class of dispersion relations for functions that are analytic in the z plane except for a branch line along the real axis.

8.3.5 PROPOSITION *If $f(z)$ is analytic in the z plane with a branch line from $z = a$ to ∞, and if $|f(z)| = O(1/z)$, then*

$$f(z) = \lim_{\epsilon \to 0+} \frac{1}{2\pi i} \int_a^\infty \frac{1}{x - z} [f(x + i\epsilon) - f(x - i\epsilon)] \, dx$$

(The notation $O(1/z)$ is explained in Sec. 7.2; $\displaystyle\lim_{\epsilon \to 0+}$ means the limit is taken through $\epsilon > 0$.)

PROOF Take the contour of Fig. 8.3.6 and apply the Cauchy theorem to $f(\zeta)/(\zeta - z)$. ∎

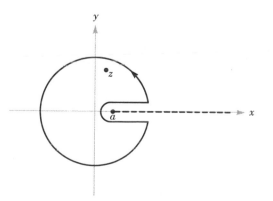

FIGURE 8.3.6 Contour used in the proof of Proposition 8.3.5.

If, in addition to the hypotheses of Proposition 8.3.5, $f(z)$ also satisfies the relation $\overline{f(\bar{z})} = f(z)$, that is, if $u(x, \epsilon) + iv(x, \epsilon) = u(x, -\epsilon) - iv(x, -\epsilon)$, so that the real part of f is continuous across the real axis whereas the imaginary part is discontinuous, then we obtain

$$f(z) = \lim_{\epsilon \to 0+} \frac{1}{\pi} \int_a^\infty \frac{1}{x - z} \, \mathrm{Im}\, f(x + i\epsilon) \, dx$$

When z actually moves onto the real axis, we can take real parts of each side of the preceding equation to obtain

$$\mathrm{Re}\, f(x_0) = \lim_{\epsilon \to 0+} \frac{\mathrm{P.V.}}{\pi} \int_a^\infty \frac{1}{x - x_0} \, \mathrm{Im}\, f(x + i\epsilon) \, dx$$

The Wave Equation in Three Dimensions

The ideas that have been developed thus far in this section for wave motion in one dimension can easily be extended to higher-dimensional problems. In two dimensions the vibrating string is replaced by a vibrating membrane. In three dimensions we can think of sound waves propagating in air. The pressure $\phi(r, t)$ then satisfies the equation of motion,

$$\frac{1}{c^2} \cdot \frac{\partial^2 \phi}{\partial t^2} (r, t) = \nabla^2 \phi(r, t) + F(r, t) \tag{14}$$

where F represents some external source of waves, $r = (x, y, z)$, and

$$\nabla^2 = \frac{\partial^2}{\partial x^2} + \frac{\partial^2}{\partial y^2} + \frac{\partial^2}{\partial z^2}$$

is the Laplace operator. When Eq. (14) is expressed in terms of rectangular coordinates, the homogeneous and inhomogeneous solutions to Eq. (14) are obtained in much the same way as they were previously. The really new and exciting features of Eq. (14) not present in one dimension arise in the scattering problem, and these are most interesting and tractable when the scattering medium has spherical symmetry.

Consider an incident plane wave $e^{i(x/c - \omega)t}$ traveling from the left down the x axis that impinges on a ball located at the origin (Fig. 8.3.7). Part of the wave may penetrate the ball, part of the wave is scattered by the surface of the ball and then travels radially outward, and the remainder of the wave simply bypasses the ball. To solve for ϕ, we proceed as previously. We first obtain the solutions for ϕ in regions I and II separately and then require that ϕ and the radial derivative $\partial \phi / \partial r$ be continuous at the surface of the ball. This procedure eventually specifies the total wave in region I that results from the "impurity" of the medium in region II.

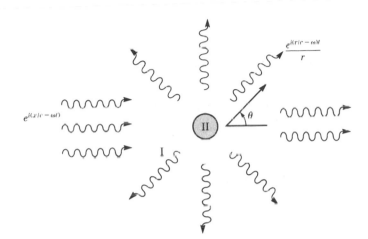

FIGURE 8.3.7 Scattering of a wave by a spherical obstacle.

These calculations are not detailed here, because such a task would take us too far afield into the subject of partial differential equations. However, the form of the final result is not too difficult to anticipate. The wave in region I will be a sum of the incident wave and the outgoing radial wave, and this will take the asymptotic form

$$\phi_1(r, t) \sim \left[e^{i x/c} + \frac{e^{i r/c}}{r} f(\omega, \theta) \right] e^{-i\omega t}$$

as $|r| = r \to \infty$. Here $f(\omega, \theta)$ is the amount of scattering wave that is traveling outward at an angle θ to the axis of symmetry (see Fig. 8.3.7); it is the three-dimensional analogue of the scattering amplitude $T(\omega)$ of the one-dimensional problem discussed earlier in the section.

Note that f is now a function of *two* complex variables, ω and θ. Physically observable scattering occurs, of course, only when ω and θ are real with $0 \le \theta \le \pi$. But by studying the properties of f for complex values of ω and θ, as we will do in the following paragraphs, we do gain a deeper understanding of the characteristics of f.

It will be convenient to change variables as follows:

$$s = \frac{4\omega^2}{c^2} \qquad t = -2 \frac{\omega^2}{c^2} (1 - \cos \theta)$$

Define the function A of the two complex variables s, t by $A(s, t) = f(\omega, \theta)$. For a large class of scattering problems, it can be shown that A has the following properties:

(i) $A(s, t)$ is analytic in the two complex variables s and t with branch lines from $s = a$ to ∞ and $t = b$ to ∞.

(ii) $A(s, t) = O(1/s)$ as $s \to \infty$ for each t.

(iii) $A(\bar{s}, t) = \overline{A(s, t)}$ for t real.

Applying Proposition 8.3.5,

$$A(s_0, t_0) = \begin{cases} \text{limit } \dfrac{1}{2\pi i} \displaystyle\int_a^\infty \dfrac{1}{s - s_0} [A(s + i\epsilon, t_0) - A(s - i\epsilon, t_0)] \, ds & s_0, t_0 \text{ in } \mathbf{C} \quad (15) \\[4mm] \text{limit } \dfrac{1}{\pi} \displaystyle\int_b^\infty \dfrac{1}{s - s_0} \text{Im } A(s + i\epsilon, t_0) \, ds & t_0 \text{ real} \quad (16) \end{cases}$$

In Eq. (15) we are integrating $A(\zeta, t)$ for ζ slightly above and below the real axis. Now consider the integrand. By property (i) of $A(s, t)$, we can write a dispersion relation in the t variable as follows:

$$A(s + i\epsilon, t_0) = \lim_{\delta \to 0+} \frac{1}{2\pi i} \int_b^\infty \frac{1}{t - t_0} [A(s + i\epsilon, t + i\delta) - A(s + i\epsilon, t - i\delta)] \, dt$$

Using a similar representation for the second term of Eq. (15), we finally obtain the *double dispersion relation*

$$A(s_0, t_0) = \frac{1}{\pi^2} \int_a^\infty \frac{1}{s - s_0} \left[\int_b^\infty \frac{1}{t - t_0} \rho(s, t) \, dt \right] ds \qquad (17)$$

where

$$\rho(s, t) = \lim_{\epsilon, \delta \to 0+} \left(\frac{1}{2i} \right)^2 [A(s + i\epsilon, t + i\delta) - A(s + i\epsilon, t - i\delta)$$
$$- A(s - i\epsilon, t + i\delta) + A(s - i\epsilon, t - i\delta)].$$

Equation (17) is the representation for $A(s, t)$ first obtained by S. Mandelstam in 1958.

Exercises

Solve the differential equations in Exercises 1 through 9 by using Laplace transforms.

1. $y'' - 4y = 0$, $y(0) = 2$, $y'(0) = 1$

2. $y'' + 6y - 7 = 0$, $y(0) = 1$, $y'(0) = 0$

3. $y'' + 9y = H(t - 1)$, $y(0) = y'(0) = 0$

4. $y' + y = e^t$, $y(0) = 0$

5. $y' + y + \int_0^t y(\tau)\, d\tau = f(t)$ where $y(0) = 1$ and

$$f(t) = \begin{cases} 0 & 0 \le t < 1,\, t \ge 2 \\ 1 & 1 \le t < 2 \end{cases}$$

6. $y'' + 9y = H(t)$, $y(0) = y'(0) = 0$

7. Solve the following systems of equations for $y_1(t)$, $y_2(t)$ by using Laplace transforms.

(a) $\begin{cases} y_1' + y_2 = 0 \\ y_2' + y_1 = 0 \end{cases}$ $y_1(0) = 1,\, y_2(0) = 0$ (b) $\begin{cases} y_1' + y_2' + y_1 = 0 \\ y_2' + y_1 = 3 \end{cases}$ $y_1(0) = 0,\, y_2(0) = 0$

8. $y' + y = \cos t$, $y(0) = 1$

9. Solve $y'' + y = t \sin t$, $y(0) = 0$, $y'(0) = 1$.

10. Study the solution of $y'' + \omega_0^2 y = \sin \omega t$, $y(0) = y'(0) = 0$, and examine the behavior of solutions for various ω, especially those near $\omega = \omega_0$. Interpret these solutions in terms of forced oscillations.

11. Apply the boundary conditions on p. 546 to obtain expressions for $T(\omega)$, $R(\omega)$, $A(\omega)$, $B(\omega)$. Verify that when ω is real,

$$|T(\omega)|^2 + |R(\omega)|^2 = 1$$

(This last relation expresses "conservation of intensity": In a medium without dissipative loss, the unit intensity of the incident wave of Fig. 8.3.5 is equal to the sum of the intensity of the wave R reflected back into region I and intensity of the wave T transmitted into region III.)

12. Since Eq. (11) solves the Dirichlet problem for the upper half plane, apply it to the problem illustrated in the right side of Fig. 5.3.5 and obtain the same solution. Thus we have two methods for solving Laplace's equation: conformal mapping, and integral relations of the type described by Eq. (11).

13. As an application of Proposition 8.3.5, let

$$f(z) = \frac{1}{\sqrt{z}} = r^{-1/2} e^{i\theta/2} \quad \text{for} \quad 0 < \theta < 2\pi$$

be defined on the z plane with a branch line along the positive real axis. What equality results?

REVIEW EXERCISES FOR CHAPTER 8

1. Compute the Laplace transform and the abscissa of convergence for $f(t) = H(t-1)\sin(t-1)$.

2. Compute the Laplace transform and the abscissa of convergence for $f(t) = H(t-1) + 3e^{-(t+6)}$.

3. Compute the Laplace transform and the abscissa of convergence for

$$f(t) = \begin{cases} t & 0 \le t \le 1 \\ 1 & t > 1 \end{cases}$$

4. Let $f(t)$ be a bounded function of t. Show that $\sigma(f) \le 0$.

5. Compute the Laplace transform and the abscissa of convergence for

$$f(t) = \frac{e^t - 1}{t}$$

6. If $f(t) = 0$ for $t < 0$, then $\hat{f}(y) = \tilde{f}(iy)$ is called the *Fourier transform* of f. Using Corollary 8.2.2, show that, under suitable conditions,

$$f(x) = \frac{1}{2\pi} \int_{-\infty}^{\infty} \hat{f}(y) e^{ixy} \, dy$$

(This result is called the *inversion theorem for fourier transforms*.)

7. Compute the inverse Laplace transform and the abscissa of convergence for

$$F(z) = \frac{e^{-z}}{z^2 + 1}$$

8. Compute the inverse Laplace transform and the abscissa of convergence for

$$F(z) = \frac{1}{(z+1)^2}$$

9. Compute the inverse Laplace transform and the abscissa of convergence for

$$F(z) = \frac{z}{(z+1)^2} + \frac{e^{-z}}{z}$$

10. (a) Let $\tilde{f}(z)$ be the Laplace transform of $f(t)$. Show that $\tilde{f}(z) \to 0$ as $\mathrm{Re}\ z \to \infty$.
(b) Use (a) to show that, under suitable conditions, $z\tilde{f}(z) \to f(0)$ as $\mathrm{Re}\ z \to \infty$.
(c) Can a nonzero polynomial be the Laplace transform of any $f(t)$?
(d) Can a nonzero entire function F be the Laplace transform of a function $f(t)$?

11. Solve the following differential equations, using Laplace transforms:

(a) $y'' + 8y + 15 = 0$, $y(0) = 1$, $y'(0) = 0$ (b) $y' + y = 3$, $y(0) = 0$

12. Suppose that $f(t) \geq 0$ and is infinitely differentiable. Prove that $(-1)^k \tilde{f}^{(k)}(z) \geq 0$, $k = 0, 1, 2, \ldots$, for $z \geq 0$. (The converse, called *Bernstein's theorem*, is also true but is more difficult to prove.)

13. Solve the following differential equations using Laplace transforms.

(a) $y'' + y = H(t - 1)$, $y(0) = 0$, $y'(0) = 0$ (b) $y'' + 2y' + y = 0$, $y(0) = 1$, $y'(0) = 1$

14. As yet another way to solve the Laplace equation, take Fourier transforms of

$$\frac{\partial^2 u}{\partial x^2}(x, y) + \frac{\partial^2 u}{\partial y^2}(x, y) = 0 \tag{a}$$

with respect to the variable x (use the method described on p. 541); show that this method gives the ordinary differential equation

$$\frac{\partial^2 \hat{u}}{\partial y^2}(k, y) - k^2 \hat{u}(k, y) = 0 \tag{b}$$

By solving, summing over all solutions, and keeping only those that decay exponentially in the limit $y \to +\infty$, obtain the formula

$$u(x, y) = \frac{1}{2\pi} \int_{-\infty}^{+\infty} A(k)e^{ikx - k/y} \, dk \tag{c}$$

With relative ease, $A(k)$ can now be evaluated in terms of the boundary data at $y = 0$. Since at $y = 0$ Eq. (b) reduces to

$$u(x, 0) = \frac{1}{2\pi} \int_{-\infty}^{+\infty} A(k)e^{ikx} \, dk$$

$A(k)$ must in fact be the Fourier transform of $u(x, 0)$. Using this result, substituting in Eq. (b), and interchanging orders of integration, obtain

$$u(x, y) = \int_{-\infty}^{+\infty} u(z, 0) \left(\frac{1}{2\pi} \int_{-\infty}^{+\infty} e^{ik(x-z) - |k|y} \, dk \right) dz$$

Perform the integral on k to get

$$u(x, y) = \int_{-\infty}^{+\infty} u(z, 0) \cdot \frac{1}{\pi} \cdot \frac{1}{(x - z)^2 + y^2} \, dz \tag{d}$$

Referring to Eq. (d), show that

$$\frac{y}{(x-z)^2 + y^2} = \frac{-\partial}{\partial w} G(x, y \,|\, z, w) \bigg|_{w=0}$$

where

$$G(x, y \,|\, z, w) = \log |r - r'|$$

with $r = x + iy$, $r' = z + iw$. (The function G is called *Green's function* and is recognized as the potential at r caused by a unit charge at the point r' in the plane.)

15. Many problems in applied mathematics include an infinite series

$$F(x) = \sum_{n=0}^{\infty} f_n(z)$$

in which the function $F(z)$ has singularities in addition to all those of each $f_n(z)$. These singularities are introduced by the failure of the series to converge. The simplest example of this type of series is

$$\frac{1}{1-z} = 1 + z + z^2 + \cdots$$

The individual terms on the right of the equation are entire functions; the function their sum defines has a simple pole at $z = 1$. As a second example, consider

$$\frac{z}{z-1} = 1 + \frac{1}{z} + \frac{1}{z^2} + \cdots$$

The terms on the right of this equation have singularities at $z = 0$; their sum is singular at $z = 1$.
  Consider

$$F(j) = \sum_{n=0}^{\infty} [g \log(j - \alpha)]^n$$

where α, g are constants.
(a) What are the analytic properties of the individual terms on the right side of the last equation?
(b) By summing the series in closed form, verify that the sum in this equation has a pole at $j = \alpha + e^{1/g}$.

ANSWERS TO ODD-NUMBERED EXERCISES

1.1 INTRODUCTION TO COMPLEX NUMBERS

1. (a) $6 + 4i$ (b) $\frac{11}{17} + i\frac{10}{17}$ (c) $\frac{3}{2} - i\frac{5}{2}$ **3.** $z = \pm(2 - i)$

5. $\operatorname{Re}\dfrac{1}{z^2} = \dfrac{x^2 \, y^2}{(x^2 + y^2)^2}$ $\operatorname{Im}\dfrac{1}{z^2} = -\dfrac{2xy}{(x^2 + y^2)^2}$

$\operatorname{Re}\dfrac{1}{3z + 2} = \dfrac{3x + 2}{(3x + 2)^2 + 9y^2}$ $\operatorname{Im}\dfrac{1}{3z + 2} = \dfrac{-3y}{(3x + 2)^2 + 9y^2}$

7. No; let $z = w = i$.

9. If $z = x + iy$, then $\operatorname{Re}(iz) = \operatorname{Re}(ix - y) = -y = -\operatorname{Im}(z)$, and $\operatorname{Im}(iz) = \operatorname{Im}(ix - y) = x = \operatorname{Re}(z)$.

11. The proof of the associative law for multiplication was outlined in the text. We show that addition is commutative and leave the others for the reader: if $z = x + iy$ and $w = u + iv$, where x, y, u, and v are real, then z and w correspond to (x, y) and (u, v) respectively and thus

$$z + w = (x, y) + (u, v) = (x + u, y + v)$$
$$= (u + x, v + y) = (u, v) + (x, y) = w + z$$

as desired.

13. First show that $a = (x^2 - y^2)/(x^2 + y^2)$ and $b = -2xy/(x^2 + y^2)$. Then show that the squares of these sum to 1.

15. A complex number z can be written as $z = x + iy$ with x and y real in only one way, corresponding to the vector (x, y). The real numbers were to correspond to vectors of the form $(x, 0)$, and so $y = 0$, and therefore $z = x = \operatorname{Re} z$.

17. (a) -4 (b) i

19. (a) $\sqrt{1+\sqrt{i}} = \pm\left(\dfrac{\sqrt{1+\sqrt{2}+\sqrt{4+2\sqrt{2}}}}{2^{3/4}} + i\,\dfrac{\sqrt{-1-\sqrt{2}+\sqrt{4+2\sqrt{2}}}}{2^{3/4}}\right)$

or $\sqrt{1+\sqrt{i}} = \pm\left(\dfrac{\sqrt{\sqrt{2}-1+\sqrt{4-2\sqrt{2}}}}{2^{3/4}} - i\,\dfrac{\sqrt{1-\sqrt{2}+\sqrt{4-2\sqrt{2}}}}{2^{3/4}}\right)$

(b) $\sqrt{1+i} = \pm\left(\sqrt{\dfrac{1+\sqrt{2}}{2}} + i\,\sqrt{\dfrac{\sqrt{2}-1}{2}}\right)$ (c) See Worked Example 1.1.6.

1.2 PROPERTIES OF COMPLEX NUMBERS

1. (a) $z = \sqrt[5]{2}\left(\cos\dfrac{2\pi k}{5} + i\sin\dfrac{2\pi k}{5}\right)$ $k = 0, 1, 2, 3, 4$

(b) $z = \cos\left(\dfrac{3\pi}{8} + \dfrac{2\pi k}{4}\right) + i\sin\left(\dfrac{3\pi}{8} + \dfrac{2\pi k}{4}\right)$ $k = 0, 1, 2, 3$

3. $(3-8i)^4/(1-i)^{10}$

5. $\cos 5x = \cos^5 x - 10\cos^3 x \cdot \sin^2 x + 5\cos x \cdot \sin^4 x$
$\sin 5x = \sin^5 x - 10\cos^2 x \cdot \sin^3 x + 5\cos^4 x \cdot \sin x$

7. $\sqrt{\frac{377}{5}}$

9. Use the identity $(1-w)(1+w+w^2+\cdots+w^{n-1}) = 1 - w^n$.

11. $|a-b|^2 + |a+b|^2 = (a-b)\overline{(a-b)} + (a+b)\overline{(a+b)}$
$= (a-b)(\bar{a}-\bar{b}) + (a+b)(\bar{a}+\bar{b})$
$= |a|^2 - a\bar{b} - b\bar{a} + |b|^2 + |a|^2 + a\bar{b} + b\bar{a} + |b|^2$
$= 2(|a|^2 + |b|^2)$

13. All the points must have the same argument. They must lie on the same ray from the origin.

15. No; take $z = i$. $z^2 = |z|^2$ if and only if z is real.

17. Each side is a positive real number whose square is $a^2 a'^2 + a^2 b'^2 + a'^2 b^2 + b^2 b'^2$.

19. $|z-(8+5i)| = 3$ **21.** The real axis.   **23.** $1 + |a|$

25. Using de Moivre's formula and the identity $1 + w + \cdots + w^n = (1-w^{n+1})/(1-w)$,

$$\sum_{k=0}^{n} \cos k\theta = \text{Re}\left[\sum_{k=0}^{n}(\cos\theta + i\sin\theta)^k\right]$$

$$= \text{Re}\,\frac{1-(\cos\theta + i\sin\theta)^{n+1}}{1-\cos\theta - i\sin\theta} = \text{Re}\,\frac{1-\cos(n+1)\theta - i\sin(n+1)\theta}{1-\cos\theta - i\sin\theta}$$

$$= \frac{1-\cos\theta - \cos(n+1)\theta + \cos(n+1)\theta\cos\theta + \sin(n+1)\theta\sin\theta}{2-2\cos\theta}$$

$$= \frac{1}{2} + \frac{\cos n\theta - \cos(n+1)\theta}{2(1-\cos\theta)}$$

$$= \frac{1}{2} + \frac{2\sin(\theta/2)\sin(n+\frac{1}{2})\theta}{2(1-\cos\theta)} = \frac{1}{2} + \frac{\sin(n+\frac{1}{2})\theta}{2\sin(\theta/2)}$$

27.   (a) $(z_2 - z_1)/(z_3 - z_1)$ is real.   (b) $\left(\dfrac{z_4 - z_1}{z_4 - z_2}\right) \cdot \left(\dfrac{z_3 - z_2}{z_3 - z_1}\right)$ is real.

29.   Multiply by $1 - w$ and use Exercise 9 to show that the sum is $-n/(1 - w)$.

1.3   SOME ELEMENTARY FUNCTIONS

1.   (a) $e^2 (\cos 1 + i \sin 1)$ (b) $\dfrac{1}{2} (\sin 1) \left(\dfrac{1}{e} + e\right) + i \dfrac{1}{2} (\cos 1) \cdot \left(e - \dfrac{1}{e}\right)$

3.   (a) $z = \pm \left(\dfrac{\pi}{4} + 2\pi n - i \dfrac{1}{2} \log 2\right)$

(b) $z = \pm [2\pi n - i \log (4 + \sqrt{15})]$. (*Note:* $\log (4 - \sqrt{15}) = -\log (4 + \sqrt{15})$.)

5.   (a) $\log 1 = 2\pi n i$ (b) $\log i = \pi i/2 + 2\pi n i$

7.   (a) $e^{\pi/2} e^{-2\pi n} = e^{-2\pi(n-1/4)}$

(b) $e^{(1/2) \log 2 - 2\pi n - \pi/4} \left[\cos \left(\dfrac{1}{2} \log 2 + \dfrac{\pi}{4}\right) + i \sin \left(\dfrac{1}{2} \log 2 + \dfrac{\pi}{4}\right) \right]$

9. $z = n\pi$ for any integer n

11.   Since $|e^z| = e^{\mathrm{Re}\, z}$, $|e^z|$ goes to 0 along rays pointing into the left half plane. It is 1 along the imaginary axis, and it goes to $+\infty$ along rays into the right half plane.

13.   (a) $e^{x^2 - y^2} (\cos 2xy + i \sin 2xy)$ (b) $e^{-y} (\cos x + i \sin x)$

(c) $e^{x/(x^2 + y^2)} \left(\cos \dfrac{y}{x^2 + y^2} - i \sin \dfrac{y}{x^2 + y^2} \right)$

15. $\sin (\pi/2 - z) = \dfrac{e^{i(\pi/2 - z)} - e^{-i(\pi/2 - z)}}{2i} = e^{\pi i/2} \dfrac{e^{-iz} - e^{iz} e^{-\pi i}}{2i}$

$= \dfrac{e^{-iz} + e^{iz}}{2} = \cos z$

The other two assertions follow in a similar way.

17. $|\sin z|^2 = \sin^2 x \cosh^2 y + \cos^2 x \sinh^2 y$
$= \sin^2 (x)(1 + \sinh^2 y) + \cos^2 x \sinh^2 y$
$= \sin^2 x + \sinh^2 y \geq \sinh^2 y$

and so $|\sin z| \geq |\sinh y|$. The other inequality follows in a similar manner.

19.   No, not even for real a and b. Let $a = 2$, $b = -1$. Then $|a^b| = |2^{-1}| = \frac{1}{2}$, but $|a|^{|b|} = 2$.

21.   If $|z| = 1$, then $z = e^{i\theta}$ for some θ, and so $z + 1/z = e^{i\theta} + e^{-i\theta} = 2 \cos \theta$. As θ varies from 0 to 2π, this covers the interval $[-2, 2]$ twice.

23.   Since $|1/z| = 1/|z|$, the map interchanges the inside and outside of the unit circle. Circles of radius r are mapped to circles of radius $1/r$. If $z = re^{i\theta}$, then $1/z = (1/r)e^{-i\theta}$, and so the ray defined by $\arg z = \theta$ is mapped to the ray with argument $-\theta$.

25. This holds iff $b \log a$ has its imaginary part in $[-\pi, \pi[$. Otherwise, the formula reads $\log a^b = b \log a + 2\pi i k$.

27. These are the nth roots of 1, since $(w^k)^n = [(e^{2\pi i/n})^k]^n = e^{2\pi k i} = 1$. They are all different, since $w^j = w^k$ implies $e^{2\pi i(k-j)/n} = 1$. By Proposition 1.3.2(vii), this forces $(k-j)/n$ to be an integer.

29. $\sin z = 0$ if and only if $e^{iz} = e^{-iz}$ or $e^{2iz} = 1$. By Proposition 1.3.2(vii), this happens exactly when $2iz = 2\pi n i$, or $z = n\pi$.

31. The maximum is $\cosh(2\pi) \approx 267$ attained at $z = 2\pi i$, $\pi + 2\pi i$, and $2\pi + 2\pi i$.

33. (a) $\approx 24 - i4.5$ (b) $\approx 1.17 - i(1.19) + 2\pi n i$ (c) $\approx 96.16 - i1644.43$

35. No, $\sin z$ is not one-to-one on $0 \le \operatorname{Re} z < 2\pi$. For example, $\sin(0) = \sin(\pi) = 0$. We know $\sin z = \sin(z + 2\pi)$. Now suppose $\sin z = \sin w$. Then

$$0 = \sin z - \sin w = 2 \sin \frac{z-w}{2} \cos \frac{z+w}{2}$$

and by Exercise 29 and a similar result for cosine,

$$\left. \begin{array}{ll} z - w = 2k\pi & k = 0, \pm 1, \pm 2, \ldots \\ z + w = n\pi & n = \pm 1, \pm 3, \pm 5, \ldots \end{array} \right\} (*)$$

or

Using Exercise 34 and this result, for each $z_0 \in \mathbf{C}$ there is *precisely one* w with $-\pi/2 \le \operatorname{Re} w \le \pi/2$ such that $\sin w = z_0$, provided, for example, that the portion of the boundary of this strip lying below the real axis is omitted. Taking this value of w defines a branch of $\sin^{-1} z_0$. The others are given by the pair of formulas (*).

The discussion for $\cos^{-1}$ is analogous.

1.4 CONTINUOUS FUNCTIONS

1. Since $|w|^2 = (\operatorname{Re} w)^2 + (\operatorname{Im} w)^2$, all three assertions follow from the observation that if $a \ge 0$ and $b \ge 0$, then $a \le \sqrt{a^2 + b^2} \le a + b$.

3. Since f is continuous, there is a $\delta > 0$ such that $|z - z_0| < \delta$ implies $|f(z) - f(z_0)| < |f(z_0)|/2$. Thus $f(z) \ne 0$, for if $f(z)$ were equal to 0, then $|f(z_0)|$ would be less than $|f(z_0)|/2$, which is absurd.

5. Let $\{a_1, a_2, \ldots, a_n\}$ be a finite set of points and let z_0 be in its complement. Let $\delta_k = |z_0 - a_k|$ and let $\delta = \min\{\delta_1/2, \ldots, \delta_n/2\}$. Then no a_k can lie in $D(z_0, \delta)$, since $\delta < |z_0 - a_k|$.

7. Let $\epsilon > 0$ and $\delta = \epsilon$. If $|z - z_0| < \delta$, then $|f(z) - f(z_0)| = |\bar{z} - \bar{z}_0| = |\overline{(z - z_0)}| = |z - z_0| < \epsilon$. Thus for any $z_0 \in \mathbf{C}$, $\lim\limits_{z \to z_0} f(z) = f(z_0)$.

9. $\mathbf{C} \backslash \{2\pi n i \mid n \text{ is an integer}\}$ **11.** $|z| < 1$

13. (a) Open, not closed   (b) Neither open nor closed   (c) Not open, closed

15. (a) Connected and compact   (b) Compact, not connected
(c) Connected, not compact   (d) Neither compact nor connected

17. $z \in \mathbb{C} \backslash f^{-1}(A) \Leftrightarrow z \notin f^{-1}(A) \Leftrightarrow f(z) \notin A \Leftrightarrow f(z) \in \mathbb{C} \backslash A \Leftrightarrow z \in f^{-1}(\mathbb{C} \backslash A)$

19. If the U_α's are open sets and $z \subset \underset{\alpha}{\cup} U_\alpha$, then $z \in U_{\alpha_0}$ for some α_0. Since U_{α_0} is open, there is an $\epsilon > 0$ such that $D(z, \epsilon) \subset U_{\alpha_0} \subset \underset{\alpha}{\cup} U_\alpha$. This shows the union is open, since it can be done for any such z.

21. $\overset{\infty}{\underset{n=1}{\cap}} D(0; 1/n) = \{0\}$, and this set is not open.

23. Let $R > 0$. We need to show there is an N such that $|z^n/n| \geq R$ whenever $n \geq N$. A little arithmetic shows this to be equivalent to $(nR)^{1/n} \leq |z|$. But L'Hôpital's rule may be used to show that $\underset{n \to \infty}{\text{limit}} (nR)^{1/n} = 1$. (Take logarithms first and use L'Hôpital's rule to show that $\log (nR)^{1/n} \to 0$.)

Since $|z| > 1$, we have the inequality we need for large enough n.

1.5 ANALYTIC FUNCTIONS

1. (a) Analytic on all of $\mathbb{C}$. The derivative is $3(z + 1)^2$.
(b) Analytic on $\mathbb{C} \backslash \{0\}$. The derivative is $1 - 1/z^2$.
(c) Analytic on $\mathbb{C} \backslash \{1\}$. The derivative is $-10[1/(z - 1)]^{11}$.
(d) Analytic on $\mathbb{C} \backslash \{e^{2\pi i/3}, e^{4\pi i/3}, 1, i\sqrt{2}, -i\sqrt{2}\}$. The derivative is

$$-[1/(z^3 - 1)^2(z^2 + 2)^2] \cdot [(z^3 - 1)2z + 3z^2(z^2 + 2)]$$

3. (a) If $n \geq 0$, it is analytic everywhere. If $n < 0$, it is analytic everywhere except at 0. The derivative is nz^{n-1}.
(b) Analytic on $\mathbb{C} \backslash \{0, i, -i\}$. The derivative is given by

$$-2 \frac{1}{(z + 1/z)^3} \left(1 - \frac{1}{z^2}\right)$$

(c) Analytic except at the nth roots of 2, $\sqrt[n]{2} e^{2\pi i k/n}$. The derivative is given by

$$\frac{(1 - n)z^n - 2}{(z^n - 2)^2}$$

5. (a) Locally, f rotates by $\theta = 0$ and multiplies lengths by 1.
(b) Locally, f rotates by an angle $\theta = 0$ and stretches lengths by a factor of 3.
(c) Locally, f rotates by an angle π and stretches lengths by a factor of 2.

7. $(f^{-1} \circ f)'(z) = (f^{-1})'(f(z))f'(z)$. But $(f^{-1} \circ f)(z) = z$, and so $(f^{-1} \circ f)'(z) = 1$. Hence $(f^{-1})'(f(z)) \cdot f'(z) = 1$.

9. $f(z) = z^2 + 3z + 2 = (x^2 - y^2 + 3x + 2) + i(2xy + 3y)$, and so $\partial u/\partial x = 2x + 3 = \partial v/\partial y$ and $\partial u/\partial y = -2y = -\partial v/\partial x$.

11. Since $x = r \cos \theta$ and $y = r \sin \theta$, the chain rule gives

$$\frac{\partial u}{\partial r} = \cos \theta \frac{\partial u}{\partial x} + \sin \theta \frac{\partial u}{\partial y} \quad \text{and} \quad \frac{\partial u}{\partial \theta} = -r \sin \theta \frac{\partial u}{\partial x} + r \cos \theta \frac{\partial u}{\partial y}$$

Solving for $\partial u/\partial x$ and $\partial u/\partial y$ gives

$$\frac{\partial u}{\partial x} = \cos\theta\,\frac{\partial u}{\partial r} - \frac{\sin\theta}{r}\,\frac{\partial u}{\partial \theta} \quad\text{and}\quad \frac{\partial u}{\partial y} = \sin\theta\,\frac{\partial u}{\partial r} + \frac{\cos\theta}{r}\,\frac{\partial u}{\partial \theta}$$

Similarly,

$$\frac{\partial v}{\partial x} = \cos\theta\,\frac{\partial v}{\partial r} - \frac{\sin\theta}{r}\,\frac{\partial v}{\partial \theta} \quad\text{and}\quad \frac{\partial v}{\partial y} = \sin\theta\,\frac{\partial v}{\partial r} + \frac{\cos\theta}{r}\,\frac{\partial v}{\partial \theta}$$

and so the Cauchy-Riemann equations become

$$\cos\theta\,\frac{\partial u}{\partial r} - \frac{\sin\theta}{r}\,\frac{\partial u}{\partial \theta} = \sin\theta\,\frac{\partial v}{\partial r} + \frac{\cos\theta}{r}\,\frac{\partial v}{\partial \theta}$$

and

$$\sin\theta\,\frac{\partial u}{\partial r} + \frac{\cos\theta}{r}\,\frac{\partial u}{\partial \theta} = -\cos\theta\,\frac{\partial v}{\partial r} + \frac{\sin\theta}{r}\,\frac{\partial v}{\partial \theta}$$

Multiplying the first by $\cos\theta$ and the second by $\sin\theta$ and adding gives $\dfrac{\partial u}{\partial r} = \dfrac{1}{r}\dfrac{\partial v}{\partial \theta}$. Similarly, $\dfrac{\partial v}{\partial r} = -\dfrac{1}{r}\dfrac{\partial u}{\partial \theta}$.

13.
$$\frac{\partial f}{\partial \bar{z}} = \frac{1}{2}\left(\frac{\partial f}{\partial x} - \frac{1}{i}\frac{\partial f}{\partial y}\right) = \frac{1}{2}\left[\frac{\partial u}{\partial x} + i\frac{\partial v}{\partial x} - \frac{1}{i}\left(\frac{\partial u}{\partial y} + i\frac{\partial v}{\partial y}\right)\right]$$
$$= \frac{1}{2}\left(\frac{\partial u}{\partial x} - \frac{\partial v}{\partial y}\right) + i\frac{1}{2}\left(\frac{\partial v}{\partial x} + \frac{\partial u}{\partial y}\right)$$

Thus the Cauchy-Riemann equations $\partial u/\partial x = \partial v/\partial y$ and $\partial u/\partial y = -\partial v/\partial x$ are equivalent to saying that the complex quantity $\partial f/\partial \bar{z}$ is zero.

15. If $f = u + iv$, then $\partial u/\partial x = 0 = \partial u/\partial y$ since u is constant. By the Cauchy-Riemann equations, $\partial v/\partial y = \partial u/\partial x = 0$ and $\partial v/\partial x = -\partial u/\partial y = 0$ also. Thus $f'(z) = \partial u/\partial x + i(\partial v/\partial x) = 0$ everywhere on A. Since A is connected, f is constant.

17. By the Cauchy-Riemann equations, $\partial u/\partial x = \partial v/\partial y$. Hence $2\partial v/\partial y = 0$ and so $\partial u/\partial x$ and $\partial v/\partial y$ are identically 0 on A. Thus u depends only on y, and v depends only on x. But then $\partial u/\partial y$ can depend only on y and $\partial v/\partial x$ only on x. Since $\partial u/\partial y = -\partial v/\partial x$ for all x and y, $\partial u/\partial y$ and $-\partial v/\partial x$ equal the same real constant c. Thus $u = cy + d_1$ and $v = -cx + d_2$. Therefore $f = u + iv = -ic(x + iy) + (d_1 + id_2)$.

19. (a) $\mathbb{C}\backslash\{1\}$.   (b) Yes.   (c) x axis$\backslash\{1\}$, unit circle$\backslash\{1\}$.   (d) $90°$.

21. $\mathbb{C}\backslash\{1, e^{2\pi i/3}, e^{-2\pi i/3}\}$

23. (a) u is the imaginary part of the function $f(z) = z^2 + 3z + 1$, which is analytic on all of $\mathbb{C}$. Thus u is harmonic on $\mathbb{C}$.
(b) Either check the second derivatives directly in Laplace's equation or notice that u is the real part of $f(z) = 1/(z - 1)$ to see that u is harmonic on $\mathbb{C}\backslash\{1\}$.

25. Locally in B, $w = \operatorname{Re} g$, where g is analytic. Then $w \circ f = \operatorname{Re}(g \circ f)$. But $g \circ f$ is analytic, and so $w \circ f$ is harmonic.

27. (a) $\dfrac{\partial^2 u}{\partial x^2} + \dfrac{\partial^2 u}{\partial y^2} = e^x \cos y - e^x \cos y = 0$ for all (x, y)

(b) We need $\partial v/\partial y = \partial u/\partial x = e^x \cos y$; thus, $v(x, y) = e^x \sin y + g(x)$. Then $e^x \sin y + g'(x) = \partial v/\partial x = -\partial u/\partial y = e^x \sin y$. Thus $g'(x) = 0$, and so g is constant. To obtain $v(0, 0) = 0$, take $v(x, y) = e^x \sin y$.

(c) $e^z = e^x \cos y + ie^x \sin y$. By parts (a) and (b), the real and imaginary parts satisfy the conditions of the Cauchy-Riemann theorem 1.5.8, and so f is analytic.

29. (a) No. Counterexample: $u(x, y) = x^2 - y^2$ and $v(x, y) = x$ are harmonic, but $u(v(x, y), 0) = x^2$ is not harmonic.

(b) No. Counterexample: $u(z) = v(z) = x$. Then $u(z) \cdot v(z) = x^2$ is not harmonic.

(c) Yes.

31. Write

$$U = \frac{\partial u}{\partial x} \quad \text{and} \quad V = -\frac{\partial u}{\partial y}$$

Then $f = U + iV$. By assumption, U and V have continuous partial derivatives. By the assumption of continuous second partials for u and v, we get

$$\frac{\partial^2 u}{\partial x\, \partial y} = \frac{\partial^2 u}{\partial y\, \partial x} \quad \text{that is,} \quad \frac{\partial U}{\partial y} = -\frac{\partial V}{\partial x}$$

one of the Cauchy-Riemann equations for f. The other equation comes from

$$\frac{\partial U}{\partial x} = \frac{\partial^2 u}{\partial x^2} = -\frac{\partial^2 u}{\partial y^2} = \frac{\partial}{\partial y}\left(-\frac{\partial u}{\partial y}\right) = \frac{\partial V}{\partial y}$$

f is thus analytic by the Cauchy-Riemann theorem (1.5.8).

1.6 DIFFERENTIATION OF THE ELEMENTARY FUNCTIONS

1. (a) Analytic for all z; the derivative is $2z + 1$.

(b) Analytic on $\mathbb{C}\backslash\{0\}$; the derivative is $-1/z^2$.

(c) Analytic on $\mathbb{C}\backslash\{z = (2k + 1)\pi/2 \mid k = 0, \pm 1, \pm 2, \pm 3, \ldots\}$; the derivative is $1/\cos^2 z$.

(d) Analytic on $\mathbb{C}\backslash\{1\}$. The derivative is

$$\frac{2z^3 - 3z^2 - 1}{(z - 1)^2} \exp \frac{z^3 + 1}{z - 1}$$

3. (a) 1.   (b) The limit does not exist.   **5.** No.   **7.** Yes.

9. (a) Analytic on $\mathbb{C}\backslash\{\pm 1\}$. The derivative is $-(z^2 + 1)/(z^2 - 1)^2$.

(b) Analytic on $\mathbb{C}\backslash\{0\}$. The derivative is $(1 - 1/z^2)e^{z + 1/z}$.

11. The minimum is $1/e$, at $z = \pm i$.

13. The map $z \mapsto 2^{z^2}$ is a composition of entire functions and so is entire. $z^{2z} = e^{2z \log z}$ is analytic on the region of analyticity of the logarithm chosen.

REVIEW EXERCISES FOR CHAPTER 1

1. $e^i = \cos(1) + i \sin(1)$ $\log(1+i) = \frac{1}{2} \log 2 + \frac{\pi i}{4} + 2\pi ni$, n an integer

$\sin i = i \frac{1}{2} \left(e - \frac{1}{e} \right)$ $\exp[2 \log(-1)] = 1$

3. $e^{\pi i/16} 1$, $e^{\pi i/16} \left(\frac{1}{\sqrt{2}} + \frac{i}{\sqrt{2}} \right)$, $e^{\pi i/16} i$, $e^{\pi i/16} \left(\frac{-1}{\sqrt{2}} + \frac{i}{\sqrt{2}} \right)$,

$e^{\pi i/16}(-1)$, $e^{\pi i/16} \left(\frac{-1}{\sqrt{2}} - \frac{i}{\sqrt{2}} \right)$, $e^{\pi i/16}(-i)$, $e^{\pi i/16} \left(\frac{1}{\sqrt{2}} - \frac{i}{\sqrt{2}} \right)$

5. $z = 2\pi n \pm i \log(\sqrt{3} + \sqrt{2})$

7. (a) The real axis   (b) A circle, centered at $(\frac{17}{8}, 0)$ of radius $\frac{3}{8}$.

9. (a) $(z^3 + 8)' = 3z^2$ on all of $\mathbf{C}$

(b) $\left(\frac{1}{z^3 + 1} \right)' = \frac{-3z^2}{(z^3 + 1)^2}$ on $\mathbf{C} \backslash \{-1, e^{\pi i/3}, e^{-\pi i/3}\}$

(c) $[\exp(z^4 - 1)]' = 4z^3 \exp(z^4 - 1)$ on all of $\mathbf{C}$

(d) $[\sin(\log z^2)]' = \frac{2}{z} \cos(\log z^2)$, on all $\mathbf{C}$ except the entire imaginary axis

11. (a) Analytic on $\mathbf{C} \backslash \{0\}$; $(e^{1/z})' = -e^{1/z}/z^2$
(b) Analytic on $\mathbf{C} \backslash \{z = (\pi/2) + 2\pi n \,|\, n = 0, \pm 1, \pm 2, \pm 3, \ldots \}$;
$(1/(1 - \sin z)^2)' = 2(\cos z)/(1 - \sin z)^3$
(c) $\mathbf{C} \backslash \{\pm a\}$; $[e^{az}/(a^2 + z^2)]' = [(a^2 + z^2)ae^{az} - 2ze^{az}]/(a^2 + z^2)^2$

13. (a) No.   (b) Yes.   (c) Yes.   **15.** If and only if f is constant.   **17.** $x = y = -1$

19. Note that $v(x, y) = 0$ on A. Therefore,

$$\frac{\partial v}{\partial x} = \frac{\partial v}{\partial y} = 0$$

and the Cauchy-Riemann equations give

$$\frac{\partial u}{\partial x} = \frac{\partial u}{\partial y} = 0$$

Thus f' is identically 0 on the connected set A. Hence, by Proposition 1.5.5, f is constant on A.

21. By hypothesis, $(d/dz)(f(z) - \log z) = 0$. Now use Proposition 1.5.5.

23. $\operatorname*{limit}_{h \to 0} \dfrac{(z_0 - h)^n - z_0^n}{h} = f'(z_0)$ where $f(z) = z^n$

25. Fix a branch of log, for example, the principal branch, which is analytic on $\mathbf{C} \setminus \{$nonpositive real axis$\}$. Then we can write $z^z = e^{z \log z}$, which is analytic on the region of analyticity of logarithm chosen. The derivative is $z^z(1 + \log z)$.

27. $z = 2e^{i\pi/2},\ 2e^{7\pi i/6},\ 2e^{11\pi i/6}$

29. They are the real and imaginary parts of z^3.

31. (a) 0.   (b) Differentiable at all points $z \in \mathbf{C}$.

33. (a) $\dfrac{\partial^2 u}{\partial x^2} = -6y,\ \dfrac{\partial^2 u}{\partial y^2} = 6y$, and so $\dfrac{\partial^2 u}{\partial x^2} + \dfrac{\partial^2 u}{\partial y^2} = 0$.   (b) $v(x, y) = x^3 - 3xy^2$.

35. $z = i, -1, -i, 1$.

37. $f(z) = f(2z) = f(4z) = \cdots = f(2^n z)$ for any z and positive integer n. Letting $w = 2^n z$, we get $f(w/2^n) = f(w)$ for all n. Letting $n \to \infty$ and using the continuity of f at 0 gives $f(0) = f(w)$. Since this can be done for any w in $\mathbf{C}$, f is constant.

2.1 CONTOUR INTEGRALS

1. (a) $2 + i\frac{1}{2}$ (b) $\frac{1}{2}[\cos(2 + 2i) - \cos(2i)]$ (c) 0

3. The principal branch of the logarithm is a function which is analytic on an open set containing γ and whose derivative is $1/z$ there. Since γ is closed, the value of the integral is 0, by the fundamental theorem of calculus for contour integrals (2.1.7).

5. No; for example, let $f(z) = z$, $\gamma(t) = it$ for $t \in [0, 1]$. Then

$$\int_\gamma \operatorname{Re} f = \int_0^1 0 \cdot i \, dt = 0$$

but

$$\operatorname{Re} \int_\gamma f = \operatorname{Re} \int_0^1 iti \, dt = \operatorname{Re}\left(-\frac{t^2}{2}\bigg|_0^1\right) = -\frac{1}{2}$$

7. (a) $2\pi i$ (b) $-\dfrac{i}{3}$

9. For $|z| = 1$ we have

$$\left|\frac{1}{2 + z^2}\right| = \frac{1}{|2 + z^2|} \leq \frac{1}{2 - |z|^2} = 1$$

since $|z_1 + z_2| \geq |z_1| - |z_2|$. Hence

$$\left|\int_\gamma \frac{dz}{2 + z^2}\right| \leq 1 \cdot l(\gamma) = \pi$$

11. (a) $\displaystyle\int_{|z|=1} \frac{dz}{z} = 2\pi i$ $\quad\displaystyle\int_{|z|=1} \frac{dz}{|z|} = 0$ $\quad\displaystyle\int_{|z|=1} \frac{|dz|}{z} = 0$ $\quad\displaystyle\int_{|z|=1} \left|\frac{dz}{z}\right| = 2\pi$ $\quad$ (b) $-\dfrac{i}{3}$

13. 0

15. If $z = e^{i\theta}$ is on γ, then

$$\left|\frac{\sin z}{z^2}\right| = |\sin z| \le \frac{|e^{iz}| + |e^{-iz}|}{2} = \frac{e^{-\sin\theta} + e^{\cos\theta}}{2} \le e$$

Since γ has length 2π, the estimate follows from Proposition 2.1.6.

2.2 CAUCHY'S THEOREM: INTUITIVE VERSION

1. (a) -6 $\quad$ (b) 0 $\quad$ (c) 0 $\quad$ (d) 0

3. 0, by Cauchy's theorem applied to $z = z_0 + re^{i\theta}$.

5. The integral will be zero iff γ encircles neither or both of the roots $-(1/2) \pm (\sqrt{3}/2)i$ of $z^2 + z + 1 = 0$.

7. No; let $f(z) = z$, $\gamma(t) = e^{it}$, $t \in [0, 2\pi]$ (the unit circle). Then $\int_\gamma \operatorname{Re} f(z)\, dz = \pi i$, while $\int_\gamma \operatorname{Im} f(z)\, dz = -\pi$.

9. $-\frac{2}{3} - \frac{2}{3}i$ $\quad$ **11.** $4\pi i$

2.3 CAUCHY'S THEOREM: PRECISE VERSION

1. If it were, then the circle $|z| = 1$ would be homotopic to a point in $\mathbf{C}\backslash\{0\}$ so that $\int_{|z|=1} dz/z = 0$. But $\int_{|z|=1} dz/z = 2\pi i$. This contradiction shows that $\mathbf{C}\backslash\{0\}$ is not simply connected.

3. Let $\gamma: [a, b] \to A$ be a closed curve in A. Define a homotopy $H: [a, b] \times [0, 1] \to A$ by $H(t, s) = s\gamma(t) + (1 - s)z_0$. H is clearly continuous and is into A since A is starlike at z_0. Thus γ is homotopic by H to the constant map at z_0.

5. G contains the segment between each of its points and 0. Sample of part of a proof: Consider a point c for which $0 < \operatorname{Re} c < 1$ and $0 < \operatorname{Im} c < 3$. If $z = sc + (1 - s)0 = sc$ with $0 \le s \le 1$ is on the segment between c and 0, then $\operatorname{Re} z = \operatorname{Re}(sc) = s \operatorname{Re} c$ and $\operatorname{Im} z = s \operatorname{Im} c$, so that $0 \le \operatorname{Re} z \le 1$ and $0 \le \operatorname{Im} z \le 3$. Thus $z \in G$. Points in other parts of G are handled similarly.

7. (a) $2\pi i$ $\quad$ (b) 0 $\quad$ (c) 0 $\quad$ (d) πi $\quad$ **9.** (a) $2\pi i$ $\quad$ (b) $2\pi i$

2.4 CAUCHY'S INTEGRAL FORMULA

1. (a) $2\pi i$ $\quad$ (b) $2\pi i$

3. The Cauchy inequalities (2.4.7) show that $f^{(k)}(z)$ is identically 0 for $k > n$. The conclusion follows from Exercise 20 of Sec. 1.5.

5. (a) 0   (b) $-\pi i/3$

7. Using the Cauchy inequalities, $|f'(0)| \le 1/R$ for every $R < 1$. Hence $|f'(0)| \le 1$. This is the best possible bound, as is clear from the example $f(z) = z$.

9. Let $\tilde{\gamma}$ be the circle $\tilde{\gamma}(t) = z_1 + re^{it}$, $0 \le t \le 2\pi$, $|z_0 - z_1| < r$. γ is homotopic to $\tilde{\gamma}$ by $H(t, s) = s(z_1 + re^{it}) + (1 - s)(z_0 + re^{it})$. Since z_1 is not in the image of the homotopy,

$$I(\gamma, z_1) = \frac{1}{2\pi i} \int_\gamma \frac{1}{z - z_1} = \frac{1}{2\pi i} \int_{\tilde{\gamma}} \frac{1}{z - z_1} = 1$$

11. By Proposition 1.5.3, f is analytic on $A \setminus \{z_0\}$, so it is continuous there. Since

$$f(z_0) = F'(z_0) = \lim_{z \to z_0} \frac{F(z) - F(z_0)}{z - z_0} = \lim_{z \to z_0} f(z)$$

f is also continuous at z_0. By the corollary to Morera's theorem (2.4.11), f is analytic on A.

13. (a) 0   (b) 0   (c) 0   (d) $\pi i/2$ **15.** $4\pi i$

17. $1/f$ is entire and $|1/f(z)| \le 1$ on C. Therefore, $1/f$ is constant by Liouville's theorem. Hence f is constant.

19. (a) $\pi/2 + i(\pi/2)$ (b) 0

21. Write down Cauchy's integral formula for $f(z_1), f(z_2)$ and $f'(z_0)$, substitute into the left-hand side, and simplify.

2.5 MAXIMUM MODULUS THEOREM AND HARMONIC FUNCTIONS

1. e

3. Let $A = C \setminus \{0\}$, $f(z) = e^z$.

5. $f - g$ is continuous on cl (A) and analytic on A. $(f - g)(z) = 0$ for $z \in$ bd (A), and so by the maximum modulus theorem, $(f - g)(z) = 0$ for all $z \in A$. In other words, $f(z) = g(z)$ for all $z \in A$. Hence $f = g$ on all of cl $(A) = A \cup$ bd (A).

7. $|e^{z^2}|$ attains a maximum value of e at ± 1.

9. (a) $v(x, y) = -\cosh x \cos y$ (b) $v = \arctan (y/x)$ or $-\arctan (x/y)$
(c) $v(x, y) = e^x \sin y$ (Note that an arbitrary constant may be added to each.)

11. If $z = x + iy$, Re $e^z = e^x \cos y$ and Im $e^z = e^x \sin y$. The normal vector to the level curves of these functions is given by the gradient vectors

$$(e^x \cos y, -e^x \sin y) \quad \text{and} \quad (e^x \sin y, e^x \cos y)$$

Since these are orthogonal, the curves are also orthogonal.

13. Since f is analytic and nonconstant on A, z_0 cannot be a relative maximum. Thus in every neighborhood of z_0, and, in particular, in $\{z$ such that $|z - z_0| < \epsilon\}$, there is a point z with $|f(z)| >$

$|f(z_0)|$. If $f(z_0) \neq 0$, then by continuity, $f(z) \neq 0$ in some small disk D centered at z_0. Thus $1/f(z)$ is analytic on D. By this last argument, there is a ζ close to z_0 such that

$$\left| \frac{1}{f(\zeta)} \right| > \left| \frac{1}{f(z_0)} \right|$$

Therefore,

$$|f(\zeta)| < |f(z_0)|$$

15. $|g(0)| = |0| = 0$, and so $g(0) = 0$. Also, $|g(z)| = |z| < 1$ for all $z \in \{z$ such that $|z| < 1\}$. Thus Schwarz's lemma applies, and so $g(z) = cz$ with $|c| = 1$.

17. 0

REVIEW EXERCISES FOR CHAPTER 2

1. (a) 0   (b) 0   (c) $2\pi i$ (d) $2\pi i \cos (1)$

3. If $r_1 > r_2 > 1$, then γ_{r_1} is homotopic in $\{z$ such that $|z| > 1\}$ to γ_{r_2}, and so the integrals are the same.

5. $-\frac{2}{3}$

7. If $z_1 \in A$, let γ be a path in A from z_0 to z_1. By the distance lemma (1.4.21), there is a $\delta > 0$ such that the set $B = \{z \mid \text{there is a point } w \text{ on } \gamma \text{ with } |z - w| < \delta\} \subset A$. The maximum modulus theorem shows that f is constant on this bounded subregion and in particular $f(z_1) = f(z_0)$. Since z_1 was arbitrary, f is constant on A.

9. $v(x, y) = \dfrac{-y}{(x - 1)^2 + y^2}$ on $\mathbb{C}\backslash\{1\}$ **11.** Consider $\displaystyle\int_{|z|=1} \frac{e^z}{z^2}\, dz$ to obtain 2π.

13. i; $\log i = i(\pi/2) + 2\pi i n$; $\log (-i) = -i(\pi/2) + 2\pi i n$; $i^{\log(-1)} = -i$.

15. By the Cauchy integral formulas, f' is analytic on A. Since f is nonzero in A, f'/f is analytic on A and the integral is 0 by the Cauchy integral theorem.

17. No; let γ be the unit circle. Then $\int_\gamma x\, dx + x\, dy = \pi$.

19. (a) No   (b) Yes   (c) Yes   **21.** $2e^{i\pi/6}$, $2e^{i5\pi/6}$, and $2e^{i3\pi/2}$

23. By the mean value property for harmonic functions (2.5.9),

$$u(0) = \frac{1}{2\pi} \int_0^{2\pi} u(Re^{i\theta})\, d\theta$$

By Poisson's formula (2.5.13),

$$u(re^{i\phi}) = \frac{R^2 - r^2}{2\pi} \int_0^{2\pi} \frac{u(Re^{i\theta})}{R^2 - 2Rr \cos (\theta - \phi) + r^2}\, d\theta$$

Using these equalities we get

$$\frac{R^2 - r^2}{2\pi} \int_0^{2\pi} \frac{u(Re^{i\theta})}{R^2 + 2Rr + r^2}\, d\theta \le u(z) \le \frac{R^2 - r^2}{2\pi} \int_0^{2\pi} \frac{u(Re^{i\theta})}{R^2 - 2Rr + r^2}\, d\theta$$

that is,

$$\frac{(R+r)(R-r)}{(R+r)(R+r)} \frac{1}{2\pi} \int_0^{2\pi} u(Re^{i\theta})\, d\theta \le u(z) \le \frac{(R-r)(R+r)}{(R-r)(R-r)} \frac{1}{2\pi} \int_0^{2\pi} u(Re^{i\theta})\, d\theta$$

Therefore,

$$\frac{R - |z|}{R + |z|} u(0) \le u(z) \le \frac{R + |z|}{R - |z|} u(0)$$

3.1 CONVERGENT SERIES OF ANALYTIC FUNCTIONS

1. (a) Does not converge. (b) 0.

3. The limit is 1 and the convergence is not uniform.

5. If $|z| \le r$, then $|z^n - 0| \le r^n$, and so $|z^n - 0| < \epsilon$ whenever $n > (\log \epsilon)/(\log r)$. These n work for all z in D_r, so the convergence is uniform. The convergence is not uniform on $D(0, 1)$. For example, if $z = r$, the required minimum n is $(\log \epsilon)/(\log r)$, which becomes arbitrarily large as r gets close to 1.

7. Neither of these series converges absolutely. However, both the real and imaginary parts of each are alternating series whose terms decrease in absolute value monotonically to zero and thus are convergent (by the alternating series test from calculus).

9. The sequence of partial sums converges uniformly; thus it converges to a continuous function and the assertion follows.

11. Let D be any closed disk in A. Then there exists an $r > 1$ such that $|z| > r$ for all $z \in D$. Hence $|1/z^n| < (1/r)^n$ for all $z \in D$. But $\sum_{n=1}^{\infty} (1/r)^n$ converges, since $1/r < 1$. Thus $\sum_{n=1}^{\infty} 1/z^n$ converges absolutely and uniformly on D. Since $1/z^n$ is analytic on A, the analytic convergence theorem (3.1.8) shows that $\sum_{n=1}^{\infty} 1/z^n$ is analytic on A.

13. Let D be a closed disk in A and let δ be its distance from the boundary Im $z = \pm 1$. For $z = x + iy \in D$, prove that $|e^{-n} \sin (nz)| \le e^{-n\delta}$.

15. By Worked Example 3.1.15, $\zeta(z) = \sum_{n=1}^{\infty} n^{-z}$ converges uniformly on closed disks in A and thus it is analytic on A with $\zeta'(z) = \sum_{n=1}^{\infty} (-\log n)n^{-z}$, which also converges uniformly on closed disks in A and thus is analytic. By induction, $\zeta^{(k)}(z) = \sum_{n=1}^{\infty} (-\log n)^k n^{-z}$ converges uniformly on closed disks in A and thus is analytic. Therefore, $(-1)^k \zeta^k(z) = \sum_{n=1}^{\infty} (\log n)^k n^{-z}$ is also analytic.

17. No; let $f_n(z) = \sum_{k=1}^{n} z^k/k^2$.

19. $\{z$ such that $|2z - 1| < 1\} = \{z$ such that $|z - \frac{1}{2}| < \frac{1}{2}\}$

3.2 POWER SERIES AND TAYLOR'S THEOREM

1. (a) 1. (b) e. (c) e. (d) 1.

3. (a) $e^z = \sum_{n=0}^{\infty} (e/n!)(z-1)^n$, which converges everywhere.
(b) $1/z = \sum_{n=0}^{\infty} (-1)^n(z-1)^n$; the series converges for $|z-1| < 1$.

5. (a) $\dfrac{\sin z}{z} = \sin(1) + [\cos(1) - \sin(1)](z-1)$

$$+ \left[\frac{\sin(1)}{2} - \cos(1)\right](z-1)^2 + [\tfrac{5}{6}\cos(1) - \tfrac{1}{2}\sin(1)](z-1)^3 + \cdots$$

(b) $z^2 e^z = \sum_{n=0}^{\infty} \dfrac{1}{n!} z^{n+2}$ (c) $e^z \sin z = z + z^2 + \tfrac{1}{3}z^3 - \tfrac{1}{30}z^5 + \cdots$

7. (a) $\sum_{n=0}^{\infty} z^{2n}/n!$ (b) $\sum_{n=0}^{\infty} [1 - 1/(2^{n+1})]z^n$

9. $\sqrt{z^2 - 1} = i - \dfrac{i}{2}z^2 - \dfrac{i}{8}z^4 + \cdots$

11. For $\sinh z$, the odd derivatives are 1 and the even derivatives are 0 at $z = 0$. Thus by Taylor's theorem,

$$\sinh z = \sum_{n=1}^{\infty} \frac{z^{2n-1}}{(2n-1)!}$$

The argument for $\cosh z$ is similar.

13. $\dfrac{1}{(1-z)^2} = \sum_{n=1}^{\infty} nz^{n-1}$ and $\dfrac{1}{(1-z)^3} = \dfrac{1}{2}\sum_{n=2}^{\infty} n(n-1)z^{n-2}$

15. If $|z| < R$, then $\Sigma\, a_n z^n$ converges absolutely; that is, $\Sigma\, |a_n||z|^n$ converges. But $|(\text{Re } a_n)z^n| \le |a_n||z^n|$, so $\Sigma\, |(\text{Re } a_n)z^n| \le \Sigma\, |a_n||z^n|$ and hence $\Sigma\, (\text{Re } a_n)z^n$ converges. Since $\Sigma\, (\text{Re } a_n)z^n$ converges for any $|z| < R$, the radius of convergence must be $\ge R$.

17. The region for the first series is $A = \{z \text{ such that } |\text{Im }(z)| < \log 2\}$. The second series is not analytic anywhere.

19. (a) Suppose that the Taylor series for f, $\sum_{n=0}^{\infty} f^{(n)}(z_0)(z - z_0)^n/n!$, has radius of convergence R and converges to a function $g(z)$ on $D(z_0; R)$. Let R_0 be the radius of D. By Taylor's theorem (3.2.7), the restriction of g to D is equal to f and $R \ge R_0$. The function g is analytic and so continuous on $D(z_0; R)$. If R were $> R_0$, then g would be continuous and hence bounded on the compact closed disk cl $(D) = \{z \text{ such that } |z - z_0| \le R_0\}$. But g and f are the same on D, and f is not bounded on D, so R is not greater than R_0 and thus $R = R_0$.
(b) Branches of $\log(1 + z)$ may be defined with the plane cut along any ray from -1 to ∞. For the principal branch this is along the negative real axis. These determinations differ by an additive constant depending on the angle between the ray and the real axis. The series expansions around $z_0 = -2 + i$ differ only in the constant terms and so have the same radius of convergence, which we may call R. If we choose the ray leading from -1 directly away from z_0, then $D(z_0; \sqrt{2})$ lies in the region of analyticity, and so $R \ge \sqrt{2}$. But $D = D(z_0; 1)$ is the largest disk centered at z_0 contained in the region of analyticity of the principal branch of $\log(1 + z)$, and $\sqrt{2} > 1$.

21. Suppose $|h(z)| \leq M$ for z in B, and let $\epsilon > 0$. If $g(z) = \Sigma_i\, g_i(z)$, then uniform convergence gives an N such that $|g(z) - \Sigma_{i=m+1}^n\, g_i(z)| < \epsilon/M$ whenever $n \geq N$ and $m \geq N$. Thus,

$$\left| h(z)g(z) - \sum_{i=m+1}^n h(z)g_i(z) \right| = |h(z)| \left| g(z) - \sum_{i=m+1}^n g_i(z) \right| < \epsilon$$

23. $\dfrac{d}{dz}\left| \displaystyle\sum_{n=0}^\infty H_n(x)\dfrac{z^n}{n!} \right| = (2x - 2z)e^{2xz-z^2} = (2x - 2z)\displaystyle\sum_{n=0}^\infty H_n(x)\dfrac{z^n}{n!}$

Therefore

$$\sum_{n=1}^\infty H_n(x)\frac{z^{n-1}}{(n-1)!} = \sum_{n=0}^\infty 2xH_n(x)\frac{z^n}{n!} - \sum_{n=1}^\infty 2H_{n-1}(x)\frac{z^n}{(n-1)!}$$

$$H_1(x) + \sum_{n=1}^\infty H_{n+1}(x)\frac{z^n}{n!} = 2xH_0(x) + \sum_{n=1}^\infty [2xH_n(x) - 2nH_{n-1}(x)]z^n/n!$$

Equating coefficients gives the desired results.

25. $f(z) = 1 + 2z + \dfrac{5}{2}z^2 + \dfrac{5}{4}\displaystyle\sum_{n=3}^\infty \dfrac{2^n}{n!}z^n = \dfrac{5e^{2z} - 2z - 1}{4}$

3.3 LAURENT'S SERIES AND CLASSIFICATION OF SINGULARITIES

1. (a) $\dfrac{1}{z} - \dfrac{1}{3!z^3} + \dfrac{1}{5!z^5} - \cdots$, $0 < |z| < \infty$

(b) $\dfrac{1}{z} - 1 + z - z^2 + z^3 - z^4 + \cdots$, $0 < |z| < 1$

(c) $z - z^2 + z^3 - z^4 + z^5 - \cdots$, $|z| < 1$

(d) $\dfrac{1}{z^2} + \dfrac{1}{z} + \dfrac{1}{2!} + \dfrac{1}{3!}z + \dfrac{1}{4!}z^2 + \cdots$, $0 < |z| < \infty$

3. $z/(z+1) = 1/(1 + 1/z) = \Sigma_{n=0}^\infty (-1)^n(1/z)^n = \Sigma_{n=0}^\infty (-1)^n z^{-n}$ for $|z| > 1$.

5. Let γ_3 be a radial segment from γ_2 to γ_1 not passing through z. Let $\gamma_4 = \gamma_2 + \gamma_3 - \gamma_1 - \gamma_3$ as indicated in Fig. A.3.3.5. Let γ_5 be a small circle about z lying between γ_1 and γ_2 and not crossing γ_3.

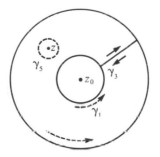

FIGURE A.3.3.5 Solution for Exercise 3.3.5.

Then γ_4 is clearly homotopic to γ_5 in the region of analyticity of $f(t)$ and of $f(t)/(t-z)$. Hence, by the Cauchy integral formula,

$$f(z) = \frac{1}{2\pi i} \int_{\gamma_5} \frac{f(t)}{t-z}\, dt = \frac{1}{2\pi i} \int_{\gamma_4} \frac{f(t)}{t-z}\, dt$$

$$= \frac{1}{2\pi i} \int_{\gamma_2} \frac{f(t)}{t-z}\, dt + \frac{1}{2\pi i} \int_{\gamma_3} \frac{f(t)}{t-z}\, dt - \frac{1}{2\pi i} \int_{\gamma_1} \frac{f(t)}{t-z}\, dt - \frac{1}{2\pi i} \int_{\gamma_3} \frac{f(t)}{t-z}\, dt$$

$$= \frac{1}{2\pi i} \int_{\gamma_2} \frac{f(t)}{t-z}\, dt - \frac{1}{2\pi i} \int_{\gamma_1} \frac{f(t)}{t-z}\, dt$$

7. Use the fact that there is an analytic function ϕ defined on a neighborhood of z_0 such that $\phi(z_0) \neq 0$ and $f(z) = \phi(z)/(z-z_0)^k$.

9. (a) No. (b) No. (c) Yes. **11.** $\dfrac{1}{e^z-1} = \dfrac{1}{z} - \dfrac{1}{2} + \dfrac{1}{12}z - \dfrac{1}{720}z^3 + \cdots$

13. $\cot z = \dfrac{1}{z} - \dfrac{1}{3}z - \dfrac{1}{45}z^3 - \dfrac{2}{945}z^5 + \cdots$

15. Since $\sum_{n=1}^{\infty} b_n/(z-z_0)^n$ converges for $|z-z_0| = (R+r)/2 > R$, $b_n/(z-z_0)^n \to 0$ for $|z-z_0| = (R+r)/2$. That is, $2^n|b_n|/(R+r)^n \to 0$, and so is bounded. Therefore, by the Abel-Weierstrass lemma, $\sum_{n=1}^{\infty} b_n(z-z_0)^n$ converges uniformly and absolutely on $A_\rho = \{z$ such that $|z-z_0| \le \rho\}$ if $\rho < 2/(R+r)$. Taking $\rho = 1/r < 2/(R+r)$, $\sum_{n=1}^{\infty} b_n(z-z_0)^n$ converges uniformly and absolutely on $\{z$ such that $|z-z_0| \le 1/r\}$. In particular, $\sum_{n=1}^{\infty} |b_n|/r^n$ converges. If $z \in F_r$, $|z-z_0| > r$, and so $\left|\dfrac{b_n}{(z-z_0)^n}\right| < \dfrac{|b_n|}{r^n}$. Thus $\sum_{n=1}^{\infty} b_n/(z-z_0)^n$ converges uniformly on F_r by the Weierstrass M test.

17. $\cos(1/z)$ has a zero of order 1 at

$$\frac{1}{z} = \frac{(2n+1)\pi}{2} \qquad n = 0, \pm 1, \pm 2, \ldots$$

that is, at

$$z = \frac{2}{(2n+1)\pi}$$

Thus, $1/\cos(1/z)$ has simple poles at these points and $z = 0$ is not an isolated singularity.

19. (a) $\frac{1}{2}$ (b) $\frac{1}{2}$ (c) 1 (d) 0

REVIEW EXERCISES FOR CHAPTER 3

1. $\sum_{n=1}^{\infty} (-1)^{n-1}(z-1)^n/n$ **3.** $\dfrac{1}{z^2} - \dfrac{1}{z} + 1 - z + z^2 - z^3 + \cdots$

5. $\displaystyle\sum_{n=1}^{\infty} \frac{(-1)^{n+1}}{(2n-1)!} z^{4n}$

7. Suppose that $w \in \mathbb{C}$, $w \neq 0$. Solve $e^{1/z} = w$ for z as follows:

$$\frac{1}{z} = \log w = \log|w| + i(\arg w + 2\pi n)$$

$$z = \frac{1}{\log|w| + i(\arg w + 2\pi n)}$$

Infinitely many of these solutions lie in any deleted neighborhood of the origin.

9. (a) $\frac{1}{2}$.   (b) 1.

11. The coefficients for either series are given by

$$a_n = \frac{1}{2\pi i} \int_\gamma \frac{f(\zeta)}{\zeta^{n+1}} \, d\zeta$$

where γ is the circle of radius 2 centered at the origin.

13. Schwarz' lemma applies to give $|f(z)| \leq |z|$ for all $|z| < 1$. Also, $|f(z)| = |z|$ for any $|z| < 1$ implies that $f(z) = cz$ for a constant c with $|c| = 1$.

15. $-2\pi i/3$

17. (a) $= \frac{1}{z} - z + z^3 - z^3 + z^7 - \cdots$ (b) $= \frac{1}{z^3} - \frac{1}{z^5} + \frac{1}{z^7} - \frac{1}{z^9} + \frac{1}{z^{11}} - \cdots$

19. Yes. You could base your argument on the fact that a composition of analytic functions is analytic.

21. Since f is bounded near z_0, z_0 is neither a pole nor an essential singularity

23. 2π

25. (a) Poles of order 1 at $z = 1$ and $z = 5$
(b) Removable singularity at $z = 0$
(c) Pole of order 1 at $z = 0$
(d) Pole of order 1 at $z = 1$

27. $2\pi i$

29. (a) $a_0 = -\frac{1}{2}$, $a_1 = 0$, $a_2 = (4 - \pi^2)/8$.
(b) The denominator has zeros of order 2 at the odd integers. The numerator has zeros of order 1 at ± 1, and so the function has simple poles at ± 1 and poles of order 2 at all the other odd integers.
(c) The closest singularities to 0 are at ± 1. The radius of convergence is 1.

31. No; $\sin z$.   **33.** (a) 2π.   (b) $B_n = \frac{n!}{2\pi} \int_0^{2\pi} \frac{e^{-i(n-1)\theta}}{e^{e^{i\theta}} - 1} \, d\theta$ **35.** 1

37. $a_0 + z + z^2 + z^4 + z^8 + z^{16} + \cdots$, where $a_0 = f(0)$. Show uniqueness by showing the coefficients are uniquely determined.

39. $|f(z)| = \lim\limits_{n \to \infty} \left| \sum\limits_{k=0}^{n} \frac{f^{(k)}(0)}{k!} z^k \right|$ for all z. But

$$\left| \sum_{k=0}^{n} \frac{f^{(k)}(0)}{k!} z^k \right| \le \sum_{k=0}^{n} \frac{|f^{(k)}(0)|}{k!} |z|^k \le \sum_{k=0}^{n} \frac{1}{k!} |z|^k \le \sum_{k=0}^{\infty} \frac{1}{k!} |z|^k = e^{|z|},$$

and so the limit is no more than $e^{|z|}$.

4.1 CALCULATION OF RESIDUES

1. (a) 0.   (b) 1.   (c) -1.   (d) $\frac{1}{6}$.   (e) 0.   **3.** Let $f(z) = 1/z$.

5. The correct residue is 2.

7. (a) Res $(f, 0) = \frac{1}{64}$, Res $(f, -4) = -\frac{1}{64}$ (b) Res $(f, -1) = 0$
(c) Res $(f, \sqrt[3]{3}) = 3^{-5/3}$, Res $(f, \sqrt[3]{3}e^{2\pi i/3}) = 3^{-5/3}e^{-4\pi i/3}$, Res $(f, \sqrt[3]{3}e^{4\pi i/3}) = 3^{-5/3}e^{-2\pi i/3}$

9. $\frac{1}{6}$

11. Res $(f_1 f_2, z_0) = a_1$ Res $(f_2, z_0) + a_2$ Res (f_1, z_0) where a_i is the constant term in the expansion of f_i.

13. (a) 0.   (b) $-e/2$.
(c) Res $(f, 0) = 1$, Res $(f, 1) = -1$.   (d) Res $(f, 0) = 1$, Res $(f, 1) = -e/2$.

4.2 THE RESIDUE THEOREM

1. (a) 0.   (b) 0.   **3.** 0   **5.** $-12\pi i$

7. (a) 0.   (b) 0.   **9.** (a) $2\pi i$.   (b) $2\pi i$.

11. $-\int_{\gamma} g\left(\frac{1}{z}\right) \frac{1}{z^2} \, dz = \int_{\tilde{\gamma}} g(w) \, dw$ where $\tilde{\gamma}$ is the curve $1/\gamma$

13. (a) -8.   (b) $2\pi i$.   **15.** $-\pi i$

4.3 EVALUATION OF DEFINITE INTEGRALS

1. $\pi/\sqrt{3}$ **3.** $\dfrac{\pi a}{(a^2 - b^2)^{3/2}}$

5. $\dfrac{\pi e^{-m/\sqrt{2}}}{2\sqrt{2}} \left(\cos \dfrac{m}{\sqrt{2}} + \sin \dfrac{m}{\sqrt{2}} \right)$ if $m > 0$ $\dfrac{\pi e^{m/\sqrt{2}}}{2\sqrt{2}} \left(\cos \dfrac{m}{\sqrt{2}} - \sin \dfrac{m}{\sqrt{2}} \right)$ if $m < 0$

7. $\pi e^{-1}/2$ **9.** $-\pi i/(a-1)^2$

11. The function is even and line 3 of Table 4.3.1 applies.   **13.** 0

15. $-\pi i/2$ (or $\pi i/2$ if a different branch is used). Construct $\sqrt{z^2-1}$ much as in Example 4.3.15, but make the branch cut for the factor $\sqrt{z-1}$ go from 1 to $-\infty$ and that for $\sqrt{z+1}$ from -1 to $-\infty$. Crossing the real axis at x with $|x| > 1$ requires crossing either both cuts or neither. The product is analytic on $\mathbf{C}\backslash\{z \mid \operatorname{Im} z = 0 \text{ and } |\operatorname{Re} z| \le 1\}$, as in Example 4.3.15.

17. $\pi e^{-ab}/a$ **19.** Use Exercise 18; Res $((-z)^{a-1}f(z), 1) = (e^{\pi i})^{a-1}$.

21. After checking that all the integrals exist and the operations are justified, compute

$$\left(\int_{-\infty}^{\infty} e^{-y^2}\, dy\right)\left(\int_{-\infty}^{\infty} e^{-x^2}\, dx\right) = \int_{-\infty}^{\infty}\int_{-\infty}^{\infty} e^{-(x^2+y^2)}\, dx\, dy = \int_0^{\pi}\int_0^{2\pi} e^{-r^2}r\, d\theta\, dr$$

$$= 2\pi \int_0^{\infty} e^{-r^2}r\, dr = -\pi e^{-r^2}\Big|_0^{\infty}$$

$$= -\pi(0-1) = \pi$$

and so $\int_{-\infty}^{\infty} e^{-x^2}\, dx = \sqrt{\pi}$.

23. $2\pi(97\sqrt{3}-168)/3$ **25.** (a) $\sqrt{\pi i}$

4.4 EVALUATION OF INFINITE SERIES AND PARTIAL-FRACTION EXPANSIONS

1. As in the proof of Proposition 4.4.2, $\displaystyle\int_{C_N} \frac{\pi \cot \pi z}{z^4}\, dz \to 0$ as $N \to \infty$. The residue at $n \ne 0$ is $1/n^4$. Compute the first few terms of the Laurent series $\cot z = 1/z - \frac{1}{3}z - \frac{1}{45}z^3 - \cdots$ to find that the residue at 0 is $-\pi^4/45$, so that

$$\lim_{N\to\infty}\left[-\frac{\pi^4}{45} + \sum_{n=1}^{N}\frac{1}{n^4} + \sum_{n=-1}^{-N}\frac{1}{(-n)^4}\right] = \lim_{N\to\infty}\left(-\frac{\pi^4}{45} + 2\sum_{n=1}^{N}\frac{1}{n^4}\right)$$

and thus $\sum_1^{\infty} 1/n^4 = \pi^4/90$.

3. Apply the summation theorem (4.4.1) and Proposition 4.4.2.

$$\operatorname{Res}\left(\frac{\pi \cot \pi z}{z^2 + a^2}, \pm ai\right) = -\frac{\pi}{2a}\coth \pi a$$

(check this), and so

$$\frac{\pi}{a}\coth \pi a = \sum_{n=-\infty}^{\infty}\frac{1}{n^2 + a^2} = \frac{1}{a^2} + 2\sum_{n=1}^{\infty}\frac{1}{n^2 + a^2}$$

Thus

$$\sum_{n=1}^{\infty}\frac{1}{n^2 + a^2} = -\frac{1}{2a^2} + \frac{\pi}{2a}\coth \pi a$$

and

$$\sum_{n=0}^{\infty}\frac{1}{n^2 + a^2} = \frac{1}{a^2} + \sum_{n=1}^{\infty}\frac{1}{n^2 + a^2} - \frac{1}{2a^2} + \frac{\pi}{2a}\coth \pi a$$

5. $\displaystyle\sum_{n=-\infty}^{\infty} (-1)^n f(n) = -[\text{sum of residues of } \pi \csc \pi z\, f(z) \text{ at the poles of } f]$

Here f should obey some conditions like: There exist an $R > 0$ and an $M > 0$ such that for $|z| > R$, $|f(z)| \le M/|z|^\alpha$ where $\alpha > 1$. If some of the poles of f should lie at the integers, the technique could still be used. After verifying that

$$\int_{C_N} \frac{\pi f(z)}{\sin \pi z}\, dz \to 0 \qquad \text{as} \qquad N \to \infty$$

we would have

$$-\sum_{\substack{n=-\infty \\ n \text{ not a pole of } f}}^{\infty} (-1)^n f(n) = [\text{the sum of the residues of } \pi \csc \pi z\, f(z) \text{ at the poles of } f]$$

7. Consider $f(z) = \cot z - 1/z$. Then $\displaystyle\lim_{z\to 0} f(z) = 0$ and f is analytic at 0. Check that it has simple poles at $z = n\pi$ for $n \ne 0$ with residue 1 at each. Let C_N be the square with corners at $(N + \tfrac{1}{2})\pi(\pm 1 \pm i)$. Along C_N we have $\cot z = -\cot(-z)$, and $-z$ is on C_N when z is. Thus it suffices to check $|\cot z|$ for $y = \operatorname{Im} z \ge 0$. If $z = x + iy$, $y > 0$, then

$$|\cot z| = \left| \frac{e^{iz} + e^{-iz}}{e^{iz} - e^{-iz}} \right|$$

$$= \left| \frac{e^{2ix-2y} + 1}{e^{2ix-2y} - 1} \right| \le \frac{2}{|e^{2ix-2y} - 1|}$$

on the upper horizontal of the square $y = (N + \tfrac{1}{2})\pi > 1$, and so $|\cot z| \le 2/(1 - e^{-2}) \le 4$. On the vertical sides, $x = \pm(N + \tfrac{1}{2})\pi$, and so $e^{2ix} = -1$, and $|\cot z| \le \left| \dfrac{2}{-e^{-2y} - 1} \right| < 2$. In any case, $|f(z)| \le 4 + 2/\pi$ for z on C_N, and so with $R = (N + \tfrac{1}{2})\pi$, $M = 4 + 2/\pi$, and $S = 8$, the conditions of the partial-fraction theorem (4.4.5) are met and the data on poles and residues may be entered to give the desired formula.

9. An exact answer to this seemingly simple problem is not known. The sum is $\zeta(3)$ where ζ is the Riemann zeta function, important in analysis and number theory and a source of several famous open problems in mathematics. The method of the summation theorem (4.4.1) may be used for summing $\zeta(p) = \sum_1^\infty (1/n^p)$ for even p as in Proposition 4.4.3 and Exercise 1. One gets

$$\zeta(2m) = (-1)^{m+1}(2\pi)^{2m} \frac{B_{2m}}{2(2m)!}$$

where the B_{2m}'s are the Bernoulli numbers involved in the expansion of the cotangent function. (See also Review Exercise 33 of Chap. 3.) This method fails for odd p basically because $1/(-n)^p + 1/n^p = 0$, not $2/n^p$. An approximate value is $\zeta(3) \approx 1.2020569$, but until recently it was not even known if $\zeta(3)$ was irrational. This was shown in 1978 by R. Apery. (See *Mathematical Intelligencer*, Vol. 1 (1979), pp. 195–203.) Even irrationality is still unknown for $\zeta(p)$ for other odd values of p.

REVIEW EXERCISES FOR CHAPTER 4

1. $2\pi/\sqrt{3}$ **3.** $(\pi\sqrt{2})/2$ **5.** $\dfrac{\pi}{2}\left[2 - \dfrac{1}{e} - \cos(1)\right]$

7. $\pi/\sqrt{5}$ **9.** 2π **11.** $2\pi i \sin(1)$

13. (a) $\dfrac{1}{2} + \dfrac{3}{4}z + \dfrac{7}{8}z^2 + \cdots + \dfrac{2^{n+1}-1}{2^{n+1}}z^n + \cdots$

(b) $\dfrac{1}{z^2} + \dfrac{3}{z^3} + \dfrac{7}{z^4} + \dfrac{15}{z^5} + \cdots + \dfrac{2^n-1}{z^{n+1}} + \cdots$

15. $\pi/2$

17. (a) Res $(f, 0) = -1$. The other residues are at z with $z^2 = 2\pi ni, n = \pm 1, \pm 2, \ldots$, and are equal to $-\frac{1}{2}$.
(b) Res $(f, n\pi) = 2\pi n \cos(n^2\pi^2)$ (c) $\cos(1)$

19. $\operatorname{Res}\left(\dfrac{f''}{f'}, z_0\right) = k - 1$ and $\operatorname{Res}\left(\dfrac{f''}{f}, z_0\right) = \dfrac{2k}{k+1} \cdot \dfrac{f^{(k+1)}(z_0)}{f^{(k)}(z_0)}$ for $k \neq 0, 1$

21. $\dfrac{1}{1 + (z-1)}$ has been expanded incorrectly.

23. $-\pi i$

25. (a) The radius of convergence is infinite.
(b) The radius of convergence is 1 (use the root test).
  Note: To use the ratio test, the following facts are used:

  (i) $\displaystyle\lim_{n\to\infty}\left(1 + \dfrac{z}{n}\right)^n = e^z$.

  (ii) In the power series $\Sigma\, a_n z^n$, if the coefficients a_n tend to a nonzero finite limit, then the radius of convergence must be 1.

27. (a) $1 + 3z + 6z^2 + 10z^3 + \cdots + \dfrac{(n+1)(n+2)}{2}z^n + \cdots$

(b) $-\dfrac{1}{z^3} - \dfrac{3}{z^4} - \dfrac{6}{z^5} - \cdots - \dfrac{(n+1)(n+2)}{2}\dfrac{1}{z^{n+3}} - \cdots$

(c) $\dfrac{1}{8} + \dfrac{3}{16}(z+1) + \dfrac{6}{32}(z+1)^2 + \cdots + \dfrac{(n+1)(n+2)}{2^{n+4}}(z+1)^n + \cdots$

(d) $\dfrac{-1}{(z-1)^3}$

29. (a) Use $\sin^3 x = (3 \sin x - \sin 3x)/4$. Use an argument like that for Cauchy principle value, checking directly that

$$\int_\gamma \frac{3e^{iz} - e^{3iz}}{z^3} \to -3\pi i \quad \text{as} \quad \rho \to 0$$

where γ is a half circle in the upper half plane from $-\rho$ to ρ.
(b) Use line 5 of Table 4.3.1.

31. Use Exercise 5 of Sec. 4.4. $F(z) = (\pi \csc \pi z)/(z + a)^2$ has a pole at $z = -a$ with residue $-\pi^2 \csc (\pi a) \cot (\pi a)$.

33. The last equality is wrong since the integral along the semicircle is omitted. We cannot conclude that the integral along the semicircle goes to 0 as $R \to \infty$ and thus must evaluate it more carefully.

35. It will suffice for f to be analytic in a region containing the real axis and the upper half plane and to be such that the integral of $f(z)/(z - x)$ along the upper semicircle of $|z| = R \to 0$ as $R \to \infty$. These conditions will hold if $|f(z)| < M/R^\alpha$ for some $M > 0$ and $\alpha > 0$ for large enough R, and for z lying in the upper half plane. Use Exercise 34 for the last part.

37. $\dfrac{\pi}{2b} \tanh \dfrac{\omega\pi}{2b}$

5.1 BASIC THEORY OF CONFORMAL MAPPINGS

1. The first three quadrants.

3. (a) Everywhere except $z = 0$ and $z = -\frac{2}{3}$ (b) Everywhere except $z = -\frac{1}{3}$

5. $v(x, y) = 1 - 2xy + \dfrac{x^2 - y^2}{(x^2 + y^2)^2}$

7. Let g and h be the functions guaranteed by the Riemann mapping theorem:

$$g: A \to D \quad \text{with} \quad g(z_0) = 0 \quad \text{and} \quad g'(z_0) > 0$$
$$h: B \to D \quad \text{with} \quad h(w_0) = 0 \quad \text{and} \quad h'(w_0) > 0$$

Set $f(z) = h^{-1}(e^{i\theta}g(z))$ for $z \in A$ and check that f takes A one-to-one onto B, that $f(z_0) = h^{-1}(e^{i\theta}g(z_0)) = h^{-1}(0) = w_0$, and that $f'(z_0) = e^{i\theta}[g'(z_0)/h'(w_0)]$.

9. From $f \circ f^{-1}(z) = z$ we get $f'(f^{-1}(z)) \cdot (f^{-1})'(z) = 1$. It follows that $f'(z) \neq 0$, so f is conformal by the conformal mapping theorem (5.1.1).

11. No for both parts. A function for (a) is a bounded entire function. Liouville's theorem says it would be constant and so certainly cannot map onto D. The inverse of a function for (b) is a function for (a).

13. $A \cup \mathrm{bd}(A)$ is closed. If A is bounded then $A \cup \mathrm{bd}(A)$ would also be bounded and so compact. If there were a continuous extension of f to this compact set its image would be compact and could not contain the unbounded set B.

5.2   FRACTIONAL LINEAR AND SCHWARZ-CHRISTOFFEL TRANSFORMATIONS

1.   (a) $\mathbf{R}\backslash\{1\}$ ($f(\infty) = 1$ and $f(-1) = \infty$)
(b) The circle cutting the real axis at right angles at 3 and $\frac{1}{3}$ (center $\frac{5}{3}$, radius $\frac{4}{3}$)
(c) {imaginary axis} $\cup$ {∞}   ($f(-1) = \infty$)
(d) {unit circle}$\backslash\{1\}$ ($f(\infty) = 1$)

3.   (a) $f(z) = (z + 1)/(z - 3)$ (b) $f(z) = z - 2$

5.   According to Fig. 5.2.11(vi), $z \mapsto -i\left(\dfrac{z + 1}{z - 1}\right)$ takes the disk to the upper half plane with
$0 \mapsto i$ The map $w \mapsto 2(w + 1)$ takes the upper half plane to itself. Thus,

$$f(z) = 2\left[1 - i\left(\frac{z + 1}{z - 1}\right)\right]$$

does what we want.

7. $z \mapsto \dfrac{z - \frac{1}{2}}{1 - \frac{1}{2}z} = \dfrac{2z - 1}{2 - z}$ takes D to D and $\frac{1}{2}$ to 0. $w \mapsto \dfrac{3w - 1}{3 - w}$ takes D to D and $\frac{1}{3}$ to 0. Solving
$\dfrac{3w - 1}{3 - w} = \dfrac{2z - 1}{2 - z}$ for w gives $w = \dfrac{5z - 1}{5 - z}$, and so $f(z) = \dfrac{5z - 1}{5 - z}$ is the map we want.

9. $f(z) = e^{-3\pi i/4}\left(\dfrac{z + i}{z - i}\right)$

11. T is the composition of a translation, inversion in the unit circle, reflection in the real axis, a rotation, a magnification, and another translation.

13. $(2z - 1)/(2 - z)$. (This may be multiplied by $e^{i\theta}$ for any real constant θ.)

15. $e^{i\theta}\dfrac{z - Rz_0}{R - \bar{z}_0 z}$

17.   Suppose that T is such a map. Define W by

$$W(z) = T\left(\frac{1}{i} \cdot \frac{z + 1}{z - 1}\right)$$

W maps the unit disk conformally onto itself. Now use Proposition 5.2.1.

19. $\dfrac{z^8 - i}{z^8 + i}$

21.   By Proposition 5.2.3, $T(\gamma_1)$ and $T(\gamma_2)$ are circles or straight lines, and since T is conformal, by Theorem 5.1.1, they intersect orthogonally.

23.
$$|\bar{z} - z_0||z - z_0| = |\bar{z} - \bar{z}_0||z - z_0|$$
$$= \left|\left(\frac{\bar{z}_0 z + R^2 - |z_0|^2}{z - z_0} - \bar{z}_0\right)(z - z_0)\right|$$
$$= |\bar{z}_0 z + R^2 - |z_0|^2 - \bar{z}_0 z + \bar{z}_0 z_0| = |R^2| = R^2$$

25. $f(z) = \dfrac{1 + e^z}{1 - e^z}$

27. Use a branch of log defined on $\mathbf{C}\backslash\{\text{nonpositive real axis}\}$ by $\log{(re^{i\theta})} = \log r + i\theta$, where $-\pi < \theta < \pi$.

29. By the Schwarz-Christoffel formula, the image is a polygon with four sides, three of whose angles are 90°. Hence the image is a rectangle.

31. No. The map

$$f(z) = \int_0^z (t - 1)^{-\alpha_1} t^{-\alpha_2}\, dt$$

takes the upper half plane to a triangle with exterior angles $\pi\alpha_1$, $\pi\alpha_2$, $\pi\alpha_3$.

33. The boundary circles both go through 0 and $f(0) = \infty$. Since f takes $\mathbf{R}$ to $\mathbf{R}$, the circles go to lines through $f(2)$ and $f(4)$ that are orthogonal to $\mathbf{R}$. These are the lines Re $z = 2$ and Re $z = 1$. Check $f(3)$ to make sure the region is right.

35. $f(b) = 0$ and $f(d) = \infty$. Thus a circle through b and d must map to a circle through 0 and ∞, that is, a line through the origin. Since

$$|f(z)| = \left| a\, \frac{z - b}{z - d} \right| = \left| \frac{z - b}{z - d} \right| |a|$$

we have

$$\left| \frac{z - b}{z - d} \right| = \frac{r}{|a|} \Longleftrightarrow |f(z)| = r$$

This establishes (a) and (b).

The easiest way to obtain the orthogonality is to notice that the images under the map f are trivially orthogonal. Since the inverse of a fractional linear transformation is of the same form, hence conformal, the same must have been true of the preimage. (To confirm this directly a straightforward but lengthy calculation is required.)

5.3 APPLICATIONS TO LAPLACE'S EQUATION, HEAT CONDUCTION, ELECTROSTATICS, AND HYDRODYNAMICS

1. $u(x, y) = \dfrac{1}{\pi} \arctan \dfrac{4x^3y - 4xy^3}{x^4 - 6x^2y^2 + y^4} = \dfrac{4}{\pi} \arctan \dfrac{y}{x}$

3. $\phi(x, y) = 1 - \dfrac{1}{\pi} \arctan \dfrac{\cos x \cdot \sinh y}{\sin x \cdot \cosh y - 1} + \dfrac{1}{\pi} \arctan \dfrac{\cos x \cdot \sinh y}{\sin x \cdot \cosh y + 1}$

5. $F(z) = |\alpha| \left(e^{-i\theta}z + \dfrac{1}{e^{-i\theta}z} \right)$

In polar coordinates,

$$\phi(r, \Theta) = |\alpha| \left[\left(r + \frac{1}{r} \right) \cos (\Theta - \theta) \right]$$

$$\psi(r, \Theta) = |\alpha| \left[\left(r - \frac{1}{r} \right) \sin (\Theta - \theta) \right]$$

7. In polar coordinates,

$$\phi(r, \theta) = \alpha r^4 \cos 4\theta \quad \text{and} \quad \psi(r, \theta) = \alpha r^4 \sin 4\theta$$

In rectangular coordinates,

$$\phi(x, y) = (x^4 - 6x^2y^2 + y^4)\alpha \quad \text{and} \quad \psi(x, y) = (4x^3y - 4xy^3)\alpha$$

REVIEW EXERCISES FOR CHAPTER 5

1. Any region not containing zero

3. $(2z - i)/(2 + iz)$. (This answer may be multipled by $e^{i\theta}$ for any real constant θ.)

5. $f(z) = (2z - 4)/z$ is one such. **7.** The first quadrant **9.** $f(z) = z + 1 - i$

11. No. The region $\{z \mid 1 < |z| < 2\}$ is not simply connected. The inverse would take the simply connected region B to it, which is impossible by Worked Example 5.1.7.

13. No. **15.** Use Proposition 5.2.2.

17. $F(z) = \sqrt{\dfrac{az^2 + b}{cz^2 + d}}$ where a, b, c, d are real and $ad > bc$.

19. $f(z) = \sqrt{z^2 + 1}$ where $\sqrt{}$ must be taken as the branch defined on $\mathbb{C}\backslash\{\text{positive real axis}\}$, which takes values in the upper half plane. The desired potential is $\phi(z) = \dfrac{1}{\pi} \arg \left(\dfrac{\sqrt{z^2 + 1} - 1}{\sqrt{z^2 + 1} + 1} \right)$.

21. The desired complex potential must be $F(z) = \alpha\sqrt{z^2 + 1}$. The formula

$$y^2 = \frac{K - 1}{2x^2 + K - 1} + \frac{K - 1}{2}$$

for $K > 1$ gives lines of constant ψ, which are streamlines.

23. $\phi(x, y) = 1 - \dfrac{1}{\pi} \arctan \dfrac{e^x \sin y}{e^x \cos y - 1}$. The values of arctan are chosen between 0 and π.

25. $f(z) = \dfrac{h}{\pi} (\sqrt{z^2 - 1} + \cosh^{-1} z)$ **27.** $T(x, y) = \dfrac{1}{2} - \dfrac{1}{\pi} \mathrm{Re} \left(\arcsin \dfrac{z}{a} \right)$

29. $\dfrac{\pi e^{-\sqrt{3}}}{2\sqrt{3}}$ **31.** $\sqrt{p}, p^2$

6.1 ANALYTIC CONTINUATION AND ELEMENTARY RIEMANN SURFACES

1. (a) No, it does not. An important condition of the identity theorem (6.1.1) is that the limit point z_0 must lie in A.

(b) No. Let A be the unit disk. Let $u_1(z) = \text{Im } z$ and $u_2(z) = \text{Im } e^z$. Both u_1 and u_2 are harmonic in A and are zero along the real axis, but they are not identically zero, nor do they agree with each other on A.

3. If f were constant on a neighborhood of z_2, it would be constant on all of A by the identity theorem. This would force $f'(z_1)$ to be 0, which is not true.

5. If $z = re^{2\pi ip/q}$, then $z^{n!} = r^{n!}$ whenever $n \geq q$. Any open set containing A contains a point $e^{2\pi ip/q} = z_0$. If f were analytically continued to include z_0 it would have to have a finite limit at z_0, but $\lim\limits_{r \to 1} f(re^{2\pi ip/q}) = \infty$. Check this using the first observation.

7. The union of the sets U_k is an open set A containing γ. The distance lemma gives a positive distance ρ from γ to the complement of A. Since the continuation is analytic on each U_k, the radius of convergence is at least ρ at each point along γ. The path covering lemma gives a finite chain of overlapping disks centered along γ, where each contains the center of the next so they can be used to implement the continuation by power series.

9. The situation is something like that for $\sqrt{z}$, except that now there are three sheets, each a copy of the plane cut along, say, the negative real axis. They are joined along these cuts so that following a path that winds once around zero carries one from the first sheet to the second. When the path winds once more around zero one is carried to the third sheet; when the path winds a third time around zero one is carried back to the first sheet.

11. $f(z) = 1/(1 + z)$ extends it to $\mathbf{C}\backslash\{-1\}$.

13. Use Worked Example 2.4.16. (The implicit function theorem can be used to ensure that most small rectangles meet γ at most twice.)

6.2 ROUCHÉ'S THEOREM AND THE PRINCIPLE OF THE ARGUMENT

1. Five. Consider $g(z) = z^6 - 4z^5 + z^2 - 1$ and $f(z) = -4z^5$ and use Rouché's theorem.

3. For large enough R, the curve γ_R shown would include any finite number of possible solutions in the right half plane. Let $g(z) = z + e^{-z} - 2$ and $f(z) = z - 2$. Along γ_R, $|f(z) - g(z)| = e^{-\text{Re}\,z} \leq 1 < |f(z)|$, and f has exactly one solution. Thus so does g.

5. Let $h(z) = f(z) - z$ and $g(z) = -z$. On the circle $|z| = 1$, $|h(z) - g(z)| = |f(z)| < 1 = |g(z)|$, and so Rouché's theorem shows h has one zero inside $\{z$ such that $|z| = 1\}$. A zero of h is a fixed point of f.

7. Let $r_n = 1 - 1/n$ and $f_n(z) = f(r_n z)$. Use Exericse 5 to get z_n with $f_n(z_n) = z_n$. (Use the maximum modulus principle to obtain $|f(r_n z)| < 1$ if $|z| = 1$.) The z_n's are all in the closed disk $D = \{z$ such that $|z| \leq 1\}$, and so there is a subsequence converging to a point $z_0 \in D$, say $z_{n_k} \to z_0$. Check the following: $r_{n_k} z_{n_k} \to z_0$ and so $f(r_{n_k} z_{n_k}) \to f(z_0)$, but $f(r_{n_k} z_{n_k}) = z_{n_k} \to z_0$ and so $f(z_0) = z_0$.

9. Let $g(z) = a_n z^n$, estimate $f(z) - g(z)$ along large circles as in the proof of the fundamental theorem of algebra (2.4.9), and apply Rouché's theorem.

11. Use the method of Theorem 6.2.1 to compute the residue of $f'(z)h(z)/f(z)$, obtaining $kh(a_j)$ if f has a zero of order k at a_j and $-kh(b_l)$ if f has a pole of order k at b_l.

13. Apply Exercise 11 with $h(z) = z$. (The zeros are repeated in the sum according to their multiplicity.)

15. Apply the fundamental theorem of algebra to the polynomial $f(z) - w$.

17. No. Let $f(z) = e^z - 1$. f has three zeros inside a circle of radius 3π, center 0, but $f'(z)$ has no zeros.

19. Any r such that $1 < r < 4$ will give the desired result. Rouché's theorem works with $r = 2$ and $g(z) = -4z^2$.

21. Suppose $e^{i\theta}$ and $e^{i\psi}$ are on the boundary circle. If $e^{i\phi} \neq e^{i\psi}$, then an equation $(e^{i\theta})^2 + 3(e^{i\theta}) = (e^{i\psi})^2 + 3(e^{i\psi})$ would become $(e^{i\theta})^2 - (e^{i\psi})^2 = 3(e^{i\theta} - e^{i\psi})$ or $(e^{i\theta} + e^{i\psi})(e^{i\theta} - e^{i\psi}) = 3(e^{i\theta} - e^{i\psi})$, requiring $e^{i\theta} + e^{i\psi} = 3$. This is not possible, since $e^{i\theta}$ and $e^{i\psi}$ both have absolute value 1. The function is one-to-one on the boundary circle and so on the whole region by the one-to-one theorem (6.2.10).

23. Consider $f(z) = 1/z$ and apply Rouché's theorem; you will get -1 "equal to" a nonnegative number.

6.3 MAPPING PROPERTIES OF ANALYTIC FUNCTIONS

1. (a) $\{z$ such that $|z| < \frac{1}{2}\}$
(b) $\{z$ such that $|z - 1| < \frac{3}{2}\}$

3. Let $f(z) = z^3$, $w_0 = z_0 = 1$, $r = 1$. The roots of $z^3 - 1$ lie at 1, $e^{2\pi i/3}$, and $e^{4\pi i/3}$. Of these only one lies in $D = \{z$ such that $|z - 1| < 1\}$. $f'(z) = 3z^2$ and is 0 only at 0, which does not lie in D. However, $f(re^{\pi i/3}) = f(re^{-\pi i/3}) = -r^3$, and for small enough r, these points lie in D.

5. Use the chain rule to show that $g(z) = f(z^n) - f(0)$ has a zero of order n at $z_0 = 0$ and apply Worked Example 6.3.7.

7. Let u be harmonic and nonconstant on a region A, $z_0 \in A$, and let U be any open neighborhood of z_0 lying in A. By Exercise 6, u is an open mapping, and so $u(U)$ is an open neighborhood of $u(z_0)$ in $\mathbf{R}$. This means that arbitrarily near z_0, u takes on values that are both larger and smaller than $u(z_0)$.

9. Let $f(z) = e^{z-a} - z$ and $g(z) = -z$. Then, for $z = x + iy$ on the unit circle, $|f(z) - g(z)| = |e^{x+iy-a}| = e^{x-a} < 1 = |g(z)|$. Rouché's theorem now applies.

11. $|f|$ has a minimum somewhere in D since it is continuous. It is not on the boundary, since $f(0) = 1 < 2$. The minimum is at an interior point z_0 of D. If $f(z)$ were never 0, then $1/f$ would be analytic with a local maximum at z_0. The maximum modulus principle would say $1/f$ and so also f were constant. But it is not, since $f(0) = 1$ and $|f(1)| = 2$.

REVIEW EXERCISES FOR CHAPTER 6

1. f is identically zero on $\{z \text{ such that } |z| < 1\}$.

3. (a) $g(z) = f(z) - \overline{f(\bar{z})}$ is entire and $g(a_k) = 0$ for all k. Since the a_k's are bounded, there is a subsequence convergent to some a_0. Thus $g = 0$ by the identity theorem, and so $\overline{f(\bar{z})} = f(z)$ for all z. Taking z real gives the result.

(b) By part (a), $f(x)$ is real for real x. Use the mean value theorem from calculus to obtain b_n with $a_{2n+1} \leq b_n \leq a_n$ and $f'(b_n) = 0$. Check that $b_n \to 0$ and so $f' = 0$ by the identity theorem. Conclude that f is constant.

5. Use the Schwarz reflection principle to define f on the upper half plane. The functions f and its reflection agree on the strip $\{z \mid 0 < \operatorname{Im} z < 1\}$ and together define a bounded entire function. Now use Liouville's theorem.

7. Use the identity theorem to show that $g(z) = f(z + 1) - f(z)$ is identically equal to 0.

9. $\sqrt{5}$ **11.** Use the root counting formula.   **13.** Use $\int_\gamma \dfrac{f'}{f} = 2\pi i \sum\limits_{j=1}^n I(\gamma, \alpha_j)$.

15. Let $0 < r < 1$, $D(0, r) = \{z \text{ such that } |z| < r\}$, $\gamma_r = \{z \text{ such that } |z| = r\}$. By assumption, f is one-to-one on γ_r, so $f(\gamma_r)$ is a simple closed curve. Since f is bounded on A, f must map $D(0, r)$ to the interior of γ_r. The one-to-one theorem (6.2.10) now shows that f is one-to-one on $D(0, r)$. Because this holds for any $r < 1$, f is one-to-one on A.

17. Let $h(z) = f(z) - g(z)$ and use the maximum principle for harmonic functions.

19. (a) True;   (b) true;   (c) true;   (d) true;   (e) false;   (f) false;   (g) true;   (h) false;
(i) true;   (j) true.   **21.** No   **23.** (a) Yes.   (b) No.   **25.** $\frac{1}{2}$

7.1 INFINITE PRODUCTS AND THE GAMMA FUNCTION

1. The partial products are

$$\prod_{n=2}^{N} \left(1 - \frac{1}{n^2}\right) = \prod_{n=2}^{N} \frac{(n-1)(n+1)}{n^2} = \frac{1}{2} \cdot \frac{3}{2} \cdot \frac{2}{3} \cdot \frac{4}{3} \cdot \frac{3}{4} \cdot \frac{5}{4} \cdots \frac{(N-1)}{N} \cdot \frac{(N+1)}{N} = \frac{1}{2} \frac{N+1}{N}$$

which converge to $\frac{1}{2}$ as $N \to \infty$.

3. Show that for small ϵ and large n,

$$0 < (1 - \epsilon)|z_n| \leq |\log(1 + z_n)| \leq (1 + \epsilon)|z_n|$$

and use part (iii) of the convergence theorem for products (7.1.2).

5. Let $z = \frac{1}{2}$ in $\sin \pi z = \pi z \prod_{n=1}^\infty (1 - z^2/n^2)$ to obtain

$$1 = \frac{\pi}{2} \prod_{n=1}^\infty \left[1 - \frac{1}{(2n)^2}\right] = \frac{\pi}{2} \prod_{n=1}^\infty \frac{(2n-1)(2n+1)}{(2n)^2} = \frac{\pi}{2} \cdot \frac{1}{2} \cdot \frac{3}{2} \cdot \frac{3}{4} \cdot \frac{5}{4} \cdot \frac{5}{6} \cdot \frac{7}{6} \cdot \frac{7}{8} \cdot \frac{9}{8} \cdots$$

7. $G(z) = \Pi_{n=1}^{\infty} (1 + z/n)e^{-z/n}$ has zeros at $-1, -2, -3, \ldots$, and so $\Gamma(z) = [ze^{\gamma z}G(z)]^{-1}$ has poles at $0, -1, -2, \ldots$. We know that $G(z - 1) = ze^{\gamma}G(z)$, so that

$$\Gamma(z + 1) = \frac{1}{(z + 1)e^{\gamma z}e^{\gamma}G(z + 1)} = \frac{1}{e^{\gamma z}G(z)} = \frac{z}{ze^{\gamma z}G(z)} = z\Gamma(z)$$

as long as we stay away from the poles.

9.
$$zG(z)G(-z) = z \prod_{n=1}^{\infty} \left(1 + \frac{z}{n}\right) e^{-z/n} \prod_{n=1}^{\infty} \left(1 - \frac{z}{n}\right) e^{z/n}$$

$$= z \prod_{n=1}^{\infty} \left(1 - \frac{z^2}{n^2}\right) = \frac{\sin \pi z}{\pi}$$

by Worked Example 7.1.10. Using Exercise 7,

$$\Gamma(z)\Gamma(1 - z) = -z\Gamma(z)\Gamma(-z) = \frac{-z}{ze^{\gamma z}G(z)(-z)e^{-\gamma z}G(-z)} = \frac{1}{zG(z)G(-z)} = \frac{\pi}{\sin \pi z}$$

11. Start with the third line of the proof of Euler's formula:

$$\frac{1}{\Gamma(z)} = z \lim_{n\to\infty} n^{-z} \prod_{k=1}^{n} \left(1 + \frac{z}{k}\right) = \lim_{n\to\infty} \frac{z}{n^z} \prod_{k=1}^{n} \frac{k + z}{k}$$

$$= \lim_{n\to\infty} \frac{z(z + 1)(z + 2) \cdots (z + n)}{n^z \cdot 1 \cdot 2 \cdot 3 \cdots n}$$

$$= \lim_{n\to\infty} \frac{z(z + 1) \cdots (z + n)}{n! n^z}$$

13. Use the Laurent expansion and Res $(\Gamma, -m) = (-1)^m / m!$.

15. One way is to proceed as follows. First, show that $1 + y \le e^y \le (1 - y)^{-1}$ for $0 \le y \le 1$ by using power series or calculus. Set $y = t/n$ to obtain $0 \le e^{-t} - (1 - t/n)^n$ and conclude that $e^{-t} - (1 - t/n)^n \le e^{-t}[1 - (1 - t^2/n^2)^n]$. Use the inequality $(1 - k)^n \ge 1 - nk$ for $0 \le k \le 1$ to get

$$1 - \left(1 - \frac{t^2}{n^2}\right)^n \le \frac{t^2}{n}$$

for $0 \le t \le n$.

17. Γ has simple poles at $0, -1, -2, \ldots$ and is analytic elsewhere. Therefore,

$$\int_{\gamma} \Gamma(z) \, dz = 2\pi i \text{ Res } (\Gamma, 0) = 2\pi i \lim_{z\to 0} z\Gamma(z)$$

$$= 2\pi i \lim_{z\to 0} \Gamma(z + 1) = 2\pi i \Gamma(1) = 2\pi i$$

19. Let the radius of the circular part of C be $r < 1$. For $n > 1$ consider $f_n(z) = \int_C (-t)^{z-1} e^{-t} \, dt$.

(a) Use Worked Example 2.4.15 to show $f_n(z)$ is entire.
(b) Estimate the part of the integral with $|t| > n$ to show that the improper integral converges and that the convergence of f_n to that integral is uniform on closed disks
(c) Conclude that the Hankel integral is an entire function.

(d) Use Cauchy's theorem to show the value is independent of r and ϵ.

(e) Use $\arg(-t) = -\pi$ on the upper side of the real axis and π on the lower side to show that the straight-line portions combine to give $-2i \sin \pi z \int_r^\infty t^{z-1} e^{-t}\, dt$.

(f) The part along the circle goes to 0 as r goes to 0.

(g) Use Euler's integral for $\Gamma(z)$ to conclude that the formula holds for $\operatorname{Re} z > 0$.

(h) Use the identity theorem to conclude that the formulas agree everywhere both sides make sense; that is, at $z \neq 0, 1, 2, \ldots$.

(i) Use $\Gamma(z)\Gamma(1-z) = \pi/(\sin \pi z)$ to get the last assertion.

21. Do the first integral for positive y with the substitution $t = y^2$. For the second, integrate by parts with $u = y$ and $dv = ye^{-y^2}\, dy$.

7.2   ASYMPTOTIC EXPANSIONS AND THE METHOD OF STEEPEST DESCENT

1. There are constants R_1, B_1, R_2, and B_2 such that $|f(z)/h(z)| < B_1$ whenever $|z| \geq R_1$ and $\alpha \leq \arg z \leq \beta$, and $|g(z)/h(z)| < B_2$ whenever $|z| \geq R_2$ and $\alpha \leq \arg z \leq \beta$, so that if $|z| \geq R = \max(R_1, R_2)$ and $\alpha \leq \arg z \leq \beta$, then $|[af(z) + bg(z)]/h(z)| \leq a|f(z)/h(z)| + b|g(z)/h(z)| \leq aB_1 + bB_2$. Thus, $af(z) + bg(z) = O(h(z))$.

3. If $S_n(x) = \Sigma_2^n(a_k/x^k)$, then $f - S_n = o(1/x^n)$, and so by Proposition 7.2.3(v), $\int_x^\infty f - \int_x^\infty S_n = o(1/x^{n-1})$. That is, $\int_x^\infty f - \Sigma_2^n [a_k/(k-1)x^{k-1}] = o(1/x^{n-1})$, or $\int_x^\infty f \sim (a_2/x) + (a_3/2x^2) + \cdots$, as desired.

5. Use the geometric series and apply Proposition 7.2.5.

7. For the even case of (a), integrate by parts repeatedly to reduce to the case $k = 0$. Then change variables by $zy^2/2 = t^2$ and use $\int_{-\infty}^\infty e^{-t^2}\, dt = \sqrt{\pi}$. In part (b) either do the same thing or put $z = 2$ in part (a). For the odd cases, the integrand is an odd function of y. Thus, if the integral converges, it must be to 0. Check that it converges.

9. For $S_4(10)$ the error is ≤ 0.00024, and for $S_5(10)$ it is 0.00012. For fixed x, the error term decreases as n increases until n becomes larger than x, at which point it begins to increase again. In fact, for fixed x, our bound on the error term goes to infinity with n. For fixed n, $\lim\limits_{x\to\infty} n!/x^{n+1} = 0$, so the error goes to zero as x increases.

11. $f(z) \sim \dfrac{\sqrt{2\pi}}{\sqrt{z}} \left(\dfrac{1}{z} - \dfrac{1 \cdot 3 \cdot 5}{3!} \dfrac{1}{z^3} + \dfrac{1 \cdot 3 \cdot 5 \cdot 7 \cdot 9}{5!} \dfrac{1}{z^5} - \cdots \right)$

13. The path of steepest descent is the real axis.

15. $y = \log \dfrac{1 - \sin x}{\cos x}$ **17.** $f(z) \sim e^z \sqrt{\dfrac{2\pi}{z}}$ **19.** $f(z) \sim e^z(1 - i)\sqrt{\dfrac{\pi}{|z|}}$

7.3 STIRLING'S FORMULA AND BESSEL FUNCTIONS

1. Differentiate any convenient formula for $J_n(z)$ twice and substitute it in the equation. For example:

$$J_n(z) = \frac{1}{2\pi i} \left(\frac{z}{2}\right)^n \int_\gamma t^{-n-1} e^{t-(z^2/4t)} \, dt$$

3. From Eq. (4), $J_0(z) = 1/\pi \int_0^\pi \cos{(z \sin \theta)} \, d\theta$, and so $J_0'(z) = -1/\pi \int_0^\pi \sin{(z \sin \theta)} \sin \theta \, d\theta$. But

$$J_1(z) = \frac{1}{\pi} \int_0^\pi \cos{(\theta - z \sin \theta)} \, d\theta$$

$$= \frac{1}{\pi} \int_0^\pi [\cos \theta \cos{(z \sin \theta)} + \sin \theta \sin{(z \sin \theta)}] \, d\theta$$

Use symmetry in $\pi/2$ to show the first term vanishes.

5. From Eq. (12) (line 8 of Table 7.2),

$$J_0'(z) = \frac{0}{z} J_0(z) - J_1(z) = -J_1(z)$$

and so

$$J_0''(z) = -J_1'(z) = -\frac{1}{z} J_1(z) + J_2(z) = J_2(z) + \frac{1}{z} J_0'(z)$$

7. By definition,

$$J_{1/2}(z) = \sum_{k=0}^\infty \frac{(-1)^k z^{(1/2)+2k}}{2^{(1/2)+2k} \Gamma(\frac{1}{2} + k + 1) k!}$$

$$= \sqrt{\frac{2}{z}} \sum_{k=0}^\infty \frac{(-1)^k z^{2k+1}}{\sqrt{\pi} \Gamma(2k + 2)} \qquad \text{(by Legendre's duplication formula)}$$

$$= \sqrt{\frac{2}{\pi z}} \sin z$$

9. Use the stationary phase theorem (7.2.10) with $\gamma = [0, \pi]$, $h(\theta) = -\sin \theta$, $g(\theta) = e^{in\theta}$, $\theta_0 = \pi/2$, and $f(z) = \int_0^\pi e^{-iz \sin \theta + ni\theta} \, d\theta$. Clearly h is real on γ with a strict minimum at $\pi/2 = \theta_0$; $h'(\theta_0) = 0$, $h''(\theta_0) = 1$. All conditions of the theorem are met, and so

$$f(z) \sim \frac{e^{izh(\theta_0)} \sqrt{2\pi} e^{\pi i/4}}{\sqrt{z} \sqrt{h''(\theta_0)}} g(\theta_0)$$

$$= \sqrt{\frac{2\pi}{z}} e^{-iz} e^{\pi i/4} e^{in\pi/2} = \sqrt{\frac{2\pi}{z}} e^{-i(z - n\pi/2 - \pi/4)}$$

11. Let $n \le 0$ and $m = -n \ge 0$. The residue in the expansion of Eq. (5) now is $1/(k-m)!$, so that

$$J_n(z) = \sum_{k=m}^\infty \frac{(-1)^k z^{n+2k}}{2^{n+2k}} \cdot \frac{1}{(k-m)!} \cdot \frac{1}{k!}$$

Put $j = k - m$ and $k = j + m$ to get

$$J_n(z) = \sum_{j=0}^{\infty} \frac{(-1)^{j+m} z^{n+2(j+m)}}{2^{n+2(j+m)}} \cdot \frac{1}{(j+m-m)!} \cdot \frac{1}{(j+m)!}$$

$$= \sum_{k=0}^{\infty} \frac{(-1)^{k-n} z^{m+2k}}{2^{m+2k}} \cdot \frac{1}{k!} \cdot \frac{1}{(k+m)!} = (-1)^n \sum_{k=0}^{\infty} \frac{(-1)^k z^{m+2k}}{2^{m+2k} k! (m+k)!}$$

$$= (-1)^n J_m(z) = (-1)^n J_{-n}(z) = \sum_{k=0}^{\infty} \frac{(-1)^{k-n} z^{-n+2k}}{2^{-n+2k} k! (k-n)!}$$

REVIEW EXERCISES FOR CHAPTER 7

1. 2

3. Use $z = \pi/4$ and $z = \pi/2$ in Worked Example 7.1.10 to obtain

$$\sqrt{2} = \frac{\displaystyle\prod_{1}^{\infty}(1 - 1/16n^2)}{\displaystyle\prod_{1}^{\infty}(1 - 1/4n^2)} = \frac{\displaystyle\lim_{N\to\infty}\prod_{1}^{N}(1 - 1/16n^2)}{\displaystyle\lim_{N\to\infty}\prod_{1}^{N}(1 - 1/4n^2)}$$

$$= \lim_{N\to\infty} \frac{\displaystyle\prod_{1}^{N}(1 - 1/16n^2)}{\displaystyle\prod_{1}^{N}(1 - 1/4n^2)} = \lim_{N\to\infty} \prod_{1}^{N} \frac{(4n-1)(4n+1)}{(4n-2)(4n+2)}$$

$$= (\tfrac{3}{2})(\tfrac{5}{6})(\tfrac{7}{6})(\tfrac{9}{10})(\tfrac{11}{10})(\tfrac{13}{14}) \cdots$$

5. (a) $\{z$ such that $|z| < 1\}$ (b) $\{z \,|\, \text{Re } z > 1\}$ **7.** $f(z) \sim e^{i(z-\pi/4)}\sqrt{2\pi}/\sqrt{z}$ **9.** $\Gamma(\tfrac{1}{3})/3$

11. First establish, for example from the product representation for $1/\Gamma$, that $\Gamma(z) = \overline{\Gamma(\bar{z})}$. With $z = \tfrac{1}{2} + iy$, this gives $|\Gamma(\tfrac{1}{2} + iy)|^2 = \Gamma(\tfrac{1}{2} + iy) \cdot \Gamma(\tfrac{1}{2} - iy) = \Gamma(\tfrac{1}{2} + iy) \cdot \Gamma(1 - \tfrac{1}{2} + iy) = \pi/\sin(\pi(\tfrac{1}{2}) + iy) = 2\pi/(e^{-\pi y} + e^{\pi y})$ which goes to 0 as $y \to \infty$.

13. $\sqrt{\dfrac{2\pi}{z}}\left(1 - \dfrac{1 \cdot 3}{2} \cdot \dfrac{1}{z^2} + \dfrac{1 \cdot 3 \cdot 5 \cdot 7}{4!} \cdot \dfrac{1}{z^4} - \cdots\right)$

15. Apply the generalized steepest descent theorem (7.2.9) after writing

$$J_n(iy) = \frac{1}{2\pi}\left[\int_0^{2\pi} e^{-y\sin\theta}(\cos n\theta + 1)\, d\theta - \int_0^{2\pi} e^{-y\sin\theta}\, d\theta\right]$$

$$- \frac{i}{2\pi}\left[\int_0^{2\pi} e^{-y\sin\theta}(\sin n\theta + 1)\, d\theta - \int_0^{2\pi} e^{-y\sin\theta}\, d\theta\right]$$

and using $h(\theta) = -\sin\theta$, $\theta_0 = 3\pi/2$, and $\gamma = [0, 2\pi]$.

17. From part (b) of Exercise 16,

$$P_n(x) = \frac{1}{2^n n!} \cdot (2n)(2n-1) \cdots (n+1) \cdot x^n + (\text{lower-order terms})$$

$$= \frac{(2n)!}{2^n (n!)^2} x^n + (\text{lower-order terms})$$

8.1 BASIC PROPERTIES OF LAPLACE TRANSFORMS

1. $\tilde{f}(z) = \dfrac{2}{z^3} + \dfrac{2}{z}$ $\sigma(f) = 0$ **3.** $\tilde{f}(z) = \dfrac{1}{z^2} + \dfrac{1}{z+1} + \dfrac{1}{z^2+1}$ $\sigma(f) = 0$

5. $\tilde{f}(z) = \dfrac{1}{z} + n \cdot \dfrac{1}{z^2} + n(n-1)\dfrac{1}{z^3} + \cdots + n!\dfrac{1}{z^{n+1}}$ $\sigma(f) = 0$

7. $\tilde{f}(z) = \dfrac{2az}{(z^2+a^2)^2}$ $\sigma(f) = |\text{Im } a|$ **9.** $\tilde{f}(z) = \dfrac{z^2 - a^2}{(z^2+a^2)^2}$ $\sigma(f) = |\text{Im } a|$

11. $\tilde{f}(z) = \int_0^\infty e^{-zt} \cos at\, dt = \frac{1}{2}\int_0^\infty e^{-t(z-ia)}\, dt + \frac{1}{2}\int_0^\infty e^{-t(z+ia)}\, dt$. For Re $(z - ia) > 0$, the first integral converges to $1/2(z - ia)$. For Re $(z + ia) > 0$ the second converges to $1/2(z + ia)$. (See Worked Example 8.1.11.) Thus, for Re $z > |\text{Im } a|$, $\tilde{f}(z)$ converges to $\frac{1}{2}[1/(z - ia) + 1/(z + ia)] = z/(z^2 + a^2)$.

13. For z real and positive, put $u = zt$ to obtain $\tilde{f}(z) = \int_0^\infty e^{-u}u^{(a+1)-1}/z^{a+1}\, du$. Since $a > -1$, this converges to $\Gamma(a+1)/z^{a+1}$, and so $f(z)$ converges on the positive real axis. Lemma 8.1.8 gives convergence on the open right half plane. The identity theorem shows that $f(z) = \Gamma(a+1)/z^{a+1}$ for Re $z > 0$ and Worked Example 8.1.13 shows $\sigma(f) = 0$.

15. $\tilde{f}(z) = \int_0^\infty e^{-zt}f(t)\, dt = \sum_{n=0}^\infty \int_{np}^{(n+1)p} e^{-zt}f(t)\, dt$. In the nth integral put $u = t - np$. Then $\tilde{f}(z) = \sum_{n=0}^\infty \int_0^p e^{-z(u+np)}f(u+np)\, du = (\sum_{n=0}^\infty e^{-zpn})\int_0^p e^{-zu}f(u)\, du$, and since $|e^{-zp}| < 1$, this is $[\int_0^p e^{-zt}f(t)\, dt]/(1 - e^{-zp})$.

17. $\tilde{g}(z) = \dfrac{1}{z} \cdot \dfrac{1}{(z+1)^2 + 1}$; $\tilde{f}(z) = -\dfrac{3z^2 + 4z + 2}{(z^3 + 2z^2 + 2z)^2}$

19. Suppose f is of exponential order ρ and let $g(T) = \int_0^T f(s)\, ds$. For $\epsilon > 0$, there is an A with $|f(s)| \le Ae^{(\rho+\epsilon)s}$ for every $s \ge 0$. Then $|g(T)| \le A\int_0^T e^{(\rho+\epsilon)} ds = A(e^{(\rho+\epsilon)T} - 1)/(\rho + \epsilon)$. If $\rho \ge 0$, then $e^{(\rho+\epsilon)T} > 1$, and so $|g(T)| \le 2Ae^{(\rho+\epsilon)T}/(\rho + \epsilon)$ and thus $\rho(g) \le 0$. In any case, $\rho(g) \le$ max $[0, \rho(f)]$. If f is piecewise continuous, then g is continuous and piecewise C^1. For Re $z \ge$ max $[0, \rho(f)]$, the above gives Re $z > \rho(g)$ and Proposition 8.1.3 gives $(g')^\sim(z) = zg(z) - g(0)$. But $g'(t) = f(t)$, and $g(0) = 0$, so $f(z) = zg(z)$ and thus $g(z) = f(z)/z$.

21. If $B > 0$ and $t \ge 0$, then $-e^t \le -1$ and $|f(t)| < e^{-1} < e^{Bt}$.
If $B < 0$, let $A =$ max $(1/e,\, e^{B - B\log(B)})$, and let $g(t) = e^t + Bt + \log A$. Then $g(0) = 1 + \log A \ge 0$. Also, $g'(t) = 0$ only at $t = \log(-B)$ and $g(\log(-B)) = -B + B\log(-B) + \log A \ge 0$, and so $-e^{-t} \le Bt + \log A$ for $t \ge 0$ and $|f(t)| \le Ae^{Bt}$. Thus $\rho(f) = -\infty$, and therefore $\sigma(f) = -\infty$.

8.2 THE COMPLEX INVERSION FORMULA

1. (a) $\cos t$ (b) $f(t) = te^{-t}$ (c) $f(t) = [e^t + 2e^{-t/2} \cos(\sqrt{3}t/2)]/3$

3. Equation (1) in the complex inversion formula (8.2.17) cannot be applied, since there are no constants M and R for which $|e^{-z}/z| < M/|z|$ whenever $|z| > R$.

5. (a) $f(t) = 2e^{-2t} - e^{-t}$.

(b) $\sinh z$ has no inverse Laplace transform. If $\tilde{f}(z) = \sinh z$, let $g(t) = tf(t)$ and $h(t) = tg(t)$. Then $\tilde{g}(z) = -\cosh z$ and $\tilde{h}(z) = \sinh z = \tilde{f}(z)$. This would force $f(t) = h(t) = t^2 f(t)$ and so $\tilde{f}(z) = \sinh z - 0$ which isn't true.

(c) $f(t) = t^2 e^{-t}/2$.

7. $g(t) = \displaystyle\int_0^t [\sin(t - s)] f(s)\, ds$ **9.** $f(t) = (6te^{-3t} - e^{-3t} + 1)/9$

8.3 APPLICATION OF LAPLACE TRANSFORMS TO ORDINARY DIFFERENTIAL EQUATIONS

1. $y(t) = (5e^{2t} + 3e^{-2t})/4$

3. $y(t) = \begin{cases} 0 & 0 \le t < 1 \\ \frac{1}{9}[1 - \cos(3t - 3)] & t \ge 1 \end{cases}$

5. $y(t) = g_1(t) + g_2(t) + g_3(t)$ where

$$g_1(t) = \frac{e^{-t/2}}{\sqrt{3}}\left(3\cos\frac{\sqrt{3}}{2}t - \sin\frac{\sqrt{3}}{2}t\right)$$

$$g_2(t) = \begin{cases} 0 & 0 \le t < 2 \\ -\dfrac{e^{-(t-2)/2}}{\sqrt{3}}\, 2\sin\dfrac{\sqrt{3}}{2}(t - 2) & t \ge 2 \end{cases}$$

and

$$g_3(t) = \begin{cases} 0 & 0 \le t < 1 \\ \dfrac{e^{-(t-1)/2}}{\sqrt{3}}\, 2\sin\dfrac{\sqrt{3}}{2}(t - 1) & t \ge 1 \end{cases}$$

7. (a) $y_1(t) = (e^t + e^{-t})/2 = \cosh t$; $y_2(t) = -(e^t - e^{-t})/2 = -\sinh t$

(b) $y_1(t) = -3t$, $y_2(t) = 3[(1 + t)^2 - 1]/2$

9. $y(t) = [(t + 4)\sin t - t^2 \cos t]/4$

11. If we substitute the expression for ϕ on p. 546 into the boundary conditions that follow, we get four simultaneous equations involving A, B, R, and T. After some algebraic manipulations, show that we get Eq. (10), as well as

$$R = \frac{T}{2ic_1 c_2}(c_2^2 - c_1^2)e^{ia\omega/c_1}\sin\frac{a\omega}{c_2}$$

Together, these give $|R|^2 + |T|^2 = 1$.

13. $\dfrac{1}{\sqrt{x_0 + iy_0}} = -\dfrac{i}{\pi} \displaystyle\int_0^\infty \dfrac{1}{x} \dfrac{1}{z_0 \sqrt{x}} dx$

(Making the change of variables $w = \sqrt{x}$, you can evaluate the right side of the last equation directly to verify the equality.)

REVIEW EXERCISES FOR CHAPTER 8

1. $\tilde{f}(z) = e^{-z}/(z^2 + 1)$, $\sigma(f) = 0$.   **3.** $\tilde{f}(z) = \dfrac{1 - e^{-z}}{z^2}$ **5.** $\tilde{f}(z) = \log\left(\dfrac{z}{z-1}\right)$, $\sigma = 1$.

7. $f(t) = \begin{cases} 0 & t \le 1 \\ \sin(t-1) & t > 1 \end{cases}$

9. $f(t) = \begin{cases} -te^{-t} + e^{-t} & t < 1 \\ -te^{-t} + e^{-t} + 1 & t > 1 \end{cases}$

11. (a) $y(t) = (-15 + 23 \cos 2\sqrt{2}t)/8$ (b) $y(t) = 3(1 - e^{-t})$

13. (a) $y(t) = \begin{cases} 0 & 0 \le t < 1 \\ 1 - \cos(t-1) & t \ge 1 \end{cases}$ (b) $y(t) = 2te^{-t} + e^{-t}$

15. Each term of the series has a logarithmic-type singularity at $j = \alpha$. But

$$F(j) = \dfrac{1}{1 - g \log(j - \alpha)}$$

Thus, in addition to the logarithmic singularity at $j = \alpha$, $F(j)$ also has a simple pole singularity when the denominator vanishes, at $j = \alpha + e^{1/g}$.

INDEX

686- 8713